Selected Aerothermodynamic Design Problems of Hypersonic Flight Vehicles

Ernst Heinrich Hirschel · Claus Weiland

Selected Aerothermodynamic Design Problems of Hypersonic Flight Vehicles

Springer

Prof. Dr. Ernst Heinrich Hirschel
Herzog-Heinrich-Weg 6
85604 Zorneding
Germany
Email: e.h.hirschel@t-online.de

Dr. Claus Weiland
Föhrenstrasse 90
83052 Bruckmühl
Germany
Email: claus.weiland@t-online.de

Jointly published with the American Institute of Aeronautics and Astronautics (AIAA)

ISBN 978-3-642-44295-7 ISBN 978-3-540-89974-7 (eBook)

DOI 10.1007/978-3-540-89974-7

Typesetting: Scientific Publishing Services Pvt. Ltd., Chennai, India.
Coverdesign: eStudio Calamar S.L

Printed in acid-free paper

9 8 7 6 5 4 3 2 1

springer.com

A serious problem showed up during the first Orbiter flight, viz. the hypersonic pitching moment anomaly, which gave rise to grave concerns in the HERMES project. This new vehicle had a shape totally different to that of the Orbiter. The question was whether similar or other problems – undetected in the vehicle design – would become manifest during flight. Because the hypersonic pitching moment anomaly was obviously a ground-facility simulation problem, much emphasis was put on the development and application of numerical methods and their validation. Consequently, an experimental vehicle was proposed, the 1 to 6 down-scaled MAIA. The first author of this book was deeply involved in the definition of its scientific payload (neither MAIA nor HERMES actually flew).

Work in the SÄNGER programme revealed that in airbreathing hypersonic flight viscosity effects dominate, rather than pressure or compressibility effects as is the case in re-entry flight. Viscous thermal surface effects of all kinds, governed by surface radiation cooling, became a major focal point in hypersonic research and design work. These are important subjects covered in a previous publication by the first author on the basics of aerothermodynamics.

The fantastic increase in computer power in the second half of the 1990s showed that it will be possible in future to treat the many strong couplings between the disciplines involved in the design of hypersonic flight vehicles in new ways. Multidisciplinary numerical simulation and optimization methods became a major focus in ASTRA and IMENS, in which the second author was strongly involved.

All these findings and developments, together with the responsibility of the first author for an initial structuring of general medium and long term technology development and verification strategies in both the SÄNGER technology programme and FESTIP, have shaped the content of the present book. It discusses selected aerothermodynamic design problems of winged and non-winged re-entry and airbreathing hypersonic flight vehicles – but not the full vehicle design.

The work and experience of the authors are reflected in the chapters on winged re-entry vehicles (RV-W), airbreathing cruise and acceleration vehicles (CAV) and non-winged re-entry vehicles (RV-NW). Besides the major aerothermodynamic phenomena and simulation problems, particular trends in aerothermodynamics of these vehicle classes are discussed.

Special attention is paid to the hypersonic pitching moment anomaly of the Space Shuttle Orbiter and to forebody aerothermodynamics of airbreathing vehicles. Furthermore, there is a comprehensive presentation of waverider design issues. For non-winged re-entry vehicles trim and dynamic stability issues are discussed. In particular, aerothermodynamic issues of stabilization, trim and control devices are also considered.

The authors' research, university teaching and industrial involvement have shown that it is important to cover topics that are usually not in the major focus of books on aerothermodynamics. These are the fundamentals of flight

Preface

Hypersonic flight and aerothermodynamics are fascinating topics. Design problems and aerothermodynamic phenomena are partly very different for the various kinds of hypersonic flight vehicles. These are – and will be in the future – winged and non-winged re-entry vehicles as well as airbreathing cruise and acceleration and also ascent and re-entry vehicles.

Both authors of the book worked for almost four decades in hypersonics: at the German aerospace research establishment (DVL/DFVLR, now DLR) to the end of the 1970s, then in industry (MBB/Dasa, now EADS). They were involved in many major technology programmes and projects. First, in the early 1970s, the German ART programme (Association for Re-Entry Technologies), and, in the 1980s, the European (ESA) HERMES project and the German Hypersonics Technology (SÄNGER) Programme. Then followed, in the 1990s, the Future European Space Transportation Investigations Programme (FESTIP), the Manned Space Transportation Programme (MSTP) with the Atmospheric Re-Entry Demonstrator (ARD), the X-CRV Project with the X-38 vehicle and, later, the German technology programmes TETRA (Technologies for Future Space Transportation Systems), ASTRA (Selected Systems and Technologies for Future Space Transportation Systems Applications), and IMENS (Integrated Multidisciplinary Design of Hot Structures for Space Vehicles).

Research in the 1960s and 1970s placed great emphasis on low-density flows, high-temperature real-gas effects in ground-simulation facilities and, already, on discrete numerical computation methods. After the first flight of the Space Shuttle Orbiter with its generally very good aerodynamic performance, interest in low-density problems diminished. In view of the layout of thermal protection systems, high-temperature real-gas effects, surface catalycity and also laminar-turbulent transition became important. Numerical methods received a large boost first during the post flight analysis of the Orbiter flights and then, in particular in Europe, during the research and development activities accompanying the HERMES project.

trajectory mechanics, including the general equations for planetary flight, the describing mathematical equations in general formulation of forces, moments, center of pressure, trim and stability, as well as multidisciplinary design aspects including the mathematical models and the coupling procedures.

Another outcome of the authors' work was the recognition that it would be useful to provide the reader with quantitative examples of the coefficients of – at least – longitudinal motion for a variety of shapes of operational vehicles, demonstrators, and studied concepts. In this way, numbers are available to compare and to check the results of the readers own work. Many of the data that we provide were generated with both numerical simulations and experimental tests in the department formerly headed by the second author.

Although the thermal state of a vehicle surface and the ensuing thermal loads and thermal surface effects are among the major topics of hypersonic vehicle design, they are not treated separately in this book. Their treatment is integrated in the corresponding chapters. However, a short overview of the basic issues, as well as a simulation compendium, is given in a separate chapter.

Both authors have advanced over many years the use of discrete numerical methods of aerothermodynamics in research and in industrial applications. These methods now permit a thorough quantification of and deep insights into design problems and the relevant aerothermodynamic phenomena. A good overall knowledge is necessary for their successful application, as is an eye for the relevant features. Consequently, this book discusses in great detail results of numerical simulations, also in view of the multidisciplinary implications of aerothermodynamics.

The book is intended for graduate students, doctoral students, design and development engineers, and technical managers. A useful prerequisite is a knowledge of the basics of aerothermodynamics.

We see presently an up and down of the different modes and vehicles of hypersonic flight. Non-winged concepts now seem to displace winged re-entry concepts. Airbreathing hypersonic flight is a concept still waiting for its time. Nevertheless, we are convinced that hypersonic flight has a bright future. We hope that our book will help the reader to make himself familiar with a number of problems regarding the aerothermodynamic design of hypersonic flight vehicles.

July 2008

Ernst Heinrich Hirschel
Claus Weiland

Acknowledgements

The authors are much indebted to several persons, who read the book or parts of it, and provided critical and constructive comments.

First of all we would like to thank G. Simeonides and W. Kordulla. They read all of the book and their input was very important and highly appreciated.

Many thanks are due also to R. Behr, T. M. Berens, W. Buhl, T. Eggers, M. Haupt, M. Hornung, E. D. Sach, G. Sachs, D. Schmitz, W. Staudacher, and O. Wagner, who read parts of the book.

Data and illustrative material was directly made available for the book by many colleagues. We wish to thank J. Ballmann, R. Bayer, R. Behr, A. Celic, J. M. Delery, T. Eggers, U. Ganzer, P. Gruhn, A. Gülhan, J. Häberle, H. Hansen, M. Haupt, R. Henke, A. Henkels, M. Hornung, D. Kliche, A. Knoll, M. Korfanty, G. Lange, H. Lüdecke, A. Mack, M. Marini, A. W. Markl, D. Schmitz, G. Simeonides, W. Staudacher, B. Thorwald, and W. Zeiss. General permissions are acknowledged at the end of the book.

Special thanks for the draft of figures are due to H. Reger and B. Thorwald, and also to J. M. A. Longo, DLR Braunschweig as well as Ch. Mundt and M. Pfitzner, Universität der Bundeswehr, Neubiberg/München, for sponsoring some of the preparatory work.

Finally we wish to thank our wives for their support and patience.

Ernst Heinrich Hirschel
Claus Weiland

Contents

altitudes $H \lessapprox 100.0\ km$ and flight velocities $v_\infty \lessapprox 8.0\ km/s$. We don't attempt to cover the issue of overall aerothermodynamic vehicle design, but to explain, demonstrate and illustrate design problems. Because of the partly very different vehicle shapes and the different flow phenomena present, a classification of the flight vehicles into winged re-entry vehicles, airbreathing cruise and acceleration vehicles, and non-winged re-entry vehicles, i. e. space capsules, like given in [2], is employed.

This classification is sketched first, then we discuss shortly aerothermodynamics and the definition and development of flight vehicles with regard to design sensitivities, design margins, data uncertainties, and the potential and deficits of simulation means. The latter also appears to be necessary, because design problems usually are coupled to simulation problems, too. Sometimes even a design problem is "only" a simulation problem. This chapter is closed with a sketch of scope and content of the book.

1.1 Three Reference Classes of Hypersonic Vehicles

Winged re-entry vehicles (RV-W), airbreathing cruise and acceleration vehicles (CAV), and non-winged re-entry vehicles or space capsules (RV-NW) are chosen to be the reference flight vehicle classes in this book. In [2] the class of ascent and re-entry vehicles (ARV) is added which in principle are single-stage-to-orbit (SSTO) space transportation systems with airbreathing (and rocket) propulsion. In a sense the design problems of these vehicles are a mixture of the problems encountered for RV-W-type and CAV-type vehicles. Therefore, we don't treat this class here separately. In [2] aeroassisted orbital transfer vehicles (AOTV) are introduced as a separate class to serve as a kind of extreme reference vehicle class. We don't treat this class either but have introduced instead the class of RV-NW-type vehicles with their operation, like that of the other classes, predominantly in the $H \lessapprox 100.0\ km$ and $v_\infty \lessapprox 8.0\ km/s$ domain.

For more details regarding vehicle classifications see [2] and for a very detailed classification, e.g., [3]. The vehicles of the three reference classes referred to in the accordant chapters are:

1. Winged re-entry vehicles (RV-W): Space Shuttle Orbiter, HERMES, HOPE-X, X-34, X-38, and HOPPER/PHOENIX. RV-W-type flight vehicles are launched typically by means of rocket boosters, but potentially also as rocket-propelled upper stages of two-stage-to-orbit (TSTO) space transportation systems.
2. Cruise and acceleration vehicles with airbreathing propulsion (CAV): the lower stage of the TSTO system SÄNGER, and also the ARV-type vehicle Scram 5. Flight Mach numbers lie in the ramjet propulsion regime up to $M_\infty = 7$, and the scramjet propulsion regime up to $M_\infty = 12$ (to 14).
3. Non-winged re-entry vehicles (RV-NW), some of them not operating in the Earth atmosphere: HUYGENS, BEAGLE2, OREX, APOLLO, ARD, SOYUZ, VIKING, AFE, CARINA and others.

1 Introduction

When studying papers discussing aspects of the aerodynamic shape definition process of the Space Shuttle Orbiter, see, e.g., [1], one is confronted with a host of different methods, correlations, simulation tools, et cetera which were employed. At that time the discrete numerical methods of aerodynamics and aerothermodynamics were just beginning to appear. In the meantime very large algorithmic achievements and fantastic developments in computer speed and storage, and in general in the information technologies, have happened and change now profoundly the aerothermodynamic design processes, but also the scientific work.

However, numerical methods, like ground-simulation facilities, are "only" tools. Basic knowledge of both aerothermodynamic phenomena and design problems are the prerequisites which must be present in order to use the tools effectively. It is important to note in this context, that extended design experience for space transport vehicles, backed by flight experience, is available only from the Space Shuttle Orbiter as well as from the capsules APOLLO and SOYUZ.

The Space Shuttle Orbiter is a winged re-entry flight vehicle which has its specific aerothermodynamic phenomena and design problems. If one looks at airbreathing hypersonic flight vehicles, the picture is radically different. Re-entry vehicles on purpose have large drag, while airbreathing flight vehicles must have a drag as low as possible, which immediately brings into play the viscous drag and many aerothermodynamic phenomena and design problems other than those present or important for the Space Shuttle Orbiter.

It is even apt to distinguish between hypersonic flight and hypersonic flow. A winged re-entry vehicle flies at hypersonic speed, but usually the flow in the shock layer is in the subsonic to low supersonic speed domain. An airbreathing flight vehicle flies at hypersonic speed, but now the flow past the vehicle is a true hypersonic flow. This implies that *hypersonics* in one case can be a topic vastly different from *hypersonics* in another case, [2].

In this book we treat selected aerothermodynamic design problems of predominantly hypersonic flight vehicles operating in the Earth atmosphere at

Table 1.1. Comparative consideration of particular aerothermodynamic features of the three reference classes of hypersonic vehicles. Features which are common to all classes are not listed.

Item	Winged re-entry vehicles (RV-W)	Cruise and acceleration vehicles (CAV)	Non-winged re-entry vehicles (RV-NW)
Mach number range	30–0	0–7(12)	30–0
Configuration	blunt	slender	very blunt, blunt
Flight time	short	long	short
Angle of attack	large	small	head on
Drag	large	small	large
Aerodynamic lift/drag ratio	small	large	small, zero
Flow field	compressibility-effects dominated	viscosity-effects dominated	compressibility-effects dominated
Thermal surface effects: 'viscous'	not important/ locally important	very important	not important
Thermal surface effects: 'thermo-chemical'	very important	important	very important

Each of the three classes has specific aerothermodynamic features which are summarized in Table 1.1, see also [2].

Without a quantification of features and effects we can say, that for CAV-type flight vehicles viscosity effects, notably laminar-turbulent transition and turbulence (which occur predominantly at altitudes below approximately 60.0 to 40.0 km) play a major role, while high-temperature real-gas (thermo-chemical) effects are very important for RV-W and RV-NW-type vehicles. Viscous thermal surface effects play a large role for CAV-type flight vehicles, while thermo-chemical thermal surface effects are very important for RV-W and RV-NW-type vehicles, [2].

The main objective of Table 1.1 is to sharpen the perception, that for instance a CAV-type flight vehicle, i. e. an airbreathing hypersonic flight vehicle, definitely poses an aerothermodynamic (and multidisciplinary) design problem quite different from that of a RV-W-type vehicle. The CAV-type vehicle is aircraft-like, slender, flies at small angles of attack, all in contrast to the RV-W-type vehicle. The RV-W-type flight vehicle is a pure re-entry vehicle, which is more or less "only" a deceleration system, however not a ballistic or quasi-ballistic one as is the RV-NW-type vehicle. Therefore it has a blunt shape, and flies at large angles of attack in order to increase the effective bluntness.

Thermal loads must always be considered together with the structure and materials concept of the respective vehicle, and its passive or active cooling concept. As discussed in [2], the major passive cooling means for outer surfaces is surface-(thermal-)radiation cooling. The thermal management of a CAV-type flight vehicle, for instance, must take into account all thermal loads (heat sources), cooling needs and cooling potential of the airframe, propulsion system, sub-systems and cryogenic fuel system.

1.2 Aerothermodynamics and the Definition and Development of Flight Vehicles

High performance and at the same time high cost efficiency of all kinds of hypersonic flight vehicles will most likely not be achieved by single large technological breakthroughs. A good chance exists that they will be achieved in the future, at least partly, by better, more accurate and more versatile disciplinary and multidisciplinary numerical simulation and optimization tools in all vehicle definition and development phases and processes. This certainly is possible, when we observe how discrete numerical methods in all involved technology areas advance, supported by the vast growth of computer power. A similar development is underway in classical aircraft design, see, e.g., [4] to [6].

In view of this prospect we think it is useful to look at some important aspects of vehicle design and development, in particular with respect to design sensitivities and margins, and accounting for data uncertainties and deficits of aerothermodynamic simulation means.

1.2.1 Design Sensitivities and Margins Versus Data Uncertainties

In [2] the term "simulation triangle" is used. The simulation triangle consists of computational simulation, ground-facility simulation and in-flight simulation. None of the simulation means in the triangle permits a full simulation of the aerothermodynamic properties and functions of hypersonic flight vehicles of RV-W-type, CAV-type, and RV-NW-type, Section 1.2.2. It is the art and experience of the engineer to arrive nevertheless at a viable aerothermodynamic design. However, in future the engineer will command much more powerful numerical simulation and optimization tools than he has available today.

In the following a short consideration of the three important entities in the definition and development processes, "sensitivities", "uncertainties", and "margins", is given, following [2]. For a discussion of design methodologies of hypersonic flight vehicles as we consider them in this book, see, e.g., [7] to [9].

The objectives of the definition and development processes are:

- the design of the flight vehicle with its performance, properties and functions according to the specifications, (coefficients of longitudinal motion) of a number of RV-W-type vehicle shapes. Finally issues of vehicle flyability and controllability, in particular the hypersonic pitching moment anomaly of the Space Shuttle Orbiter are discussed.

A similar range of topics is treated in Chapter 4 for winged airbreathing (CAV-type) flight vehicles, there especially also issues of aerothermodynamic airframe/propulsion integration and of waverider design, and in Chapter 5 for re-entry capsules (RV-NW-type vehicles). For a couple of the capsules aerodynamic data sets of longitudinal motion are given, together with an in-depth discussion of the role of the z-offset of the center-of-gravity and nominal and parasite trim.

Chapter 6 is devoted to a thorough presentation of aerothermodynamic design problems of stabilization, trim and control devices, which to a large extent are the same for all vehicle classes considered in this book. In Chapter 7 general formulations are given of forces, moments, center of pressure, trim, and stability of flight vehicles. These, like the flight mechanical equations in Chapter 2 usually are not found in a general form in the literature.

Chapter 8 is devoted to a discussion of multidisciplinary design aspects which are of interest for hypersonic vehicle design. Also in this evolving discipline the describing equations in a general form are seldom found in the literature, which also holds for the coupling procedures. Few application examples are available so far from hypersonic vehicle design.

The thermal state of a vehicle surface and the ensuing thermal loads and thermal surface effects are of large interest in aerothermodynamic vehicle design. Nevertheless, we do not treat them separately in this book, but integrate them in the corresponding chapters. However, a short overview of the basic issues, as well as a simulation compendium, is given in Chapter 9, together with a compilation of the most important approximate relations used or referred to in the book in Chapter 10. The full governing equations for flow with high-temperature real-gas effects are presented then in Chapter 11.

Finally in Chapter 12 properties of the Earth atmosphere are given which should provide the reader quickly with data especially also for the solution of the problems which are provided at the end of several of the chapters. The book closes with constants, units and conversions, Chapter 13, symbols, Chapter 14, a glossary, abbreviations and acronyms, Chapter 15, and a solution guide to the problems, Chapter 16, followed by the acknowledgement of copyright permissions, the author and the subject index.

References

1. Woods, W.C., Arrington, J.P., Hamilton II, H.H.: A Review of Preflight Estimates of Real-Gas Effects on Space Shuttle Aerodynamic Characteristics. In: Arrington, J.P., Jones, J.J. (eds.) Shuttle Performance: Lessons Learned. NASA CP-2283, Part 1, pp. 309–346 (1983)
2. Hirschel, E.H.: Basics of Aerothermodynamics. Progress in Astronautics and Aeronautics, AIAA, Reston, Va, vol. 204. Springer, Heidelberg (2004)

3. Bushnell, D.M.: Hypersonic Ground Test Requirements. In: Lu, F.K., Marren, D.E. (eds.) Advanced Hypersonic Test Facilities. Progress in Astronautics and Aeronautics, AIAA, Reston, Va, vol. 198, pp. 1–15 (2002)
4. Hirschel, E.H.: Present and Future Aerodynamic Process Technologies at DASA Military Aircraft. In: ERCOFTAC Industrial Technology Topic Meeting, Florence, Italy (1999)
5. Vos, J.B., Rizzi, A., Darracq, D., Hirschel, E.H.: Navier-Stokes Solvers in European Aircraft Design. Progress in Aerospace Sciences 38, 601–697 (2002)
6. Hirschel, E.H.: Towards the Virtual Product in Aircraft Design? In: Periaux, J., Champion, M., Gagnepain, J.-J., Pironneau, O., Stouflet, B., Thomas, P. (eds.) Fluid Dynamics and Aeronautics New Challenges. CIMNE Handbooks on Theory and Engineering Applications of Computational Methods, Barcelona, Spain, pp. 453–464 (2003)
7. Neumann, R.D.: Defining the Aerothermodynamic Methodology. In: Bertin, J.J., Glowinski, R., Periaux, J. (eds.) Hypersonics. Defining the Hypersonic Environment, vol. 1, pp. 125–204. Birkhäuser, Boston (1989)
8. Hunt, J.L.: Hypersonic Airbreathing Vehicle Design (Focus on Aero-Space Plane). In: Bertin, J.J., Glowinski, R., Periaux, J. (eds.) Hypersonics. Defining the Hypersonic Environment, vol. 1, pp. 205–262. Birkhäuser, Boston (1989)
9. Hammond, W.E.: Design Methodologies for Space Transportation Systems. Education Series, AIAA, Reston, Va (2001)
10. Hirschel, E.H.: CFD - from Solitary Tools to Components of the Virtual Product. In: Proceedings Basel World CFD User Days 1996. Third World Conference in Applied Computational Fluid Dynamics, Freiburg, Germany, pp. 24.1–24.9 (1996)
11. Arrington, J.P., Jones, J.J. (eds.): Shuttle Performance: Lessons Learned. NASA CP-2283 (1983)
12. Throckmorton, D.A. (ed.): Orbiter Experiments (OEX) Aerothermodynamics Symposium. NASA CP-3248 (1995)
13. Woods, W.C., Watson, R.D.: Shuttle Orbiter Aerodynamics – Comparison Between Hypersonic Ground-Facility Results and STS-1 Flight-Derived Results. In: Throckmorton, D.A. (ed.) Orbiter Experiments (OEX) Aerothermodynamics Symposium. NASA CP-3248, Part 1, pp. 371–409 (1995)
14. Lu, F.K., Marren, D.E. (eds.): Advanced Hypersonic Test Facilities. Progress in Astronautics and Aeronautics, AIAA, Reston, Va, vol. 198 (2002)
15. Haney, J.W.: Orbiter (Pre STS-1) Aeroheating Design Data Base Development Methodology – Comparison of Wind Tunnel and Flight Test Data. In: Throckmorton, D.A. (ed.) Orbiter Experiments (OEX) Aerothermodynamics Symposium. NASA CP-3248, Part 2, pp. 607–675 (1995)

2 Short Introduction to Flight Trajectories for Aerothermodynamicists

Aerothermodynamic design, aerothermodynamic phenomena, and the choice of flight trajectories of either re-entry vehicles, space-transportation systems or hypersonic aircraft depend mutually on each other. We give here a short introduction to issues of flight trajectories in order to provide basic knowledge about these dependencies.

The very fast flight of hypersonic vehicles, partly with vast changes of the flight altitude, makes a precise flight guidance necessary. This is especially a problem with CAV-type space transportation systems because of their very small pay-load fractions. The basic problem is to find a flight trajectory which permits the vehicle to fulfill its mission with minimum demands on the vehicle system. However, different from classical aircraft design, the physical properties and the functions of a hypersonic vehicle and its components must be extremely closely tailored to the flight trajectory and vice versa.

To design and to optimize a vehicle's flight trajectory in a sense is to solve a guidance problem. While the fulfillment of the basic mission is the primary objective of the trajectory definition, other, secondary objectives may exist. In the multi-objective design and optimization of a trajectory, these must be identified as guidance objectives. It is further necessary to define and to describe the trajectory control variables, which permit the vehicle to fly the trajectory. Finally, a system reduction is necessary to identify a few characteristic physical loads and vehicle properties/functions, whose limitations and/or fulfillments are introduced as systems and operational constraints in the trajectory design and optimization process. The eventual outcome are guidance laws, which in general have a rather small number of free parameters to fulfill the mission objectives under the given conditions.

Prerequisites for trajectory design and optimization are flyability and controllability of the considered vehicle on the sought trajectory. Under flyability we understand longitudinal trimmability, and static and dynamic stability, which, with a few exceptions, is the rule for both the longitudinal and the lateral motion of the vehicle. Controllability is the ability to steer the vehicle around all relevant vehicle and air-path related axes with the help of control devices. For RV-W-type flight vehicles these are aerodynamic control surfaces and usually, in addition, reaction control systems (RCS) in the form of small rocket thrusters located appropriately around the vehicle, for RV-NW-type vehicles they are in general solely reaction control systems[1]. We stress the fact that only a "trimmed" trajectory is a viable trajectory. For airbreathing (CAV-type) flight vehicles the influence of the thrust vector of the propulsion system in the lift-drag plane on the longitudinal force and the moment balance must be taken into account.

Trajectory design and optimization must allow for uncertainties in the describing data of the vehicle, its sub-systems, and the flight environment, for a RV-W-type vehicle see, for instance, [1]. The uncertainties concern the aerodynamic model – the aerodynamic data set – of the vehicle including uncertainties in the performance data of the control devices, and also uncertainties in the performance data of the propulsion system in the case of CAV-type flight vehicles. Uncertainties of other kinds are present as a rule regarding the vehicle mass, the location of the center-of-gravity of the flight vehicle and its moments of inertia. This holds especially for RV-NW-type vehicles with ablation cooling. With CAV-type flight vehicles all these are anyway not constant because of the fuel consumption during flight, and, in the case of TSTO-systems, also because of the separation of the upper stage.

Still other uncertainties come in from the sensor systems (air data, acceleration data) et cetera, and are also given in the form of deviations from the, for the trajectory design chosen, standard atmosphere during the actual mission, especially regarding the density ρ_∞, and the possible presence of wind. The latter concerns in particular CAV-type vehicles, because these fly predominantly in the troposphere and the stratosphere, Chapter 12.

In the following sections we look at the trajectory design and optimization elements which have close connections to aerothermodynamics (guidance objectives, trajectory control variables, systems and operational constraints). We consider the forces acting on a vehicle, discuss the equilibrium glide trajectory of RV-W-type and RV-NW-type vehicles, give qualitative results, and show in case studies some examples of trajectories. Finally the compact and frame-consistent derivation of the general equations for planetary flight is given. We refrain from discussing guidance laws, and refer the reader the reader instead to, e.g., [2]. We begin with RV-NW-type vehicles, where considerable flight experience is available,[2] and proceed with CAV-type flight vehicles, where, however, flight experience is not available.

---

[1] The major role, however, of the RCS of a flight vehicle leaving the atmosphere (above $H \approx 80.0$ to $100.0\ km$) and/or performing orbital flight, is to carry out orbital manoeuvering.

[2] The reader is especially referred to [1] about the Space Shuttle Orbiter's re-entry guidance.

2.1 Flight Trajectories of Winged and Non-Winged Re-entry Vehicles

2.1.1 General Aspects

RV-W-type and RV-NW-type vehicles have in common that their re-entry flight as decelerating flight is actually a braking mission. Their large initial total air-path energy:

$$E_{t,i} = m\left(gH_i + \frac{1}{2}v_i^2\right), \tag{2.1}$$

is dissipated exclusively by means of the aerodynamic drag. In eq. (2.1) m is the vehicle mass, g the gravitational acceleration as function of the altitude, Section 13.1, H_i the initial altitude, and v_i the initial speed.

The dissipation of the large initial total energy requires specific systems constraints of which the dynamic pressure, the thermal surface loads and the aerodynamic load factor belong to the most important ones. The result is an usually very narrow re-entry flight corridor. We show in Fig. 2.1 as an example the flight corridor of the Space Shuttle Orbiter for the operational angle of attack profile, [1].

The minimum weight of the Space Shuttle Orbiter's thermal protection system (TPS) is achieved by flying on a large part of the trajectory the maximum angle of attack, consistent with the cross-range requirements, in order to minimize the thermal loads. During the initial five flights, which served as test flights, this angle was $\alpha = \alpha_{max} = 40°$, during the following operational flights $\alpha = \alpha_{max} = 38°$, [1].

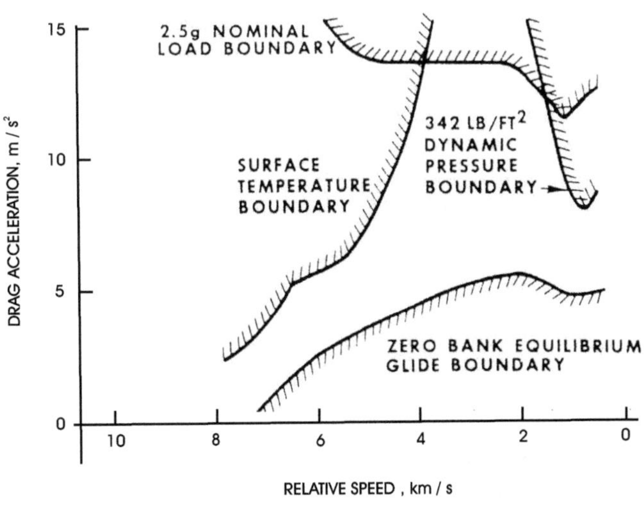

Fig. 2.1. Flight corridor of the Space Shuttle Orbiter (operational flights), [1].

A RV-W-type vehicle flies, with basically fixed configuration, in a large Mach number and altitude range. During the high speed re-entry it flies at large, and at low Mach numbers at small angles of attack. With increasing angle of attack the effective longitudinal "nose" radius in the stagnation point region of the vehicle increases (rise of effective bluntness)[3]. With increasing nose radius, at constant flight speed and altitude, the boundary layer thickness increases and the thermal loads, both the heat flux in the gas at the wall, q_{gw}, and the surface temperature T_w (which without slip-flow effects is equal to the temperature in the gas at the wall T_{gw}, Section 9.1) of the radiation cooled TPS surface, decrease, [5].

Increased effective bluntness also increases the portions of the bow shock with large inclination against the free-stream, and hence the wave drag and with that the deceleration[4] of the vehicle along the flight path. The blunt vehicle shape at large angle of attack thus serves both low thermal loads and high drag (and deceleration), [5].

In the altitude-velocity map the flight trajectories of RV-W-type and RV-NW-type vehicles can be distinguished, Fig. 2.2. The lift parameter $\alpha_W = W/(A_{ref}C_L)$ and the ballistic parameter $\beta_W = W/(A_{ref}C_D)$ are derived in Sub-Section 2.1.4. They can be related to each other by the lift-to-drag ratio L/D. The "lifting" re-entry trajectory of RV-W-type vehicles is much "higher" than that of RV-NW-type vehicles. Our intuition tells us that the higher the trajectory, the smaller the thermal loads, but the lower the effectiveness of aerodynamic stabilization, trim, and control surfaces. The ballistic or semi-ballistic re-entry of RV-NW-type vehicles thus is marked by much larger thermal loads than the lifting re-entry of RV-W-type vehicles.

Cross-range capabilities of RV-W-type and especially RV-NW-type vehicles are limited because of their small lift-to-drag ratios. Usually RV-W-type flight vehicles have in the high speed domain a $L/D = O(1)$ due to the blunt vehicle shape and the large angles of attack. The Space Shuttle Orbiter has a trimmed $L/D \approx 1.0$ at $\alpha \approx 40°$, Fig. 2.3[5], [7]. For the upper stage HORUS of the TSTO reference concept SÄNGER of the former German Hypersonics Technology Programme, [8], $L/D \approx 1.9$ was envisaged at that angle of attack. For RV-NW-type vehicles we find $L/D = 0.1$ to 0.3, [9], which is achieved by an offset of the center-of-gravity from the centerline, and hence is the trimmed L/D. For purely ballistic re-entry capsules $L/D = 0$. All these vehicles can be considered as compressibility or pressure effects dominated flight vehicles, Section 1.1.

[3] For the Space Shuttle Orbiter's equivalent axisymmetric body, [3], the "nose" radius rises almost linearly from $R_N = 0.493\ m$ at $\alpha = 21.8°$ to $R_N = 1.368\ m$ at $\alpha = 42.75°$, [4].

[4] In trajectory design and optimization the "drag acceleration" is used instead of the "deceleration", Sub-Section 2.1.4.

[5] The agreement between the flown L/D data of the trimmed vehicle and the predicted data is very good. The flight data show Mach number independence, Section 3.6.

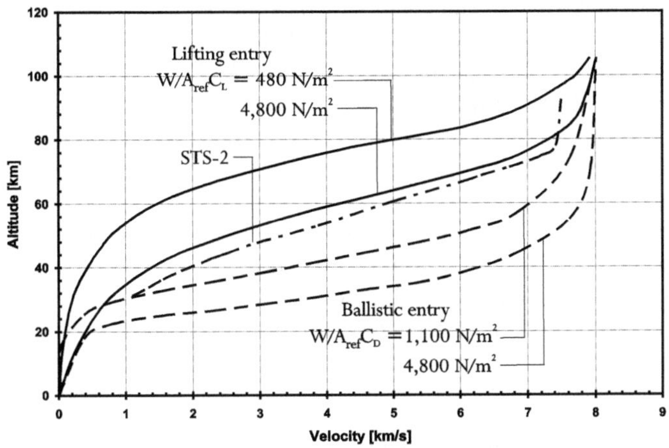

Fig. 2.2. Trajectories of RV-W-type and RV-NW-type vehicles with typical values of lift parameters and ballistic parameters in the altitude-velocity map (STS-2: second flight of the Space Shuttle Orbiter, data from [6]).

Lift-to-drag ratios of O(1) of RV-W-type flight vehicles are due to the blunt, although elongated shape of the vehicles – usually with large portions of the lower side being approximately flat, Sub-Section 3.2.2 – in combination with large angles of attack.

During a re-entry flight, which is performed at large angles of attack, L/D can be increased by both reducing the angle of attack α (reduction of the "effective" bluntness of the configuration) and the actual nose bluntness (nose

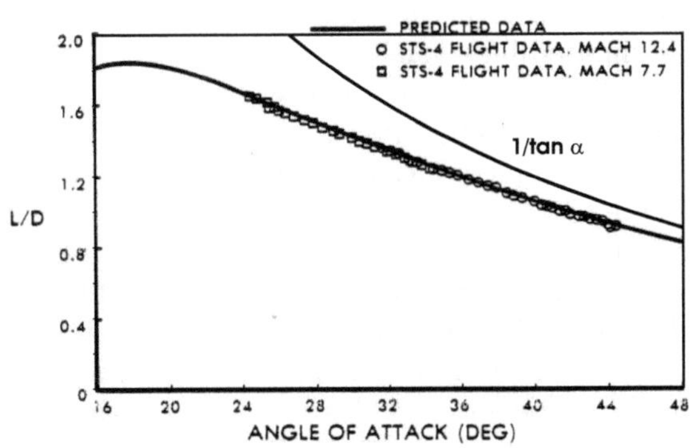

Fig. 2.3. Trimmed lift-to-drag ratio L/D of the Space Shuttle Orbiter in the hypersonic domain as function of the angle of attack α, [7]. The large angle of attack interval of the flight data was achieved by transient pushover-pull-up maneuvers around the actual flight angle of attack.

radius R_N). If we approximate the lower side of a RV-W-type flight vehicle by an equivalent flat plate, the RV-W-type RHPM-flyer, Section 10.1, and apply Newton's theory, we find for the lift-to-drag ratio $L/D = 1/tan\, \alpha$, which in the case of the Space Shuttle Orbiter is a fair approximation for $\alpha \gtrapprox 25°$, Fig. 2.3. Thus reducing the angle of attack, also for realistic vehicle shapes, is an effective means to increase the lift-to-drag ratio, as is amply demonstrated by Fig. 2.3.

We note, however, that in reality a reduction of L/D of a given flight vehicle is undertaken on appreciable parts of the trajectory via a reduction of L. With the bank angle μ_a of the vehicle an effective lift $L_{eff} \leqq L$ and/or a side force is achieved, which serve as trajectory-control means, Sub-Section 2.1.2.

2.1.2 Guidance Objectives, Trajectory Control Variables, and Systems and Operational Constraints

We discuss now some issues of the above mentioned guidance objectives, trajectory control variables, and systems and operational constraints.

Guidance objectives: For RV-W-type and RV-NW-type vehicles the most important guidance objectives are:

- Minimization of the time-integrated heat flux in the gas at the wall $\overline{q}_{gw}$ at selected reference locations:

$$\overline{q}_{gw} = \int_{t_0}^{t_{flight}} q_{gw} dt, \tag{2.2}$$

  which is used as a measure of the thickness and hence the weight of the heat protecting or insulating structure. The reference locations are at least the nose cap, approximated by a sphere, where q_{gw} would be the forward stagnation point heat flux, and usually parts of the TPS, where the heat fluxes q_{gw} can be approximated by those of a flat plate or a swept cylinder et cetera, see, e.g., [10]. Simple relations for the determination of q_{gw} are provided in Chapter 10.

  The use of the time integral of q_{gw} in trajectory design and optimization has historical roots. In reality, it is the time integral of q_w, the heat flux which actually enters the TPS or the hot primary structure, which is of importance. In presence of radiation cooled surfaces, Section 9.1, which are the rule for hypersonic vehicles in the velocity and altitude range considered in this book, this is $\overline{q}_w = \overline{q}_{gw}$ - $\overline{q}_{rad}$, where $\overline{q}_{rad}$ is the time-integrated radiation cooling heat flux q_{rad}.
- Cross range achievement. The cross range is the lateral distance of the prescribed landing site from the exit orbital plane. The down range – in direction of the trace of the exit orbit plane – is to be achieved by a proper timing of the de-orbiting impulse. For both ranges possible cross and

down winds must be compensated. In general a multi-objective problem exists, because possible contingency returns of the flight vehicle must be considered, too. In all cases the vehicle with its mass m must arrive at the terminal area (TA) with an adequate total air-path energy state – $E_{t,TA} = m_{TA}(g H_{TA} + 0.5 v_{TA}^2) \geqq E_{TA,min}$ – and a reasonable vehicle attitude.

Trajectory control variables are the attitude variables which permit to fly a RV-W- or RV-NW-type vehicle on its re-entry trajectory. For RV-W-type vehicles most experience about vehicle control is available from the flights of the Space Shuttle Orbiter, see, e.g., [1]. In principle only three vehicle trajectory control variables are at one's disposal:

- The angle of attack α is the trajectory control variable of RV-W-type flight vehicles which first of all governs the thermal loads on the vehicle's structure. In general it holds, the larger α, the smaller are these loads. The angle of attack further governs the drag and hence the deceleration of the vehicle (Sub-Section 2.1.4). The larger α, the larger is the total drag D of the vehicle. The Space Shuttle Orbiter flies at large Earth-relative speed ($4.0\ km/s \lessapprox v_\infty \lessapprox 8.0\ km/s$, equivalent to $12 \lessapprox M_\infty \lessapprox 28$) at $\alpha = O(40°)$, [1]. A mission usually is flown with a predefined function $\alpha(v_\infty)$. If bank angle changes or reversals are performed on the trajectory, small deviations around $\alpha(v_\infty)$ are the rule.

  For (axisymmetric) RV-NW-type vehicles thermal loads on the front shield are smallest, if the angle of attack, Chapter 5, is zero (ballistic entry). Then also the drag is largest. However, in this case $L/D = 0$, and the cross range – without attitude-change capability – is zero, too.
- To modulate the effective aerodynamic lift L_{eff} ($= L \cos \mu_a$, Sub-Section 2.1.4), and the effective lift-to-drag ratio L_{eff}/D of RV-W-type and RV-NW-type flight vehicles, a bank angle μ_a is employed[6]. On a large (initial) part of the trajectory, changes of the angle of attack are usually not available to modulate the lift, because α must be large in order to minimize the thermal loads (see above). A large α is further needed for sufficient drag and hence vehicle deceleration.

  The bank angle in addition is the means to control the cross range of a vehicle. Because it directs part of the lift sidewards ($L \sin \mu_a$), it induces the lateral motion. We see from this, that RV-NW-type vehicles need to have a $L/D > 0$, i. e., to fly at angle of attack, if a cross range capability is wanted. If the bank angle is used to reduce the flight time, Sub-Section 2.1.5, and the lateral motion is unwanted, bank angle reversals are used to compensate for it. The Space Shuttle Orbiter employs bank angles up to $\mu_a = \pm\ 80°$, [1].

---

[6] Although RV-NW-type vehicles usually have axisymmetric shapes, a bank angle is an option for them too, if it means roll around the velocity vector, see Fig. 2.22. However, it implies a non-zero angle of attack which anyway is necessary in order to achieve a $L/D > 0$, Chapter 5.

The bank angle μ_a is defined as roll angle around the velocity vector $\underline{v}_\infty$, Fig. 2.22, of the RV-W-type or RV-NW-type vehicle (also called air-path vector, [11], or wind vector), and not as roll angle around the longitudinal axis of the vehicle. Roll must be made around the vehicle's velocity vector, otherwise the lift is not effectively modulated or directed sidewards. Further, regarding RV-W-type vehicles, the lower side of the vehicle with its TPS must always remain the vehicle's windward side. For such vehicles, initially on the trajectory the bank angle is zero: $\mu_a = 0$, i. e. the wings are in horizontal position. $\mu_a = 180°$ would mean that the vehicle flies fully inverted.

- The sideslip (yaw) angle β of a flight vehicle is a potential trajectory control variable. But because it would induce unwanted increments of thermal loads, at least in the high Mach number segment of the trajectory, sideslip is kept zero on a large part of the trajectory, i. e. the vehicle flies at high speed with $\beta = 0$.

Systems and operational constraints: For both RV-W-type and RV-NW-type vehicles several flight path systems and operational constraints exist. They may influence only parts of the trajectory or the whole trajectory[7]. We list the most important flight path systems and operational constraints of both vehicle types. For details regarding especially RV-W-type vehicles we refer the reader to, e.g., [1].

- The dynamic pressure q_∞ is one of the most important systems and operational constraints with:

$$q_\infty = \frac{1}{2}\rho_\infty v_\infty^2 \leqq q_{\infty,max}. \tag{2.3}$$

The dynamic pressure is a measure of the mechanical (pressure and shear stress) loads on the vehicle structure, mainly the surface pressure. Important is that also all aerodynamic forces and moments are proportional to it. This holds also for performance and efficiency of aerodynamic stabilization and control devices and the sizes of the hinge moments of aerodynamic trim and control surfaces, which govern the actuator performance.

For RV-W-type flight vehicles the maximum allowed dynamic pressure typically chosen is about $q_{\infty,max} \approx 14.0\ kPa$, which amounts to the maximum value of the Space Shuttle Orbiter, [12]. Initially on a re-entry trajectory q_∞ is much lower due to the low density, see the X-38 example in Sub-Section 2.1.7. This leads to the problem of reduced and insufficient effectiveness of aerodynamic stabilization, trim and control surfaces. To overcome this problem, reaction control devices, as in the case of the Space Shuttle Orbiter, see, e.g., [13], are employed. There, the vertical RCS jets operate in roll for $q_\infty \lessapprox 0.48\ kPa$ ($H \gtrapprox 81.0\ km$, $M_\infty \gtrapprox 27$), and in pitch

[7] In any case not only the nominal flight trajectory must be considered, but conceivable alternative trajectories, and especially abort trajectories.

for $q_\infty \lessapprox 1.9\ kPa$ ($H \gtrapprox 69.0\ km$, $M_\infty \gtrapprox 22$). The yaw RCS jets are active as long as $M_\infty > 1$. The body flap and the elevons are active at $q_\infty \gtrapprox 0.1\ kPa$ ($H \lessapprox 90.0\ km$, $M_\infty \lessapprox 27$), the rudder only at Mach numbers $M_\infty \lessapprox 5$, where at angles of attack $\alpha \lessapprox 20°$ to $25°$ the vertical stabilizer begins to leave the "shadow" of the fuselage[8].

For RV-NW-type flight vehicles a large range of maximum dynamic pressures can be found from low values like 6.2 kPa for OREX up to 25.0 kPa for the lunar return of APOLLO and even larger ones.

- Thermal loads[9] must be sufficiently low so that the thermal protection system can cope with them, and functions and integrity of all other structural elements are not deteriorated.

  The maximum permissible thermal loads must be seen relative to the considered locations on the flight vehicle. The structure and materials concept of, e.g., the TPS, or of a control surface et cetera, always is tailored closely to the expected and actual thermal (and mechanical) loads in order to save weight and cost. The thermal loads are very different at the different locations of the vehicle. This holds even for the TPS, whose thickness decreases from the stagnation point region (maximum) towards the aft part of the vehicle (minimum). Behind this fact is the behavior, Chapter 9, of the heat flux in the gas at the wall, q_{gw}, and the wall temperature T_w in the presence of radiation cooling, which in general is quite close to the radiation-adiabatic temperature T_{ra}, Section 9.1.

  Trajectory design and optimization in general is made using a heat flux as constraint, either the total heat flux Q_∞, eq. (2.4), or, as the rule, the heat flux in the gas at the wall q_{gw} (see the guidance objectives above) at one or more reference locations as a result of the mentioned system reduction. The reference locations are usually and primarily the nose cap, which is approximated by a sphere (forward stagnation point of the vehicle), and in addition other configuration parts[10], which can be approximated by swept cylinders (leading edges) or flat plates (flat portions of the lower side), et cetera, see, e.g., [10]. At the sphere, for instance, it is demanded that $q_{gw} \lessapprox q_{gw,max}$, with, depending on the overall structure and material layout of the vehicle, $q_{gw,max} = 400.0$ to 1,200.0 kW/m^2.

  The requirement that q_{gw} at the reference locations remains within given constraints $q_{gw,max}$, holds for the whole configuration on the

---

[8] The vertical stabilizer has the drawback that it is not effective at the high angles of attack which are flown on a large part of the trajectory. However, it must be seen in the frame of the overall vehicle and mission layout, which is a result of numerous design trades.

[9] We define thermal loads, Chapter 9, in the sense that they encompass both surface temperatures T_w and the heat flux into the surface q_w, [5].

[10] In [5] it is shown that, due to strong interaction phenomena, thermal loads at wing leading edges, represented in that case by the radiation-adiabatic temperature, can be as large as at the forward stagnation point of the vehicle.

nominal flight trajectory and on all conceivable alternative trajectories, including abort trajectories. What actually is demanded, and that is the problem of a proper system reduction, is that everywhere at the vehicle and at any time the thermal loads in the form of the wall temperature T_w (in view of material strength and endurance, also with regard to erosion, where the wall shear stress τ_w plays a role, too), and the heat flux q_w into the wall, as well as the time integrated wall heat flux $\overline{q}_w$, eq. (2.2), are within the limits of the given vehicle layout[11]. To check whether this requirement is met, appropriate post-optimization analyses of the trajectory must be made.

The total heat flux Q_∞ is the heat transported per unit area and unit time towards a flight vehicle:

$$Q_\infty = \rho_\infty v_\infty \left(h_\infty + \frac{v_\infty^2}{2} \right), \tag{2.4}$$

with ρ_∞ and v_∞ being the free-stream density and speed (their product is the mass flux per unit area towards the flight vehicle), and h_∞ the enthalpy of the free-stream, i. e., of the undisturbed atmosphere.

At hypersonic speed the kinetic energy is dominant, and hence the transported heat is approximately proportional to the flight velocity squared, and we note:

$$Q_\infty \sim \rho_\infty v_\infty \frac{v_\infty^2}{2} \sim q_\infty v_\infty. \tag{2.5}$$

The heat flux in the gas at the wall, q_{gw}, of a sphere usually is approximated with the help of the relation of Fay and Riddell, Section 10.3. For the computation of large amounts of trajectory points, simpler relations for q_{gw} of the type:

$$q_{gw} = C_{gw} \rho_\infty^n v_\infty^m, \tag{2.6}$$

for spheres, swept cylinders, flat plates, the latter two for both laminar and turbulent (flight below approximately 60.0 to 40.0 km altitude, [5]) flow, can be found, Section 10.3. For relations for flat surfaces see, e.g., Section 10.4. For spheres $n = 0.5$ to 1, and $m \gtrapprox 3$. C_{gw} must be chosen accordingly. For a sphere, for instance, it includes the inverse of the square root of the radius, $(1/\sqrt{R})$, Section 10.3.

We remind the reader that for the actual structure and materials layout the heat flux into the wall q_w, respectively the time-integrated value $\overline{q}_w$ is the relevant one. The wall temperature T_w is of equal importance as q_w, [5], because it determines the choice of the material. It must be below the

[11] For the development of the Aeroheating Design Data Base of the Space Shuttle Orbiter the following major reference locations were distinguished [10]: the fuselage lower side including the nose area, the wing lower surface, the wing leading edges, control surfaces, the wing lee side, the fuselage sides and its upper surface.

maximum permissible one: $T_w \leqq T_{w,max}$. For radiation cooled surfaces the radiation-adiabatic temperature T_{ra} in general is a good approximation of the real T_w. However, in any case possible non-convex situations, Section 9.1, but also possible transverse, Sub-Section 6.3.3, and/or tangential heat transport, Sub-Section 8.4.3, through the structure must be taken into account. With the approximate relations for T_{ra} given in Chapter 10, it can easily and with acceptable cost be checked, whether on the trajectory $T_w \leqq T_{w,max}$ is fulfilled at the reference locations.

- The normal load factor is defined as the ratio of the normal aerodynamic force $N = q_\infty C_N A_{ref}$ to the vehicle weight $W = m\,g$:

$$n_z = \frac{q_\infty C_N A_{ref}}{m\,g},$$

where, following [11], g should be taken as function of the distance of the actual flight path to the center of Earth: $g = g(R_E + H)$, where R_E is the mean Earth radius and H the flight altitude, Chapter 13. It concerns loads on the flight vehicle structure but also on passengers. The constraint is $n_z \lessapprox n_{z,max}$, and the maximum value is usually defined for RV-W-type vehicles as $n_{z,max} = 2$ to 2.5, while for RV-NW-type vehicles it can reach $n_{z,max} = 8$ to 10.

- Equilibrium glide, Sub-Sections 2.1.4 and 2.1.5, is associated with the minimum drag acceleration if the bank angle is zero, and is used as operational constraint, too.
- Trimmability can be considered as indirect operational constraint which partly is covered by the dynamic-pressure constraint. It must be assured on the whole trajectory. It can have severe implications for the trajectory. If, for instance, a large trim-surface (e.g. body flap) deflection is needed on a large part of the trajectory, the ensuing trim drag will influence the cross range of the vehicle, not to mention the resulting hinge moments (demands on actuator performance), and the mechanical and thermal loads, which then pose additional systems and operational constraints.
- Stability and controllability, like trimmability are also indirect operational constraints. Forces exerted by aerodynamic stabilization and control surfaces can pose systems and operational constraints in both directions, either by being too large or too small. If, for instance, on a high trajectory aerodynamic stabilization and control surfaces are not effective, a RCS must be foreseen, see the "dynamic pressure" constraint.

2.1.3 Forces Acting on a Re-entry Vehicle

We consider a RV-W-type flight vehicle on its flight path. We summarize the axes, forces and moments in Table 2.1, following [11]. The general axis convention is: the x-axis in both the body-axis and the air-path axis system points forward, the z-axis downward, and the y-axis to the right, when looking

Table 2.1. Axes, and aerodynamic forces (in brackets the "aerodynamic" notation) and moments (all are right-hand orthogonal triples) in the body-axis system and the air-path system, [11].

Axis system	Axis	Force	Moment
Body axis			
	longitudinal axis x	axial force X (A)	rolling moment L^A
	lateral axis y	side force Y	pitching moment M^A
	normal axis z	normal force Z (N)	yawing moment N^A
Air-path axis			
	air-path axis x_a	drag X_a (D)	rolling moment L_a
	air-path lateral axis y_a	lateral force Y_a (C)	pitching moment M_a
	air-path normal axis z_a	lift Z_a (L)	yawing moment N_a

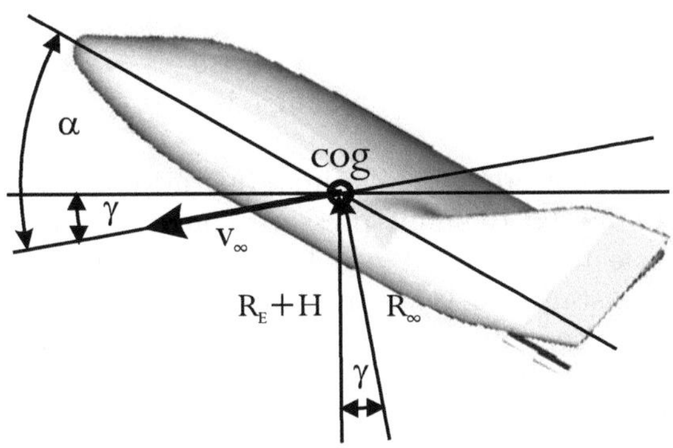

Fig. 2.4. Schematic of a re-entry vehicle on its flight path (cog: center-of-gravity). $R_E + H$ is the distance to the center of Earth, R_∞ is the local radius of the curved flight path.

in flight direction (see Fig. 7.1 and [14] to [17]). Both axis systems are right-handed systems. The definition of the forces is given in Fig. 7.2.

The aerodynamicists usually have the x-axis pointing backward, the z-axis upward, and the y-axis to the right, when looking in flight direction, see, e.g., Fig. 7.3. We use this convention throughout the book[12].

[12] The reader is generally warned, that it is mandatory and absolutely necessary, when using data and literature from other disciplines, or when dealing with representatives of other disciplines, that a full understanding and agreement is reached about axis conventions, symbols, signs, and nomenclature!

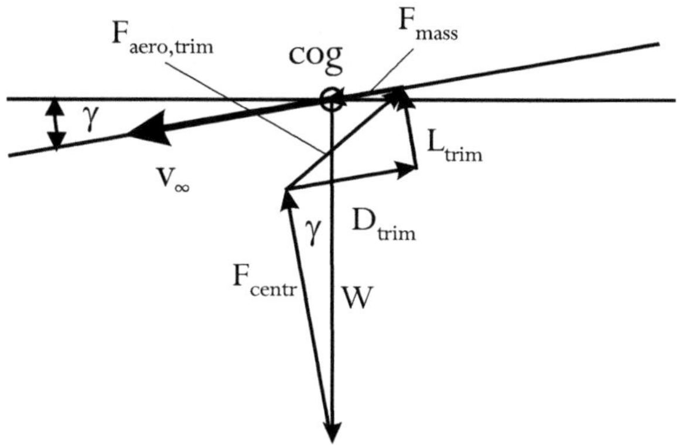

Fig. 2.5. Schematic point-mass force polygon at re-entry flight (the aerodynamic forces are those of trimmed flight, $F_{aero,trim}$ is the resultant aerodynamic force of trimmed flight).

For the pitching moment in the body-axis system usually and also in this book the symbol M is used instead of M^A. In Chapter 5 the more convenient notation $[L, M, N]$ is used for the components of the moment vector.

The vehicle has a weight W, acting on it is the aerodynamic lift L, the aerodynamic drag D, and the centrifugal force F_{centr} ("g-reduction" due to the high speed flight on a curved trajectory). At a CAV-type vehicle also the inlet drag force and the thrust force of the propulsion system act on the vehicle, Section 2.2.3. In non-steady flight an inertial or mass force F_{mass} complements these forces. The flight path is inclined by the flight path angle γ against the local horizontal plane, which is defined positive in upward direction and negative in downward direction, Fig. 2.4. The angle γ varies along the flight path. The flight vehicle is approximated as a point mass, i. e., it is trimmed, but moments are not examined[13].

The forces acting on the vehicle are shown in Fig. 2.5. The flight path angle γ initially is small ($\gamma = O(\text{-}1°)$) (danger of phugoid motion of the vehicle[14]), and varies only little down to approximately 40.0 km altitude.

2.1.4 The Equilibrium Glide Trajectory

The equilibrium glide trajectory is not a viable trajectory. However, it is very useful, because it permits to obtain a closed solution for the flight speed as function of the flight time. From this several important, though in general qualitative results can be found, which we discuss in Sub-Section 2.1.5.

In Section 2.3 we present a compact and frame-consistent derivation of the general equation for unpowered planetary flight, which results in three scalar

[13] For trim and stability considerations see Sub-Section 3.4.2

[14] Long-period lightly damped oscillatory modes are called phugoid motion, see, e.g., [18].

equations, eq. (2.55), see also [19]. By assuming the planet as non-rotating ($\omega = 0$) and the flight path azimuth angle χ as constant, we have for planar flight[15], Fig. 2.5:

$$\frac{dv_\infty}{dt} = -\frac{D}{m} - \frac{W}{m}\sin\gamma = -\frac{D}{m} - g\sin\gamma = -\frac{F_{mass}}{m}, \tag{2.7}$$

$$v_\infty\frac{d\gamma}{dt} = \frac{1}{m}L - \frac{W}{m}\cos\gamma + \frac{v_\infty^2}{R_\infty} = \frac{1}{m}L - g\cos\gamma + \frac{v_\infty^2}{R_\infty}, \tag{2.8}$$

where $m = W/g$ is the vehicle mass[16], g the gravitational acceleration, which is a function of the flight altitude, Chapter 13, and $R_\infty = (R_E + H)/\cos\gamma \approx R_E + H$, the local radius of the flight path. R_E is the mean Earth radius, H the flight altitude, and, compared to eq. (2.55), we have the identities $v_\infty \equiv V$ and $R_E + H \equiv r$.

For the sake of completeness of eqs. (2.7) and (2.8) for the description and evaluation of flight trajectories, we give the kinematic relations for the altitude H and the longitudinal angle θ, see eq. (2.52):

$$\frac{dr}{dt} \equiv \frac{d(R_E + H)}{dt} = \frac{dH}{dt} = V\sin\gamma, \tag{2.9}$$

and

$$r\frac{d\theta}{dt} \equiv (R_E + H)\frac{d\theta}{dt} = V\cos\gamma. \tag{2.10}$$

The third term on the right-hand side of eq. (2.8) times the vehicle mass m is the centrifugal force F_{centr} due to the curved trajectory of the flight vehicle:

$$F_{centr} = m\frac{v_\infty^2}{R_\infty}. \tag{2.11}$$

For flight control and guidance purposes the lift vector $\underline{L}$ can be rotated out of the planar surface by the bank angle μ_a, Section 2.3 and Sub-Section 2.1.2. In this case we replace L by $L\cos\mu_a$, which is then the component of the lift acting against the Earth gravitation. We call this the effective lift L_{eff}:

$$L_{eff} = L\,cos\mu_a. \tag{2.12}$$

The bank angle also serves to create the side force $L\,sin\mu_a$ needed for cross range control, see Sub-Section 2.1.2 and third equation of eq. (2.55).

Introducing the aerodynamic force coefficients C_L and C_D, and the reference area A_{ref}, yields from the above relations after rearrangement the with g normalized acceleration along the flight path:

$$n_t = \frac{1}{g}\frac{dv_\infty}{dt} = -sin\gamma - \frac{\rho_\infty v_\infty^2}{2}\left(\frac{W}{A_{ref}C_D}\right)^{-1}, \tag{2.13}$$

[15] The subscript 'trim' of L and D in Fig. 2.5 for convenience is omitted in this and in the following equations.

[16] The reader should note, that W is not the weight of the vehicle at sea level, because $g = g(H)$, Chapter 13.

and normal to it:

$$n_n = \frac{v_\infty}{g}\frac{d\gamma}{dt} = \frac{\rho_\infty v_\infty^2}{2}\left(\frac{W}{A_{ref}C_L}\right)^{-1} - cos\gamma + \frac{1}{g}\frac{v_\infty^2}{R_\infty}. \tag{2.14}$$

The terms in brackets are the ballistic parameter, also called ballistic factor:

$$\beta_W = \frac{W}{A_{ref}C_D}, \tag{2.15}$$

and the lift or glide parameter:

$$\alpha_W = \frac{W}{A_{ref}C_L}. \tag{2.16}$$

The two parameters are given in terms of the vehicle weight $W = mg$. Their dimensions are $[M/L\,t^2]$. In the literature they are often expressed in English units: $[lb_f/ft^2]$, [20]. In metric units they read $[N/m^2] = [Pa]$.

If the parameters are defined in terms of the vehicle mass $m = W/g$, the ballistic parameter is:

$$\beta_m = \frac{m}{A_{ref}C_D}, \tag{2.17}$$

and the lift or glide parameter:

$$\alpha_m = \frac{m}{A_{ref}C_L}. \tag{2.18}$$

Their dimensions are in this case $[M/L^2]$. In the literature they are expressed usually in metric units: $[kg/m^2]$.

Ballistic and glide parameter are related to each other via the lift-to-drag ratio:

$$\frac{W}{A_{ref}C_D} = \frac{W}{A_{ref}C_L}\frac{C_L}{C_D}. \tag{2.19}$$

For RV-W-type flight vehicles they are around[17] $W/A_{ref}C_D = 3{,}600.0$ to $4{,}000.0\ Pa$, and if, for instance, $C_L/C_D \approx 0.8$ in the hypersonic domain at high angle of attack ($\alpha = O(40°)$), $W/A_{ref}C_L \approx 4{,}500.0$ to $5{,}000.0\ Pa$, Fig. 2.2. RV-NW-type vehicles have ballistic parameters between $W/A_{ref}C_D =$ 300.0 and 4,300.0 Pa, Tab. 5.1, see also Fig. 2.2.

The velocity change (deceleration) follows from eq. (2.7) as:

$$\frac{dv_\infty}{dt} = -\frac{D + W sin\gamma}{m} = -\frac{D}{m} - g sin\gamma, \tag{2.20}$$

where D/m is called "drag acceleration".

The flight path angle γ is small on a considerable portion of a lifting entry trajectory and, in addition, changes only slowly. Equilibrium glide flight hence

[17] This is the approximate value for the Space Shuttle Orbiter. For L/D of the Space Shuttle Orbiter as function of α see Fig. 2.3.

is defined as flight in the limit $d\gamma/dt = 0$, with γ being small, such that sinγ ≈ 0, and cos$\gamma \approx 1$, *and* by zero bank angle μ_a.

We assume now $R_\infty = R_E$ and $g = g_0$. With the introduction of the circular or orbital speed, i. e. – disregarding aerodynamic forces – the speed of a body near the surface of Earth, which keeps it in orbit:

$$v_c = \sqrt{g_0 R_E}, \tag{2.21}$$

and after rearrangement of the above eqs. (2.13) and (2.14), we find:

$$\frac{dv_\infty}{dt} = -g_0 \frac{D}{L}\left(1 - \frac{v_\infty^2}{v_c^2}\right), \tag{2.22}$$

where

$$1 - \frac{v_\infty^2}{v_c^2} = \frac{L}{W} = \frac{\rho_\infty v_\infty^2}{2}\frac{A_{ref} C_L}{W}. \tag{2.23}$$

Combining eqs. (2.22) and (2.20), the drag acceleration of equilibrium glide flight reads:

$$\frac{D}{m}|_{equilibrium\ glide} = \frac{g_0(1 - v_\infty^2/v_c^2)}{L/D} \quad (= -\frac{dv_\infty}{dt}), \tag{2.24}$$

indicating that an equilibrium glide trajectory has minimum drag level, and hence can be associated with the maximum down range capability.

As we have seen in Sub-Section 2.1.2, several systems and operational constraints are to be regarded. The drag acceleration of equilibrium glide therefore is considered as the lower limit of drag acceleration on the trajectory:

$$\frac{D}{m}|_{trajectory} \geqq \frac{D}{m}|_{equilibrium\ glide}. \tag{2.25}$$

If L/D is fixed, for instance, due to a prescribed angle of attack, a larger drag acceleration can only be attained with a non-zero bank angle: $\mu_a \neq 0$, which means $L_{eff} < L$. This is the commonly used praxis in trajectory design and optimization[18].

Eq. (2.22) can be integrated by separation of variables, see, e.g., [18], to yield the flight speed v_∞ on the equilibrium glide trajectory as function of time t :

$$\frac{v_\infty}{v_c} = tanh\left(-\frac{g_0}{v_c}\frac{t}{L/D}\right) + C. \tag{2.26}$$

In [18] it is chosen $v_\infty/v_c = 0$ at $t = 0$, which yields $C = 0$. With that choice, the time must be counted negative.

[18] The concept of pseudo-equilibrium glide takes into account a time-varying lift due to a finite bank angle.

These considerations hold for RV-W-type and RV-NW-type flight vehicles, for the latter, if $L/D > 0$. For purely ballistic RV-NW-type vehicles with $L/D = 0$, similarly elements of the re-entry trajectory can be described, see, e.g., [18].

2.1.5 Equilibrium Glide Trajectory: Qualitative Results

Reentry flight is not made on an equilibrium glide trajectory. Initially a trajectory with a small and time-dependent flight path angle $\gamma = O(-1°)$ may be flown, [9], later a pseudo-equilibrium glide may be attained, [18].

Nevertheless, the above considerations allow to gain a number of qualitative insights regarding the trajectory and the aerothermodynamic phenomena and design problems. The following results must be understood in the sense that small increments of a variable lead to bounded increments of the dependent variable, while the other involved variables remain in principle unchanged. Therefore, for instance, the dependence of the lift L and the drag D on the dynamic pressure q_∞ is not explicitly noted. Large increments of a variable may lead to fundamental changes of the trajectory, which would change the whole picture.

- Combining eq. (2.23), for instance, with eq. (2.6), gives a relation for the heat flux in the gas at the wall at an appropriate reference location of the vehicle:

$$q_{gw} = C_{gw} \left(\frac{2W}{A_{ref} C_L} \right)^n \left(1 - \frac{v_\infty^2}{v_c^2} \right)^n \left(\frac{v_\infty}{v_c} \right)^{m-2n} v_c^{m-2n}. \tag{2.27}$$

  We see that q_{gw} depends inversely on some power of the lift coefficient C_L. This coefficient increases in the considered angle of attack domain at all Mach numbers with α, hence increasing α decreases q_{gw}, and a RV-W-type vehicle on a large part of the trajectory will fly at large angle of attack[19], Sub-Section 2.1.1. The heat flux depends also on some power of the wing loading W/A_{ref}. The larger the wing loading, the larger is q_{gw}.
- Eq. (2.27) can be used to guess the flight speed at which maximum heating occurs. Setting its differential with respect to the velocity ratio v_∞/v_c to zero:

$$\frac{dq_{gw}}{d(v_\infty/v_c)} = 0,$$

  we obtain:

$$\frac{v_\infty}{v_c}\Big|_{q_{gw,max}} = \sqrt{\frac{m-2n}{m}}. \tag{2.28}$$

[19] We remember that also with a finite bank angle $\mu_a \neq 0$, the angle of attack α is kept, because the roll motion is made around the air-path or velocity vector.

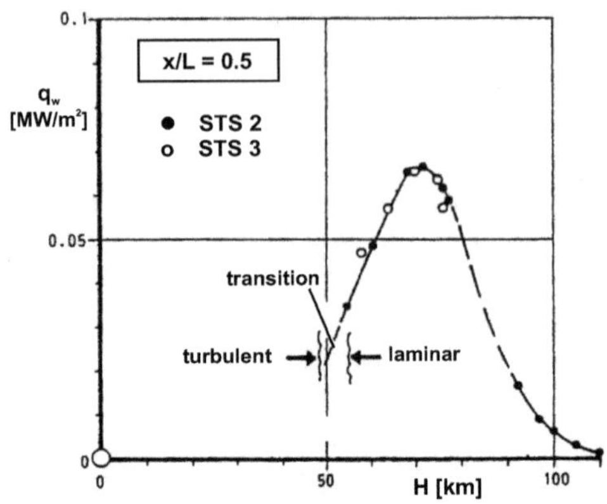

Fig. 2.6. Flight data of the heat flux in the gas at the wall, $q_w \equiv q_{gw}$, at $x/L = 0.5$ in the lower symmetry line of the Space Shuttle Orbiter, [21]. Sources of data: 90 $km \lessapprox H \lessapprox 110$ km: [22], 60 $km \lessapprox H \lessapprox 80$ km: [23], laminar-turbulent transition domain: [24], $H \lessapprox 50$ km: [25].

If we chose $n = 0.5$ and $m = 3$, the resulting velocity is with eq. (2.21) $v_\infty = 0.816\, v_c = 6{,}451.55\ m/s$. This amounts, Fig. 2.2, to a flight altitude (STS-2) of approximately 70.0 km. Comparing this with flight data of the Space Shuttle Orbiter, Fig. 2.6, we find quite a good agreement.

This result must not be generalized. It holds for the lower side of the vehicle including the nose region at the trajectory part above 40.0 to 60.0 km altitude, where the attached viscous flow is laminar. Below that altitude range the viscous flow becomes turbulent[20] which leads to a strong increase of thermal loads. Depending on the considered trajectory, these can be equal or larger than those at approximately 70.0 km altitude with laminar flow. Data are available in this regard for the Space Shuttle Orbiter, see, e.g., [7], where it is also shown that thermal loads on the orbital maneuvering system (OMS) pod (heat flux in the gas at the wall q_{gw}) due to vortex interaction are largest for $\alpha \approx 20°$, which is an angle of attack typical for flight at an altitude of about 40.0 to 30.0 km.

- For the part of the trajectory with laminar flow, eq. (2.6) tells us directly, that flight at the same speed, but on a higher trajectory point ("lifting entry", Fig. 2.2), reduces the heating q_{gw} (and the dynamic pressure q_∞), because ρ_∞ decreases with increasing altitude. A lower trajectory point ("ballistic entry") at the same speed increases heating (and dynamic pressure).

  Rearranging eq. (2.8) and assuming small γ and zero μ_a yields:

$$v_\infty^2 = \frac{R_\infty}{m}(W - L). \tag{2.29}$$

[20] The accurate and reliable prediction of laminar-turbulent transition is still one of the large unsolved problems of fluid mechanics, see, e.g., [5].

If the aerodynamic lift L would be increased (visualize it with the help of Fig. 2.5), the flight altitude, here the radius of the flight path R_∞, must be increased, in order to keep v_∞ constant, and the vehicle flies at smaller density ρ_∞ and hence dynamic pressure q_∞ than before.

In general, increasing the lift, of course via $C_L(\alpha)$ and finally α, leads from a ballistic re-entry trajectory to a "higher" flight trajectory and hence can be used to reduce thermal loads on the flight vehicle[21], and to reduce the deceleration, via q_∞, which the vehicle undergoes during re-entry. It thus permits in addition larger down ranges and cross ranges.

- From eq. (2.26) a guess of the flight time can be made. Rearranging it yields:

$$t = -\frac{v_c}{2g_0}\frac{L}{D} ln \left(\frac{1 + v_\infty/v_c}{1 - v_\infty/v_c}\right). \tag{2.30}$$

Hence (with the time counted positive):

$$t \sim \frac{L}{D},$$

which indicates, that the larger L/D, the larger the flight time and, in addition, also the time-integrated heat flux $\overline{q}_{gw}$, eq. (2.2), if the effect is not compensated by fundamental trajectory changes.

Reducing the flight time, and hence the time-integrated heat flux, can be achieved by reducing L/D. As we have seen above, this is not made via the angle of attack, but via the bank angle of the flight vehicle. Then the effective lift, eq. (2.12), is $L_{eff} = \cos\mu_a \; L < L$, and hence $L_{eff}/D < L/D$.

- Not only the flight time is proportional to the effective lift-to-drag ratio, but also the down range:

$$\Delta x \sim \frac{L}{D}\cos\mu_a, \tag{2.31}$$

and the cross range:

$$\Delta y \sim \frac{L}{D}\sin\mu_a. \tag{2.32}$$

We see from these relations, that at given lift and drag the bank angle $\mu_a = 0$ gives the largest down range. To get a cross range, a bank angle $\mu_a \neq 0$ is necessary. From eq. (2.32) $\mu_a = 90°$ appears to yield the largest cross range. However, actually that is achieved at $|\mu_a| = 45°$. We don't derive the exact relations, but refer the reader to, e.g., [18].

[21] This holds in any case for the nose region, but this effect can be configuration-dependent. Even higher thermal loads than at the nose can be present at other parts of a configuration, due to, for instance, thin boundary layers in strong expansion regions such as edges and shoulders. Hence a careful analysis is necessary in each case.

- The larger the drag D, the larger is the drag acceleration, eq. (2.20). The drag, like the lift, increases with increasing angle of attack α (rise of the effective bluntness), but stronger. This is the reason, why, at the large angles of attack flown during re-entry, L/D decreases with increasing α, Fig. 2.3.
- From eq. (2.8), also Fig. 2.5, we see that, due to the high speed of the vehicle and the curved flight path, on the initial trajectory – where γ is small – the aerodynamic lift L is small compared to the centrifugal force F_{centr}. The deceleration, eq. (2.20), see also eq. (2.7), however, is almost exclusively governed by the aerodynamic drag D.

  Hence, if density uncertainties exist on a part of the trajectory, and $W \sin\gamma \ll D$, eq. (2.7), they will affect first of all the deceleration force, drag D, and hence dv_∞/dt. If the density on a larger part of the trajectory is smaller than assumed, this would lead, without corrective measures, to an increase of the down range.

  The effect in general is large for RV-W-type flight vehicles, which fly for a rather long time at small flight path γ, see, e.g., Fig. 2.10. In [26] it is indicated that in such a situation a density, which is 25 per cent smaller than initially assumed, would lead to an increase of the down range of approximately 100.0 km.

  For RV-NW-type vehicles, density uncertainties in general have smaller impact, because the flight interval with small γ is not so large, see, e.g., Fig. 2.9. However, even here we get for a 25 per cent uncertainty in ρ_∞ a down range change of about 50.0 km. This is large for a capsule, which has a small L/D and hence a restricted trajectory correction potential.

We have seen so far, that the basic aerodynamic lift-to-drag ratio L/D is mostly the important parameter. An increase of L/D in general can be desirable, for instance for arbitrary recall from orbit (requirement of flexible down range) or large cross range demand, [9], even if constraints considerations (flight time, time-integrated heat flux) ask for a reduction of L to $L_{eff} < L$. However, in view of the small design experience base available, being in principle only that of the Space Shuttle Orbiter, one should be aware of the consequences of an increase of L/D regarding the choice and use of aerothermodynamic simulation means. This holds despite of the advances of the discrete numerical methods of aerothermodynamics which now seemingly allow to quantify more and more the aerothermodynamic properties of flight vehicles. The point is that ground-facility simulation remains for quite some time to come the major tool for data set generation, Sub-Section 12.3. This makes some further considerations necessary.

In the high angle of attack domain of re-entry flight, L/D can be increased by reducing α. If with the application of an advanced material, respectively thermal protection system, the nose radius R_N of a given RV-W-type flight vehicle could be reduced too, a further improvement of L/D would be possible, because this also reduces the wave drag. If pure surface radiation cooling of the nose region is demanded, of course, a trade-off becomes necessary between wave drag reduction and cooling effectiveness. Both wave drag and radiation cooling are reduced if the nose radius R_N is reduced, [5].

Decreasing α and R_N thus can be used to optimize the aerodynamic performance of a re-entry vehicle within the limits of its structure and materials concept, as well as to reduce the thermal loads on the airframe via trajectory shaping, [9]. However, if we reduce α on the high Mach number part of the trajectory, we move away from the benign wall (boundary-layer edge) Mach number interval, Sub-Section 3.6.4, on the windward side of the flight vehicle, and towards a Mach number and especially high-temperature real-gas sensitivity of aerodynamic coefficients, which is important in view of ground-facility simulation, Sub-Section 3.6.4.

We are also moving from the compressibility or pressure effects dominated RV-W-type flight vehicle in direction of the viscosity-effects dominated CAV-type vehicle. The more this happens, the more the proper simulation of viscous effects, including hypersonic viscous interaction phenomena, becomes important during the vehicle's definition and development process. Also thermo-chemical effects including finite-rate effects change their character. While thermo-chemical freezing phenomena in high-enthalpy ground-simulation facilities are in general not a major problem for a RV-W-type shape, like that of the Space Shuttle Orbiter, [5], this may change gradually for RV-W-type flight vehicles with reduced α and R_N compared to that of the Space Shuttle Orbiter.

Increasing the aerodynamic performance L/D of a RV-W-type flight vehicle thus increases down range and cross range capabilities and in addition, via a large L, can decrease thermal loads, though not necessarily the time-integrated ones. However, it changes the demands on the aerothermodynamic design with a kind of snow-ball effect which can be vicious. At the core of this lies the shape of the chosen trajectory.

2.1.6 Case Study 1: Trajectories of RV-NW-Type Flight Vehicles

So far we have considered qualitative results derived from the relations for the equilibrium glide trajectory. We present now quantitative results for two RV-NW-type vehicles. They are found from the numerical integration of the dynamic eqs. (2.7) and (2.8) together with the kinematic relations eqs. (2.9) and (2.10) corresponding to planar flight. We use simple relations for the gravitational acceleration $g(H)$, Chapter 13:

$$g(H) = g_0 \left(\frac{R_E}{R_E + H} \right)^2 , \tag{2.33}$$

and for the atmospheric density $\rho(H)$, Chapter 12:

$$\rho(H) = \rho_0 e^{-\beta H} , \tag{2.34}$$

with $\beta = 1.40845 \cdot 10^{-4}\ m^{-1}$.

Table 2.2. Characteristic quantities of APOLLO and SOYUZ capsules for the calculation of the trajectory in the planar limit.

Symbol	Quantity	APOLLO	SOYUZ
A_{ref}	reference area $[m^2]$	12.02	3.80
L/D	aerodynamic performance	0.30	0.26
C_L	lift coefficient	0.374	0.349
C_D	drag coefficient	1.247	1.341
m	total vehicle mass [kg]	5,470.0	2,400.0
V_e	flight velocity at entry [m/s]	7,670.0	7,900.0
γ_e	flight path angle at entry [°]	- 0.75, - 1.50, - 3.50	- 1.50
R_N	radius at stagnation point [m]	4.694	2.235

Further we assume that the lift and the drag coefficients are constant along the whole trajectory. The stagnation point heat flux is calculated with eq. (10.76).

Investigated is the entry flight behavior of the APOLLO and the SOYUZ capsule. In Table 2.2 the characteristic quantities of both capsules, necessary for the computation of the trajectories, are listed. They are taken from Tables 5.1, 5.3, 5.4 and Figs. 5.11, 5.12.

In the first example the APOLLO trajectory for three values of the flight path angle at entry (γ_e = - 0.75°, - 1.5°, - 3.5°) is considered. In Fig. 2.7 are plotted the altitude as function of the inertial velocity, a), and the inertial velocity, the altitude, the normalized acceleration n_t, the flight range and the stagnation point heat flux as functions of time, b) to f).

The lowest entry flight path angle, γ_e = - 0.75°, a value which is usually employed, leads to the smallest n_t and the smallest stagnation point heat flux, a), f), whereas the flight (down) range is largest, e). The largest angle, γ_e = - 3.5°, produces a skip trajectory, which means that the flight path angle γ becomes positive over a certain distance with the consequence that the flight altitude grows, a), c). Further the g-loads and the stagnation point heat flux increase remarkable, d), f) while the flight range is lowest, e). Of course, the flight time is much smaller than for the lower γ_e values.

In the second example the trajectories of the APOLLO and the SOYUZ capsules are compared for the same flight path angle at entry (γ_e = - 1.5°), Fig. 2.8. It seems that the tendency of SOYUZ for skipping is lower than for APOLLO, a). The maximum g-loads are reached at a later time, b), but the stagnation point heat flux is at the maximum higher and has a high value over a longer time interval leading to larger time-integrated thermal loads, d).

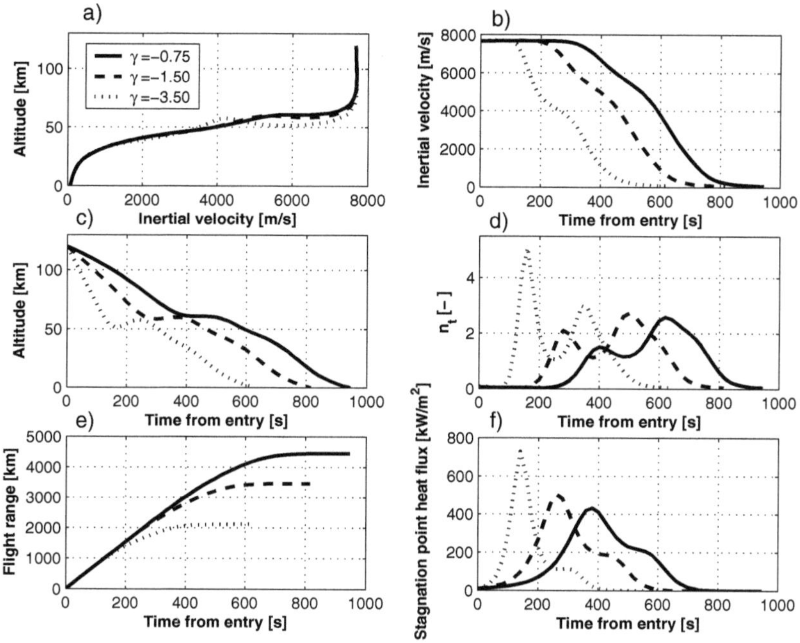

Fig. 2.7. APOLLO capsule: entry trajectory as function of the initial flight path angle γ_e based on planar equations.

As mentioned in Chapter 12, the density distribution in the atmosphere as function of the altitude is not constant but depends on the local and global weather conditions. Therefore the question arises how density uncertainties on a part of the trajectory will affect the trajectory, see also Sub-Section 2.1.5. We test this with the APOLLO trajectory, Fig. 2.7, by multiplying the density function, eq. (2.33), once with 0.75 and then with 1.25.

Fig. 2.9 shows the result in terms of the flight path angle and the vehicle's deceleration. We can perceive that the density influence on the trajectory of a lifting capsule appears to be low, since the flight path angle becomes soon large. The flight range in both directions is only changed by approximately 1.2 per cent which, however, is large for a capsule.

2.1.7 Case Study 2: Trajectory of a RV-W-Type Flight Vehicle (X-38)

In this case study we discuss the optimized trajectory of a RV-W-type re-entry vehicle, the X-38, [2]. The vehicle is to fly from an entry altitude of $H = 121.9\ km$ to the terminal area (TA) in the southwest of France, Fig. 2.10, upper right, arriving there at $24.1\ km \leqq H_{TA} \leqq 25.1\ km$ with $645.0\ m/s \leqq v_{TA} \leqq 845.0\ m/s$. The guidance objective is minimization of the time-integrated heat flux $\overline{q}_{gw}$. The determination of q_{gw} was made with eq. (10.76)

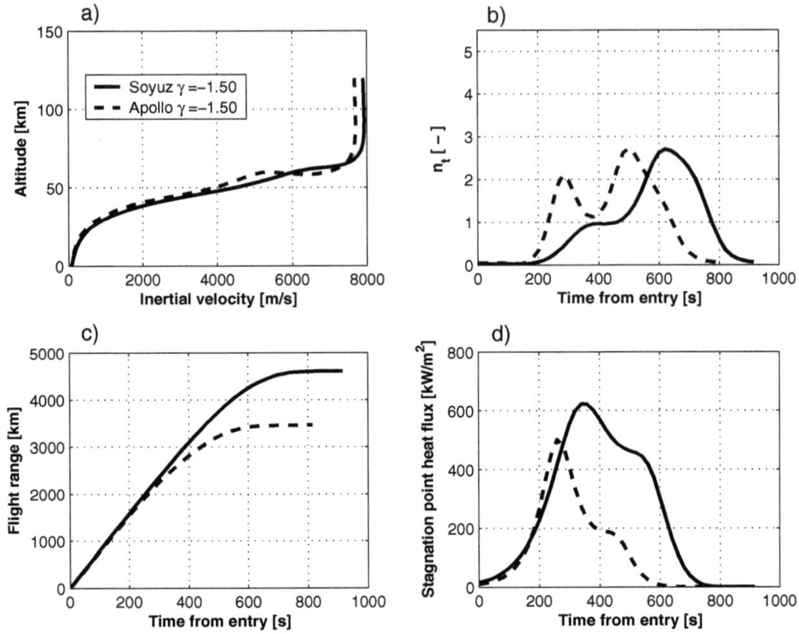

Fig. 2.8. APOLLO and SOYUZ capsule: comparison of entry trajectories based on planar equations.

Table 2.3. Systems and operational constraints of the X-38 flight, [2].

q_∞	q_{gw}	n_z	γ	α
$\leqq$ 14.346 kPa	$\leqq$ 1,175.0 kW/m^2	$\leqq$ 2	$\leqq$ 0°	35° $\leqq \alpha \leqq$ 45°

in Section 10.3 for a nose radius R_N = 0.3048 m (= 1.0 ft). The control variables are the angle of attack α and the bank angle μ_a. The systems and operational constraints are given in Table 2.3.

We discuss summarily only some of the results. For the whole picture, including the definition of the guidance law, the reader is referred to [2].

The de-orbit maneuver generates an initial flight path angle of $\gamma \approx$ - 1.6°, Fig. 2.10, lower right. With a pull-up maneuver this angle is almost zeroed out[22] at the flight time t = 322.0 s. Then it drops to again $\gamma \approx$ - 1.6° at $t \approx$ 880.0 s. In this time interval the trajectory follows the $q_{gw} = q_{gw,max}$ = 1,750.0 kW/m^2 constraint, Fig. 2.11. After a short rise the flight path angle

[22] An initial flight path angle of that magnitude would result in a trajectory initially too steep, with too large thermal loads and a large drag acceleration.

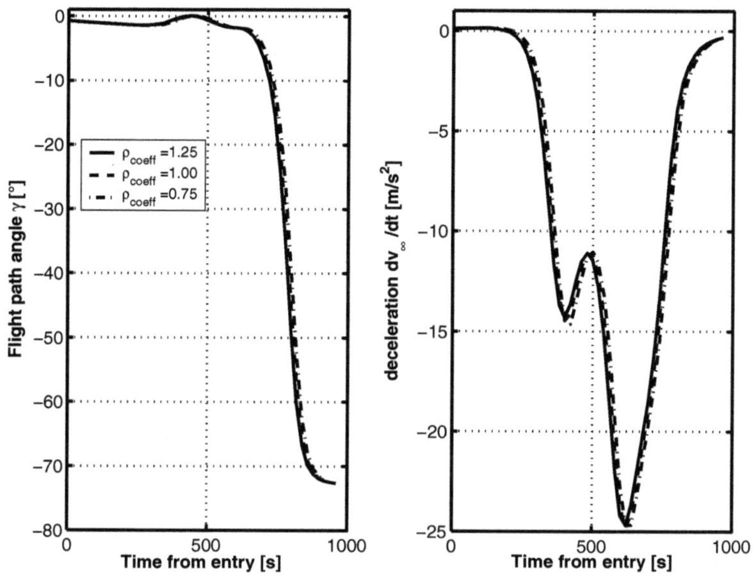

Fig. 2.9. APOLLO capsule: influence of atmospheric density variations on the flight trajectory. Flight path angle γ (left), vehicle deceleration dv_∞/dt (right).

drops to values around $\gamma \approx$ - 10°, and the trajectory follows the $n_z = n_{z,max}$ = 2 constraint.

In Fig. 2.11, upper part, this is reflected in the velocity/drag-acceleration map[23]. The trajectory, after the initial pull-up maneuver, down to $v_\infty \approx$ 5.0 km/s follows the thermal load constraint and then down to $v_\infty \approx$ 2.0 km/s that of the normal load factor. The dynamic pressure constraint $q_\infty = q_{\infty,max}$ = 14.346 kPa is approximately followed only for $v_\infty \lessapprox$ 2.0 km/s. The drag acceleration of the resulting trajectory lies clearly above that of the equilibrium glide trajectory, the latter being characterized also by a much smaller dynamic pressure, Fig. 2.11, lower part.

The angle of attack, Fig. 2.12 (upper left, also to be read from the right to the left) down to $M_\infty \approx 10$ ($v_\infty \approx$ 3.0 km/s) is the maximum angle $\alpha = \alpha_{max}$ = 45°. This high angle permits the large deceleration at minimum thermal loads while meeting the cross range demand. The bank angle μ_a, Fig. 2.12 (upper right) up to t = 322.0 s is zero in order to provide the lift needed for the pull-up maneuver. After that the effective aerodynamic lift is reduced by a large bank angle down to $\mu_a \approx$ - 80°, although without bank reversal, in order to reduce the flight time and thus the time-integrated heat flux $\overline{q}_{gw}$.

[23] The re-entry trajectory is beginning at a flight path speed of approximately 7.6 km/s, hence the figure must be read from the right to the left.

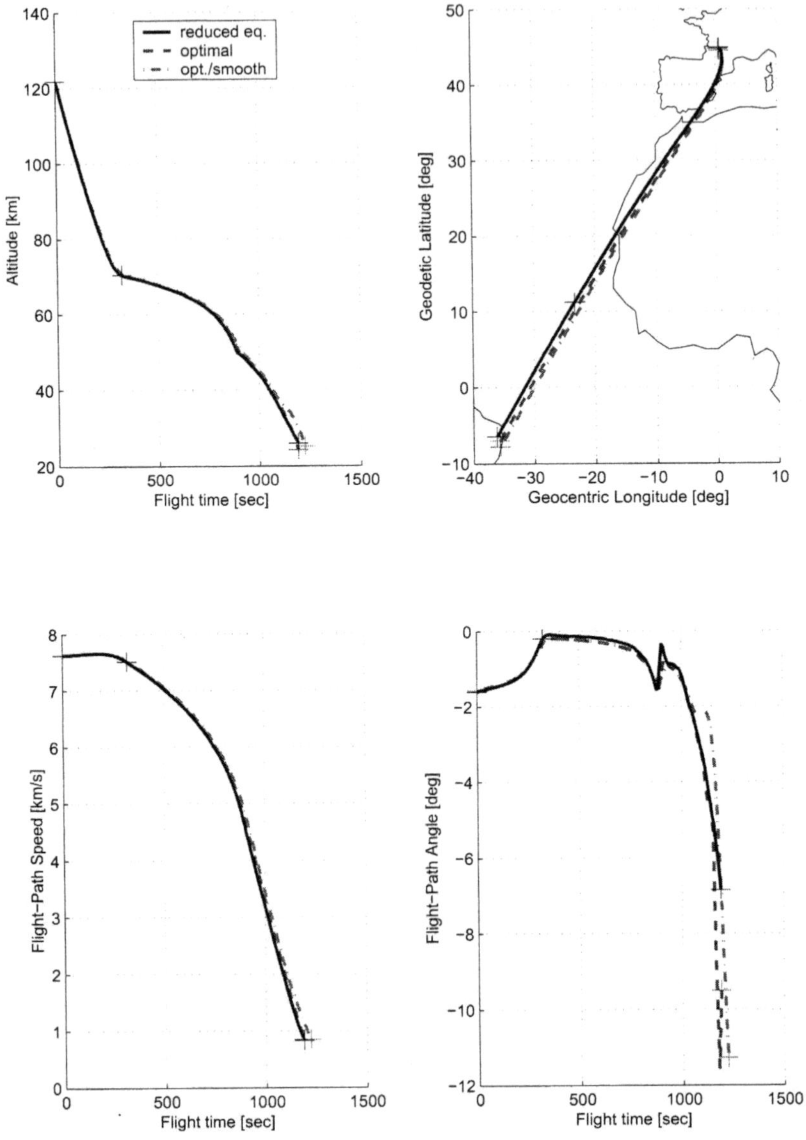

Fig. 2.10. Elements of the re-entry trajectory of the X-38, [2]. Upper left: flight altitude (H) as function of flight time (t). Upper right: ground track (the trajectory begins in this figure at the lower left and ends at the upper right). Lower left: flight path speed (v_∞) as function of flight time (t). Lower right: flight path angle (γ) as function of flight time (t).

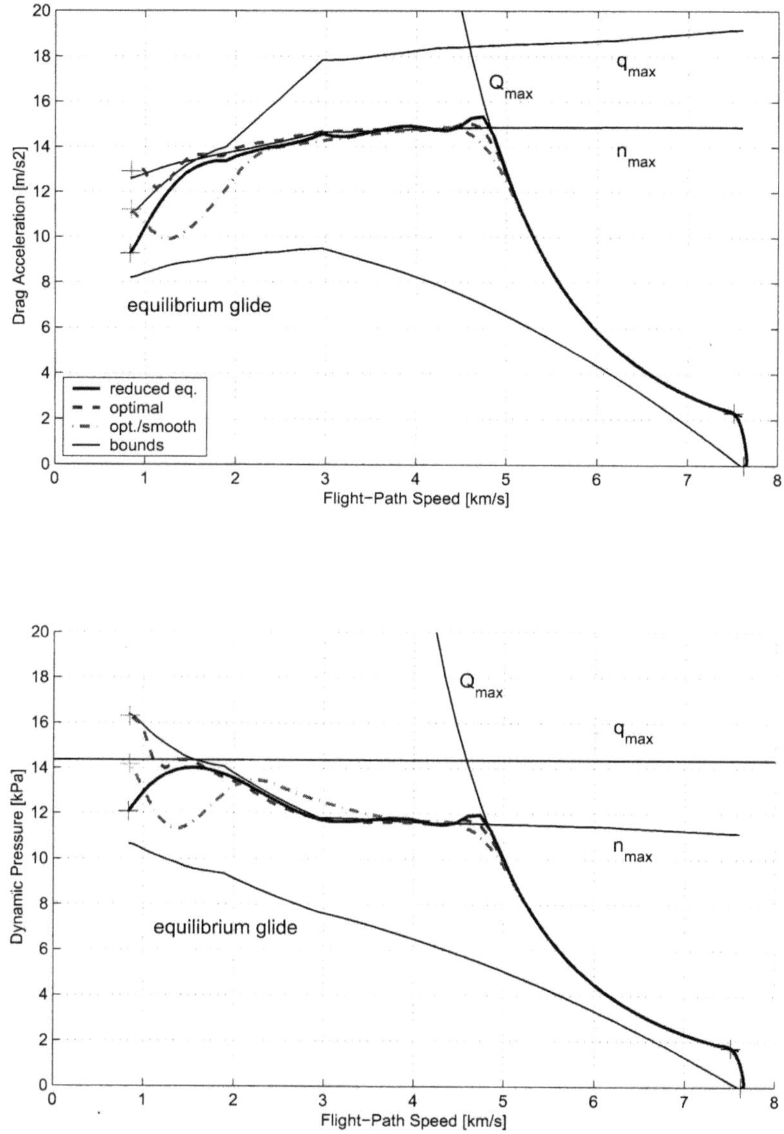

Fig. 2.11. Resulting trajectory representations, [2]. Drag acceleration (D/m) as function of flight path speed (v_∞), and dynamic pressure (q_∞) as function of flight path speed (v_∞) (the trajectories begin at the lower right). The constraints are maximum values of dynamic pressure ($q_{max} \equiv q_{\infty,max}$), of heat flux in the gas at the wall ($Q_{max} \equiv q_{gw,max}$), and of the normal load factor ($n_{max} \equiv n_{z,max}$).

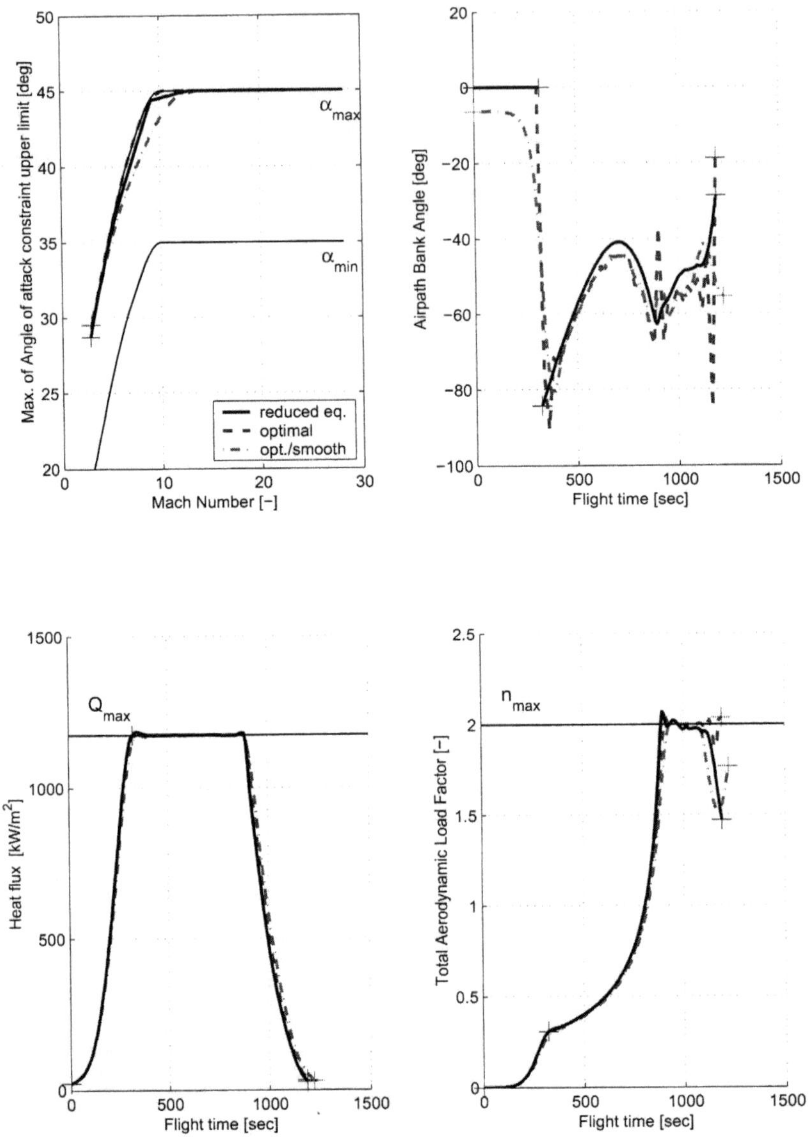

Fig. 2.12. Elements of the re-entry trajectory of the X-38, [2]. Upper left: maximum angle of attack (α_{max}) as function of flight Mach number (M_∞). Upper right: bank angle (μ_a) as function of flight time (t). Lower left: actual heat flux in the gas at the wall (q_{gw}) as function of flight time (t), $Q_{max} \equiv q_{gw,max}$. Lower right: actual normal load factor (n_z) as function of flight time (t), $n_{max} \equiv n_{z,max}$.

2.2 Flight Trajectories of Cruise and Acceleration Vehicles

2.2.1 General Aspects

In contrast to RV-W-type and RV-NW-type vehicles, CAV-type vehicles have scarcely been flown. They are in general hypothetical vehicles and very little flight experience is available. Airbreathing CAV-type vehicles are to fly like ordinary airplanes. They are drag sensitive in contrast to RV-W-type and RV-NW-type vehicles, which need to employ a large aerodynamic drag to achieve their mission. Cruise vehicles and acceleration vehicles have different missions, Chapter 4, and hence different trajectory demands. We concentrate here somewhat on cruise-type vehicles, because several concepts of this kind have been studied in the recent past, mainly TSTO space transportation systems.

A CAV-type flight vehicle, if being the lower stage of a TSTO space-transportation system, usually is thought to perform a return-to-base mission ($A \rightarrow A$ flight). Hypersonic transport aircraft, in a sense also SSTO (ARV-type) vehicles, perform $A \rightarrow B$ flights. CAV-type flight vehicles typically first employ turbojet propulsion (up to $M_\infty \approx 4$), then, depending on the maximum flight Mach number, ramjet propulsion (up to $M_\infty \approx 7$) or combined ramjet/scramjet propulsion (up to $M_\infty \approx 12$ (to 14)). SSTO vehicles finally employ rocket propulsion to reach orbit. These different propulsion modes are an important issue in trajectory design and optimization, since at least for ramjet and scramjet propulsion a strong coupling of the propulsion forces, Fig. 2.14, into the force and moment balance of the vehicle exists, Sub-Section 2.2.3.

With TSTO or multistage systems another important issue is stage separation. It is characterized by the shedding of a large percentage of the system mass in a very short time interval, which leads to a large lift surplus of the carrier stage. This problem concerns flight dynamics, guidance and control and many other systems aspects.

A particular problem especially for large CAV-type flight vehicles is the influence of aerothermoelasticity of the airframe and the aerodynamic control surfaces on the propulsion system performance, Sub-Section 4.5.4, and on the control properties of the flight vehicle. RV-W-type (and RV-NW-type) flight vehicles have rather stiff airframes, because of their cold primary (load-carrying) structures in combination with a TPS. This is not the case for large CAV-type vehicles, where the airframe and the aerodynamic control surfaces may have, in addition to their large size, hot primary structures. The resulting additional couplings of aerodynamic, propulsion and control features will be highly dynamical and pose enormous design challenges, see, e.g., [27], and also the present Sub-Section 4.5.4.

We restrict our discussion in the following sub-sections mostly to the issues of cruise-type vehicle trajectories in the turbojet/ramjet-propulsion domain without consideration of possible stage separation. We look at the guidance objectives, the trajectory control variables and the systems and operational constraints, extending the considerations of RV-W-type and RV-NW-type flight vehicles. The forces acting on a CAV-type vehicle are discussed, but, since no counterpart exists to the equilibrium glide of RV-W-type vehicles, after that only a SÄNGER trajectory is considered as an example and some qualitative and quantitative results are presented.

2.2.2 Guidance Objectives, Trajectory Control Variables, and Systems and Operational Constraints

Guidance objectives. Guidance objectives are the optimization of vehicle performance, for instance:

- Minimization of fuel consumption for a given mission.
- Maximization of total air-path energy at upper stage separation of a TSTO space transportation system.
- Maximization of pay load, i. e., of the mass inserted into Earth orbit with a TSTO space transportation system.

Trajectory control variables. The number of trajectory control variables is, like that of RV-W-type and RV-NW-type vehicles, rather limited. The control variables are basically only:

- Angle of attack α, which governs aerodynamic lift L, drag D, and pitching moment M of the vehicle. It governs further, if forebody pre-compression is employed, Section 4.5, strongly the net thrust of the (airbreathing) propulsion system.
- Power setting of the propulsion system with its different engine modes.
- Bank angle μ_a of the vehicle (again, like for RV-W-type flight vehicles, around the velocity vector $\underline{v}_\infty$). Banking is necessary for curved (in general $A \rightarrow A$) flight, and, in case of TSTO space transportation systems, to dump the surplus lift during stage separation[24].

Systems and operational constraints. The basic systems and operational constraints are the same as, or similar to those of RV-W-type and RV-NW-type vehicles:

- The dynamic pressure q_∞ now is also a measure of the demands of the airbreathing propulsion system. The ascent trajectory of the SÄNGER TSTO system for instance was studied as a 50.0 kPa trajectory for both

[24] The mass of the upper stage of, for instance, the SÄNGER space transportation system, is at upper stage separation about one third of the total mass of the system. Hence when it leaves the lower stage, the latter has much too much lift, which immediately must be reduced in order to avoid collision with the upper stage, and to insure controllability of the flight vehicle.

the turbojet and the ramjet mode [28]. If both the ramjet and the scramjet mode are to be employed, dynamic-pressure ranges of 25.0 $kPa \lessapprox q_\infty \lessapprox$ 95.0 kPa have been considered, e.g., [29], [27].

- Flight of CAV-type vehicles below 40.0 to 60.0 km altitude means that the attached viscous flow predominantly is turbulent. The forward stagnation point however again is the primary reference location where thermal loads are constrained. In addition flat portions downstream of the nose region with turbulent flow may be chosen as reference locations.
- Besides the normal load factor, again with $n_{z,max} = 2$ to 2.5, the axial load factor with $n_{x,max} = q_\infty C_A A_{ref}/(m\, g) = 3$ to 3.5 is a constraint during the stage separation process of TSTO space transportation systems.
- Due to the influence of the thrust vector on the longitudinal forces and the pitching moment, Sub-Section 2.2.3, trimmability and stability and controllability of the flight vehicle are critical operational constraints. If the thrust-force angle cannot be restricted or controlled mechanically, the only available degree of freedom to trim the vehicle is the symmetrical elevon deflection. If this angle is large, a substantial total drag D increase, due to the trim drag, will result. At the same time, depending on the overall forces and moment balance, a possible decrease of the lift L will happen, and an even stronger decrease of the lift-to-drag ratio L/D. Besides that the longitudinal stability characteristics of the vehicle will be influenced. Other issues in this regard are the resulting hinge moments (actuator performance) together with the mechanical and thermal loads on the elevons.

Besides these basic systems and operational constraints others may need to be prescribed, for instance regarding the airbreathing propulsion system. A graphical presentation of the hypothetical flight corridor in the velocity-altitude map is given in Fig. 2.13, where Areas 1 and 3 concern for instance the lower stages of TSTO space transportation systems and hypersonic aircraft, whereas Area 2 concerns SSTO (ARV-type) systems.

The shaded flight corridor in Fig. 2.13 is bounded on the lower side by structural pressure limits and thermal loads of the airframe, the propulsion system and the control surfaces (hinge moment limit). On the upper side lift and combustion limits play a role and also the limit of the aerodynamic control authority. We have met the latter limit already with RV-W-type flight vehicles. Control aspects possibly will make the flight corridor much narrower in reality, there might even be larger excluded regions due to propulsion issues, [27].

In conclusion it can be stated that flight trajectory design and optimization for future large airbreathing CAV-type flight vehicles poses extremely large challenges, much larger than those for RV-W-type vehicles. Because Cayley's design paradigm, [30], see also prologue to Chapter 8, is completely invalid for these vehicles, the highly non-linear aerodynamics/structure dynamics/propulsion/flight dynamics/flight control couplings make new vehicle

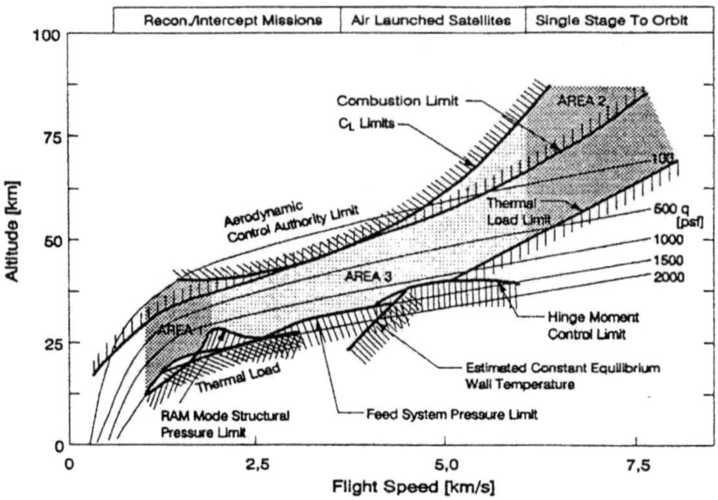

Fig. 2.13. Flight corridors and system and operational constraints of CAV-type flight vehicles in the altitude-velocity map, [27].

design approaches necessary, but also new trajectory design and optimization approaches.

2.2.3 Forces Acting on a Cruise and Acceleration Vehicle

Again we approximate the flight vehicle as a point mass. A schematic point-mass force polygon of a propelled CAV-type flight vehicle is given in Fig. 2.14. Assumed is steady level flight, for non-steady flight the figure changes accordingly.

The lower side of a CAV-type vehicle is a highly integrated lift and propulsion system, [31]. Aerodynamic lift and propulsion provision are strongly coupled, Section 4.4. Depending on the thrust vector angle, a large net thrust lift component will exist. The net thrust is the – vectorial – difference between the inlet drag, i. e. the flow momentum entering the inlet of the propulsion system, and the thrust of the nozzle.

Since the nozzle is an asymmetric external nozzle[25], the thrust vector will change in magnitude and direction in a considerable range depending on flight speed, altitude, angle of attack and power setting of the propulsion system. Fig. 2.15 shows, from the SÄNGER lower-stage studies, as an example on the right-hand side the thrust vector angle as function of the flight Mach number. The jumps in the full lines in Fig. 2.15 of both the thrust coefficient and the thrust vector angle are due to the switch-over from the turbojet to

[25] The classical bell nozzle of rocket propulsion cannot be integrated into the airframe of a CAV-type flight vehicle for thrust control (large flight Mach number span at not too large altitudes) and configurational reasons, see, e.g., [32].

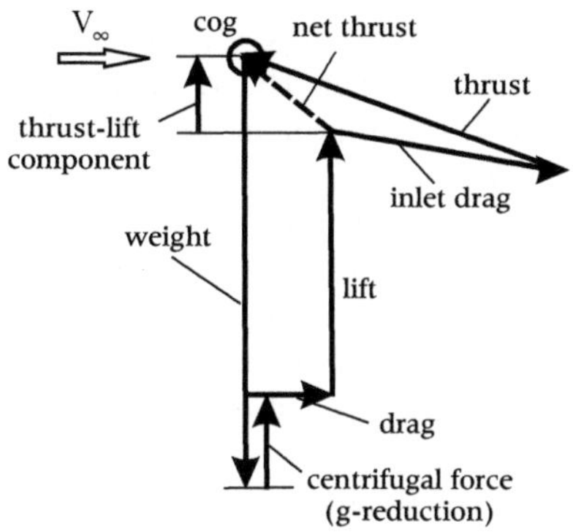

Fig. 2.14. Schematic point-mass force polygon at steady level flight of a propelled CAV-type flight vehicle, [31].

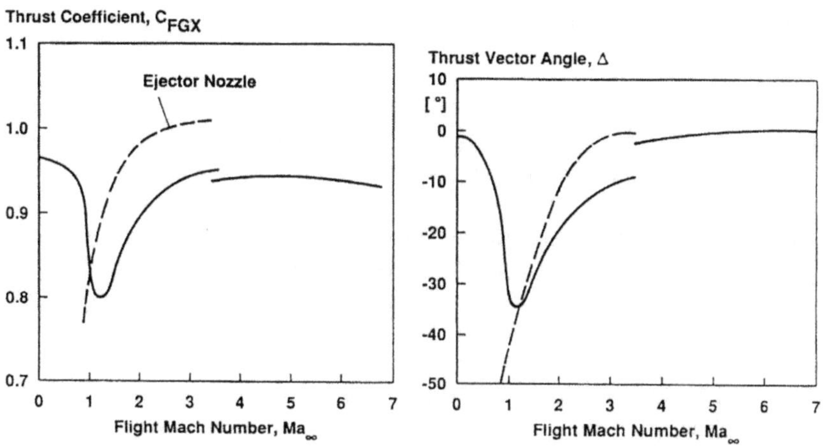

Fig. 2.15. Axial thrust coefficient C_{FGX} (left) and thrust vector angle $\triangle$ (right) as function of the flight Mach number Ma_∞ ($\equiv M_\infty$) of the SÄNGER lower stage, [28] (nominal flight trajectory).

the ramjet propulsion mode, which has not been smoothed out. The broken lines in Fig. 2.15 depict the data of the ejector nozzle. This nozzle dumps in the turbojet mode (up to $M_\infty = 3.5$) the forebody boundary layer material, which has been removed by the boundary layer diverter, see, e.g., [5], into the main nozzle flow[26].

On the left-hand side of Fig. 2.15 the axial thrust coefficient C_{FGX} is given. We see that around $M_\infty = 1$, in the transonic drag-rise domain, the C_{FGX} is smallest and the thrust vector angle $\triangle$ largest (negative). The latter

[26] It is not yet clear whether a boundary layer diverter is needed also in the ramjet and the scramjet propulsion mode.

leads (due to the need for trim) to a large elevon deflection[27] and hence to increased trim drag. To compensate in the transonic domain for both the decreased thrust (the thrust-lift component possibly is increased) and the increased drag, a transonic bunt maneuver[28] can be flown, where potential energy of the vehicle is exchanged for kinetic energy, Sub-Section 2.2.4.

Because usually forebody pre-compression is employed in order to reduce the inlet capture area, Section 4.5, a large dependence of the net thrust on the angle of attack is present. In reality this net thrust sensitivity will be enhanced by the forebody aerothermoelasticity, see, e.g., [33], a problem, for which so far no solution has e.g., [33], a problem, for which so far no solution has been found and which has not yet been accounted for in trajectory studies. It would require the quantification of the above mentioned highly non-linear aerodynamics/structure dynamics/propulsion/flight dynamics/flight control couplings of the vehicle.

2.2.4 Case Study 3: Trajectory of a CAV-Type Flight Vehicle (SÄNGER)

We discuss now as example the optimized trajectory of the TSTO space transportation system SÄNGER, [34]. The system is to fly from Southern Europe to the upper stage release at a northern latitude of 25° to 30° and back (return-to-base cruise). Again we look only at selected results.

The guidance objective is minimization of the fuel consumption for the whole mission, while the stage separation is to happen at $H = 33.35\ km$ altitude at $v_\infty = 2{,}085.0\ m/s$ with a flight path angle $\gamma = +\ 8.5°$. The control variables are angle of attack α, the power (throttle) setting δ_T, and the bank angle μ_a. During the study the power setting δ_T for the ramjet mode ($M_\infty > 3.5$) was substituted by the fuel/air equivalence ratio[29] Φ_f, because overfueling (fuel-rich mixture) of ramjet combustion was found to have a favorable effect on the stage separation maneuver performance. The systems and operational constraints are given in Table 2.4. We note that also minimum constraints are prescribed, for instance for the dynamic pressure q_∞. This is necessary to assure minimum vehicle flyability and controllability and propulsion effectiveness.

A presentation of the resulting flight trajectory is given in Fig. 2.16. We observe some particularities of the trajectory, [34]:

---

27 Whether a downward or an upward deflection is necessary depends on the actual location of the thrust vector acting line relative to the vehicle's center-of-gravity.

28 The vehicle performs part of an inverted (or outside) loop.

29 The fuel/air equivalence ratio, usually called only equivalence ratio, is the ratio of the actual fuel/air ratio to the stoichiometric fuel/air ratio, see, e.g., [35]. The stoichiometric mixture hence has the equivalence ratio $\Phi_f = 1$. For hydrogen-fueled ramjets (and scramjets) a fuel-rich mixture ($\Phi_f > 1$) can be used to increase the thrust, which, of course, leads to an increase of the fuel consumption. A fuel-lean mixture ($\Phi_f < 1$) results in a decrease of fuel consumption.

Table 2.4. Systems and operational constraints of the SÄNGER flight, [34].

	q_∞ $[kPa]$	α [°]	$\lvert\mu_a\rvert$ [°]	n_z	$\delta_{T,turbo}$	$\delta_{T,ram}$	$\Phi_{f,ram}$
Min:	10.0	- 1.5	free	0	0	0	$\Phi_f(M_\infty, \delta_{T,min})$
Max:	50.0	20.0	prescribed	2	1.0	$\delta_T(M_\infty, \Phi_{f,max})$	3

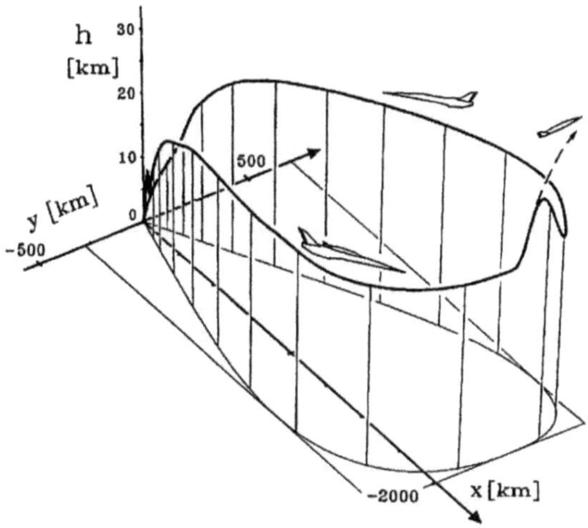

Fig. 2.16. The resulting return-to-base trajectory of the SÄNGER space transportation system in three-dimensions including upper stage separation, [34].

- The trajectory is a highly three-dimensional one with significant motion in longitudinal, vertical and lateral direction.
- The trajectory is throughout curved.
- In the transonic regime the drag increase, the large negative thrust vector angle $\triangle$, and the decrease of the axial thrust coefficient C_{FGX}, Sub-Section 2.2.3, make a bunt maneuver necessary (indicated at the far left of the ascent part of the trajectory, see also Fig. 2.17).
- The upper stage separation maneuver is characterized by a steep climb followed by a bunt, all reflected strongly in the control variables.

Fig. 2.17 shows that the flight up to the upper stage separation is an accelerated climb, and barely a cruise flight. The transonic bunt at $t \approx 200.0$ s is well discernible, similarly the stage separation maneuver at $t \gtrapprox 1{,}850.0$ s. The altitude oscillations in the trajectory parts before and after the stage separation are not due to numerical problems, but seem to be related to a phugoid-like motion, [34].

The control variables are shown in Fig. 2.18. The angle of attack α does not change much before and after stage separation. The transonic bunt is

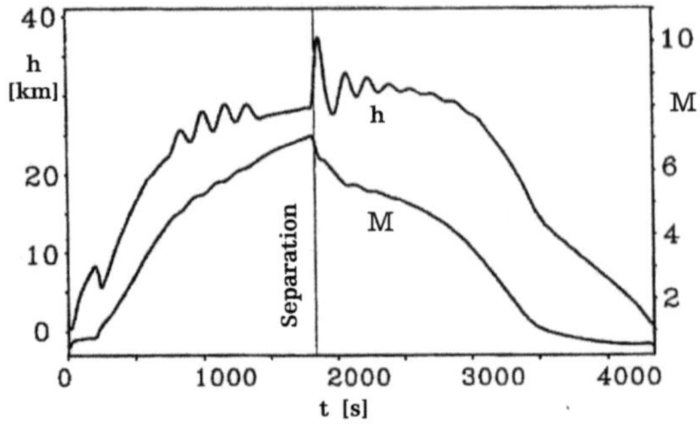

Fig. 2.17. Flight altitude h ($\equiv H$) and flight Mach number M ($\equiv M_\infty$) as function of flight time, [34]. The separation of the upper stage is initiated at $t \approx 1{,}850.0\ s$.

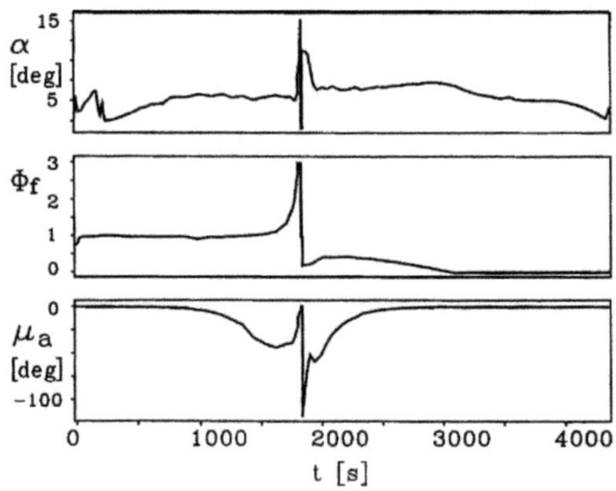

Fig. 2.18. The resulting trajectory control variables (from above): angle of attack α, equivalence ratio (power setting) Φ_f, bank angle μ_a as function of the flight time t, [34].

reflected, the pull-up for the stage separation maneuver demands a strong increase of α.

The power setting, respectively the equivalence ratio, is $\Phi_f \approx 1$ before stage separation. Transition from turbojet to ramjet propulsion – hardly observed in Fig. 2.18 – is made between $M_\infty = 3$ to 3.5 ($t \approx 540.0\ s$ to $600.0\ s$). A strong overfueling occurs prior to stage separation. On the return part of the trajectory Φ_f is much reduced, eventually to approximately zero, which amounts to a glide flight to base.

The bank angle μ_a (defined as negative to the left) increases steadily up to - 45°, is zero during stage separation (wings level condition), and then drops to $\mu_a \approx$ - 115°. This happens in order to dump the excess lift after upper stage release. Despite this the normal load factor (n_z) constraint is never violated.

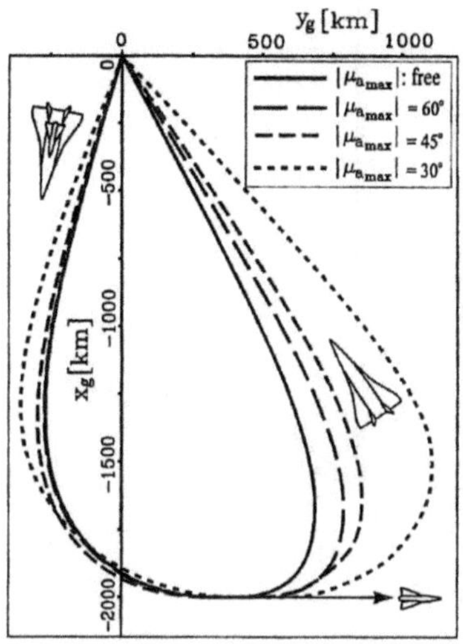

Fig. 2.19. The influence of the bank angle μ_a constraint on the resulting optimal cruise ground track, [34].

Because the resulting large bank angle might not be acceptable for a large TSTO system, trajectories with bank angle limitation were also studied in [34]. The smaller the maximum permitted bank angle, the larger becomes the ground track of the trajectory, Fig. 2.19. Accordingly rises the fuel consumption, while on the return part of the trajectory the minimum dynamic-pressure constraint becomes active. The latter is needed to assure flyability and controllability of the vehicle including sufficient residual propulsion effectiveness.

2.3 General Equations for Planetary Flight

Newton's second law is the basic equation valid for an inertial system, for example $\mathbf{O}, x_0, y_0, z_0$, Fig. 2.20, and reads:

$$m \left. \frac{d\underline{V}}{dt} \right|_0 = \underline{F}_{|0} = \underline{F}_{A\,|0} + m\underline{G}_{|0}, \tag{2.35}$$

where $\underline{F}_A$ contains the aerodynamic forces and $\underline{G}$ represents the gravity[30]. The planeto-centric coordinate system $\mathbf{p}$ rotates with the angular velocity $\underline{\bar{\Omega}}_{|p}$ around the origin $\mathbf{O}$ of the inertial system, Fig. 2.20. Therefore, applying the general rule for time derivatives of vectors in rotating systems, we find, [16], [19]:

[30] Only unpowered flight is considered.

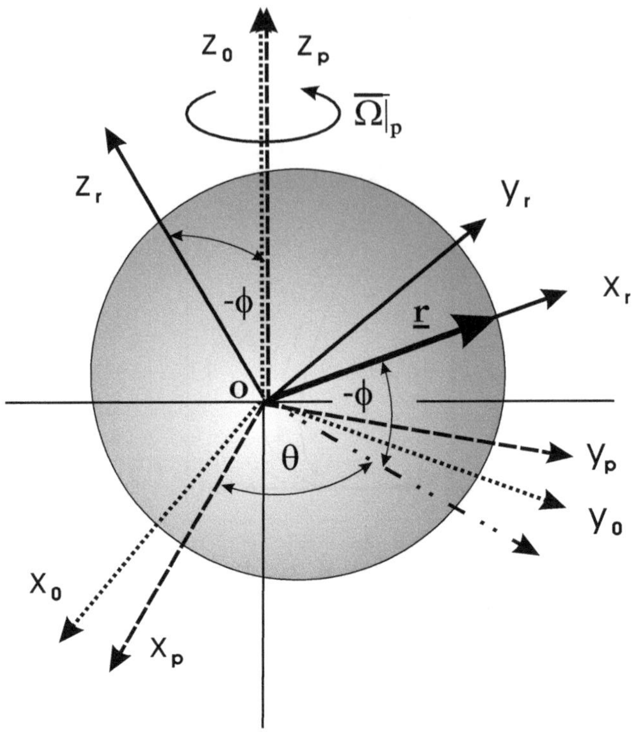

Fig. 2.20. Inertial coordinate system $\mathbf{O}, x_0, y_0, z_0$, planet fixed coordinate system $\mathbf{O}, x_p, y_p, z_p$ and rotating coordinate system $\mathbf{O}, x_r, y_r, z_r$.

$$\left.\frac{d\underline{r}}{dt}\right|_0 = \left.\frac{d\underline{r}}{dt}\right|_p + \underline{\bar{\Omega}}_{|p} \times \underline{r}_{|p}, \quad \underline{\bar{\Omega}}_{|p} = \omega \begin{pmatrix} 0 \\ 0 \\ 1 \end{pmatrix}, \tag{2.36}$$

and

$$\left.\frac{d\underline{V}}{dt}\right|_0 = \frac{d}{dt}(\underline{V}_{|p} + \underline{\bar{\Omega}}_{|p} \times \underline{r}_{|p}) + \underline{\bar{\Omega}}_{|p} \times (\underline{V}_{|p} + \underline{\bar{\Omega}}_{|p} \times \underline{r}_{|p}). \tag{2.37}$$

Since the angular velocity of the planet, $\underline{\bar{\Omega}}_{|p}$, is nearly a constant, we get with $d\underline{\bar{\Omega}}_{|p}/dt = 0$:

$$\left.\frac{d\underline{V}}{dt}\right|_0 = \left.\frac{d\underline{V}}{dt}\right|_p + 2\underline{\bar{\Omega}}_{|p} \times \underline{V}_{|p} + \underline{\bar{\Omega}}_{|p} \times (\underline{\bar{\Omega}}_{|p} \times \underline{r}_{|p}). \tag{2.38}$$

Combining eqs. (2.35) and (2.38) we obtain:

$$m\left.\frac{d\underline{V}}{dt}\right|_p = \underline{F}_{A\,|0} + m\underline{G}_{|0} - 2m\underline{\bar{\Omega}}_{|p} \times \underline{V}_{|p} - m\underline{\bar{\Omega}}_{|p} \times (\underline{\bar{\Omega}}_{|p} \times \underline{r}_{|p}). \tag{2.39}$$

We define the coordinate system **r**, which results from a rotation of the **p** coordinate system around the z_p axis, by θ and around the negative y_p axis by ϕ[31] such that x_r coincides with the position vector $\underline{r}$, Fig. 2.20. Further we need the coordinate system **g**, which is parallel to the **r** system, but with the origin **O'**, Fig. 2.21.

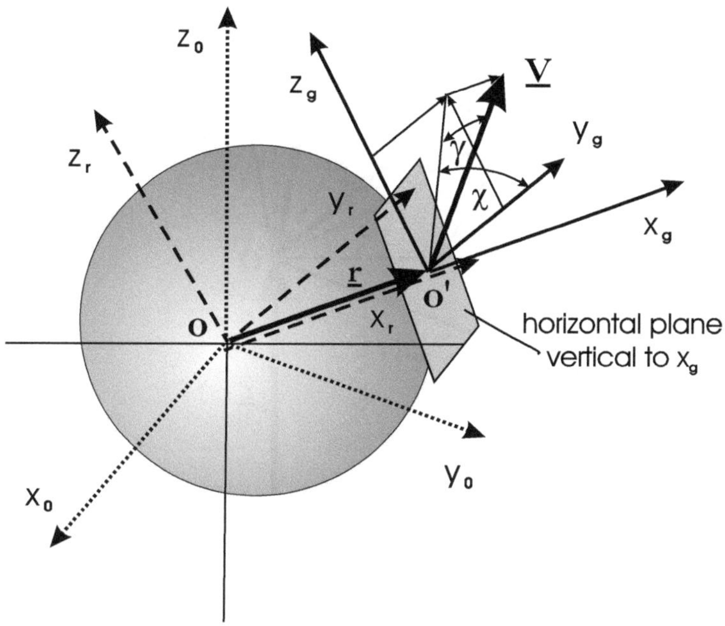

Fig. 2.21. Coordinate system **O'**, x_g, y_g, z_g parallel to coordinate system **O**, x_r, y_r, z_r.

It is now our intention to formulate all the vectors used in the **g** coordinate system. The position vector $\underline{r}$ reads in the **p** coordinate system:

$$\underline{r}_{|p} = r \begin{pmatrix} \cos\phi\cos\theta \\ \cos\phi\sin\theta \\ \sin\phi \end{pmatrix} . \tag{2.40}$$

With the matrices

$$M_{pg}^{-\phi} = \begin{pmatrix} \cos\phi & 0 & \sin\phi \\ 0 & 1 & 0 \\ -\sin\phi & 0 & \cos\phi \end{pmatrix} , \quad M_{pg}^{\theta} = \begin{pmatrix} \cos\theta & \sin\theta & 0 \\ -\sin\theta & \cos\theta & 0 \\ 0 & 0 & 1 \end{pmatrix} , \tag{2.41}$$

we transform $\underline{r}_{|p}$ and $\underline{\bar{\Omega}}_{|p}$ into the **g** coordinate system:

$$\underline{r}_{|g} = M_{pg}^{-\phi} M_{pg}^{\theta} \, \underline{r}_{|p} = r \begin{pmatrix} 1 \\ 0 \\ 0 \end{pmatrix} , \quad \underline{\bar{\Omega}}_{|g} = M_{pg}^{-\phi} M_{pg}^{\theta} \, \underline{\bar{\Omega}}_{|p} = \omega \begin{pmatrix} \sin\phi \\ 0 \\ \cos\phi \end{pmatrix} . \tag{2.42}$$

[31] The rotation around the negative y_p axis is identical with a rotation with $-\phi$.

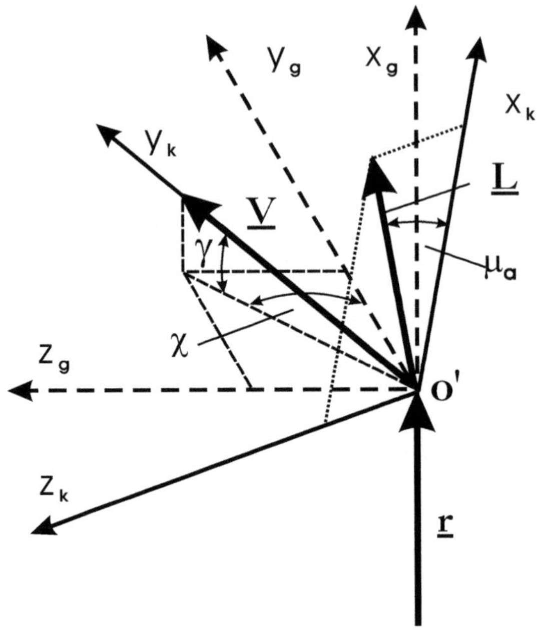

Fig. 2.22. Coordinate system **O'**, x_g, y_g, z_g and coordinate system **O'**, x_k, y_k, z_k (flight path) with the definition of bank angle μ_a and lift $\underline{L}$ outside the $\underline{r}$, $\underline{V}$ plane.

The gravitational force in the **g** coordinate system is given by:

$$m\underline{G}_{|g} = mg \begin{pmatrix} -1 \\ 0 \\ 0 \end{pmatrix} . \tag{2.43}$$

The aerodynamic force $\underline{F}_A$ is composed of the drag $\underline{D}$, which is in opposite direction to the flight velocity $\underline{V}$, and the lift $\underline{L}$ which is perpendicular to this direction.

The velocity and the drag in the **g** coordinate system follow directly from Fig. 2.21, namely:

$$\underline{V}_{|g} = V \begin{pmatrix} \sin\gamma \\ \cos\gamma\cos\chi \\ \cos\gamma\sin\chi \end{pmatrix}, \quad \underline{D}_{|g} = -D \begin{pmatrix} \sin\gamma \\ \cos\gamma\cos\chi \\ \cos\gamma\sin\chi \end{pmatrix} . \tag{2.44}$$

We define now the coordinate system **k** (flight path system), where the velocity vector $\underline{V}_{|k}$ coincides with the y_k coordinate, Fig. 2.22.

Remark: The definitions of the coordinates x_g, y_g, z_g (**g**-frame) and x_k, y_k, z_k (**k**-frame), Figs. 2.21 and 2.22, are not in agreement with the tradition of notation associated with them. We apply this exception in order to make the derivations and transformations easier to understand, having in mind that the scalar sets of eqs. (2.55) and (2.56) are independent of these definitions. Traditionally in the **g**-frame the coordinate z_g is directed vertically downwards, i.e. along the local g vector, and x_g and y_g are specified in any convenient way in the Earth horizontal plane. Further, in the **k**-frame the coordinate x_k is directed along the velocity vector $\underline{V}_{|k}$ and y_k lies in the Earth horizontal plane, [16], [17].

With the matrices:

$$M^{\gamma}_{gk} = \begin{pmatrix} \cos\gamma & -\sin\gamma & 0 \\ \sin\gamma & \cos\gamma & 0 \\ 0 & 0 & 1 \end{pmatrix}, \quad M^{\chi}_{gk} = \begin{pmatrix} 1 & 0 & 0 \\ 0 & \cos\chi & \sin\chi \\ 0 & -\sin\chi & \cos\chi \end{pmatrix}, \tag{2.45}$$

we find

$$\underline{V}_{|k} = M^{\gamma}_{gk} M^{\chi}_{gk} \, \underline{V}_{|g} = V \begin{pmatrix} 0 \\ 1 \\ 0 \end{pmatrix}. \tag{2.46}$$

In planar flight the lift force $\underline{L}$ is in the $\underline{r}, \underline{V}$ plane, but for flight control and guidance purposes $\underline{L}$ is rotated out of this plane by an angle μ_a, Fig. 2.22. μ_a is the bank angle.

Therefore the directions of the coordinates of the **k** system coincide with $x_k \Rightarrow L \sin\mu_a$, $y_k \Rightarrow \underline{V}_{|k}$ and $z_k \Rightarrow L\cos\mu_a$, Fig. 2.22.

To transform the vector $\underline{L}_{|k}$ to the **g** coordinate system we have:

$$\underline{L}_{|g} = M^{\chi}_{kg} M^{\gamma}_{kg} \, \underline{L}_{|k}, \quad \underline{L}_{|k} = L \begin{pmatrix} \cos\mu_a \\ 0 \\ \sin\mu_a \end{pmatrix}, \tag{2.47}$$

where the matrix $M^{\chi}_{kg} M^{\gamma}_{kg}$ is the inverse of the matrix $M^{\gamma}_{gk} M^{\chi}_{gk}$ with

$$M^{\gamma}_{kg} = \begin{pmatrix} \cos\gamma & \sin\gamma & 0 \\ -\sin\gamma & \cos\gamma & 0 \\ 0 & 0 & 1 \end{pmatrix}, \quad M^{\chi}_{kg} = \begin{pmatrix} 1 & 0 & 0 \\ 0 & \cos\chi & -\sin\chi \\ 0 & \sin\chi & \cos\chi \end{pmatrix}. \tag{2.48}$$

In order to be able to formulate eq. (2.39) completely in the **g** coordinate system, we need finally for $d\underline{r}\,/dt_{|p}$ and $d\underline{V}\,/dt_{|p}$ the transformations from the **p** system to the **g** system, which have the forms:

$$\left.\frac{d\underline{r}}{dt}\right|_p = \left.\frac{d\underline{r}}{dt}\right|_g + \underline{\Omega}_{|g} \times \underline{r}_{|g}, \qquad \left.\frac{d\underline{V}}{dt}\right|_p = \left.\frac{d\underline{V}}{dt}\right|_g + \underline{\Omega}_{|g} \times \underline{V}_{|g}. \tag{2.49}$$

Here $\underline{\Omega}_{|g}$ denotes the vector of the rotation velocity of the **g** system relatively to the **p** system. Since the **g** coordinate system was obtained by rotations around the z_p and y_p axes with the Euler angles[32] θ and $-\phi$, Fig. 2.21,

[32] For the definition of the Euler angles see [17] and Sub-Section 5.4.2.

we find:

$$\underline{\Omega}_{|g} = M_{pg}^{-\phi} \begin{pmatrix} 0 \\ 0 \\ \dot{\theta} \end{pmatrix} + \begin{pmatrix} 0 \\ -\dot{\phi} \\ 0 \end{pmatrix} = \begin{pmatrix} \dot{\theta}\sin\phi \\ -\dot{\phi} \\ \dot{\theta}\cos\phi \end{pmatrix}, \tag{2.50}$$

$$\left.\frac{d\underline{r}}{dt}\right|_p = \frac{dr}{dt} \begin{pmatrix} 1 \\ 0 \\ 0 \end{pmatrix}_{|g} + \begin{pmatrix} 0 \\ \dot{\theta}\, r\cos\phi \\ \dot{\phi}\, r \end{pmatrix}_{|g} = \begin{pmatrix} \dot{r} \\ \dot{\theta}\, r\cos\phi \\ \dot{\phi}\, r \end{pmatrix}_{|g} = \underline{V}_{|g}. \tag{2.51}$$

Eqs. (2.44) and (2.51) represent the kinematic equations in the form (see also [19]):

$$\begin{aligned} \dot{r} &= V\sin\gamma, \\ \dot{\theta} &= \frac{V\cos\gamma\cos\chi}{r\cos\phi}, \\ \dot{\phi} &= \frac{V\cos\gamma\sin\chi}{r}. \end{aligned} \tag{2.52}$$

All together we can write a compact form of the general flight mechanical equations for space applications in the **g** coordinate system:

$$\begin{aligned} m\left(\left.\frac{d\underline{V}}{dt}\right|_g + \underline{\Omega}_{|g} \times \underline{V}_{|g}\right) &= \underline{D}_{|g} + M_{kg}^{\chi} M_{kg}^{\gamma} \underline{L}_{|k} + m\,\underline{G}_{|g} - \\ &\quad -2m\bar{\underline{\Omega}}_{|g} \times \underline{V}_{|g} - m\bar{\underline{\Omega}}_{|g} \times (\bar{\underline{\Omega}}_{|g} \times \underline{r}_{|g}). \end{aligned} \tag{2.53}$$

By substitution of eqs. (2.42) to (2.44), (2.47), (2.50) into eq. (2.53) we obtain:

$$\begin{aligned} &\frac{d}{dt} V \begin{pmatrix} \sin\gamma \\ \cos\gamma\cos\chi \\ \cos\gamma\sin\chi \end{pmatrix} + \begin{pmatrix} \dot{\theta}\sin\phi \\ -\dot{\phi} \\ \dot{\theta}\cos\phi \end{pmatrix} \times V \begin{pmatrix} \sin\gamma \\ \cos\gamma\cos\chi \\ \cos\gamma\sin\chi \end{pmatrix} = \\ &-\frac{1}{m} D \begin{pmatrix} \sin\gamma \\ \cos\gamma\cos\chi \\ \cos\gamma\sin\chi \end{pmatrix} + \frac{1}{m} L \begin{pmatrix} \cos\gamma\cos\mu_a \\ -\sin\gamma\cos\chi\cos\mu_a - \sin\chi\sin\mu_a \\ -\sin\gamma\sin\chi\cos\mu_a + \cos\chi\sin\mu_a \end{pmatrix} + \\ &+ g \begin{pmatrix} -1 \\ 0 \\ 0 \end{pmatrix} - 2\omega \begin{pmatrix} \sin\phi \\ 0 \\ \cos\phi \end{pmatrix} \times V \begin{pmatrix} \sin\gamma \\ \cos\gamma\cos\chi \\ \cos\gamma\sin\chi \end{pmatrix} - \\ &- \omega \begin{pmatrix} \sin\phi \\ 0 \\ \cos\phi \end{pmatrix} \times \left\{ \omega \begin{pmatrix} \sin\phi \\ 0 \\ \cos\phi \end{pmatrix} \times r \begin{pmatrix} 1 \\ 0 \\ 0 \end{pmatrix} \right\}. \end{aligned} \tag{2.54}$$

Resolving eq. (2.54) for dV/dt, $Vd\gamma/dt$, $Vd\chi/dt$ by using the relations of eq. (2.52) leads to, see also [19]:

$$\begin{aligned}
\frac{dV}{dt} &= -\frac{1}{m}D - g\sin\gamma + \omega^2 r\cos^2\phi(\sin\gamma - \cos\gamma\tan\phi\sin\chi),\\
V\frac{d\gamma}{dt} &= \frac{1}{m}L\cos\mu_a - g\cos\gamma + \frac{V^2}{r}\cos\gamma + 2\omega V\cos\phi\cos\chi +\\
&\quad + \omega^2 r\cos^2\phi(\cos\gamma + \sin\gamma\tan\phi\sin\chi),\\
V\frac{d\chi}{dt} &= \frac{1}{m}\frac{L\sin\mu_a}{\cos\gamma} - \frac{V^2}{r}\cos\gamma\cos\chi\tan\phi +\\
&\quad + 2\omega V(\tan\gamma\cos\phi\sin\chi - \sin\phi) - \frac{\omega^2 r}{\cos\gamma}\sin\phi\cos\phi\cos\chi. \quad (2.55)
\end{aligned}$$

These are the force equations which describe the unpowered motion of aerospace vehicles during space flight including ascent and descent. Since ω is a small value, generally the influence of the term $\omega^2 r$ is low. More important is the quantity $2\omega V$, which describes the influence of the Coriolis forces, which aerospace vehicles experience moving relative to a rotating system. The impact of this term diminishes for flights along descent trajectories with strong deceleration.

By considering the planet as non-rotating we have $\omega = 0$, and the **p** coordinate system takes over the role of the inertial system. Eq. (2.55) then reduces to:

$$\begin{aligned}
\frac{dV}{dt} &= -\frac{1}{m}D - g\sin\gamma,\\
V\frac{d\gamma}{dt} &= \frac{1}{m}L\cos\mu_a - g\cos\gamma + \frac{V^2}{r}\cos\gamma,\\
V\frac{d\chi}{dt} &= \frac{1}{m}\frac{L\sin\mu_a}{\cos\gamma} - \frac{V^2}{r}\cos\gamma\cos\chi\tan\phi. \quad (2.56)
\end{aligned}$$

For flight, where the flight path azimuth angle χ is not changed, the third of the scalar equations, eq. (2.56), vanishes and we have the relations given at the beginning of Sub-Section 2.1.4.

Remark: It should be noted that eqs. (2.55) with the $2\omega V$- and the $\omega^2 r$-terms are only valid in the inertial **O**-frame, while eqs. (2.56) are to be applied on the basis of the **p**-frame as inertial system. Actually the magnitude of V, describing the same planetary or orbital flight situation, is different in both frames and hence in both sets of equations.

For clarification we define an example. Let us assume that a space vehicle moves along an equatorially circular orbit around the Earth, which is determined – as is well known – by the balance of the gravitational and the centrifugal forces. The velocity of the space vehicle in the **p**-frame is then given by:

$$V_{circ} = \left[(R_E + H)g(H)\right]^{1/2}. \quad (2.57)$$

The velocity in a point on this circular orbit due to the Earth rotation is:

$$V_E = \omega(R_E + H). \tag{2.58}$$

This means that the determination of the velocity of a vehicle along this circular orbit, using eq. (2.56), has to be made with:

$$V^p_{|g} = V_{circ},$$

and, using eq. (2.55), with:

$$V^0_{|g} = V_{circ} - V_E,$$

see Problem 2.5.

2.4 Problems

Problem 2.1. Calculate flight velocity and Mach number of the Space Shuttle Orbiter in connection with the RCS employment during re-entry flight for a) $q_\infty = 0.48\ kPa$ at 81.0 km altitude and b) $q_\infty = 1.9\ kPa$ at 69.0 km altitude.

Problem 2.2. The maximum heat flux at the lower symmetry line at x/L = 0.5 of the Space Shuttle Orbiter given in Fig. 2.6 is $(q_w =)\ q_{gw} = 0.067$ MW/m^2. Assume zero heat flux into the wall and no other heat-flux contributions. How large is the radiation-adiabatic temperature belonging to that heat flux, if $\varepsilon = 0.85$?

Problem 2.3. How large is the heat flux in the gas at the wall, q_{gw}, at the vehicle nose for the flight conditions of Problem 2, assuming $H = 70.0\ km$ and $v_\infty = 6{,}450.0\ m/s$? Assume an effective nose radius of $R_N = 1.3\ m$. Use the simplest relation for q_{gw} given in Section 10.3. Is the heat flux within the usual constraints? How large is T_{ra}, if $\varepsilon = 0.85$? Is it acceptable from the material point-of-view?

Problem 2.4. Repeat Problem 2.3 for an effective nose radius $R_N = 0.25$ m. Are the heat flux and T_{ra} acceptable?

Problem 2.5. a) At what altitude H is an equatorially circular orbit synchronized with the Earth rotation, saying that the position of a space vehicle is fixed with respect to an observer on the Earth surface? b) How large is the velocity magnitude V, when the simulation of the flight of a vehicle in this orbit is made with eqs. (2.55) and alternatively with eqs. (2.56)?

Assume that the gravitational acceleration is given by $g = g_0[R_E/(R_E + H)]^2$.

References

1. Harpold, J.C., Graves Jr., C.A.: Shuttle Entry Guidance. In: Proc. 25th Anniversary Conference of the American Astronomical Society (AAS 78-147), pp. 99–132 (1978)
2. Markl, A.W.: An Initial Guess Generator for Launch and Reentry Vehicle Trajectory Optimization. Doctoral Thesis, Universität Stuttgart, Germany (2001)
3. Adams, J.C., Martindale, W.R., Mayne, A.W., Marchand, E.O.: Real Gas Scale Effects on Hypersonic Laminar Boundary-layer Parameters Including Effects of Entropy-Layer Swallowing. AIAA-Paper 76-358 (1976)
4. Wüthrich, S., Sawley, M.L., Perruchoud, G.: The Coupled Euler/ Boundary-Layer Method as a Design Tool for Hypersonic Re-Entry Vehicles. Zeitschrift für Flugwissenschaften und Weltraumforschung (ZFW) 20(3), 137–144 (1996)
5. Hirschel, E.H.: Basics of Aerothermodynamics. Progress in Astronautics and Aeronautics, AIAA, Reston, Va, vol. 204. Springer, Heidelberg (2004)
6. Williams, S.D.: Columbia, the First Five Flights Entry Heating Data Series, an Overview, vol. 1. NASA CR-171 820 (1984)
7. Hoey, R.G.: AFFTC Overview of Orbiter-Reentry Flight-Test Results. In: Arrington, J.P., Jones, J.J. (eds.) Shuttle Performance: Lessons Learned. NASA CP-2283, Part 2, pp. 1303–1334 (1983)
8. Koelle, D.E., Kuczera, H.: SÄNGER – An Advanced Launcher System for Europe. IAF-Paper 88-207 (1988)
9. Neumann, R.D.: Defining the Aerothermodynamic Methodology. In: Bertin, J.J., Glowinski, R., Periaux, J. (eds.) Hypersonics. Defining the Hypersonic Environment, vol. 1, pp. 125–204. Birkhäuser, Boston (1989)
10. Haney, J.W.: Orbiter (Pre STS-1) Aeroheating Design Data Base Development Methodology: Comparison of Wind Tunnel and Flight Test Data. In: Throckmorton, D.A. (ed.) Orbiter Experiments (OEX) Aerothermodynamics Symposium. NASA CP-3248, Part 2, pp. 607–675 (1995)
11. American National Standards Institute. Recommended Practice for Atmosphere and Space Flight Vehicle Coordinate Systems. American National Standard ANSI/AIAA R-0004-1992 (1992)
12. Romere, P.O.: Orbiter (Pre STS-1) Aerodynamic Design Data Book Development and Methodology. In: Throckmorton, D.A. (ed.) Orbiter Experiments (OEX) Aerothermodynamics Symposium. NASA CP-3248, Part 1, pp. 249–280 (1995)
13. Iliff, K.W., Shafer, M.F.: Extraction of Stability and Control Derivatives from Orbiter Flight Data. In: Throckmorton, D.A. (ed.) Orbiter Experiments (OEX) Aerothermodynamics Symposium. NASA CP-3248, Part 1, pp. 299–344 (1995)
14. McCormick, B.W.: Aerodynamics, Aeronautics, and Flight Mechanics. John Wiley & Sons, New York (1979)
15. Schlichting, H., Truckenbrodt, E.: Aerodynamik des Flugzeuges. Teil I. Springer, Berlin (1967)
16. Etkin, B.: Dynamics of Atmospheric Flight. John Wiley & Sons, New York (1972)
17. Brockhaus, R.: Flugregelung. Springer, Heidelberg (2001)
18. Hankey, W.L.: Re-Entry Aerodynamics. AIAA Education Series, Washington, D.C (1988)
19. Vinh, N.X., Busemann, A., Culp, R.D.: Hypersonic and Planetary Entry Flight Mechanics. The University of Michigan Press, Ann Arbor (1980)

20. Regan, F.J.: Re-Entry Vehicle Dynamics. AIAA Education Series, Washington, D. C (1984)
21. Hirschel, E.H.: Aerodynamische Probleme bei Hyperschall-Fluggeräten. Jahrestagung der DGL (October 9 and 10, 1986) München, Germany, also MBB/S/PUB/270, MBB, München/Ottobrunn, Germany (1986)
22. Shinn, J.L., Simmonds, A.L.: Comparison of Viscous Shock-Layer Heating Analysis with Shuttle Flight Data in Slip Flow Regime. AIAA-Paper 84-0226 (1984)
23. Zoby, E.V., Gupta, R.N., Simmonds, A.L.: Temperature-Dependent Reaction-Rate Expression for Oxygen Recombination at Shuttle Entry Conditions. AIAA-Paper 84-0224 (1984)
24. Goodrich, W.D., Derry, S.M., Bertin, J.J.: Shuttle Orbiter Boundary-Layer Transition: A Comparison of Flight and Wind Tunnel Data. AIAA-Paper 83-0485 (1983)
25. Gupta, R.N., Moss, J.N., Simmonds, A.L., Shinn, J.L., Zoby, E.V.: Space Shuttle Heating Analysis with Variation in Angle of Attack and Surface Condition. AIAA-Paper 83-0486 (1983)
26. Kuczera, H.: Bemannte europäische Rückkehrsysteme – ein Überblick. Space Course, Vol. 1, Universität Stuttgart, Germany, pp. 161–172 (1995)
27. McRuer, D.: Design and Modelling Issues for Integrated Airframe/Propulsion Control of Hypersonic Flight Vehicles. In: Proc. American Control Conference, Boston, Mass., pp. 729–735 (1991)
28. Hauck, H.: Leitkonzept SÄNGER – Referenz-Daten-Buch. Issue 1, Revision 2, Dasa, München/Ottobrunn, Germany (1993)
29. Hunt, J.L.: Hypersonic Airbreathing Vehicle Design (Focus on Aero-Space Plane). In: Bertin, J.J., Glowinski, R., Periaux, J. (eds.) Hypersonics. Defining the Hypersonic Environment, vol. 1, pp. 205–262. Birkhäuser, Boston (1989)
30. Hirschel, E.H.: Towards the Virtual Product in Aircraft Design? In: Periaux, J., Champion, M., Gagnepain, J.-J., Pironneau, O., Stouflet, B., Thomas, P. (eds.) Fluid Dynamics and Aeronautics New Challenges. CIMNE Handbooks on Theory and Engineering Applications of Computational Methods, Barcelona, Spain, pp. 453–464 (2003)
31. Hirschel, E.H.: Aerothermodynamic Phenomena and the Design of Atmospheric Hypersonic Airplanes. In: Bertin, J.J., Periaux, J., Ballmann, J. (eds.) Advances in Hypersonics. Defining the Hypersonic Environment, vol. 1, pp. 1–39. Birkhäuser, Boston (1992)
32. Edwards, C.L.W., Small, W.J., Weidner, J.P.: Studies of Scramjet/Airframe Integration Techniques for Hypersonic Aircraft. AIAA-Paper 75-58 (1975)
33. Hirschel, E.H.: The Technology Development and Verification Concept of the German Hypersonics Technology Programme. AGARD R-813, pp. 12-1–12-15 (1986)
34. Bayer, R., Sachs, G.: Optimal Return-to-Base Cruise of Hypersonic Carrier Vehicles. Zeitschrift für Flugwissenschaften und Weltraumforschung (ZFW) 19(1), 47–54 (1992)
35. Heiser, W.H., Pratt, D.T.: Hypersonic Airbreathing Propulsion. AIAA Education Series, Washington, D.C (1994)

3 Aerothermodynamic Design Problems of Winged Re-Entry Vehicles

In this chapter we consider selected aerothermodynamic design problems of winged re-entry vehicles (RV-W-type vehicles, Section 1.1). Of these vehicles the Space Shuttle Orbiter so far is the only operational vehicle. The other ones are conceptual studies or projects, which have reached different degrees of maturity.

RV-W-type vehicles are either launched vertically with the help of rockets, or – in the case of TSTO-systems – horizontally from a carrier vehicle, the lower stage of the system. Other launch modes have been considered, viz. the horizontal launch from a sled. The return to the Earth surface in any case is made with an unpowered gliding flight, followed by the horizontal landing on a runway.

The winged re-entry vehicle is heavier and more complex than a non-winged vehicle (capsule), Chapter 5, but in principle is a re-usable vehicle. Its relatively high lift-to-drag (L/D) ratio results in a large cross-range capability. The aerodynamic design of RV-W-type vehicles is driven by the wide Mach number and altitude range of these vehicles, the structural design is driven by the large thermal loads, which are present on the atmospheric high speed segment of the re-entry trajectory.

In this chapter we give an overview of aerothermodynamic phenomena found on RV-W-type flight vehicles, look at a major simulation problem, viz. that due to high Mach number and enthalpy effects, and discuss particular aerothermodynamic trends. This is followed by a presentation of aerodynamic coefficients of the longitudinal motion of several vehicles. Finally we study implications of Oswatitsch's Mach number independence principle in view of the hypersonic pitching moment anomaly observed during the first re-entry flight of the Space Shuttle Orbiter.

3.1 Overview of Aerothermodynamic Issues of Winged Re-Entry Vehicles

3.1.1 Aerothermodynamic Phenomena

Winged re-entry vehicles basically fly a braking mission during return from orbit or sub-orbit to the surface of Earth. They are, therefore, on purpose

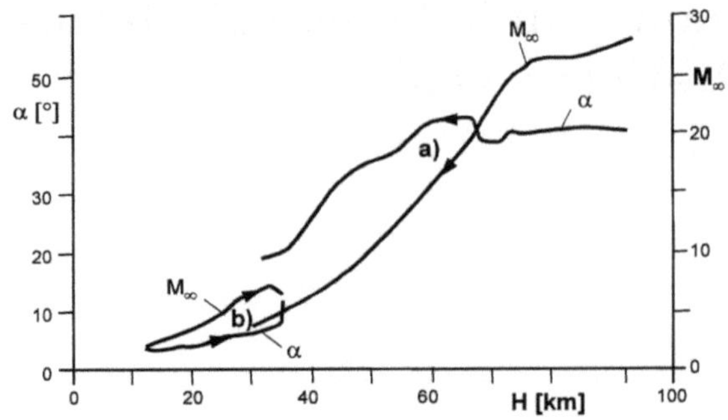

Fig. 3.1. Angle of attack α and flight Mach number M_∞ of a) the Space Shuttle Orbiter, and b) the TSTO space transportation system SÄNGER up to stage separation, as function of the flight altitude, [1].

blunt and compact vehicles. They fly on the largest part of their trajectory at high angle of attack, which is in contrast to airbreathing cruise and acceleration (CAV-type, Chapter 4) vehicles, Fig. 3.1, but also to RV-NW-type vehicles, Chapter 5.

The high angle of attack of the vehicle with a more or less flat windward side, Sub-Section 3.2.2, increases the "effective" bluntness, and thus increases further the (wave) drag of the vehicle. The large nose and wing leading edge radii, the wing with strong backward sweep, permit an effective surface radiation cooling, [1].

A problem of these vehicles is that they have very small aspect ratios. Their low speed capabilities are not very suitable for approach and landing. This holds for all hypersonic flight vehicles. A remedy is to provide a double-delta, or strake-delta wing[1], Fig. 3.2. In the case of the HERMES vehicle, winglets are employed, see also Chapter 6.

In the aerothermodynamic design of RV-W-type vehicles flow and high-temperature real-gas phenomena must be taken into account which do not occur in subsonic, transonic and low supersonic flight. Major phenomena of the flow past RV-W-type vehicles in the continuum-flow regime (below approximately 100.0 to 90.0 km altitude) are indicated in Fig. 3.3.

As mentioned, high drag demand and thermal loads minimization call for a blunt nose and large angle of attack of the more or less flat lower side of the vehicle. This gives rise to a detached bow shock and strong compressibility effects on the windward side. Due to the high angles of attack on the re-entry trajectory down to small altitudes, the flow past the windward side of the vehicle is of major interest. The leeward side is in the hypersonic shadow, which, however, is important in view of vehicle control, Section 2.1 and Chapter 6.

[1] The effective wing surface includes in any case the whole lower side of the airframe. With airbreathing CAV-type vehicles, Chapter 4, the lower side of the airframe is at the same time also part of the propulsion system.

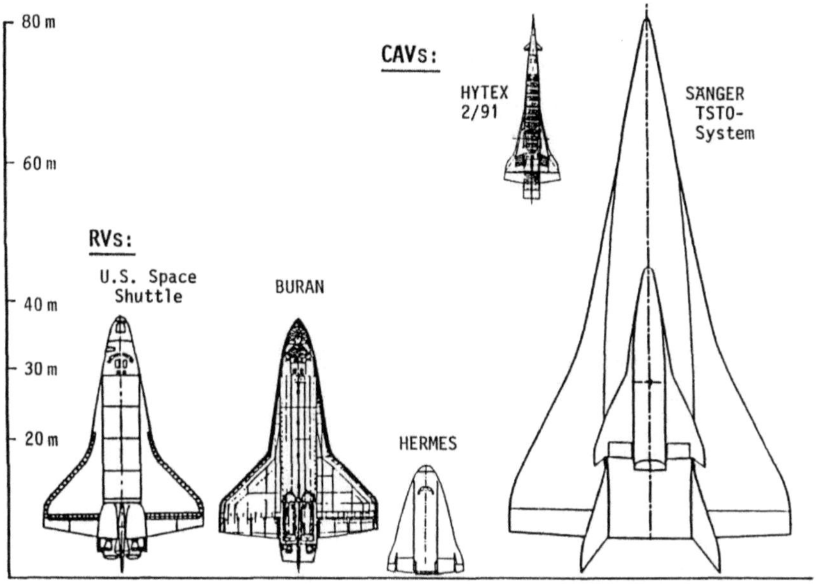

Fig. 3.2. Shape (planform) and size of hypersonic flight vehicles of RV-W-type and CAV-type, [2]. HYTEX: airbreathing experimental vehicle studied in the German Hypersonics Technology Programme, [3].

On the windward side of the vehicle we point out first the presence of the entropy layer. Two forms of the entropy layer are possible, [1]. One, typical for symmetric flow, has a (inviscid) velocity profile normal to the body surface like a slip-flow boundary layer. The second form is typical for asymmetric flow, where the streamline hitting the forward stagnation point does not cross the normal portion of the bow shock surface. In this case the maximum entropy streamline lies off the body surface and is at the same time the minimum inviscid velocity streamline. The entropy layer hence has a wake-like structure. We discuss this form, which in general seems to be present at the windward side of a RV-W-type configuration at high angle of attack, in some detail in Sub-Section 6.2.2.

The windward side is marked further by strong high-temperature real-gas effects. Thermo-chemical equilibrium and non-equilibrium flow is present at the forward stagnation point region and downstream, depending on flight speed and altitude. The strength of surface catalytic recombination depends much on the properties of the surface (coating) material. As indicated in Fig. 3.3, the general trend is that oxygen recombines rather at rearward locations, because the temperature is too high at forward locations. There nitrogen may recombine.

The static temperature at the edge of the boundary layer is very large. The attached viscous flow is in general of second-order boundary layer type, see, e.g., [1]. Above approximately 40.0 to 60.0 km altitude the boundary

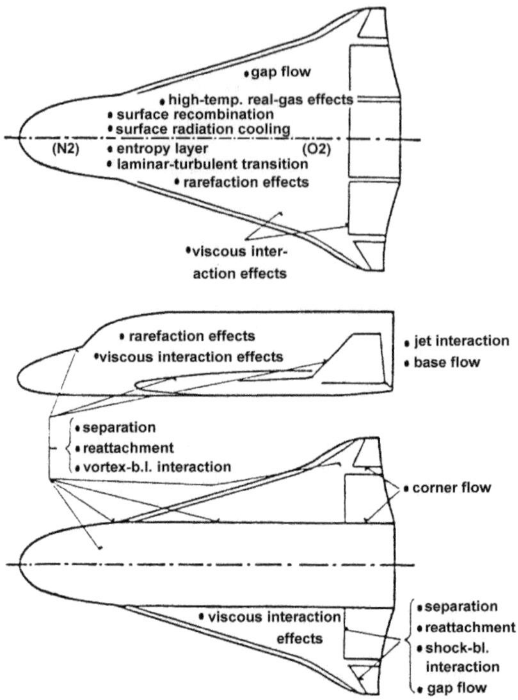

Fig. 3.3. Locations of flow and high-temperature real-gas phenomena on a RV-W-type vehicle (HERMES) at hypersonic flight, [4]. Upper figure: view from below, middle figure: side view, lower figure: view from above. ($N2$), ($O2$): location of possible/expected catalytic surface recombination of nitrogen, oxygen. Note that all indicated phenomena depend on flight speed and altitude, and on shape and attitude of the given RV-W-type vehicle.

layerusually is laminar, which eases some of the aerothermodynamic simulation problems. Below that altitude it becomes turbulent. The wall temperature, due to surface radiation cooling, is rather small, decreasing at $M_\infty = 24$ and 70.0 km altitude (laminar flow) from around 2,000.0 K at the stagnation point to around 1,000.0 K at a downstream distance of about 30.0 m.

The radiation-adiabatic temperature T_{ra} in general is a good approximation of the actual wall temperature T_w. The heat flux q_w into the thermal protection system (TPS) is small, compared to the heat flux in the gas at the wall, q_{gw}, Section 9.1. Due to the large temperature and heat flux gradients in downstream, partly also in lateral direction, the thickness of the TPS varies with the location in order to reduce its mass. The TPS of the Space Shuttle Orbiter has about 30,000 tiles of different thickness, shape and size.

Because of the large angle of attack, the boundary layer (inviscid) edge Mach number M_e at the vehicle's windward side is small compared to the flight Mach number[2]. Hence the boundary layer is a transonic/supersonic boundary layer. When it becomes turbulent, both wall temperature and heat flux in the gas at the wall increase markedly, Section 9.2. Hence thermal loads are also of concern on the lower segments of the trajectory.

Hypersonic viscous interaction and also rarefaction effects are mainly confined to the initial re-entry trajectory segment. Strong interaction phenomena, like separation, attachment, reattachment, shock/boundary layer as well as vortex/boundary layer interaction may occur, depending on the shape and attitude of the flight vehicle, at several locations, but especially in the hinge-line neighborhood of aerodynamic control surfaces, Section 6.3. Jet/boundary layer interaction occurs at the location of the thrusters of the reaction control system, Section 6.5. Base flow and base drag in general are of minor interest on the high speed segments of the trajectory, because the base pressure is close to zero there.

Summarizing we note that we have, mainly in the flow path at the windward (lower) side of RV-W-type flight vehicles:

- strong compressibility effects,
- strong high-temperature real-gas effects,
- surface radiation cooling without and with non-convex effects, radiation cooling related hot-spot situations,
- complex surface heat transfer mechanisms (thermal surface radiation, non-equilibrium effects, catalytic surface recombination, slip flow and temperature-jump effects),
- strong (shock/shock and shock/boundary layer) interaction, corner flow, boundary layer separation,
- viscous and thermo-chemical thermal surface effects,
- hypersonic viscous interaction and rarefaction (slip flow) effects at high altitudes,
- gap flow (trim and control surfaces, thermal protection system),
- reaction control system related strong flow interactions,
- base flow, plume spreading,
- laminar flow at altitudes above approximately 40.0 to 60.0 km,
- Mach number and surface property effects on laminar-turbulent transition and on turbulent flow phenomena.

It is important to note, and this holds also for the aerothermodynamic phenomena at CAV- and ARV-type as well as RV-NW-type vehicles, that the listed phenomena usually do not occur isolated. Strong synergistic effects may take place which can result in a highly non-linear behavior. Then it might be difficult to understand each effect and its influence on the flight vehicle properties and performance. To simulate in such cases the effects properly in ground-simulation facilities and/or computationally is very difficult, if not impossible.

[2] Of influence are also the large static temperature at the windward side and high-temperature real-gas effects.

3.1.2 Major Simulation Problem: High Mach Number and Total Enthalpy Effects

In view of the dominating compressibility effects, Mach number and high-temperature real-gas effects are extremely important, but difficult to treat in ground-simulation, Section 9.3, see also the discussions in [1]. This holds for both ground-facility and computational simulation.

Ground-simulation in modern facilities, in principle, is possible for the whole Mach number range of RV-W-type vehicles. The problem is to have a sufficiently large wind-tunnel or shock-tunnel model and further, regarding especially data set generation, a high enough productivity of the ground-facility (polars per time unit). The latter holds for computational simulation, too. However, the growing computer capabilities will help to overcome this problem rather soon.

The fact, that force coefficients, shock wave shapes et cetera become independent of the flight Mach number above a – configuration-dependent – Mach number (Oswatitsch's Mach number independence principle, Sub-Section 3.6.1), allows to make measurements at much lower Mach numbers than actual flight Mach numbers. This works, but, as was seen with the hypersonic pitching moment anomaly observed at the Space Shuttle Orbiter, Section 3.5, has its dangers.

High-enthalpy facilities allow to achieve the real flight enthalpy but reach, because of the inherently large static temperature in the working section, too small Mach numbers. However, if the Mach number is large enough, so that independence is present, this is no principal problem[3]. The problem is the role of high-temperature real-gas effects in the Mach number independence principle. We discuss this problem in Section 3.6.

In view of the still growing computation capabilities, one is tempted to assume, that at least pure data set generation in ground-simulation facilities will become obsolete in some years. This assumption requires that flow-physics, thermodynamics and surface-property models are improved accordingly.

3.2 Particular Trends in RV-W Aerothermodynamics

3.2.1 Stagnation Pressure

The stagnation pressure coefficient for incompressible flow is, with $p_{stag} = p_\infty + q_\infty$ (the dynamic pressure is $q_\infty = 0.5\,\rho_\infty\, v_\infty^2$):

$$c_{p_{stag}} = \frac{p_{stag} - p_\infty}{q_\infty} = 1. \tag{3.1}$$

[3] Of course, there are other problems in such facilities, which are discussed in, e.g., [1].

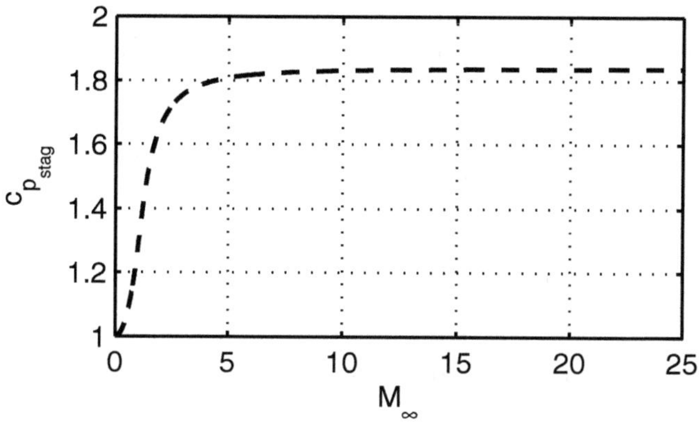

Fig. 3.4. Stagnation pressure coefficient $c_{p_{stag}}$ as function of the flight Mach number M_∞, $\gamma = 1.4$.

For compressible subsonic flow we get at the stagnation point with $p_{stag} \equiv p_t$ (isentropic compression), see, e.g., [1]:

$$c_{p_{stag}} = \frac{2}{\gamma M_\infty^2} \left[\left(1 + \frac{\gamma - 1}{2} M_\infty^2 \right)^{\frac{\gamma}{\gamma-1}} - 1 \right]. \tag{3.2}$$

This means that for $0 < M_\infty \leqq 1$ and $\gamma = 1.4$ we have $1 \leqq c_{p_{stag}} \leqq 1.2756$.

In supersonic flow we must specify the flow situation because ahead of a blunt body a bow shock exists which causes a loss of total pressure. In the case of a symmetric blunt body at zero angle of attack, the total pressure loss is that across the normal shock portion of the bow shock, see, e.g., [1]. Taking that into account, we have:

$$c_{p_{stag}} = c_{p_{t_2}} = \frac{2}{\gamma M_\infty^2} \left\{ \left[\frac{(\gamma+1)^2 M_\infty^2}{4\gamma M_\infty^2 - 2(\gamma - 1)} \right]^{\frac{\gamma}{\gamma-1}} \left[\frac{2\gamma M_\infty^2 - (\gamma - 1)}{\gamma + 1} \right] - 1 \right\}. \tag{3.3}$$

For very large Mach numbers, $M_\infty \to \infty$, this relation reduces to:

$$c_{p_{stag}} = c_{p_{t_2}} \to \left[\frac{(\gamma+1)^2}{4\gamma} \right]^{\frac{\gamma}{\gamma-1}} \frac{4}{\gamma + 1}. \tag{3.4}$$

With $\gamma = 1.4$ we obtain in this case $c_{p_{stag}} = 1.8394$. Hence in supersonic and hypersonic perfect gas flow, $M_\infty > 1$, we find: $1.2756 < c_{p_{stag}} < 1.8394$, Fig. 3.4. This means that we get at hypersonic Mach numbers only a finite, even not very large pressure coefficient at the forward stagnation point of a blunt body.

However, the question is, how high-temperature real-gas effects change this picture. In Table 3.1 we have listed in this respect a number of computation

Table 3.1. Configuration, flight data and computation details of the stagnation pressure data given in Fig. 3.5. For the Nos. 6 and 7 mean values from several contributions to [5] (proceedings of the 1991 Antibes workshop) are given in Fig. 3.5.

No.	Configuration	M_∞ [–]	H [km]	α [°]	Eqs.	Gas Model	Ref.	Remarks
1	double ellipse	6.6	-	0	NS	non-eq.	[5]	WT cond.
2	sphere	10	30.0	0	Euler	eq.	[6]	
3	sphere	15	50.0	0	Euler	eq.	[6]	
4	sphere	20	50.0	0	Euler	eq.	[6]	
5	Space Shuttle Orb.	23.86	73.1	40	NS	non-eq.	[7]	flight cond.
6	double ellipsoid	25	75.0	40	Euler, NS	non-eq.	[5]	
7	double ellipsoid	25	75.0	40	Euler, NS	eq.	[5]	

cases for which the stagnation pressure coefficients as function of the Mach number are shown in Fig. 3.5.

In Fig. 3.5 the stagnation pressure coefficients, actually the c_{p_w} at the forward stagnation points, of the computation cases are marked by symbols. In addition curves $c_{p_{stag}}(M_\infty)$ are given which were found with the normal-shock relations with different values of γ_{eff}, Sub-Section 10.1.3. They represent the stagnation pressure coefficient $c_{p_{t_2}}$ just behind the normal shock.

Concerning the curves in Fig. 3.5, the reader should remember, that in reality γ is not constant in the layer between the bow shock and the body surface. Further, if the body is asymmetric or at angle of attack, the streamline hitting the forward stagnation point does not cross the normal portion of the bow shock surface. In this case the maximum entropy streamline lies off the body surface. Nevertheless, the curves give the general trend of c_{p_w} in terms of $c_{p_{stag}}$ at the forward stagnation point of blunt bodies at hypersonic flight Mach numbers. For $\gamma \rightarrow 1$ we have for all Mach numbers $c_{p_{stag}} = 2$, which, of course, represents the Newton limit ($M_\infty \rightarrow \infty$), see, e.g., [1].

We note, that $c_{p_{stag}}$ increases for all given γ_{eff} with increasing Mach number M_∞. The gradient $d\,c_{p_{stag}}/d\,M_\infty$ becomes small and then approximately zero, Figs. 3.4 and 3.5. Decreasing γ_{eff} increases $c_{p_{stag}}$ at all Mach numbers.

As we will see in Section 3.6, this behavior is reversed, at least for blunt body shapes, at some location downstream of the stagnation point area. The cross-over location in general lies at the beginning of the flat surface portion of the windward side, see also next sub-section. Behind it the inviscid wall pressure coefficient c_{p_w} decreases with increasing Mach number as well as with decreasing γ_{eff}, both independently of each other. However, there exists presumably – for sufficiently high flight Mach numbers – a wall (boundary-layer edge) Mach

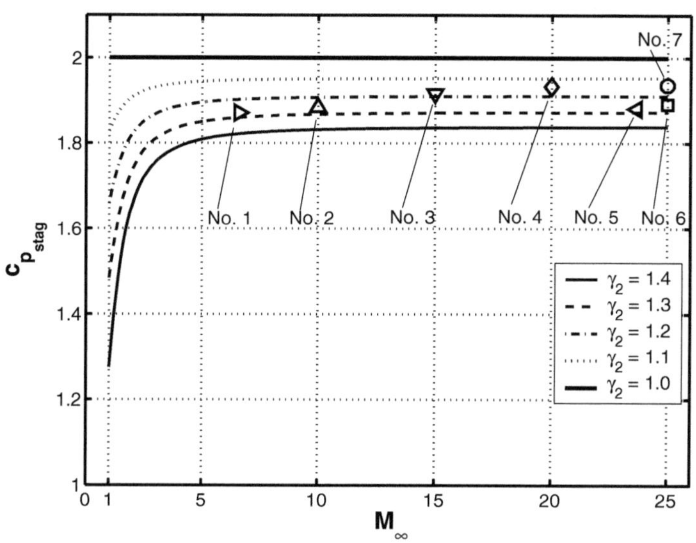

Fig. 3.5. Computed stagnation pressure coefficient $c_{p_{stag}}$ of different, mainly inviscid cases, Table 3.1, as function of the flight Mach number M_∞. Included are curves $c_{p_{stag}}(M_\infty)$ for different γ_{eff} ($\equiv \gamma_2$).

number interval, the benign M_w interval, where these effects are so small, that we have virtually a flight Mach number and real-gas effects independence[4].

Important is also the observation, that the small gradient $d\,c_{p_{stag}}/d\,M_\infty$ at high supersonic/hypersonic Mach numbers would make flight speed and Mach number determination with a conventional Pitot tube, for instance during re-entry of a RV-W-type vehicle, difficult if not impossible at high Mach numbers. This effect is also a consequence of Oswatitsch's Mach number independence principle.

There are, however, situations possible at a vehicle configuration, where a much larger wall pressure than the forward stagnation point pressure can arise. This happens when a multiple shock system is present, which leads, for instance, to the Edney type IV shock/shock interaction, see, e.g., [1]. There very locally the wall pressure can be up to ten times and more larger than the stagnation pressure. Unfortunately in such strong interaction situations also the thermal loads locally become very large.

We can find such a situation at locations where the vehicle bow shock interacts with an embedded bow shock, for instance that of a pylon, which is the famous X-15 case, which was the motivation for Edney's work, [8]. Other locations, where such situations can arise, are at wings (second – outboard – delta

[4] The curves in Fig. 3.5 suggest that with increasing high-temperature real-gas effects the Mach number independence occurs at smaller Mach numbers. However, such a conclusion should be drawn with caution.

part) and at stabilizer surfaces, if the sweep angle locally is small or at the inlet of CAV/ARV-type vehicles.

3.2.2 Topology of the Windward Side Velocity Field

From an aerothermodynamic point of view the windward side of a RV-W-type vehicle ideally should be predominantly flat. This means also that fuselage and wing combined should have a plane lower side. A flat windward side would lead to a larger bow shock portion with high shock angle θ against the free-stream. In this way the wave drag can be enhanced and the total drag maximized. We find such lower side shapes on most of the flight vehicles considered in Section 3.3.

A flat or approximately flat windward side leads to a more or less two-dimensional flow field. This has the advantage of an optimal or near-optimal onset flow geometry of the body flap and also of other trim and control surfaces, Section 6.2. See in this respect also Figs. 6.3 and 6.4 in that section.

If the windward side of a RV-W-type vehicle is sufficiently flat, we have another large advantage, viz. that in a large angle of attack domain the flow there is flight Mach number and high-temperature real-gas effects independent, Section 3.6. This, as is discussed there, would already no more be the case, if a boattailing is present. In the boattailing region especially the real-gas independence is lost. This was causing the hypersonic pitching moment anomaly experienced during the first flight of the Space Shuttle Orbiter.

Last but not least, another interesting advantage of the full or approximately two-dimensionality of the flow is, that flow field studies, both inviscid and viscous, can be made with the help of plane or axisymmetric two-dimensional equivalent shapes of the original configuration, Sub-Section 10.1.1 and also Section 3.6.

Because the (blunt) wing leading edges of RV-W-type vehicles are strongly swept, the vehicle planform and hence the windward side in many cases resembles more or less the lower side of a delta wing, or double-delta wing, see, e.g., Figs. 3.2 and 3.3. On such shapes we find the desired topology of the surface velocity/skin-friction line field: downstream of the nose area, we have on the flat part a more or less two-dimensional (plane) flow between two primary attachment lines.

In reality the windward sides of RV-W-type vehicles are only approximately flat, and that not necessarily everywhere. Take for instance the Space Shuttle Orbiter, Figs. 3.40 and 3.41. Large portions of its windward side are approximately flat for $x/L \gtrapprox 0.25$. Ahead of this location the fuselage resembles more a blunt cone.

This is reflected by the computed skin-friction lines on the windward side of HALIS in Fig. 3.6. The computations for this figure and Fig. 3.7 were made in the frame of the the Manned Space Transportation Programme (MSTP) for the conditions of the ONERA tunnel F4 and flight conditions with the parameters given in Table 3.2.

Table 3.2. Flight parameters of the numerical simulation of the flow past the HALIS configuration, [9].

Case	M_∞	H $[km]$	α	ρ_∞ $[kg/m^3]$	T_∞ $[K]$	v_∞ $[m/s]$	T_w $[K]$	η_{bf}
F4 conditions	8.86	-	40°	$0.545{\cdot}10^{-3}$	795.0	4,930.0	300.0	15°
Flight conditions	24	70.0	40°	$0.55{\cdot}10^{-4}$	212.65	7,028.0	1,300.0	15°

The forward stagnation point, Fig. 3.6, is located at the lower side of the configuration at $x \approx 0.4\ m$, compare with Fig. 6.5 in Sub-Section 6.2.2. Up to $x/L \approx 0.25$ we have an attachment line[5] at the lower symmetry line (blunt cone flow). Downstream of this location the central attachment line branches into the left and right attachment line. They are located below the strake (first delta of the wing) and then, at the second delta, below and close to the leading edge. We see that between the two attachment lines the flow is indeed approximately two-dimensional, with larger deviations at the outer part of the second delta due to its dihedral, Fig. 3.41.

This is basically the same pattern which we see also in Fig. 3.7. We have included this figure, too, because it shows several other interesting features of the topology of the skin-friction line field. First of all we see differences in the skin-friction line pattern compared to that in Fig. 3.6. Although the flight cases are the same, the two-dimensionality is now partly less pronounced. The separation region ahead of the deflected body flap is considerably larger in Fig. 3.6. We don't discuss the differences in the two solutions further.

In Fig. 3.7 for the two cases the influence of the wall temperature on the extent of the separation region can be clearly seen. In the F4 case (upper part), the temperature is much lower than in the flight case[6] (lower part), consequently the boundary layer has a higher flow momentum and can negotiate better the pressure increase at the flap, see also Fig. 6.8 in Sub-Section 6.2.3.

The differences in the skin-friction line patterns in Figs. 3.6 and 3.7 are possibly due to different grid structures and flow field (grid) resolutions. This is not the case for the two patterns shown in Fig. 3.7. The differences between the F4 case (upper part) and the flight case (lower part) are small, except for the vicinity of the body flap and at locations where the flow is

[5] Note that in this figure, in contrast to Fig. 3.7, the skin-friction lines were computed from the aft part of the configuration towards the forward stagnation point. This gives a richer pattern, however partly with a somewhat not so nice accumulation of skin-friction lines.

[6] The assumed constant wall temperature $T_w = 1{,}300.0\ K$ in the flight case is not a good approximation of the real wall temperature. That is better approximated by the radiation-adiabatic temperature, which is at the position of the hinge line of the body flap approximately $1{,}000.0\ K$. The fact, that hypersonic vehicles have radiation cooled surfaces was not yet fully accepted by the aerothermodynamic community at the time of the computations.

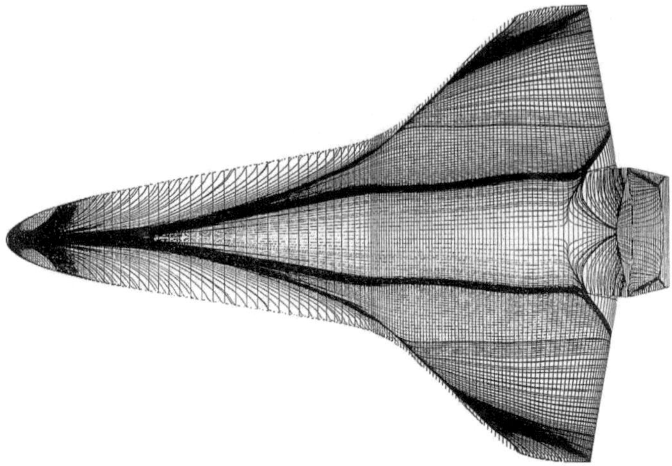

Fig. 3.6. HALIS flow field computation, flight conditions, Table 3.2: skin-friction lines at the windward side. Laminar Navier-Stokes solution with chemical non-equilibrium, [10].

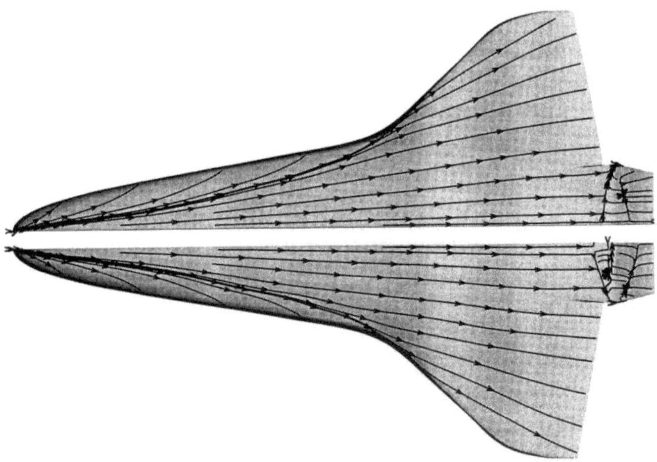

Fig. 3.7. HALIS flow field computation, upper part F4 conditions, lower part flight conditions, Table 3.2: skin-friction lines at the windward side. Laminar Navier-Stokes solutions with chemical non-equilibrium, [9].

more three-dimensional. A larger wall temperature enhances anyway a given three-dimensionality of a boundary layer, [1].

What we basically see, is an illustration of the Mach number independence principle, Section 3.6. The F4 as an high-enthalpy ground facility – the total enthalpy is still smaller than that of flight – has the typical large free-stream temperature and the also typical low free-stream Mach number. Nevertheless,

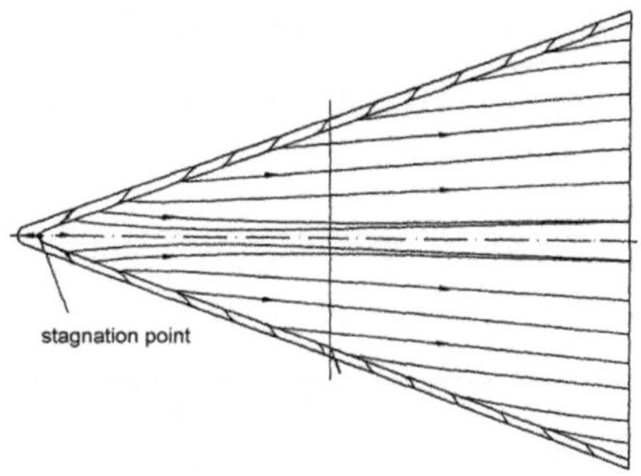

Fig. 3.8. Skin-friction lines on the windward side of the Blunt Delta Wing (BDW), [11]. Navier-Stokes solution, laminar flow, perfect gas, $M_\infty = 7.15$, $H = 30.0\ km$, $L_{BDW} = 10.0\ m$, $\alpha = 15°$.

the Mach number appears to be large enough, so that Mach number independence is more or less reached. Mainly therefore we see only small differences in the skin-friction line patterns.

Quite another skin-friction line pattern is present at the windward side of the Blunt Delta Wing (BDW), Fig. 3.8, [11]. It resembles that usually found at the windward side of true delta wings. The BDW was derived as simple generic configuration for aerothermodynamic studies in the research activities of the HERMES project, [12].

Fig. 3.8 shows the forward stagnation point and the two primary attachment lines, which all lie at the large angle of attack at the lower side of the vehicle. The lower side of the BDW has a rather large dihedral angle of 15°. Despite the strong dihedral we see between the primary attachment lines a more or less two-dimensional pattern of the skin-friction lines. For a detailed discussion of the topology of the flow field past the BDW see [1].

We note finally, that the leeside of a RV-W-type vehicle at hypersonic speed and large angle of attack does not contribute very much to the aerodynamic forces acting on the vehicle, Sub-Section 3.2.3. The vacuum pressure is reached soon. Of course, the leeside pressure depends on flight Mach number, altitude and angle of attack. At the BDW, with the flight parameters given in Fig. 3.8, we have at the location $x/L = 0.5$ a computed wall pressure coefficient $c_{p_w} = 0.24$ to 0.32 on the windward side and $c_{p_w} =$ - 0.022 to - 0.025 on the leeside[7], [11].

This is reflected by the thermal loads on the radiation cooled vehicle surface. The levels of the wall temperature $T_w = T_{ra}$ and the heat flux in the gas at the wall q_{gw} are significantly lower on the leeside than on the windward side. Because they are so low at the lee side, this side usually does not need a coating with a high surface emissivity. The lee sides of RV-W-type vehicles, see e.g. the Space Shuttle Orbiter, actually have white surfaces. In orbit the

[7] The vacuum pressure coefficient, eq. (3.6), in this case is $c_{p_{vac}} =$ - 0.028.

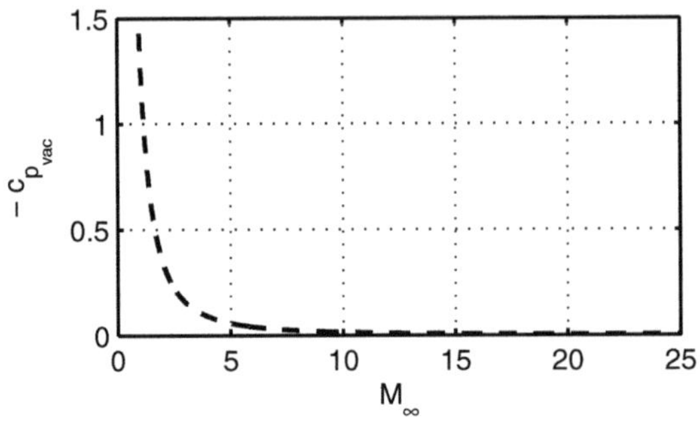

Fig. 3.9. Vacuum pressure coefficient $c_{p_{vac}}$ as function of the flight Mach number M_∞, $\gamma = 1.4$.

vehicle turns this side towards the sun, and the low emissivity prevents the heating of the vehicle due to the sun's thermal radiation. In contrast to this, CAV- and ARV-type vehicles need, because they fly at very low angles of attack (ARV-type vehicles only at ascent), a surface with high emissivity all around.

However, RV-W-type vehicles may have at the leeside – or at the fuselage side walls – hot-spot and cold-spot situations due to vortex interaction phenomena (separation and attachment lines), see, e.g., [11], and the discussion in Section 3.3 of [1].

3.2.3 Lift Generation

Another trend, typical for high supersonic and hypersonic flight, is that with increasing flight Mach number the leeward side of a wing ceases to contribute to lift. This is due to the limited expansion angle at supersonic flow turning (Prandtl-Meyer expansion), see, e.g., [1].

The pressure coefficient c_p can be written:

$$c_p = \frac{p - p_\infty}{q_\infty} = \frac{2}{\gamma M_\infty^2}\left(\frac{p}{p_\infty} - 1\right). \tag{3.5}$$

If we define the expansion limit as the static pressure p approaching zero, we obtain from this the so called vacuum pressure coefficient $c_{p_{vac}}$ as function of the Mach number:

$$c_{p_{vac}} = -\frac{2}{\gamma M_\infty^2}. \tag{3.6}$$

Fig. 3.9 shows how fast this pressure coefficient falls with increasing Mach number. Effectively zero $c_{p_{vac}}$ is reached for $M_\infty \approx 10$, thereafter independent of the Mach number.

For practical purposes, usually the value $c_{p_{min}} = 0.8\ c_{p_{vac}}$ is taken. At large Mach numbers the required expansion turning angles to reach $c_{p_{min}}$ are small.

The Prandtl-Meyer theory does not permit to determine the zero-pressure state. It can be found from the condition, that due to the finite total enthalpy of the flow, the maximum attainable velocity is finite, see, e.g., [1]. For a perfect gas we have $v_{max} = \sqrt{2\, c_p T_t}$ at $T = 0$, and hence $p = 0$. Here we use a result from thin airfoil theory, see, e.g., [13]. On the upper (lee) side of a flat plate, being at negative angle of inclination against the free-stream, $\overline{\alpha}$, the pressure coefficient is approximately:

$$c_p = \frac{2\,\overline{\alpha}}{\sqrt{M_\infty^2 - 1}}. \tag{3.7}$$

We equate this to the vacuum pressure coefficient, eq. (3.6), and find:

$$c_p = c_{p_{vac}} \quad \Rightarrow \quad \overline{\alpha}_{vac} = -\frac{\sqrt{M_\infty^2 - 1}}{\gamma M_\infty^2}. \tag{3.8}$$

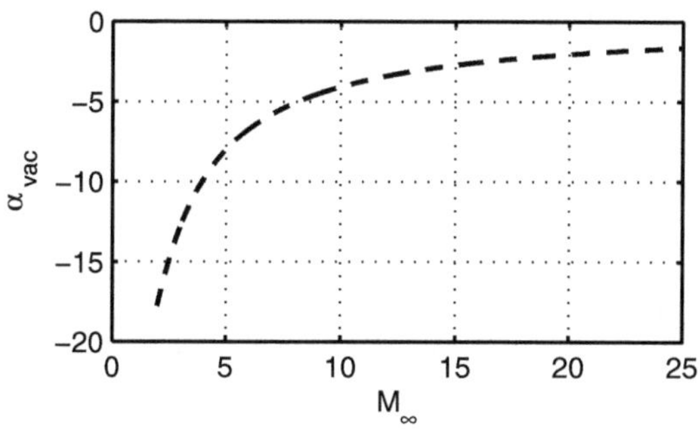

Fig. 3.10. Inclination angle α_{vac} ($\equiv \overline{\alpha}_{vac}$) needed to reach the vacuum pressure coefficient $c_{p_{vac}}$ as function of the flight Mach number M_∞, $\gamma = 1.4$.

Fig. 3.10 shows us, that indeed small negative inclination angles $\overline{\alpha}$ suffice to reach $c_{p_{vac}}$. This is an approximative result, valid for the RHPM flyer. Nevertheless, we can use it for our considerations. In reality the flow in any case first has passed the bow shock, caused by the blunt nose or a blunt leading edge, hence the local Mach number must be regarded, and the total pressure loss must be taken into account. Remember in this context the very different flow situations at RV-W-type and CAV/ARV-type flight vehicles, [1].

The "disappearance" of pressure at the leeside of a configuration at high Mach numbers is also called hypersonic shadow effect. This phenomenon permits the use of the so called impact methods, especially the Newton method, in design work. Further, because of it, the HALIS configuration, [14], see below, as a simplified Space Shuttle Orbiter configuration and also other similar configurations – at sufficiently large angle of attack – can be used for numerical investigations.

3.2.4 Base Pressure and Drag

Directly related to the expansion turning effect is the base pressure. At a base the flow turns up to 90°, hence in any case $c_{p_{vac}}$ is reached. Base pressure drops and base drag rises with blunt afterbody portions and blunt wing/flap trailing edges.

So far any RV-W-type vehicle has a blunt afterbody because there either the propulsion unit is located, or the connection apparatus to the launch vehicle, see the images given in Section 3.3. The blunt afterbody drag is no problem, because a large drag is anyway desired on most of the re-entry trajectory. On the low Mach number trajectory parts, especially in the transonic regime, however, the reduction of L/D usually is tolerated. The situation is different for CAV/ARV-type vehicles, Sub-Section 4.1.4. For them the base drag, especially in the transonic flight regime, is a major problem[8].

Blunt wing/flap trailing edges are usually employed at RV-W-type vehicles in order to alleviate the large thermal loads there. They can have an interesting side effect. During the development of the X-15 it was seen that very large (all-movable[9]) horizontal and vertical stabilizer surfaces were needed in order to assure directional and pitch stability, [15]. To reduce the vertical stabilizer surface, the design was changed from the original thin airfoil to the wedge design, which was proposed in [16]. The wedge with an angle of 10°, leading to a considerable trailing-edge bluntness, improved the effectiveness of the stabilizer surface and allowed a smaller size. The effect is due to the initial load on both sides of the wedge surface, and the fact that larger setting angles are possible before the vacuum limit is reached on the respective lee side. It seems, however, that no sizeable use of the proposal of [16] was made later with RV-W-type vehicles.

3.3 Aerodynamic Performance Data of RV-W-Type Vehicles

In this section we give extracts from aerodynamic data sets of longitudinal motion of some RV-W-type flight vehicles, which were established in recent decades. Two of these vehicles, which were never built, are the European HERMES and the Japanese HOPE-X. Three are demonstrators, viz. the American X-38[10] and X-34, and the German PHOENIX. These were manufactured and have conducted at least some drop tests. Finally it is the Orbiter of the Space Shuttle system, which is the only RV-W vehicle which became

[8] We note in this context the tail cone, which is attached to the Space Shuttle Orbiter for ferry flights on top of the Boeing 747 Shuttle Carrier Aircraft. It reduces drag and buffeting due to flow separation.

[9] All-movable stabilizer surfaces have the benefit, that no extra control surfaces are employed and hence the strong interaction phenomena around the hinge lines, Chapter 6, are avoided.

[10] The X-38 is only approximately a RV-W-type vehicle, Sub-Section 3.3.5.

Table 3.3. Geometrical data of the considered RV-W-type flight vehicles.

Vehicle	L_{ref} $[m]$	A_{ref} $[m^2]$	Wing span $[m]$	Sweep angle [°]	Strake angle [°]	x_{cog}/L_{ref} [–]
Space Shuttle Orbiter	32.774	249.91	23.790	45.0	81.0	0.650
HERMES	15.500	73.000	9.375	74.0	–	0.600
HOPE-X	15.249	50.220	9.670	55.0	75.0	0.635
X-34	17.678	33.213	8.534	45.0	80.0	0.600
X-38	8.410	21.670	3.658	–	–	0.570
HOPPER	43.650	545.120	27.200	55.0	–	0.680
PHOENIX	6.900	15.14	3.84	62.5	–	0.700

truly operational. In Table 3.3 we have listed some geometrical data of the configurations, chronologically ordered.

In the following sub-sections we discuss selected aerodynamic coefficients of these configurations which are available from own industrial design work and the literature. We consider in general the untrimmed state, i. e. the data found, for instance, in a ground-simulation facility. The pitching moment in all cases, except for the X-38, is given with all aerodynamic control surfaces in neutral position ($\eta_{body\ flap\ and\ elevons} = 0°$).

Note, that in general in the trimmed state the lift-to-drag ratio is smaller than in the untrimmed state, regardless of whether the vehicle flies stable or unstable, Sub-Section 3.4.2. In the unstable case, which is characterized by downward deflected trim surfaces, in the trimmed state the resulting aero force $\underline{F}_{aero,trim}$ is larger than the force $\underline{F}_{aero}$ in the untrimmed state, Fig. 3.34. For the stable case it is the other way around.

The pitching moment charts are given for provisionally chosen center-of-gravity positions x_{cog}/L_{ref}, Table 3.3, which are not necessarily the actual positions. Therefore they in general do not reflect the state of trim and stability of the considered vehicle configuration.

3.3.1 Space Shuttle Orbiter

The Space Shuttle system consists of the three elements:

- (two) solid rocket boosters,
- external tank,
- Orbiter vehicle.

The Space Shuttle is a semi-reusable system of which the burned-out boosters are recovered from the Atlantic ocean, refurbished and then refilled with solid propellant. The external tank is expended. The Orbiter vehicle is reused after inspection and refurbishment. This system is the only winged and manned system to reach orbit and to land horizontally. So far it has performed more than 120 flights.

The missions of the Space Shuttle Orbiter involve carrying large and heavy payloads to various orbits including elements of the International Space Station (ISS), performing service missions, also to satellites, e.g. Hubble, and serving as crew transport system for the ISS.

The Space Shuttle program was officially started by the Nixon administration in January 1972. The first launch has taken place on April 12, 1981, followed by the first re-entry flight on April 14, 1981. Detailed accounts regarding flight experience, aerothermodynamic performance and problems of this flight were published in, for instance, [17] and [18].

Five Orbiter vehicles were built and flown (first flight):

- Columbia (1981),
- Challenger (1983),
- Discovery (1984),
- Atlantis (1985),
- Endeavour (1992).

The Orbiters Challenger (1986) and Columbia (2003) were lost by accidents. It is planned to retire the Space Shuttle system from service in 2010.

We are concerned here with some aerodynamic data of the Orbiter vehicle, Fig. 3.11.

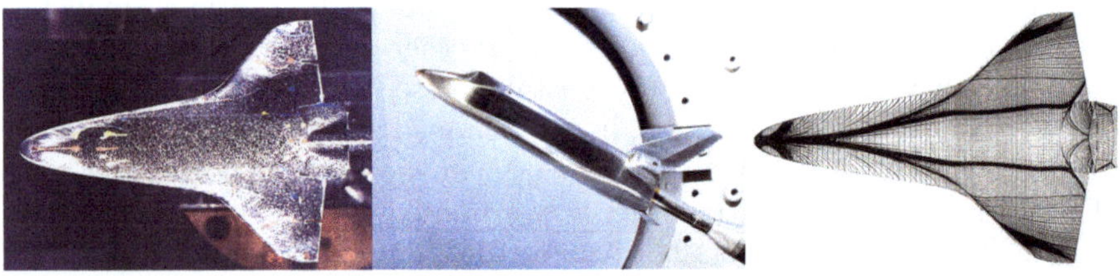

Fig. 3.11. Space Shuttle Orbiter: oil flow pattern on the leeward side of a wind tunnel model (left), model in wind tunnel (middle), [19], skin-friction lines on the windward side found with numerical flow simulation (right), [10].

The selected aerodynamic data, Fig. 3.12, are taken from [20]. The lift coefficient behaves linear for the Mach numbers up to $M = 1.5$, but shows a positive lift-curve break (change of the gradient $dC_L/d\alpha$ with α) for the high Mach numbers. The drag coefficient has its minimum for all Mach numbers at small angles of attack, as expected, and rises then fast to large values. The drag coefficient for the transonic Mach numbers is always the largest. Mach

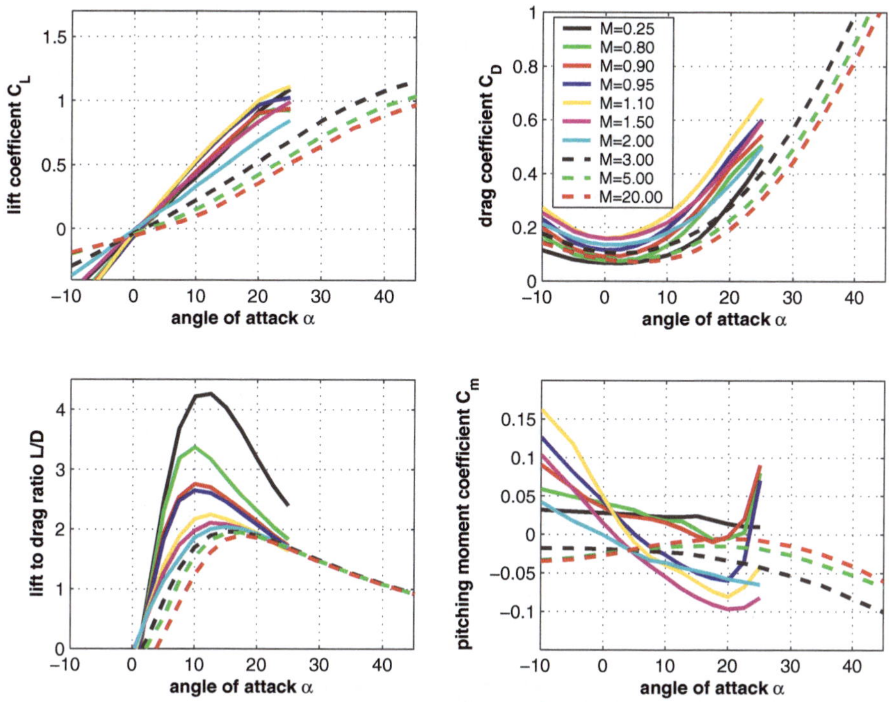

Fig. 3.12. Selected aerodynamic data of the Space Shuttle Orbiter for particular subsonic to hypersonic Mach numbers as function of the angle of attack α, moment reference point $x_{ref} = 0.65\ L_{ref}$. Data source: [20].

number independence is present for $M_\infty \gtrsim 5$ to 10, although not directly visible due to the selected Mach numbers (see also Figs. 3.32 and 2.3).

The maximum lift-to-drag ratio comes to $L/D|_{max} \approx 4.5$ in the subsonic regime and reduces to $L/D \lesssim 2$ in the hypersonic regime. This is somewhat lower than that of the HOPPER/PHOENIX and HOPE-X shapes, Figs. 3.29 and 2.3.

The behavior of the pitching moment with respect to the Mach number is very similar to the one of the HOPPER/PHOENIX shape, Fig. 3.29. The largest (negative) derivative $dC_m/d\alpha$, found at $M_\infty = 1.1$, indicates strong static stability for $\alpha \lesssim 18°$. The configuration turns to become unstable for $M_\infty \gtrsim 3$, Fig. 3.12, lower right. For $\alpha \gtrsim 20°$ we see a stable but un-trimmed behavior. The forward part of the upper side of the configuration contributes no more to the nose-up moment, because $c_{p_{vac}}$ is reached there.

Further one can observe for $0.8 \leqq M_\infty \leqq 1.5$ a sudden change of the derivative $dC_m/d\alpha$ from negative to positive at approximately $\alpha \approx 20°$, indicating static instability there. This is the same for the HOPPER/PHOENIX shape and to some extent also for the HOPE-X shape, Fig. 2.3. The reason is the onset of leading edge vortex break-down above the wing, travelling continuously forward with increasing α. There are no marks of this behavior in the HERMES aerodynamics since non-linearity of the lift up to $\alpha < 30°$ can not be detected for it, Fig. 3.14. Remember, however, that the Orbiter data are data for the chosen $x_{cog}/L_{ref} = 0.65$. The actual longitudinal center-of-gravity[11] envelope of the Orbiter at the entry interface of 121,920.0 m (400,000.0 ft) altitude up to the year 1995 was $0.65 < x_{cog}/L_{ref} < 0.675$, however, during STS-1 it was $0.667 \lessapprox x_{cog}/L_{ref} \lessapprox 0.671$, [21].

3.3.2 HERMES Configuration

In 1984 the French government launched a proposal for the development of a space transportation system in order to guarantee Europe an autonomous and manned access to space. Key parts of this system were the space plane HERMES, responsible for a gliding re-entry from space to an Earth landing site, and the launch system ARIANE V, which at that time was a completely new to be developed rocket system.

In the following the HERMES space plane project, Fig. 3.13, originally a pure French project, became a European project under the supervision of the European Space Agency (ESA). Officially ESA started the programme in November 1987.

Fig. 3.13. HERMES shape 1.0: pressure distribution resulting from a numerical flow field simulation (left), synthetic image (middle), planform view (right), [22].

The RV-W vehicle HERMES was conceived to have the following features:

- initially, the transportation of six astronauts and 4,500.0 kg payload into space, and after a later reorientation a reduction of the transport capacity to three astronauts and 3,000.0 kg payload,
- ascent to near Earth orbit (up to 800.0 km) on top of the ARIANE V rocket,
- 30 to 90 days mission duration,
- total launch mass 21,000.0 kg,
- full reusability.

[11] Note that in [21], and in other contributions to [18], the x-locations of the center-of-gravity are given in terms of the design coordinate system. It has its origin 5.9944 m (236.0 in) ahead of the nose point, which usually in aerodynamics lies at $x/L = 0$.

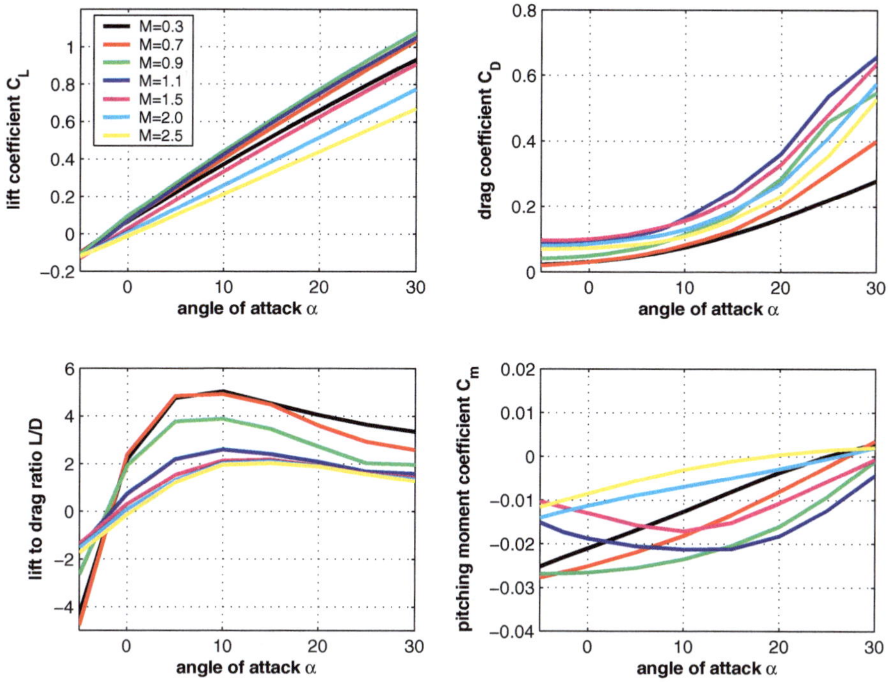

Fig. 3.14. Aerodynamic data of the HERMES shape 1.0 for the subsonic, transonic and low supersonic Mach number regimes, moment reference point $x_{ref} = 0.6\ L_{ref}$. Data source: [23].

In 1993 the HERMES project was cancelled due to the new political environment (end of the cold war) and budget constraints. No HERMES vehicle, nor the proposed sub-scale experimental vehicle MAIA, see, e.g., [4], was ever built.

The aerodynamic data presented here for the subsonic, transonic and low supersonic flight regime in Fig. 3.14, as well as for the hypersonic flight regime in Fig. 3.15, were composed with results from wind tunnel tests, approximate design methods and numerical simulations, [22], [23].

When we have a look at the lift coefficient C_L for subsonic Mach numbers, Fig. 3.14 upper left, the linear behavior all along the plotted angle of attack range is conspicuous compared to the HOPPER and HOPE-X data, where nonlinear effects for $\alpha \approx 15°$ to $20°$ can be observed. The drag coefficient around $\alpha \approx 0°$ is very small for all Mach numbers, and rises then for increasing angle of attack as expected. The drag coefficient for the transonic Mach numbers, Fig. 3.14, is the largest.

We find the maximum lift-to-drag ratio with $L/D_{max} \approx 5$ in the vicinity of $\alpha \approx 10°$ for subsonic Mach numbers, Fig. 3.14, and somewhat below 2 for the supersonic, Fig. 3.14, and hypersonic Mach numbers, Fig. 3.15, each

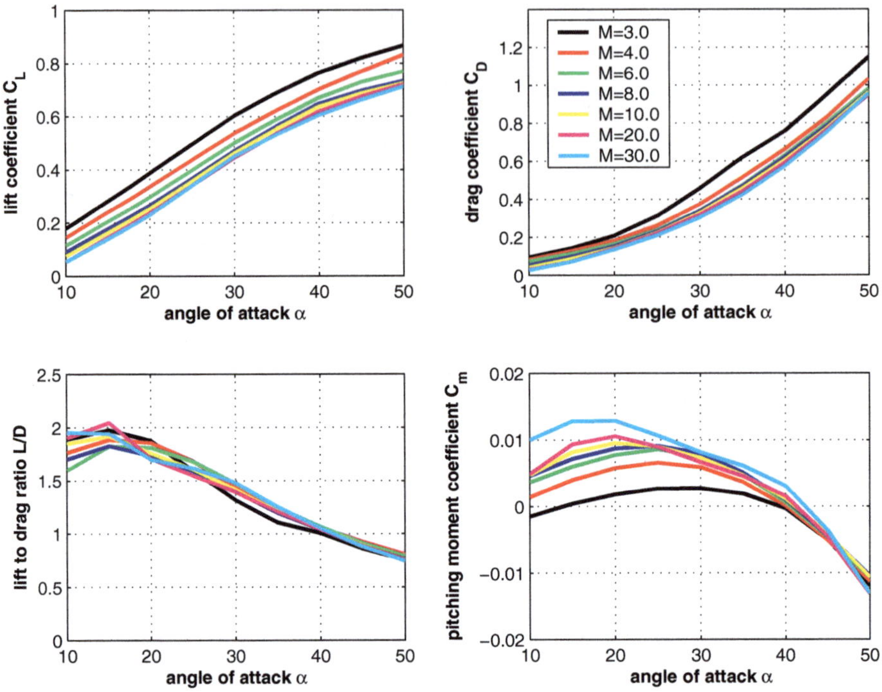

Fig. 3.15. Aerodynamic data of the HERMES shape 1.0 for the hypersonic Mach number regime, moment reference point $x_{ref} = 0.6\ L_{ref}$. Data source: [23].

lower left. These values are somewhat lower than the ones of the HOPPER and HOPE-X shapes.

The pitching moment diagram, Fig. 3.14 lower right, indicates static stability for transonic/low supersonic Mach numbers ($1.1 \leqq M_\infty \leqq 1.5$) up to $\alpha \approx 10°$, but no trim point, and instability for subsonic and low supersonic Mach numbers. This is essentially in agreement with the data of other vehicles discussed in this Section, see Figs. 3.29, 2.3 and 3.19. For higher supersonic and hypersonic Mach numbers static stability and trim for $\alpha \approx 40°$ are observed, Fig. 3.15 lower right. Mach number independence of C_L and C_D is seen for $M \gtrapprox 6$ to 8.

Considering the example of $M_\infty = 10$, we have a view, Fig. 3.16, at the effectiveness of the body flap and in one case of the elevator. With a positive (downward) body flap deflection one can shift the trim point to lower α values if the flight trajectory would require this. By applying a body flap deflection of $\eta_{bf} = 10°$, for example, the trim point is shifted from $\alpha \approx 38°$ to $\alpha \approx 27°$.

As we will also see in Fig. 3.31 for the HOPPER/PHOENIX shape, it is not unusual that a re-entry vehicle operates with negative body flap deflections in order to trim the vehicle, see in this respect Sub-Section 3.4.2 and also Sub-Section 6.1.2. This can be necessary since the free-stream conditions, for

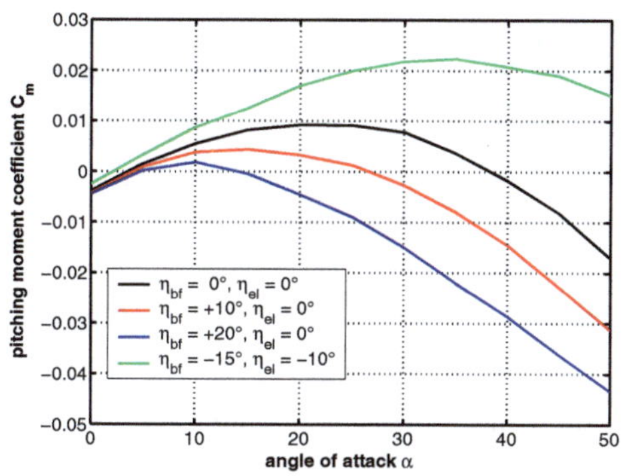

Fig. 3.16. HERMES shape 1.0, body flap and elevator efficiency for $M_\infty = 10$, moment reference point: $x_{ref} = 0.6\ L_{ref}$. Data source: [23].

example the Mach number, and the angle of attack, vary strongly in a wide range. In the HERMES project, investigations in this regard were carried out. Deflecting the body flap by η_{bf} = - 15° and the elevators by η_{el} = - 10°, the pitching moment increases strongly, which can help to trim the vehicle, Fig. 3.16, especially at high angle of attack. The effect of flap-force reduction at negative flap deflection angle, Sub-Section 6.1.2, increases with increasing angle of attack.

3.3.3 HOPE-X Configuration

In the 1980s Japan joined the community of nations striving for an autonomous access to space. Besides their rocket activities, they developed a concept of a re-entry vehicle, called HOPE, which was planned to take over payload transportation to and from the ISS. Due to budget constraints in the 1990s this program was re-oriented in order to develop a smaller, lighter and cheaper vehicle, called HOPE-X, which was to operate unmanned, Fig. 3.17. The project was cancelled in 2003.

For subsonic and transonic Mach numbers Fig. 2.3 shows aerodynamic data of longitudinal motion. The lift coefficient of this vehicle with a slender wing is rather independent of the Mach number, indicating a behavior according to slender-body theory, see, e.g., [13]. The L/D values in the subsonic regime are very similar to those of the HOPPER vehicle, Fig. 3.29, with a maximum of about 6 at an angle of attack of approximately 10°. Also the agreement of the pitching moment with its distinct static stability and trim behavior at $M_\infty = 1.1$ of the two shapes is noteworthy. Further the marginally static stability for $0.8 \leqq M_\infty \leqq 0.9$ with the strong trend to become unstable for higher angles of attack should be emphasized. Beyond the transonic regime, say for $M_\infty \gtrapprox 1.1$, the aerodynamic performance L/D_{max} drops for HOPE-X, like for all re-entry vehicles, to values not very much greater than 2, see also Figs. 3.29, 2.3, 3.19, 3.14, and 3.15.

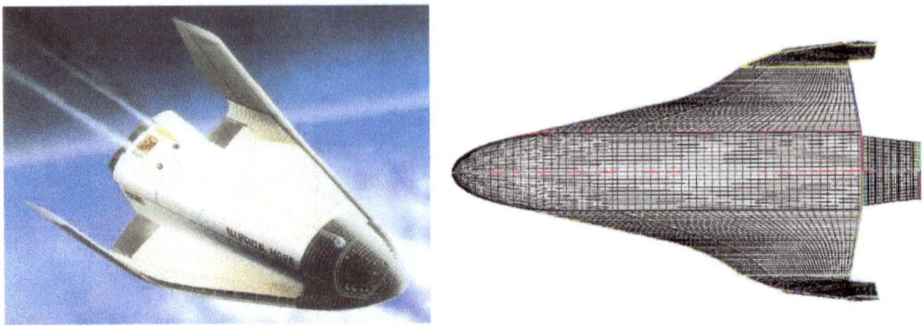

Fig. 3.17. HOPE-X shape: synthetic image, [24],[25]. Note that the shape of the synthetic image differs somewhat from that of the panel model.

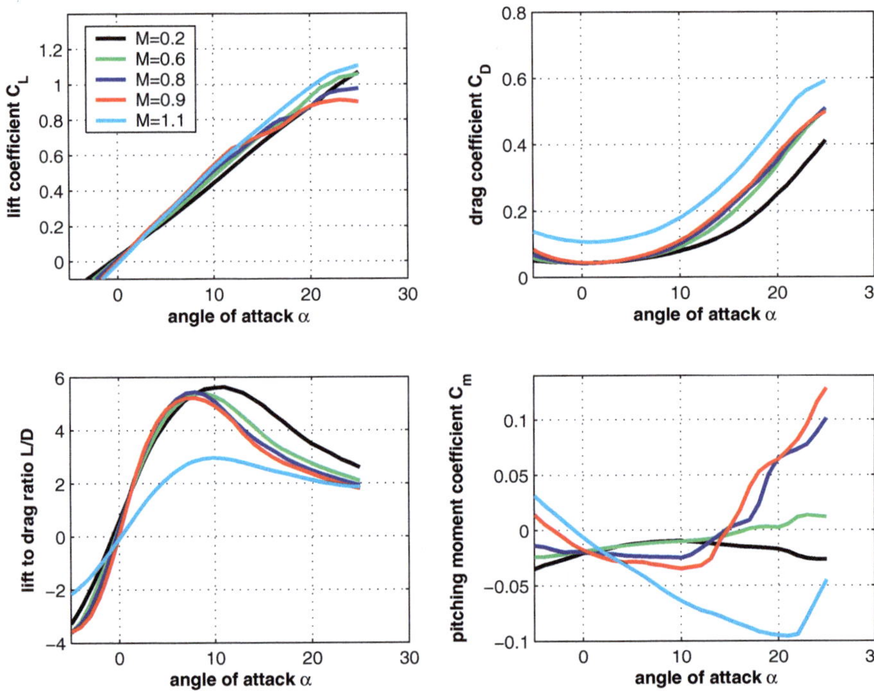

Fig. 3.18. Aerodynamic data of the HOPE-X shape for the subsonic and transonic Mach number regime. Data source: [24].

The aerodynamic data for the supersonic and hypersonic Mach number regime, shown in Fig. 3.19, is the outcome of a set of approximate design methods, [24]. Mach number independence is present above $M_\infty \approx 6$. Again in this case a comparison with the data of the HOPPER configuration, Fig. 3.30, indicates the same trends.

Further we concern ourselves with the trim feasibility and the static stability of the vehicle. With body flap deflection $\eta_{bf} = 0°$, trim and static

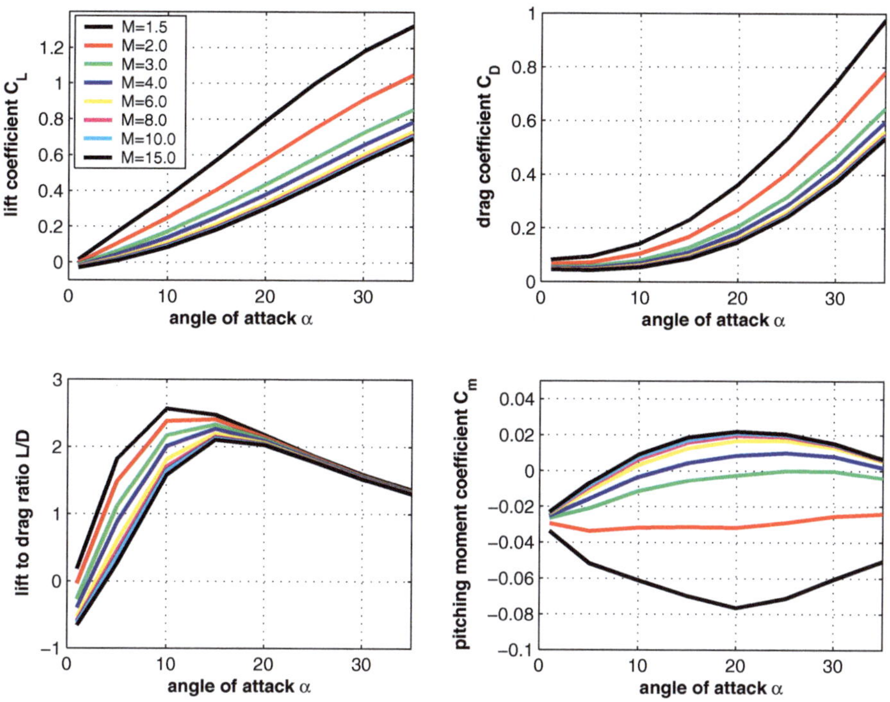

Fig. 3.19. Aerodynamic data of the HOPE-X vehicle for the supersonic and hypersonic Mach number regime. Data source: [24].

stability can be achieved for $M_\infty \geqq 4.0$ and $\alpha \geqq 30°$, Fig. 3.19. Application of a large body flap deflection of $\eta_{bf} = 30°$ leads to a corresponding nose-down pitching moment and a slight increase of the static stability, but causes a problem to trim the vehicle, Fig. 3.20 (below). The same is still true when the center-of-gravity is shifted forward by 2.5 per cent, Fig. 3.20 (above).

3.3.4 X-34 Configuration

In July 1996 NASA started a program with the goal to develop key technologies for the transportation of payloads and crews to space by using reusable launch vehicles. It was planned to prove these technologies in a realistic flight environment. For that reason two demonstrators, the X-33 and the X-34, Fig. 3.21, were developed. For the X-33 no data are available, therefore we discuss only available data of the X-34.

The mission profile of X-34 was as follows:

- captive carriage under the belly of a L-1011 aircraft up to an altitude of 11.5 km at a Mach number $M_\infty = 0.7$,
- drop-separation from the L-1011 aircraft,

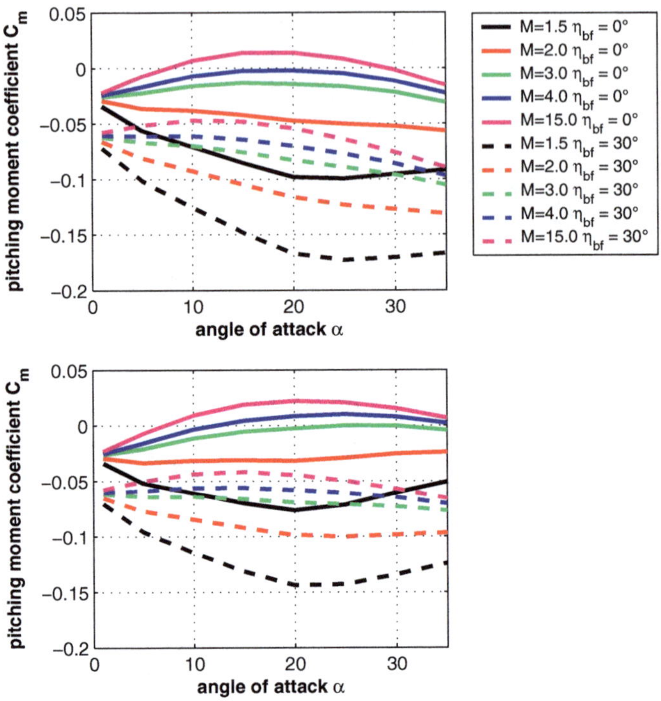

Fig. 3.20. Pitching moment behavior of the HOPE-X shape. Influence of the variation of the body flap angle η_{bf} for the nominal moment reference point, (below), and for a moment-reference point shifted 2.5 per cent forward, (above). Data source: [24].

Fig. 3.21. X-34 shape, planform view (left) [26], synthetic image (middle), X-34 on NASA Dryden ramp (right), [27].

- ignition of the rocket engine and acceleration to $M_\infty = 8$ at an altitude of 76.2 km,
- gliding back to Earth after engine burn-out, and execution of an autonomous landing on a conventional runway.

Some of the key technologies initially planned for demonstration were:

- lightweight composite airframe structures,
- advanced thermal protection systems,

- low cost avionics,
- automatic landing techniques.

In March 2001 NASA decided to terminate the technology program including test flying the X-34 vehicle.

We present in Fig. 3.22 aerodynamic data of the X-34 vehicle, i. e. essentially results from wind tunnel tests, [26], [28], [29].

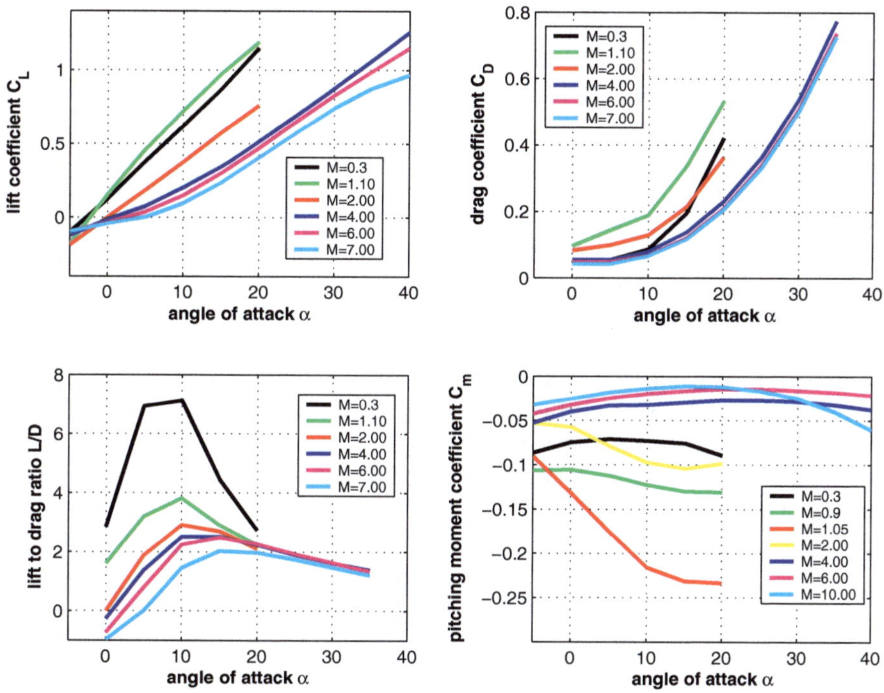

Fig. 3.22. Aerodynamic data of the X-34 shape for the subsonic to the hypersonic regime, moment reference point $x_{ref} = 0.6\ L_{ref}$. Data sources: [26], [28], [29].

The first thing, which is salient, is that X-34 does not encounter stall for the angles of attack and Mach numbers considered, Fig. 3.22 upper left. This is similar to the behavior of HERMES, Fig. 3.14. Secondly, the slender X-34 vehicle reaches the largest maximum aerodynamic performance of all the vehicles considered in this section with $L/D \approx 7.2$ at $M_\infty = 0.3$.

The pitching moment follows the general trend of RV-W-type vehicles with this kind of shape, namely that – for the given center-of-gravity location – an instability is present at subsonic Mach numbers (for small angles of attack) which turns to a strong static stability in the transonic and low supersonic regime. Further for hypersonic Mach numbers we observe instability again which is due to the increasing lift in the forward part of the vehicle, see Fig. 3.22 lower right. In all cases trim is not found due to the strong negative zero-lift pitching moments. Mach number independence of the coefficients is seen for $M_\infty \gtrapprox 6$.

3.3.5 X-38 Configuration

In the 1990s NASA had envisaged to develop a Crew Return Vehicle (CRV) for the International Space Station. In case of illness of crew members or any other emergency, the CRV should be able to bring back to Earth up to seven astronauts. The essential point is that the crew return is done unpiloted, which means that the vehicle operates automatically. For this mission a configuration on the basis of a lifting body, viz. the X-24A shape, was chosen and named X-38.

The X-38 is not a RV-W-type vehicle in the strict sense. As a lifting body it glides, like a winged re-entry vehicle, unpowered from an orbit along a given trajectory down to a specified altitude, but conducts then the final descent and landing by a steerable parafoil system. The parafoil system is a must, because the aerodynamic performance L/D of such a lifting body in the subsonic flight regime is too low for an aero-assisted (winged) terminal approach and landing.

In order to provide enough space for seven crew members the X-24A shape on the one hand was scaled up by a factor of 1.2, and on the other hand the fuselage was redesigned on the leeward side for getting more volume. The resulting shape, named X-38 Rev. 8.3, Fig. 3.23, has served as a technology demonstrator for the prototype CRV.

Fig. 3.23. X-38 shape, Rev. 8.3, planform view (left), vehicle #2 in free flight (middle), [27], model in wind tunnel (right), [30].

The vehicle had to have the following features:

- accurate and soft landing to allow for the transportation of injured persons,
- load factor (n_t, n_n) minimization,
- sufficient cross range capability for reaching the selected landing site.

In 2002 NASA changed its strategy and desired to pursue a multi-purpose vehicle, which could include both crew transport and crew return capabilities, instead of the single-purpose vehicle X-38. In June 2002 the X-38 project was cancelled.

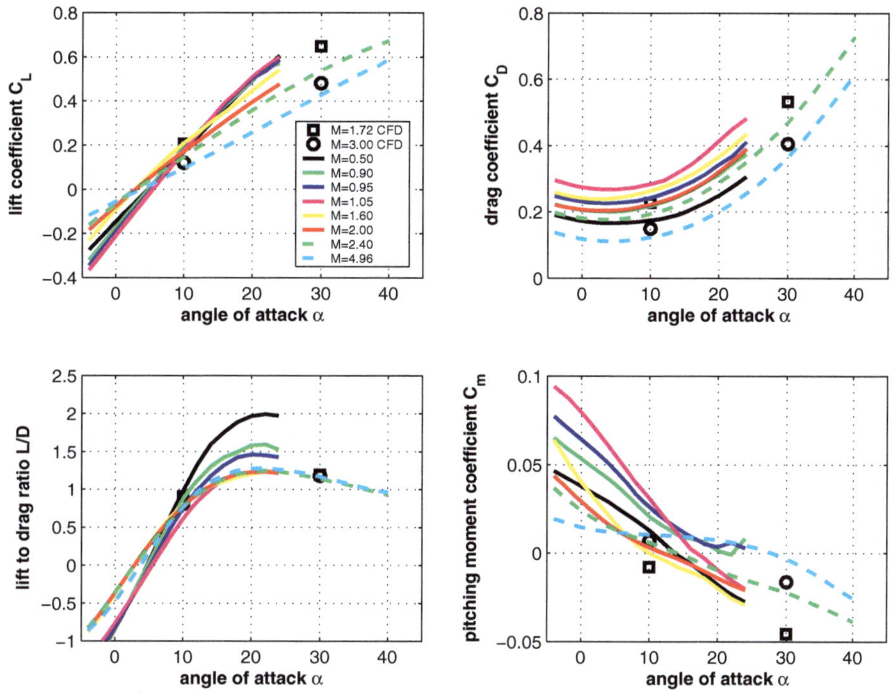

Fig. 3.24. Aerodynamic data of the X-38 shape, Rev. 8.3, for the subsonic, transonic and supersonic flight regime, moment reference point $x_{ref} = 0.57\ L_{ref}$, body flap deflection $\eta_{bf} = 20°$. Data sources: [32], [33].

The aerodynamic and aerothermodynamic characterization of the X-38 shape was made co-operatively by NASA, DLR (in the frame of the German technology programme TETRA[12]), and ESA, with Dassault Aviation as integrator of the activities, [31].

All vehicles considered so far in this section have in common that they are able to conduct an aero-assisted landing. This requires for low subsonic Mach numbers a minimum $L/D|_{max}$ value between 4.5 and 5.

As mentioned above, the X-38 as a lifting body does not strictly belong to the class of RV-W-type vehicles. When we look at its aerodynamic performance in Fig. 3.24, we find for the X-38 in the subsonic regime a maximum L/D of ≈ 2 and for higher supersonic Mach numbers of ≈ 1.3. The large pitching moment coefficients at small angles of attack are due to the strong boattailing of the lower side of the flight vehicle which can be seen at the right in Fig. 3.23.

Obviously that is due to the larger drag coefficient of the lifting body, Fig. 3.24 (upper right), compared to HOPPER/PHOENIX and HERMES. The

[12] Technologies for Future Space Transportation Systems, 1998-2001.

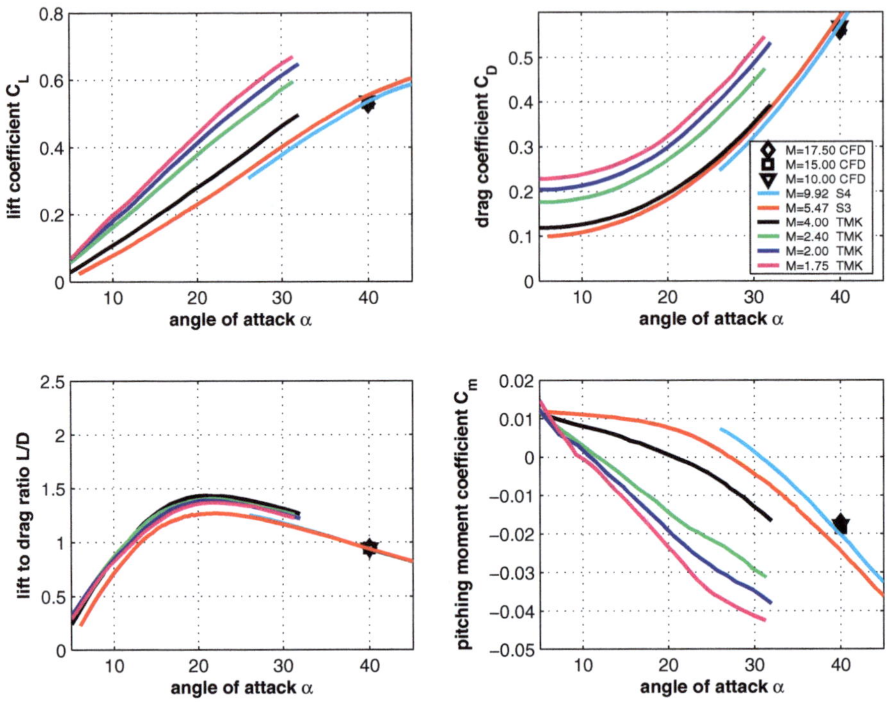

Fig. 3.25. Aerodynamic data of the X-38 shape, Rev. 8.3, for the supersonic and hypersonic flight regimes, moment reference point $x_{ref} = 0.57\ L_{ref}$, body flap deflection $\eta_{bf} = 20°$. Data sources: [32], [34] to [36].

data are taken from the Aerodynamic Data Book (ADB)[13] assembled by Dassault Aviation, [32]. The results of some Euler computations for verification purposes, [33], are also plotted in the figure. The agreement with the data from the ADB is rather good except for the pitching moment for $M_\infty = 1.72$, where some deviations occur. For low supersonic to hypersonic Mach numbers the aerodynamic data received from three different wind tunnels are plotted in Fig. 3.25.

Again we compare these data with some results extracted this time from laminar Navier-Stokes solutions. Three different Mach numbers ($M_\infty = 10, 15, 17.5$) at constant angle of attack $\alpha = 40°$ were considered. From Fig. 3.25 we can draw the following conclusions:

- Firstly, the wind tunnel data are remarkably well confirmed by the data from the numerical flow simulation.
- Secondly, as expected, the aerodynamic coefficients are asymptotically independent of the Mach number above $M_\infty \approx 6$.

[13] We present the data for a body flap deflection $\eta_{bf} = 20°$ since these describe more likely the nominal case than the data for $\eta_{bf} = 0°$.

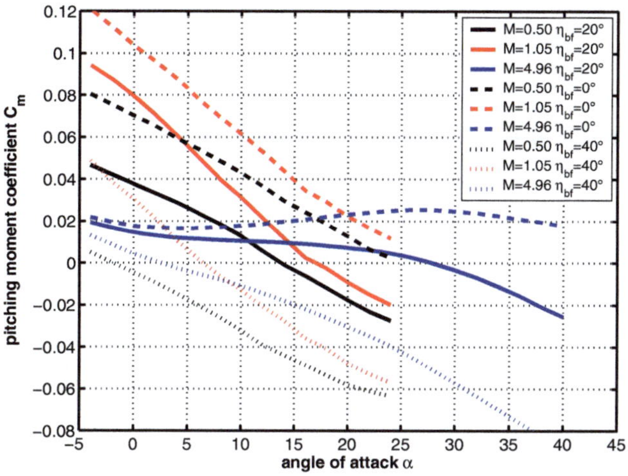

Fig. 3.26. X-38 shape, Rev. 8.3, pitching moment coefficient behavior for various body flap deflection angles η_{bf}, moment reference point $x_{ref} = 0.57\ L_{ref}$. Data source: [32].

A peculiarity is the behavior seen in the trim cross-plot, Fig. 3.26. In the whole Mach number range the vehicle behaves stable and can be trimmed, which is different from the pitching moment characteristics of the RV-W-type vehicles discussed before[14]. Further it is interesting to have a glance at the effectiveness of the body flap. Since the body flap deflection $\eta_{bf} = 20°$ is the nominal case, we concentrate on the two other cases $\eta_{bf} = 0°$ and $\eta_{bf} = 40°$.

In the first case, $\eta_{bf} = 0°$, the pitching moment coefficient increases strongly compared to the nominal case, so that for the two lower Mach numbers ($M_\infty = 0.5$ and 1.05) trim can be achieved only for $\alpha \gtrapprox 25°$ which is not a realistic value. For the higher supersonic Mach number ($M_\infty = 4.96$) no trim and a reduced static stability can be observed. On the other hand a body flap deflection $\eta_{bf} = 40°$ (second case) reduces strongly the pitching moment, whereas the static stability grows and the trim points are shifted to much lower angle of attack values.

3.3.6 HOPPER/PHOENIX Configuration

In the frame of the "Future European Space Transportation Investigation Programme (FESTIP)" several concepts of Reusable Launch Vehicles (RLV) were designed and developed with the goal to make the access to space more reliable

[14] Note that this is a consideration in a sense different from those with the previously discussed RV-W-type vehicles. There arbitrary pitching moment data were discussed, i. e. for a given η_{bf}, here we study the actual trim capabilities of the vehicle.

Fig. 3.27. HOPPER/PHOENIX shape. Synthetic images: planform view (left), 3-D view (right), [38], [39].

and, in particular, cost effective. One of these concepts is the HOPPER vehicle, Fig. 3.27, which was foreseen to fly along a suborbital trajectory, [37].

This means that this vehicle does not reach the velocity necessary for moving into an Earth target orbit. That also is the reason why the vehicle is not able to return to its launch base, but has to fly to another landing ground. The main features of the system are the horizontal take-off (sled launch) and landing capability and reusability. To demonstrate the low speed and landing properties of the vehicle shape, a 1 : 6 down-scaled flight demonstrator, called PHOENIX, was manufactured. PHOENIX was successfully flown[15] in 2004 in Vidsel, Sweden, Fig. 3.28.

Fig. 3.28. PHOENIX demonstrator: Test of subsonic free flight (helicopter drop test) and landing capability in Vidsel, Sweden in May 2004.

The aerodynamic data set was established on the basis of tests in several wind tunnels and on numerical flow field simulations employing the Navier-Stokes equations. In Fig. 3.29 we have plotted the lift, drag and pitching moment coefficients as well as the lift-to-drag ratio for subsonic, transonic and

[15] The Italian (CIRA) PRORA-USV Experimental Vehicle, [40], [41], made its first balloon-launched transonic/subsonic flight in February 2007.

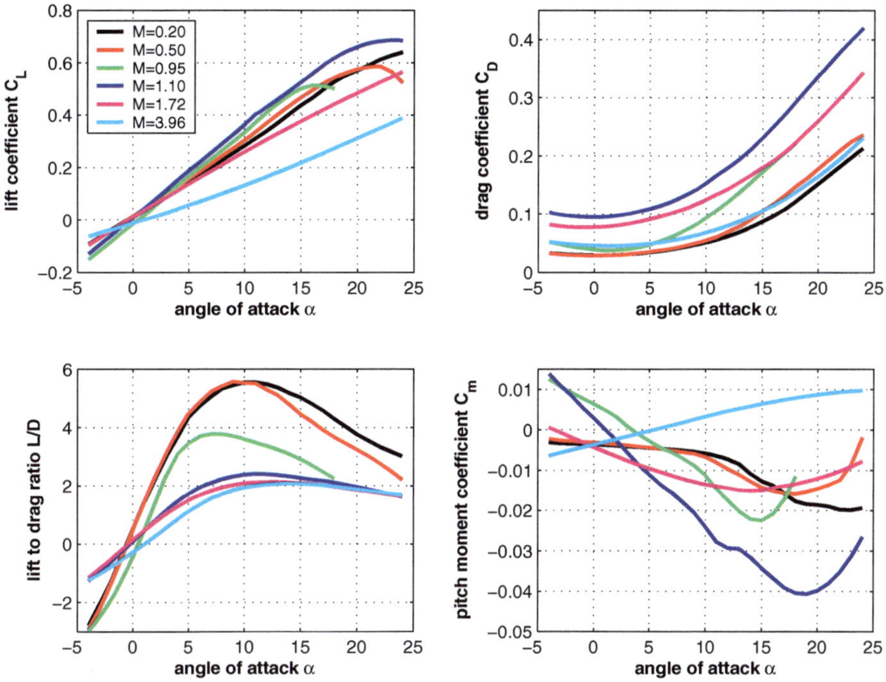

Fig. 3.29. Aerodynamic data of the HOPPER/PHOENIX vehicle for subsonic, transonic and supersonic Mach numbers based on wind tunnel measurements. Moment reference point $x_{ref} = 0.68\ L_{ref}$. Data source: [38].

supersonic flight Mach numbers. These data stem exclusively from wind tunnel tests.

As expected the highest L/D values are found in the subsonic regime with a maximum of $L/D|_{max} \approx 6$, which drops to $L/D|_{max} \approx 2$ in the supersonic regime. The pitching moment diagram reveals – for the selected center-of-gravity position – for all Mach numbers, except $M_\infty = 3.96$, static stability. But trim for small positive angles of attack is only achieved for $M_\infty = 0.95$ and 1.1. The moment reference point with $x_{ref} = 0.68\,L_{ref}$ was found to be a realistic center-of-gravity position as a consideration of the internal layout of the vehicle had shown.

The aerodynamic data for supersonic and in particular hypersonic Mach numbers, which were derived from Navier-Stokes solutions, are displayed in Fig. 3.30. For Mach numbers $M \gtrapprox 10$ lift and drag coefficients are nearly Mach number independent, but not the aerodynamic performance L/D. Further, as expected, the vehicle behaves statically unstable in the high Mach number regime for the neutral aerodynamic controls setting.

We describe now by which measures a vehicle can be trimmed and longitudinal static stability can be achieved, if, for example, the pitching moment

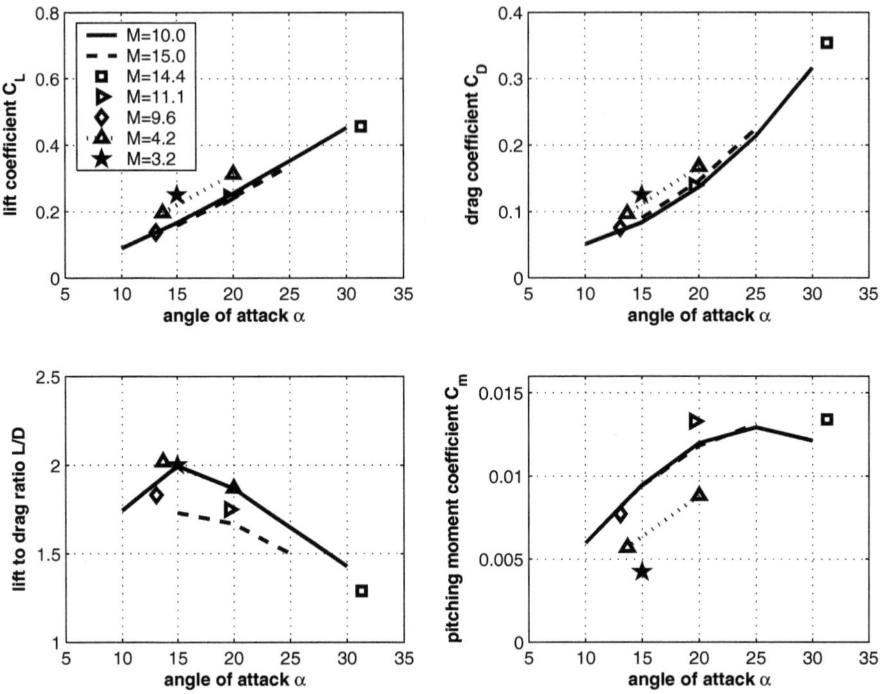

Fig. 3.30. Aerodynamic data of the HOPPER/PHOENIX vehicle for the supersonic and hypersonic Mach number regime based on CFD data. Moment reference point $x_{ref} = 0.68\ L_{ref}$. Data sources: [38], [39], [42].

coefficient behaves like the one of the HOPPER shape for $M_\infty = 3.96$, Fig. 3.29. Generally a positive (downwards) deflected body flap ($\eta_{bf} > 0$) causes an additional nose-down moment, whereas a negative one ($\eta_{bf} < 0$) produces an additional nose-up moment, as can be seen in Fig. 3.31, right.

Usually for re-entry vehicles flying along classical entry trajectories the angle of attack amounts to $20° \lessapprox \alpha \lessapprox 30°$ for flight Mach numbers $M_\infty \approx 4$, Fig. 3.1. It seems that only for $\alpha \gtrapprox 30°$ a body flap deflection $\eta_{bf} = +\ 10°$ leads to a trimmed and stable flight situation. The other possibility to attain static stability consists in a forward shift of the center-of-gravity, Fig. 3.31 (left), but in the case we consider here, the additional nose-down moment prevents trim for $\eta_{bf} = +\ 10°$. However, the vehicle behaves statically stable for $\alpha \gtrapprox 20°$ and can be trimmed with a body flap deflection $\eta_{bf} = -\ 10°$ at $\alpha \approx 28°$.

3.3.7 Summary

We summarize now what the aerodynamic data of the considered RV-W-type vehicles have in common and which differences exist.

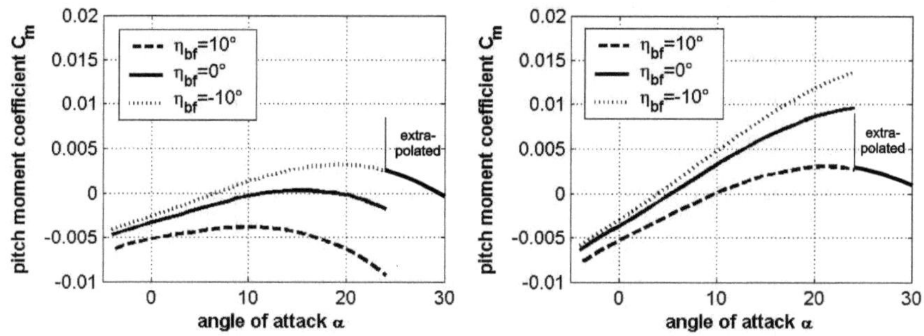

Fig. 3.31. Pitching moment behavior of the HOPPER/PHOENIX shape. Variation of the body flap angle η_{bf} for the nominal moment reference point $x_{ref} = 0.680\ L_{ref}$ (right), and for the moment reference point $x_{ref} = 0.655\ L_{ref}$ (left), $M_\infty = 3.96$. Data source: [38].

- The lift coefficient around the transonic Mach numbers ($0.8 \lessapprox M_\infty \lessapprox 1.2$) is – to different degrees – nonlinear for the Space Shuttle Orbiter, HOPE-X, X-38, and HOPPER/PHOENIX, Figs. 3.12, 2.3, 3.24, 3.29. Surprisingly, this is not the case for HERMES, Fig. 3.14 and X-34, Fig. 3.22. This means that leeward side vortices either are not produced or do not play any role (vortex break-down phenomena) for the pressure field on the upper body/ wing surface.
- The drag and the lift coefficients are not monotonic with respect to the Mach number, but have a clearly marked maximum at $M_\infty \approx 1$. Fig. 3.32 shows as example the behavior of the drag coefficients of all configurations at $\alpha = 20°$ as function of M_∞. They rise in the compressible subsonic domain, reach the transonic maxima, and then decrease with increasing M_∞, attaining finally Mach number independence at Mach numbers which are different for the different shapes.
- The aerodynamic performance L/D is highest for low subsonic Mach numbers and reaches values well beyond $L/D = 4$ for the Space Shuttle Orbiter, HERMES, HOPE-X, X-34, and HOPPER/PHOENIX, Figs. 3.12, 3.14, 2.3, 3.22, 3.29. In the hypersonic regime the maximum values diminish to $L/D \approx 2$. These observations are not true for the X-38, since this is a lifting body which reaches a maximum $L/D \approx 2$ at medium subsonic Mach numbers, which drops to a value of $L/D \approx 1.3$ for hypersonic Mach numbers.
- The magnitude of the derivative of the pitching moment with respect to the angle of attack, $dC_m/d\alpha$, has low positive to moderate negative values for the subsonic Mach numbers, grows on to maximum negative values in the transonic regime and drops back to low negative or even positive values in the supersonic and hypersonic regime for $\alpha \lessapprox 20°$. This is more or less true for all shapes.

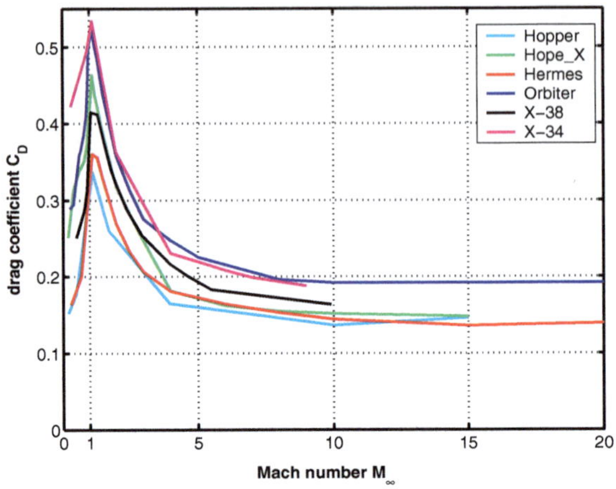

Fig. 3.32. Drag coefficient as function of the Mach number M_∞ at $\alpha = 20°$ for the vehicle shapes discussed in this section.

- Of course, the pitching moment depends on the position of the center-of-gravity and the deflection of elevators and body flaps[16]. But we can observe a clear tendency towards static stability in the transonic flight regime, Figs. 3.12, 3.14, 2.3, 3.22, 3.24, 3.29. For hypersonic Mach numbers, Space Shuttle Orbiter, HERMES, HOPE-X, X-34, and HOPPER/PHOENIX are statically unstable for lower angles of attack ($\alpha \lessapprox 20°$), but turn to static stability for higher angles of attack, Figs. 3.12, 3.15, 3.19, 3.22, 3.30. That is in contrast to the X-38 vehicle, which is statically stable throughout, Figs. 3.24, 3.25.
- It is remarkable that the drag coefficients of HERMES, HOPE-X, and HOPPER/PHOENIX attain values at high supersonic and hypersonic Mach numbers which are very close together. The drag coefficients along the Mach number span of Space Shuttle Orbiter and X-34 are very similar which is not so surprising as a look at the planform of both vehicles indicates, Figs. 3.11, 3.21.

3.4 Vehicle Flyability and Controllability

3.4.1 General Considerations

Under flyability we understand the fact that an air vehicle must be able to fly in a determinate way. Flyability means on the one hand that the vehicle can be trimmed. On the other hand it concerns especially the stability about the lateral axis, which is called longitudinal stability, the directional stability about the vertical axis, and sufficient damping of roll motion around the longitudinal

[16] Also of speed brakes, if available in appropriate configuration.

axis. The longitudinal and the directional stability can be considered as weathercock stabilities, made possible in general by the provision of horizontal and vertical, respectively, tail surfaces (stabilizers).

Regarding the lateral stability of RV-W-type flight vehicles we mention that a vertical stabilizer, like that of the Space Shuttle Orbiter, during the high angle of attack re-entry flight phase, Fig. 3.1, is in the "shadow" of the fuselage. Hence its effectiveness is severely curtailed. A possibility to avoid this effect is to provide winglets, Fig. 3.3, as was foreseen for HERMES, [43]. With the Orbiter the problem was alleviated by the use of the reaction control system, which is in any case needed for manoeuvering in orbit. For further discussion of this and the above topics see Section 2.1 and Chapter 6.

In aircraft motion of course the movements around the axes are coupled to a certain degree, depending on the inertial properties of the airframe. One also has to distinguish between static and dynamic stability. In the frame of our considerations we take no detailed look at these topics, but refer the reader to, e.g., [44].

Under controllability we understand that intended movements around all axes must be possible. This implies control surfaces of sufficient effectiveness and sufficient "volume", the latter being the product of the area of the surface times the distance (lever arm) to the respective axis of the flight vehicle[17], all in view of the requirements which are imposed by the respective moments of inertia of the vehicle. Detailed topics related to controllability, such as control surface aerothermodynamics, are treated in Chapter 6.

Regarding the matter of flyability we consider in the next sub-section trim and static stability of RV-W-type vehicles at large angles of attack.

3.4.2 Trim and Stability of RV-W-Type Flight Vehicles

A RV-W-type flight vehicle is longitudinally trimmed by a deflection of one or more trim surfaces. These can be elevons and/or a body flap, Sub-Section 6.6.1. Downward deflection (η positive) results in an upward flap force and a nose-down (negative) increment of the pitching moment[18]. Upward deflection (η negative) accordingly gives a downward flap force and a nose-up (positive) increment.

We study now aspects of these two trim modes in detail, with the vehicles flying at high angles of attack, and geometrically approximated as RHPM flyers, Figs. 3.33 (case 1, $\eta > 0$) and 3.35 (case 2, $\eta < 0$). In this approximation the z-offsets (distances to the vehicle axis x, Chapter 7) of both the center-of-gravity and the center-of-pressure, Section 7.1, are zero in the untrimmed case and small and negligible in the trimmed case. We note that the force $\underline{F}_{aero,trim}$ in the figures is equal to the sum of all forces acting on the vehicle on the flight path, Fig. 2.5, found with a point-mass consideration. Here we consider the moment balance, and nothing else.

[17] The axes have their origin in the center-of-gravity of the vehicle.

[18] For the definition of aerodynamic forces and moments see Fig. 7.3, Section 7.1.

Although actually elevons and/or a body flap may be employed, we use simply the term 'body flap' when speaking of the aerodynamic trim surface. The figures are schematics, the flight path angles γ and the deflection angles of the body flap η_{bf} are exaggerated. Exaggerated too are the sizes of the body flap forces $\underline{F}_{bf}$.

Similarity was not aimed for of the magnitudes of the trimmed forces $\underline{F}_{aero,trim}$ and the lift L and drag D components in both cases. Important is to remember that we consider in the untrimmed situation the body flap as fixed in neutral position, $\eta_{bf} = 0°$, Section 3.3. This means that it contributes with $\underline{F}_{bf,\eta=0}$ to the force $\underline{F}_{aero}$. Hence it must be vectorially subtracted from the flap force $\underline{F}_{bf}$, when the body flap is deflected, but also from $\underline{F}_{aero}$. We call the resulting flap force $\underline{F}_{bf,net}$.

In Figs. 3.33 and 3.35 we illustrate this schematically. We don't include the correct corresponding force polygons. The force $\underline{F}_{aero}$ is considered as reduced by $\underline{F}_{bf,\eta=0}$. The forces $\underline{F}_{bf}$ are assumed to act in the middle of the trim surfaces in direction normal to them.

In Fig. 3.33 we show now schematically how longitudinal trim is achieved with the downward deflection of the body flap, $\eta_{bf} > 0$. The action line a_1 of the untrimmed lift force $\underline{F}_{aero}$ lies somewhat forward of the center-of-gravity where the trimmed force $\underline{F}_{aero,trim}$ acts. The action line a_2 of the net force of the body flap, $\underline{F}_{bf,net} = \underline{F}_{bf} + \underline{F}_{bf,\eta=0}$, crosses a_1 below the center-of-gravity. Important is the observation, that trim is achieved – for this specific configuration – only if the action line a_1 of $\underline{F}_{aero}$, which goes through the center-of-pressure, lies forward of the center-of-gravity which then results in the moment balance $l_{aero}\ |\underline{F}_{aero}| = l_{bf}\ |\underline{F}_{bf,net}|$.

We observe also, Fig. 3.33, how the downward deflected trim flap creates the aerodynamic trim force $\underline{F}_{bf,net}$. This force added to $\underline{F}_{aero}$ gives the trimmed aero force: $\underline{F}_{aero,trim} = \underline{F}_{aero} + \underline{F}_{bf,net}$, Fig. 3.34 a). The action line a_3 of $F_{aero,trim}$ crosses the center-of-gravity, which is now also the center-of-pressure. Now the resultant moment around the center-of-gravity is zero, the vehicle is trimmed.

Because the original untrimmed force $\underline{F}_{aero}$ acts ahead of the center-of-gravity, the vehicle flies trimmed, but statically unstable[19]. The static margin, Chapter 7, is negative[20] and approximately equal to l_{aero} in Fig. 3.33.

However, a situation can also be found at RV-W-type vehicles, where the body flap/elevons are deflected upwards, Sub-Section 6.1.2. We show this situation without a detailed discussion in Figs. 3.35 and 3.34 b). Because now the original untrimmed force $\underline{F}_{aero}$ acts downstream of the center-of-gravity, with the upward deflected body flap the vehicle flies trimmed and statically stable.

---

[19] This is in contrast to classical aircraft which fly stable. There the longitudinal trim is achieved by horizontal surfaces (elevator unit), which have a negative angle of attack and exert a downward force.

[20] In the literature one finds sometimes that this case, i. e. the untrimmed force $\underline{F}_{aero}$ acting ahead of the center-of-gravity, is considered to have a positive static margin.

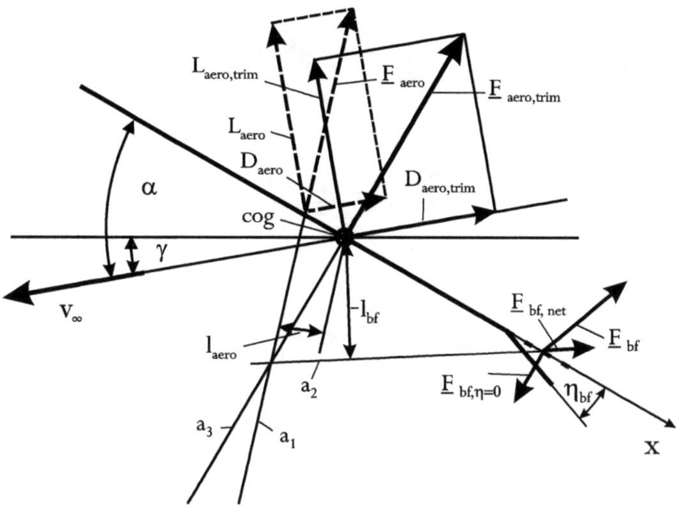

Fig. 3.33. Schematic of longitudinal trim of a RV-W-type flight vehicle at large angle of attack (geometrical RHPM approximation) with downward deflected body flap: involved force vectors $\underline{F}$ and lever arms l as well as lift L and drag D in untrimmed ('aero', broken lines) and trimmed ('aero,trim', full lines) situation. The action lines of the forces $\underline{F}$ are denoted by a_1 to a_3.

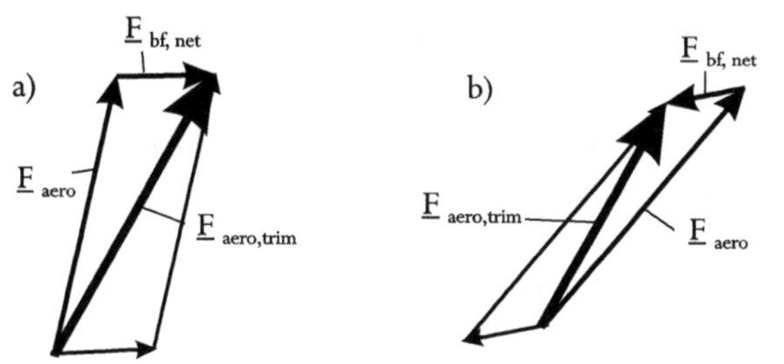

Fig. 3.34. Force polygons: cases a) with downward, b) with upward deflected body flap.

The static margin, Chapter 7, is positive and approximately equal to l_{aero} in Fig. 3.35.

Although we have considered the RV-W-type vehicle at large angle of attack only in the geometrical RHPM approximation, the results can be generalized:

- Downward deflection of body flap/elevons means trimmed but statically unstable, upward deflection trimmed but statically stable flight of the vehicle.
- In the statically unstable case $L/D|_{aero,trim} < L/D|_{aero}$, Fig. 3.33 (compare also Figs. 3.12 and 2.3). This holds also for the statically stable case, Fig. 3.35.

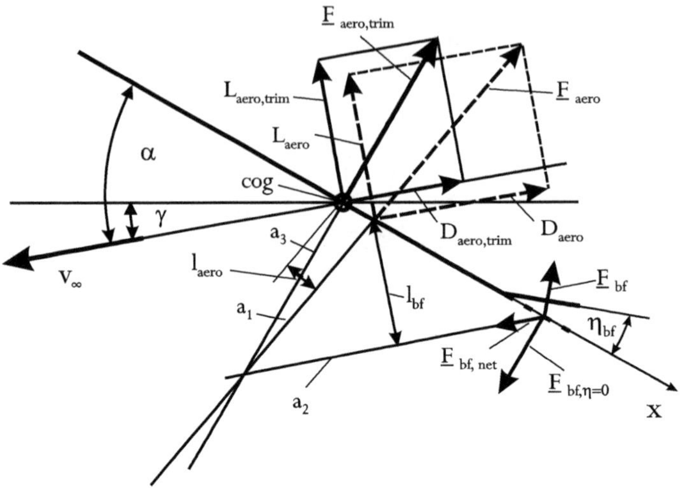

Fig. 3.35. Schematic of longitudinal trim of a RV-W-type flight vehicle (geometrical RHPM approximation) at large angle of attack with upward deflected body flap. For details see Fig. 3.33.

- If the vehicle is statically unstable, lift L and drag D are larger in the trimmed than in the untrimmed case, Figs. 3.33 and 3.34 a), but smaller, if the vehicle is statically stable, Figs. 3.35 and 3.34 b).

3.5 The Hypersonic Pitching Moment Anomaly of the Space Shuttle Orbiter

During the first flight (STS-1) of the Space Shuttle Orbiter the so-called hypersonic pitching moment anomaly showed up. The notion is due to a pitch-up moment being larger than predicted in the high flight Mach number domain. This situation required a (downward) body flap[21] deflection, which was more than twice as large as predicted, Fig. 3.36, [45]. Together with the pitching moment discrepancy also a certain discrepancy between the predicted and the actual normal force was observed.

Angles of attack corresponding to the flight Mach numbers in Fig. 3.36 are, with small oscillations, $\alpha = 40°$ from $M_\infty \approx 24$ down to $M_\infty \approx 10$, and $\alpha = 25°$ at $M_\infty \approx 5$, see, e.g., [46], and also Fig. 3.1 of this book.

The largest deviation (factor approximately two) from the predicted body flap deflection angle is found in the Mach number range $M_\infty = 24$ to $M_\infty \approx 17$, Fig. 3.36. Below $M_\infty \approx 8$ predicted and flight data are, with some noteworthy exceptions, close to each other.

[21] Originally the body flap was to act as a heat shield for the nozzles of the main engines of the Orbiter, now it is the primary longitudinal trim device, [48].

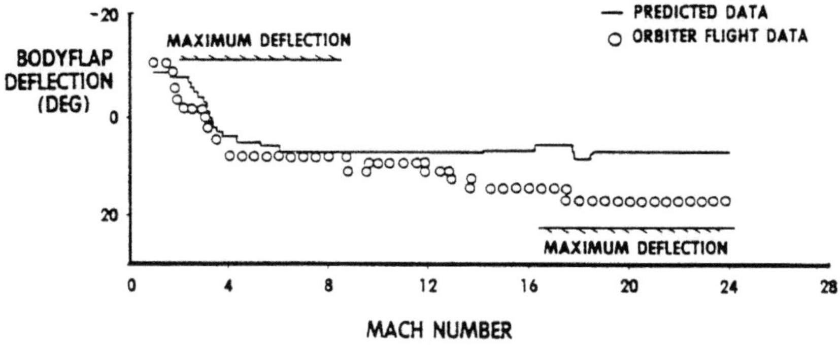

Fig. 3.36. Predicted data and flight data of body flap deflection for trim as function of the flight Mach number of the first flight (STS-1) of the Space Shuttle Orbiter, [45].

The free-stream data can be deduced, at least approximately, from the altitude-velocity map, Fig. 2.2 and the atmospheric data given in Chapter 12, or from, for instance, [47].

What did cause this "anomaly", which subsequently was extensively studied, see, e.g., [17], [18]? Before we discuss the problem in detail, we look at the trim situation and list possible causes of the anomaly.

3.5.1 Trim Situation and Possible Causes of the Pitching Moment Anomaly

Investigations after the first flight mission of the Space Shuttle Orbiter revealed that the actual center-of-pressure location was more forward (at $\alpha = 40°$ approximately 0.3 m in the numerical simulation of [14], which amounts to a little bit less than 1 per cent body length) than derived from ground simulation before flight. This means that the action line a_1 in Fig. 3.33 of the untrimmed lift force $\underline{F}_{aero}$ would lie even further forward than shown in that figure. It is evident, that this induces an additional nose-up moment. It is also evident, that a larger downward deflection angle of the trim flap was necessary to trim the vehicle, Fig. 3.36.

We now consider the aerodynamic surface forces acting on the vehicle and causing forces and moments, regarding only the longitudinal motion. These surface forces are basically the wall pressure p_w and the skin-friction τ_w, Fig. 3.37. It is important to note that the wall pressure can be increased by hypersonic viscous interaction, see, e.g., [1]. Hypersonic viscous interaction can occur during hypersonic flight at high altitudes, where large local Mach numbers in combination with low unit Reynolds numbers (both defined at the boundary layer edge) lead to thick boundary layers which then mainly induce an additional wall pressure.

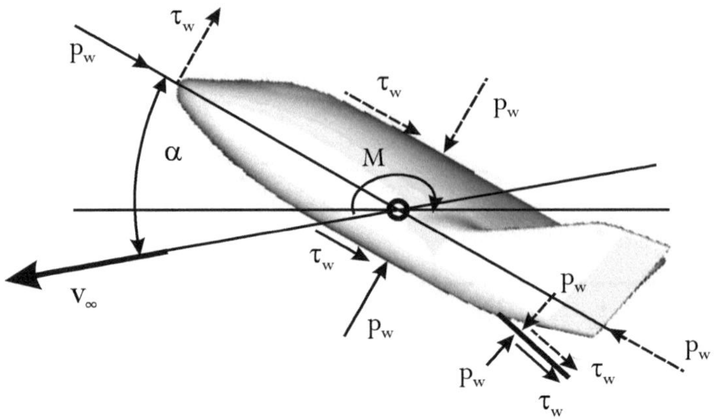

Fig. 3.37. Longitudinal motion: schematic of the wall pressure p_w and the skin-friction τ_w acting on the surface of a RV-W-type flight vehicle (broken lines: small forces which can be neglected in the considerations). Not indicated are the integral forces lift, drag et cetera.

Neglecting p_w and τ_w in the "shadow" parts of the configuration (indicated by broken lines), the integral forces on the vehicle and the contributions of p_w and τ_w to them are basically, though very approximately:

- the force in direction normal to the flight path, the lift $L = L(p_w, \tau_w)$,
- the force in direction of the flight path, the drag $D = D(p_w, \tau_w)$,
- the force in direction of the vehicle axis, the axial force $A = A(p_w, \tau_w)$,
- the force in direction normal to the vehicle axis, the normal force $N = N(p_w, \tau_w)$,
- the trim force on elevons/body flap, the force $F_{trim} = F_{trim}(p_w, \tau_w)$,
- the longitudinal moment, the pitching moment $M = M(p_w, \tau_w)$, which of course is zero, if the vehicle is trimmed.

Important is that the first flight of the Orbiter did show, that in the hypersonic flight domain the integral aerodynamic performance of the vehicle was sufficiently well predicted, except for the normal force N and especially the longitudinal trim characteristics, [45].

As major possible effects, that were not adequately taken into account in the aerodynamic data base, were seen and studied[22]:

- High-temperature real-gas effects and Mach number effects[23], [49], [14], [50] to [52], [7], [53].
- Viscous forces and effects, including hypersonic viscous interaction, regarding both airframe and trim surface (body flap) aerodynamics, [14], [54], [55].

---

22 Only a few papers are cited.

23 This is connected to the question to what extent Oswatitsch's Mach number independence principle for blunt bodies is valid in ground-facility simulations.

Trim surface and elevon effectiveness can also be adversely affected by erroneous bow shock shapes in ground-facility simulations[24], [14].

Other factors potentially having played a role are, see, e.g., [14]:

- Insufficient knowledge of actual atmospheric properties, especially air density, during re-entry.
- Uncertainties in the actual location of the center-of-gravity of the flight vehicle.
- Possible small aerodynamic shape difference between the actual flight vehicle and the ground-facility simulation models.

3.5.2 Pressure Coefficient Distribution at the Windward Side of the Orbiter

Remember, that except for the pitching moment and the normal force, all aerodynamic forces and moments were predicted sufficiently well. We assume therefore that it is sufficient to study only the wall pressure distribution at the lower side (windward side) of the Orbiter in view of the pitching moment anomaly, taking into account the configurational particularities of the lower side of the vehicle. The wall pressure, Fig. 3.37, is the most important force acting on the windward side of the vehicle. Unfortunately no detailed information about the other forces is available.

Early viscous computations on simple shapes had shown that high-temperature real-gas effects reduce aerodynamic forces, and the (nose-up) pitching moment, too, [49]. Viscous forces, i. e. the wall shear stress, acting mainly on the windward side of the flight vehicle, will induce a pitch-down increment, as can be deduced from Fig. 3.37.

Regarding the high-temperature real-gas effects, numerical studies in the aftermath of the discovery of the hypersonic pitching moment anomaly did show a different picture, which possibly is due to the particular shape of the lower side of the vehicle.

We investigate in the following the pressure-coefficient distributions in the lower symmetry line of the vehicle for two angles of attack. The computations for the two cases[25] taken from [49] ($\alpha = 25°$, inviscid, Fig. 3.38) and [51] ($\alpha = 40°$, viscous, Fig. 3.39) were made with a modified vehicle configuration[26].

---

[24] Trim surface effectiveness, however, obviously was sufficient, see the discussion on page 306 of this book.

[25] We have noted above, that the flight Mach number at $\alpha = 25°$ is $M_\infty \approx 5$. In [49] results are presented – obviously due to the limited computation capabilities at that time – for this angle of attack, though for flight Mach numbers which belong to the $\alpha = 40°$ regime. However, for high flight Mach numbers $\alpha = 25°$ is included in the aerodynamic design data book of STS-1, [20]. We keep the $\alpha = 25°$ case in the following investigations in order to widen our data base.

[26] This was necessary because of the weak computer resources in 1983 and still in 1994. Later investigations with the full geometry were made, see, e.g., [53].

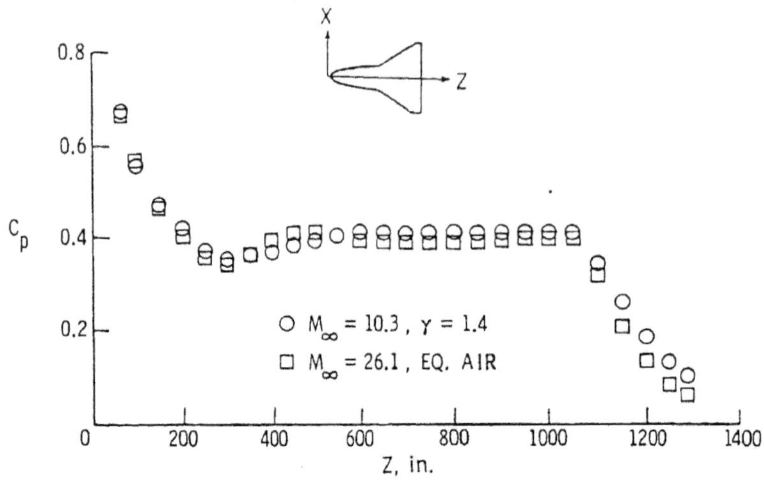

Fig. 3.38. Wall pressure coefficients c_{p_w} along the windward symmetry line of the Orbiter configuration, [49]. Angle of attack $\alpha = 25°$, inviscid computations.

We use $M_\infty = 10$ with perfect gas ($\gamma = 1.4$) as reference pre-flight case and $M_\infty = 24$ with equilibrium/non-equilibrium gas model as reference post-flight case.

During the aerodynamic shape definition process of the Orbiter it was observed, that high-temperature real-gas effects reduce c_{p_w} on the lower surface (windward side) of the vehicle. This does not hold for the part, where a re-compression after an overexpansion takes place. In the $\alpha = 25°$ case this is at 300.0 $in. \lessapprox Z \lessapprox$ 450.0 $in.$ on the center line, Fig. 3.38. If we change our notation from Z to x, this amounts to[27] $0.23 \lessapprox x/L \lessapprox 0.35$. Note that at this location the longitudinally flat surface part of the windward side of the vehicle begins. Approximately here also the cross-over of c_{p_w} as function of M_∞ and high-temperature real-gas effects, mentioned in Sub-Section 3.2.1, happens.

In the $\alpha = 40°$ case the re-compression occurs slightly ahead of that region, Fig. 3.39. The reduction of c_{p_w}, however, is strongest in the region 1,060.0 $in.$ $\lessapprox Z \leqq$ 1,290.3 $in.$ ($0.82 \lessapprox x/L \leqq 1.0$). The authors of [49] remark that other unpublished results indicate only little Mach number effects, i. e. Mach number independence, so that the differences in c_{p_w} probably can be attributed mainly to high-temperature real-gas effects.

We look at the $\alpha = 40°$ case, Fig. 3.39, and analyze then both cases. Fig. 3.39 was made with another axis convention, such that the stagnation point lies at $z = 0.0$ in. ($x/L = 0$) at the right-hand side and the end of the vehicle at $z =$ -1,290.0 in. ($x/L = 1.0$) at the left-hand side.

[27] L = 32.7736 m = 1,290.3 $in.$ is the length of the Orbiter without the 2.21 m (87.0 $in.$) long body flap, [48].

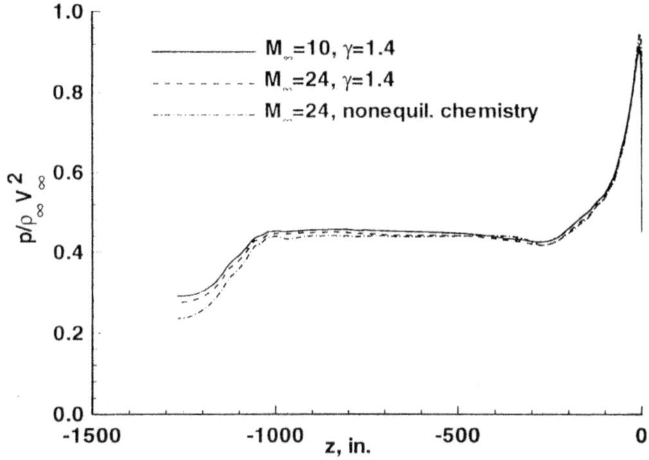

Fig. 3.39. Wall pressure $p/\rho_\infty v_\infty^2$ ($\equiv p_w/\rho_\infty v_\infty^2$) along the windward symmetry line of the Orbiter configuration, [51]. Angle of attack $\alpha = 40°$, viscous computations for wind tunnel and for flight at $H = 70.0\ km$ altitude.

The non-dimensional pressure $p/\rho_\infty v_\infty^2 = p_w/(2q_\infty)$ in Fig. 3.39 can be converted into the pressure coefficient c_p by means of the relation:

$$c_p = c_{p_w} = \frac{p - p_\infty}{q_\infty} = \frac{2\,p}{\rho_\infty v_\infty^2} - \frac{2}{\gamma_\infty M_\infty^2}. \tag{3.9}$$

The overall pattern of the pressure distribution on the windward symmetry line in the $\alpha = 40°$ case (Fig. 3.39) is quite similar to that in the $\alpha = 25°$ case (Fig. 3.38).

The reader is now asked to note in both figures the pressure plateau[28] at $0.4 \lessapprox x/L \lessapprox 0.82$, and the strong drop of c_{p_w} at $0.82 \lessapprox x/L \lesssim 1.0$. In order to understand these properties of the pressure field, consider Figs. 3.40 and 3.41.

We see from the first of these figures that the lower surface of the flight vehicle is flat in longitudinal direction in the range $0.12 \lessapprox x/L \lessapprox 0.82$. At $x/L \approx 0.82$ begins a boattailing with an angle of approximately 6°. Boattailing of a fuselage usually is applied in order to increase the scrape angle for take-off and landing, to influence – at low speed flight – the vehicles pitching moment, and to reduce the base area, hence the base drag.

The second of the two figures, Fig. 3.41, shows a slight positive (upward) dihedral of the lower side of the vehicle. Such a dihedral shape improves rolling and also lateral/directional stability, see, e.g., [44]. Behind $x/L \approx 0.62$ the lower side, approximately below the fuselage, is rather flat.

Hence for $0.4 \lessapprox x/L \lessapprox 0.82$ the lower side of the flight vehicle can be considered as approximately flat, except for the dihedral area of the outer wing,

[28] We switch completely over from the Z notation to the x or, respectively, x/L notation.

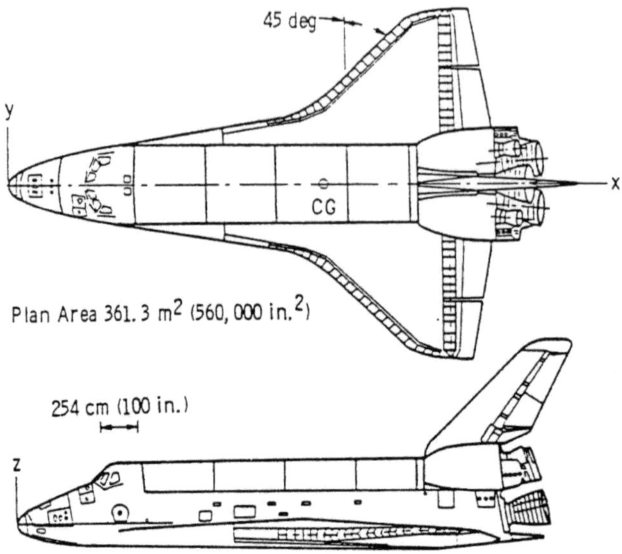

Fig. 3.40. Plan view and side view of the Orbiter, [14].

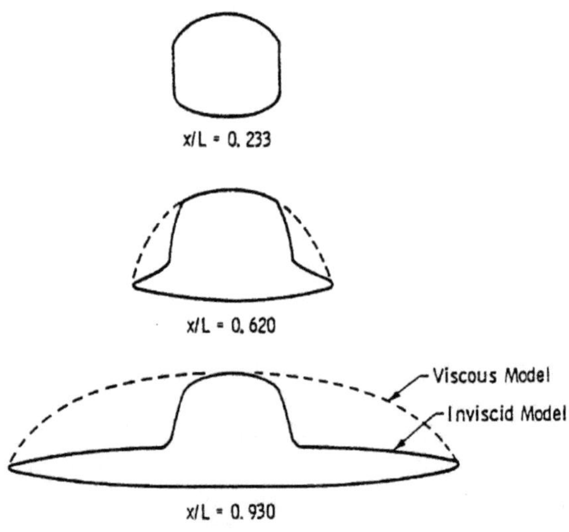

Fig. 3.41. Cross sections of the simplified Orbiter geometry (Viscous Model = HALIS configuration), [14].

Fig. 3.6. This means that, for the angle of attack regime considered in this section, the flow there – initially between the primary attachment lines – is approximately two-dimensional in nature, as is discussed in Sub-Section 3.2.2. This can also be seen from earlier computational and oil-flow data, see, e.g., [51], [55], [53].

Table 3.4. Wall pressure coefficients c_{p_w} in the lower symmetry line of the Orbiter at $x/L = 0.62$ ($\pm$ 800.0 *in.*) for $\alpha = 25°$ (Fig. 3.38) and $\alpha = 40°$ (Fig. 3.39).

Case	Fig.	α	M_∞	γ	x/L	c_{p_w}
1.a	*Fig.*3.38 (*invisc.*)	25°	10.3	1.4	0.62	0.41
1.b	*Fig.*3.38 (*invisc.*)	25°	26.1	*equil.*	0.62	0.38
2.a	*Fig.*3.39 (*visc.*)	40°	10	1.4	0.62	0.89
2.b	*Fig.*3.39 (*visc.*)	40°	24	1.4	0.62	0.90
2.c	*Fig.*3.39 (*visc.*)	40°	24	*non-eq.*	0.62	0.88

We come now back to the behavior of c_{p_w} in Figs. 3.38 and 3.39. The pressure plateaus at $0.4 \lessapprox x/L \lessapprox 0.82$ obviously are due to the longitudinal flatness of the lower side of the flight vehicle. We assemble the c_{p_w} data (case 1.a and 1.b from Fig. 3.38 and case 2.a to 2.c from Fig. 3.39) at the location $x/L = 0.62$ ($\pm$ 800.0 *in.*) in Table 3.4.

We see that in case 2.b c_{p_w} is slightly larger than that in case 2.a. It is not quite clear what is to be expected. Note that the data from Fig. 3.39 had to be converted with the help of eq. (3.9). Usually we have in this part of the configuration for a given angle of attack a c_{p_w}, which decreases with higher M_∞ and also due to high-temperature real-gas effects.

At the forward stagnation point (the data are not shown in the table), we expect, according to Fig. 3.5 in Sub-Section 3.2.1, that c_{p_w} is largest for the highest Mach number and with high-temperature real-gas effects present. This is indicated for the $\alpha = 40°$ case in Fig. 3.39.

In [1] it is discussed that the windward side boundary layer on a RV-W-type flight vehicle at large angle of attack is initially a subsonic, then a transonic, and finally a low supersonic boundary layer[29]. The boundary layer edge Mach numbers are rather small. In [56] it is reported that during an Orbiter re-entry these are typically at most $M_e \approx 2.5$, and mostly below $M_e \approx 2$. Hence we assume that we have supersonic flow in the $\alpha = 25°$ case, but low supersonic flow in the $\alpha = 40°$ case.

This is important, because the boattailing, which begins at $x/L \approx 0.82$, leads to a flow expansion, only if the flow ahead of it is supersonic (supersonic isentropic expansion turning). In our case this is given, even at the large angles of attack during the initial re-entry flight phase. The effect can be understood and approximated basically with the help of the Prandtl-Meyer expansion, see,

[29] It is, however, not an ordinary boundary layer. It is characterized by very high temperatures and hence high-temperature real-gas effects, by strong wall-normal temperature gradients due to surface radiation cooling, and is, depending on the type of the TPS, influenced by surface roughness.

Table 3.5. Wall pressure coefficients c_{p_w} in the lower symmetry line of the Orbiter in the boattailing region at $x/L = 0.91$ ($\pm$ 1,173.9 *in.*) for $\alpha = 25°$ (Fig. 3.38) and $\alpha = 40°$ (Fig. 3.39).

Case	Fig.	α	M_∞	γ	x/L	c_{p_w}
1.a	*Fig.*3.38 (*inviscid*)	25°	10.3	1.4	0.91	0.26
1.b	*Fig.*3.38 (*inviscid*)	25°	26.1	*equil.*	0.91	0.20
2.a	*Fig.*3.39 (*viscous*)	40°	10	1.4	0.91	0.63
2.b	*Fig.*3.39 (*viscous*)	40°	24	1.4	0.91	0.61
2.c	*Fig.*3.39 (*viscous*)	40°	24	*non-equil.*	0.91	0.55

e.g., [1], or understood qualitatively in terms of the stream-tube behavior, see, e.g., [13].

However, behind $x/L \approx 0.82$ the pressure coefficient drops further, therefore we have neither a "centered" nor a "simple" expansion, see, e.g., [1]. One could speak of a "sliding simple" expansion due to the initially curved boattailing, Fig. 3.40. Possibly also a deviation from the more or less two-dimensionality of the flow in the boattailing region plays a role, see, e.g., [51].

In cases where the plateau wall Mach numbers are subsonic (angles of attack larger than considered here), one can assume that the pressure would increase in the boattailing region and induce a nose-down increment of the pitching moment. A "pitching moment reversal" is indeed present in case of the Orbiter for $\alpha \gtrapprox 50°$, see, e.g., [57]. In [53], for instance, it is not attributed to a boattailing re-compression due to a subsonic plateau Mach number.

We consider now the situation in the boattailing region $0.82 \lessapprox x/L \leqq 1.0$. In Figs. 3.38 ($\alpha = 25°$) and 3.39 ($\alpha = 40°$) we see at once, that there are quantitative differences in the c_{p_w} behavior compared to that in the upstream pressure plateau region. In the boattailing region the differences in c_{p_w} become much larger, when we go from the lower flight Mach number cases with perfect gas to the higher flight Mach number cases which take into account high-temperature real-gas effects. Remembering the remark in [49] that Mach number effects were found to be small (Mach number independence), these differences are probably to be attributed mostly to high-temperature real-gas effects.

Whether the assumption of equilibrium or non-equilibrium behavior of the gas plays a role, cannot be decided with the available data. The $M_\infty = 24$ case corresponds to a flight altitude of about 72.0 km. Considering the length of the pressure plateau on the windward side of the flight vehicle, $0.4 \lessapprox x/L \lessapprox 0.82$, which amounts to approximately 14.0 m, one can assume that there thermo-chemical equilibrium is attained in the inviscid flow part of the shock layer. We come back to the influence of flight Mach number and high-temperature real-gas effects in Section 3.6, considering then also the influence of the two different angles of attack.

For the quantitative study of the situation in the boattailing region, data are collected in Table 3.5. There c_{p_w} data (case 1.a and 1.b from Fig. 3.38 and case 2.a to 2.c from Fig. 3.39) are given for the location $x/L = 0.91$ ($\pm$ 1,173.9 $in.$) which is approximately the middle of the boattailing region.

We see in the boattailing region at $x/L = 0.91$ a much stronger influence of high-temperature real-gas and Mach number effects than in the pressure-plateau region at $x/L = 0.62$. The pressure coefficient c_{p_w} drops in all cases appreciably. Related to the $M_\infty = 10.3$ and $M_\infty = 10$ cases, the reduction of c_{p_w} at $x/L = 0.91$ is about 22 per cent (case 1.b) and 13 per cent (case 2.c).

This appears not to be so much[30], however, considering the area of the boattailing region (length approximately 5.9 m, width approximately 23.0 m, hence an area of approximately 135.7 m^2), and the distance of this region to the center-of-gravity, approximately 9.04 m, this amounts to an appreciable pitch-up moment increment in both angle of attack cases[31], see Problem 3.5. Obviously it was mainly just this not so very large reduction of c_{p_w} in the boattailing region which caused the hypersonic pitching moment anomaly experienced with STS-1.

Remembering that the pressure coefficient at the forward stagnation point becomes larger with larger flight Mach number, and also when high-temperature real-gas effects are present, Fig. 3.5, we have another potential pitch-up moment increment to take into account. In this case the affected surface portion is small, maybe about 1.0 m^2, but the lever arm is rather large with approximately 21.3 m. Here, the potential increment is not large, but should be appreciated, see Problem 3.7.

3.5.3 The Forward Shift of the Center-of-Pressure

We show in a simple way, that the reduction of the wall pressure in the boattailing region, Figs. 3.38 and 3.39, as well as Table 3.5, leads to a forward shift of the center of pressure and to a reduction of the normal force, too.

Consider the action lines of the forces in Fig. 3.42. That of the force on the boattailing region, F_{bt}, crosses that of the remaining portion (main surface) of the windward side of the vehicle, F_{ms}, in the upper part of the figure. The action line a_{res} of the resulting force, F_{res}, crosses the vehicle surface in the center of pressure at x_{cp}.

If now F_{bt} is reduced, F_{res} is reduced to F'_{res}, and the new action line a'_{res} as well as the new center of pressure, x'_{cp}, are shifted forward. An increase of the

[30] The reader should note, that these are values found in the lower symmetry line. They are only approximately constant in span-wise direction. Probably the effect diminishes in that direction (three-dimensionality effects due to the dihedral of the outer wing, Fig. 3.6).

[31] The identification of this effect on the Russian BURAN vehicle (first and only flight on November 15, 1988) obviously was achieved already in 1980, before the first Space Shuttle Orbiter re-entry flight in April 1981, [50].

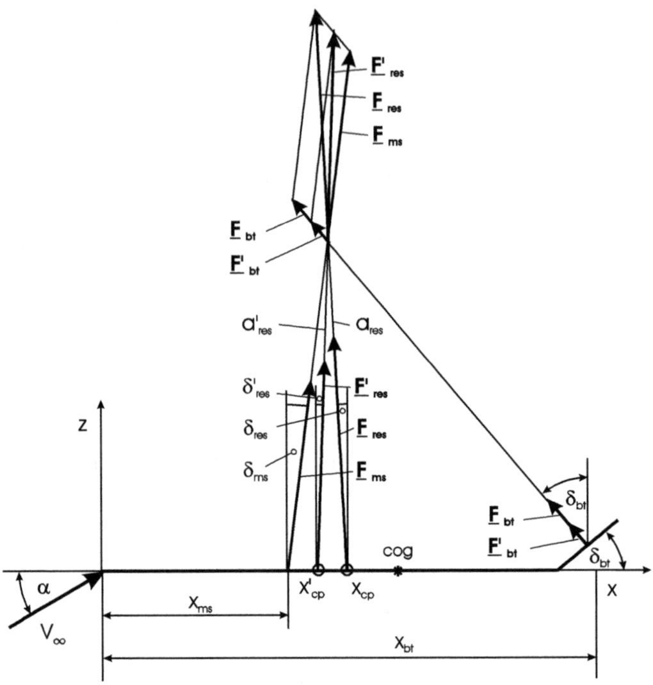

Fig. 3.42. Geometrical RHPM approximation of the windward side of the Space Shuttle Orbiter: aerodynamic forces, their action lines, and the centers of pressure. Note that the boattailing angle is exaggerated.

pressure in the boattailing region would shift the center of pressure in downstream direction.

If the boattailing angle δ_{bt} and all other angles δ are very small[32], we can write the simple balance:

$$x_{cp}\, F_{res} = x_{ms}\, F_{ms} + x_{bt}\, F_{bt}, \tag{3.10}$$

and find with $F_{res} = F_{ms} + F_{bt}$ for x_{cp}:

$$x_{cp} = \frac{x_{ms} + x_{bt}\frac{F_{bt}}{F_{ms}}}{1 + \frac{F_{bt}}{F_{ms}}}. \tag{3.11}$$

If $F_{bt} \leqq F_{ms}$ we get:

$$x_{cp} \approx (x_{ms} + x_{bt}\frac{F_{bt}}{F_{ms}})(1 - \frac{F_{bt}}{F_{ms}}), \tag{3.12}$$

and finally:

$$x_{cp} \approx x_{ms} + (x_{bt} - x_{ms})\frac{F_{bt}}{F_{ms}}. \tag{3.13}$$

[32] In this case the resultant force is equal to the normal force: $F_{res} = N$.

This relation reflects directly the observation which we made with the help of Fig. 3.42: the reduction of the wall pressure in the boattailing region and hence of F_{bt} results in a forward shift of the center of pressure. The pitch-up moment is increased, the normal force reduced, the other forces – axial, lift, drag – are, due to the given constellation, only a little changed.

Problem 3.8 yields despite the simple approach for $\alpha = 40°$ a forward shift of approximately 0.3 m which compares well with the data given in [14]. The original data were: center-of-gravity located at $x_{cog} = 21.30\ m$ ($x_{cog}/L_{ref} = 0.65$), center of pressure at $x_{cp} \approx 21.71\ m$. This means $x_{cp} > x_{cog}$ and a positive static stability margin, Sub-Section 3.4.2. Even with the forward shift of $|\triangle x_{cp}| \approx 0.3\ m$ the margin would remain positive.

However, the actually flown center-of-gravity was $0.667 \lessapprox x_{cog}/L_{ref} \lessapprox 0.671$, [21], which amounts to $x_{cog} = 21.86$ to $21.99\ m$. The stability margin was negative from the beginning. The hence predicted downward deflection of the body flap as main trim surface, was at the high flight Mach numbers $\eta_{bf} \approx 7°$, Fig. 3.36. The – unexpected – forward shift $|\triangle x_{cp}| \approx 0.3\ m$ then made $\eta_{bf} \approx 17°$ necessary in flight. Note that most of these numbers are not measured, but post-flight computed numbers.

3.6 The Hypersonic Pitching Moment Anomaly in View of Oswatitsch's Mach Number Independence Principle

3.6.1 Introduction

When the Orbiter was developed, the major aerodynamic design tools were approximate (impact) methods, and ground-simulation facilities. The discrete numerical methods of aerothermodynamics were in their infancy, see, e.g., [49]. What was available and was used thoroughly were similarity rules, correlations and parameters, see, e.g., [58]. Considered were weak and strong viscous interaction, rarefaction effects, high-temperature real-gas effects et cetera. Also employed was, usually not explicitly mentioned, Oswatitsch's Mach number independence principle for blunt bodies in hypersonic flow, [59], see also [1].

This principle says basically, that above a certain flight Mach number several properties of the flow field in the shock layer become independent[33] of the flight Mach number[34]. This holds for the shape of the bow shock surface, the pattern of the streamlines, the sonic surface, and the Mach lines in the supersonic part of the flow field. The density ratio ρ/ρ_∞, the pressure coefficient c_p, and with the latter the force and moment coefficients, are also independent of M_∞. These items thus do not depend on the free-stream or flight Mach number, but only on the body shape and also on the ratio of specific heats γ. The latter means,

[33] Independence is reached asymptotically, see for instance the graphs in Section 3.2. Independence hence has an approximative character. For certain applications it may be appropriate to define an error bound.

[34] Actually this is a precise definition of hypersonic flow.

that gas flow with constant γ must be present everywhere for Mach number independence.

In practice this means, if we obtain for a given body shape experimentally for instance the force coefficients at a large enough Mach number M'_∞ (saturation Mach number), they are valid then for all larger Mach numbers $M_\infty > M'_\infty$. This happens, depending on the body shape, already for free-stream Mach numbers as low as $M_\infty = 4$ to 5. In [58] it is argued, without going in detail, that Oswatitsch's independence principle for blunt bodies also holds for non-perfect gas flow and also for viscous flow.

The independence principle, although seldom cited in the literature, was underlying in the aerodynamic shape definition and the data set generation work for the Orbiter. The base-line Mach number range in ground-facility simulation for the data set generation was $M \approx 8$ to 10, 13. In this Mach number range the available facilities are large enough that models can be employed, which have a size that permits a high-fidelity modelling of the original vehicle shape, and they have a high productivity.

Due to the thorough and risk-conscious work in the aerothermodynamic design process the aerodynamic performance of the Orbiter was largely found as predicted. The only very critical misprediction was that of the pitching moment, which, as we have seen in the last sub-section, can be traced back mostly to the flow behavior in the boattailing region.

That the blunt body Mach number independence principle holds for perfect gas flow is a matter of fact, see also the simple examples in Section 3.2. In view of the practical application, two questions arise:

1. What is – for perfect gas flow – the saturation Mach number M'_∞ for a given blunt body shape?
2. How do high-temperature real-gas effects affect the principle?

The first of these questions can only be answered experimentally or computationally by trial and error, taking into account that Mach number independence is reached asymptotically. The second question, whose answer can give us a clue why the pitching moment anomaly was there, of course could also be approached by trial and error. In the following sub-sections we try, however, to answer the question with inviscid flow data found by computational and analytical means.

We study the influence of the flight Mach number and high-temperature real-gas effects on the wall pressure for the two angles of attack $\alpha = 25°$ (Fig. 3.38) and $\alpha = 40°$ (Fig. 3.39) with blunt-cone approximations, Section 10.1, of the lower symmetry line of the Space Shuttle Orbiter.

The two geometries with boattailing are shown in Fig. 3.43. The generatrices of the blunted cones are the lower side symmetry lines of the Orbiter. The forward stagnation points are located at $x = 0$. The nose radii were taken from [60], see also [61], where they were introduced as effective nose radii, depending on the angle of attack. The semi-vertex angles are identical with the angles of attack. The locations X at the lower side symmetry line of the Orbiter,

Fig. 3.40, are found with $x_{bc} = X/\cos\alpha$. The location $x/L = 0.82$, where the $6°$ boattailing, Sub-Section 3.5.2, begins, are $x_{bc} = 24.35\ m$ and $20.58\ m$, respectively.

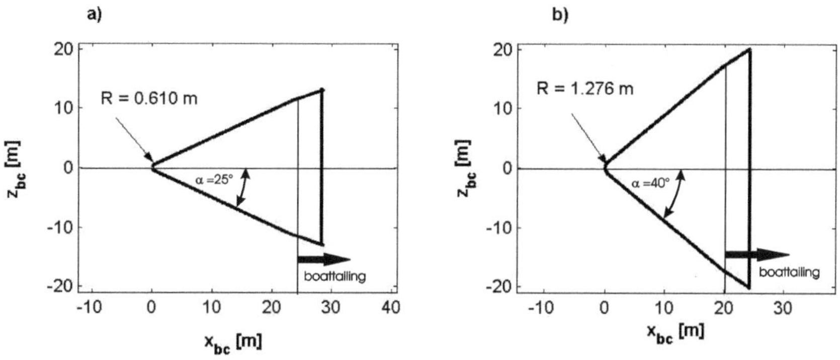

Fig. 3.43. Blunt cone approximation of the lower symmetry line of the Space Shuttle Orbiter for two angles of attack: a) $\alpha = 25°$, b) $\alpha = 40°$. The semi-vertex angles of the blunt cones are identical with the angles of attack.

The distributions of the wall pressure coefficient $c_{p_w}(x)$, discussed in the next sub-section, were found with a time marching finite-difference method for the solution of the non-conservative Euler equations with bow shock fitting, [62]. This method gives, besides an accurate bow shock location, an excellent entropy conservation along streamlines, particularly along the body surface. An advanced shock fitting method based on the quasi-conservatively formulated Euler equations, [63], was employed for the determination of the data given in Fig. 3.44, [64].

3.6.2 Wall Pressure Coefficient Distribution

Fig. 3.44 shows for the blunt cone approximations of the windward symmetry line of the Space Shuttle Orbiter at the two angles of attack the computed wall pressure coefficients for four flight Mach numbers for perfect (ID) and equilibrium (EQ) real gas. Note that the distributions are given along the generatrices of the blunt-cone configurations, Fig. 3.43, i. e. on the real windward symmetry line of the Orbiter, Fig. 3.40.

The results are in good qualitative and quantitative agreement with the data of Figs. 3.38 and 3.39. In the boattailing region, however, we see, compared to Figs. 3.38 and Fig. 3.39, a different qualitative behavior. Here we reach $c_{p_w}(x)$ plateaus, which is not the case there. This is caused by the geometrical modelling of the onset of the boattailing at the blunt cones, which is different from that at the Orbiter.

In general we can say, that ahead of the boattailing region the influence of M_∞ is small for both angles of attack, also that of high-temperature real-gas

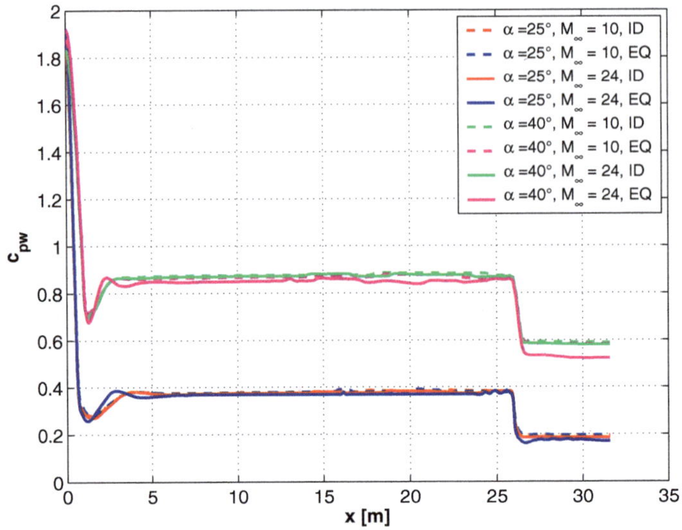

Fig. 3.44. Space Shuttle Orbiter windward symmetry line (blunt cone approximation) with boattailing: $c_{p_w}(x)$ for four flight Mach numbers M_∞, inviscid flow, $\alpha = 25°$, $\alpha = 40°$, [64]. ID: perfect gas, EQ: equilibrium real gas.

effects. Although the curves are not very smooth, we see that $c_{p_w}(x)$ is slightly smaller for larger M_∞ and in the equilibrium real gas case compared to the perfect gas case.

The picture changes in the boattailing region. There the influence of M_∞ is still small, but the real-gas effect is strong. All is more pronounced in the 40° cases.

To gain more insight, we investigate now parametrically with the method [62] the influence of flight Mach number M_∞ and high-temperature real-gas effects – the latter in terms of the effective ratio of specific heats γ_{eff} – on $c_{p_w}(x)$. We do it again with the blunt cone approximation of the Orbiter windward side, but include also a case without boattailing.

First we have a look at the pressure-coefficient distributions of the configuration without boattailing for four flight Mach numbers M_∞ and perfect gas.

In Fig. 3.45 we see for both the $\alpha = 25°$ and 40° case for the three larger flight Mach numbers a very good qualitative and quantitative agreement with the c_{p_w} distributions in Figs. 3.38 and 3.39. In the region, where the boattailing is not present, the pressure coefficient remains the same as upstream of it. The difference of the $M_\infty = 3.5$ curve to the curves of the other flight Mach numbers is somewhat larger for the $\alpha = 40°$ case than for the $\alpha = 25°$ case. Mach number independence is present for $M_\infty \geqq 10$. The pressure coefficients decrease only very slightly. The cross-over region is not discernible. We note the typical small undershoot under the plateau pressure ahead of the plateau and little kinks due to the curvature jump at the junction of the blunt nose and the cone.

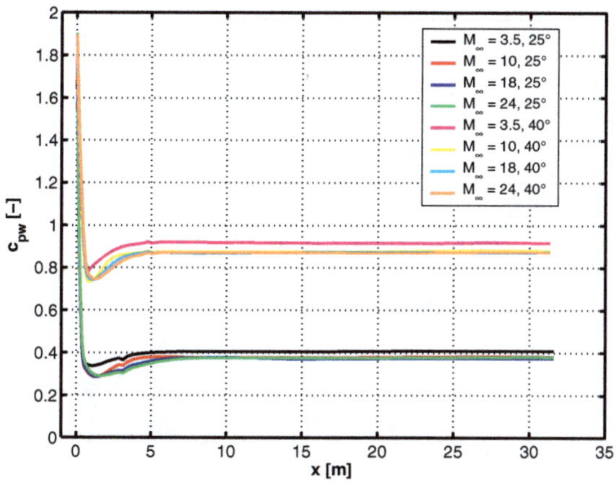

3

Fig. 3.45. Space Shuttle Orbiter windward symmetry line (blunt cone approximation) **without** boattailing: $c_{p_w}(x)$ for four flight Mach numbers, inviscid flow, perfect gas, $\alpha = 25°$ and $40°$.

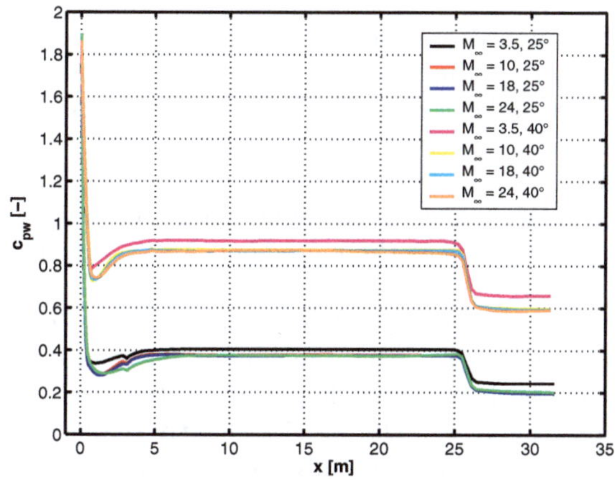

Fig. 3.46. Space Shuttle Orbiter windward symmetry line (blunt cone approximation) **with** boattailing: $c_{p_w}(x)$ for four flight Mach numbers, inviscid flow, perfect gas, $\alpha = 25°$ and $\alpha = 40°$.

Next we study results for the configuration with boattailing. In the boattailing region we now observe a spreading of the $c_{p_w}(x)$ curves. It is very small for the $\alpha = 25°$ case, and a little larger for the $\alpha = 40°$ case, Fig. 3.46. Once Mach number independence is reached, it is again very small.

Before we concern ourselves with the influence of the high-temperature real-gas effects, we look at aspects of the Mach number independence principle different from the $c_{p_w}(x)$ behavior. We have mentioned that the independence also pertains to the shape of the bow shock surface, the pattern of the streamlines, et cetera.

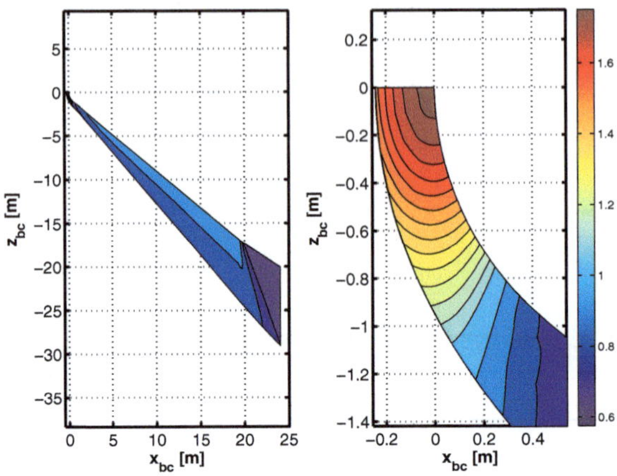

Fig. 3.47. Space Shuttle Orbiter windward symmetry plane (blunt cone approximation) **with** boattailing: $M_\infty = 3.5$, $\alpha = 40°$, bow shock shape and iso-c_p lines, inviscid flow, perfect gas. The color bar at the right side concerns c_p.

In the following three figures we show for the Space Shuttle Orbiter windward symmetry plane in the blunt cone approximation the traces of the bow shock surface, iso-c_p lines and color bands. The angle of attack in all cases is $\alpha = 40°$. The flight Mach numbers are $M_\infty = 3.5$, Fig. 3.47, $M_\infty = 10$, Fig. 3.48, and $M_\infty = 24$, Fig. 3.49.

The iso-lines in the enlargements of the figures begin each with $c_p = 0.8$ to the right, then we have three times the increment $\triangle c_p = 0.1$, and beginning with $c_p = 1.1$ the increment $\triangle c_p = 0.05$.

We observe first that – for all three Mach numbers – the bow shock lies quite close to the body surface, and that much more for the two higher Mach numbers. For the latter the bow shock shapes are virtually identical, except for the boattailing area. There for $M_\infty = 24$ a slightly stronger expansion is indicated than for $M_\infty = 10$, which is weakly seen also in Fig. 3.46. Very small differences are also present in the stagnation point region.

The effective Mach number independence for the two larger Mach numbers is also evident from the isolines of the pressure coefficient. In the stagnation point region we see small differences, see also Fig. 3.5. We note, that in general Mach number independence, if present, appears to be more pronounced somewhat away from the stagnation point area – downstream of the cross-over point – in the presumably existing "benign" wall Mach number interval, Sub-Section 3.6.3.

This is well indicated by the data in Table 3.6. At the stagnation point, in terms of the Orbiter geometry $x/L = 0$, we find a $\triangle c_p = +\ 0.031$ from $M_\infty = 10$ to $M_\infty = 24$. At the location $x/L = 0.62$, compare Table 3.4, we find $\triangle c_{p_w} = -\ 0.006$, and finally in the middle of the boattailing region, $x/L = 0.91$, compare Table 3.5, $\triangle c_{p_w} = -\ 0.017$. The data at the two latter points illustrate the cross-over of the c_{p_w} curves.

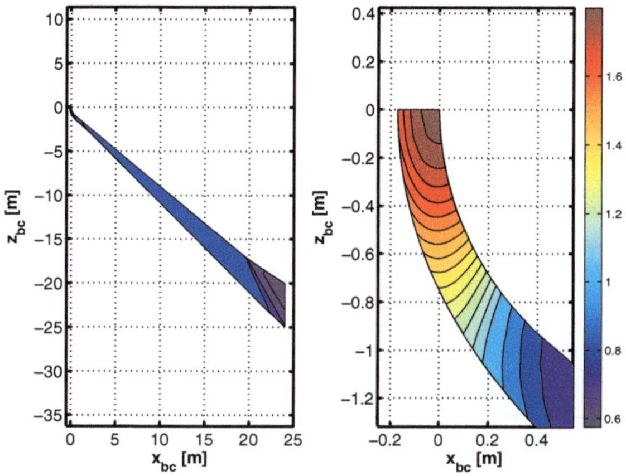

Fig. 3.48. Space Shuttle Orbiter windward symmetry plane (blunt cone approximation) **with** boattailing: $M_\infty = 10$, $\alpha = 40°$, bow shock shape and iso-c_p lines, inviscid flow, perfect gas. The color bar at the right side concerns c_p.

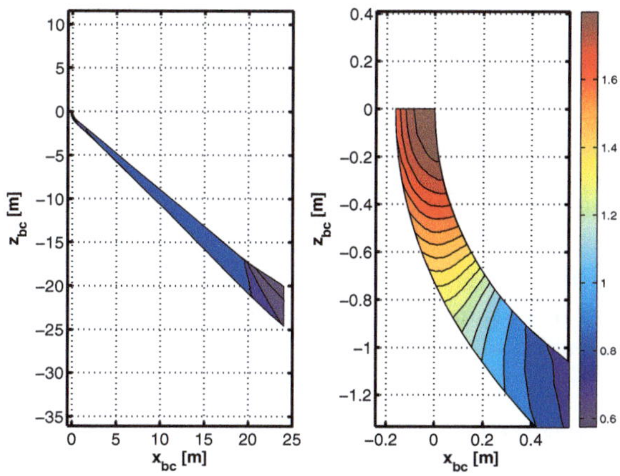

Fig. 3.49. Space Shuttle Orbiter windward symmetry plane (blunt cone approximation) **with** boattailing: $M_\infty = 24$, $\alpha = 40°$, bow shock shape and iso-c_p lines, inviscid flow, perfect gas. The color bar at the right side concerns c_p.

Now to the influence of high-temperature real-gas effects. They are studied for the $\alpha = 25°$ and the $\alpha = 40°$ case for $M_\infty = 10$ with the help of three γ_{eff}. The $c_{p_w}(x)$ distributions in Fig. 3.50 for the configuration without boattailing show very little influence of γ_{eff} for $\alpha = 25°$, and somewhat more for $\alpha = 40°$. The pressure coefficient decreases slightly in any case with decreasing γ_{eff}.

For the configuration with boattailing we see in the boattailing region the spreading of the c_{p_w} curves similar to that we have seen for different Mach numbers M_∞, Fig. 3.46, but much stronger. It is also stronger for 40° than for $\alpha =$

Table 3.6. Space Shuttle Orbiter windward symmetry line (blunt cone approximation) **with** boattailing: wall pressure coefficient c_{p_w} at three locations x_{bc} (blunt cone), x/L (Orbiter), $\alpha = 40°$, inviscid flow, perfect gas.

x_{bc} $[m]$	x/L $[-]$	$c_{p_w}\|_{M_\infty=3.5}$	$c_{p_w}\|_{M_\infty=10}$	$c_{p_w}\|_{M_\infty=24}$
0.0	0	1.790	1.833	1.864
14.37	0.62	0.917	0.873	0.867
21.09	0.91	0.659	0.607	0.590

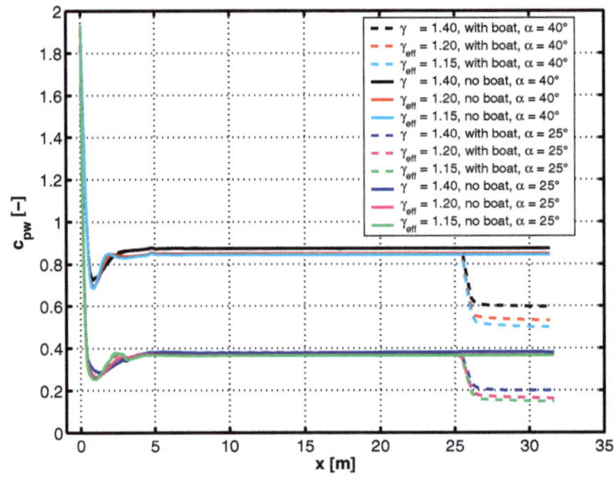

Fig. 3.50. Space Shuttle Orbiter windward symmetry line (blunt cone approximation) **without** and **with** boattailing: $c_{p_w}(x)$ for $M_\infty = 10$, and three different γ_{eff}, inviscid flow, $\alpha = 25°$ and $\alpha = 40°$.

25°. In any case we observe quantitatively the same behavior as seen in Figs. 3.38 and Fig. 3.39.

These parametric studies, performed for the $\alpha = 25°$ and the $\alpha = 40°$ case, and with γ_{eff} only for $M_\infty = 10$, show for large flight Mach numbers in the boattailing region high-temperature real-gas effects having a strong effect on $c_{p_w}(x)$, which is much stronger than that of the flight Mach number. At the configuration without boattailing, neither of the two effects has an appreciable influence on $c_{p_w}(x)$.

3.6.3 A Simple Analysis of Flight Mach Number and High-Temperature Real-Gas Effects

With Oswatitsch's Mach number independence principle for blunt bodies in the background, we investigate now the expansion history at the lower symmetry line of the flight vehicle with the help of a simple analysis, assuming one-dimensional inviscid flow. We connect only the wall pressure coefficient c_{p_w} with the Mach number of the inviscid flow at the wall (boundary-layer edge) M_w, using M_∞, $\gamma_{\inf}$, and γ_{eff} across the bow shock and along the body surface as parameters.

The problem is that we can do this only for planar flow and not, as needed in our case for conical flow. Nevertheless, we can gain helpful insight from it.

Assuming adiabatic expansion along the windward side of the vehicle (one-dimensional consideration along the lower symmetry line), beginning at the stagnation point, we find:

$$\frac{p_w}{p_\infty} = \frac{p_w}{p_{t_2}}\frac{p_{t_2}}{p_\infty} = \left(1 + \frac{\gamma - 1}{2} M_w^2\right)^{-\frac{\gamma}{\gamma-1}} \frac{p_{t_2}}{p_\infty}. \tag{3.14}$$

The wall pressure coefficient c_{pw} then reads:

$$c_{p_w} = \frac{p_w - p_\infty}{q_\infty} = \frac{p_\infty}{q_\infty}\left(\frac{p_w}{p_\infty} - 1\right) = \frac{2}{\gamma_\infty M_\infty^2}\left(\frac{p_w}{p_\infty} - 1\right). \tag{3.15}$$

We assume that the forward stagnation point streamline crosses the normal-shock portion of the bow shock, and find p_{t_2} for the chosen γ_{eff} from the relations in Sub-Section 10.1.3. We choose two free-stream Mach numbers close to those in Figs. 3.38 and 3.39, viz. $M_\infty = 10$ and 24, and add $M_\infty = \infty$.

In [55] for the flight situation computed ratios of the specific heats are $\gamma \approx$ 1.3 just behind the bow shock in the nose region, $\gamma_{eff} \approx 1.12$ close to the body in the nose region, and $\gamma_{eff} \approx 1.14$ along the lower body surface. We choose hence for each Mach number the four $\gamma_{eff} = 1.4, 1.3, 1.2, 1.1$.

The chosen wall Mach number interval is $0 \leqq M_w \leqq 5$. $M_w = 0$ represents the stagnation point and $M_w = 1$, of course, the sonic point at the wall.

In Fig. 3.51 we show computed wall pressure coefficients for $M_\infty = 24$ and the four chosen γ_{eff}. The c_{p_w} curves drop from their maxima in the stagnation point monotonously with growing M_w. At the stagnation point we see the behavior shown in Fig. 3.5: the stagnation pressure coefficient increases with decreasing γ_{eff}.

Remember, however, the remark above, that these data are not quantitatively representative for the Space Shuttle Orbiter. In [49] results from γ_{eff} studies for a blunt 30° cone are given, unfortunately without the wall Mach number. They show also the cross-over of the c_{p_w} curves, which happens in Fig. 3.51 around $M_w \approx 2.2$. At the stagnation point, the curves for the smallest γ_{eff} are above and switch then places with the curves for the larger ones.

In order to get a better impression of the differences between the different results, we consider normalized wall pressure coefficients $c^\star_{p_w}$. The c_{p_w} data are normalized with $c_{p_w, M_\infty=\infty, \gamma=1.4}$. In this way we can visualize the increments which exist in view of the Mach number independence principle.

The results in Fig. 3.52 show that ahead of the cross-over region, i. e. for $M_w \lessapprox 2.2$, the differences $c^\star_{p_w}$ - 1 between the respective results and the ones for $M_\infty = \infty$, $\gamma = 1.4$ are bounded and relatively small. The influence of M_∞

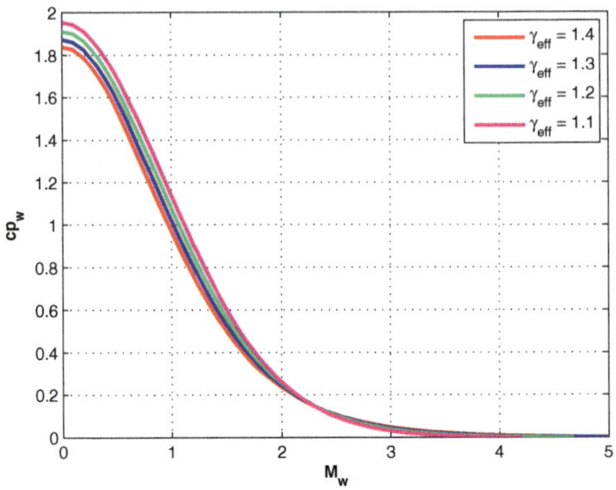

Fig. 3.51. Wall pressure coefficient c_{p_w} as function of wall Mach numbers M_w for $M_\infty = 24$, and four γ_{eff}.

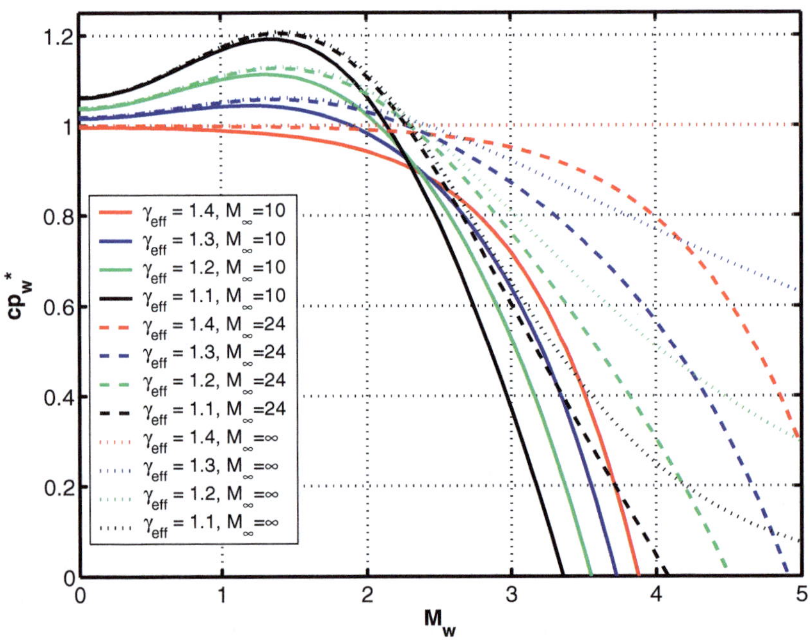

Fig. 3.52. Normalized wall pressure coefficient $c^\star_{p_w} = c_{p_w}/c_{p_{w,M_\infty=\infty,\gamma=1.4}}$ as function of wall Mach numbers M_w for selected pairs of M_∞ and γ_{eff}.

is smaller in general than that of γ_{eff}. For $M_w \gtrapprox 2.2$ the differences are unbounded and becoming large with increasing M_w and that in general stronger for smaller γ_{eff}.

These are results for the plane case. The results for the blunt cone case in Sub-Section 3.6.2 – more relevant for the Orbiter – exhibit exactly the same properties, however, with much smaller $c^{\star}_{p_w}$ - 1 differences. Unfortunately, we have no M_w results available belonging to the c_{p_w} results of Sub-Section 3.6.2. Initial results show for the $\alpha = 40°$ case a wall Mach number rise from levels of $M_w \approx 1.2$ ahead of the boattailing to ≈ 1.5 in the boattailing region. Data for the $\alpha = 25°$ are not consistent. For perfect gas and the given boattailing angle of 6° we find with the data of the Prandtl-Meyer expansion, see, e.g., [65], at small wall Mach numbers a $\triangle M_w \approx 0.2$. A dependence of $c^{\star}_{p_w}$ on the angle of attack and on one or more other parameters may be possible. In any case, is appears, that the cross-over region lies at wall Mach numbers smaller than $M_w \lessapprox 2.2$.

However, the general results of this sub-section and of Sub-Section 3.6.2 indicate, like the data in Figs. 3.38 and 3.39, that presumably a "benign" wall Mach number interval exists, where the Mach number independence principle is valid despite the presence of high-temperature real-gas effects.

With other words: *If for a given body shape and a given flight Mach number interval (flight) Mach number independence exists, presumably a "benign" wall (boundary-layer edge) Mach number interval is present, where also high-temperature real-gas effects independence exists.* A quantification of the interval would be desirable.

We close this sub-section remembering, that not only the wall pressure is of influence regarding the hypersonic pitching moment anomaly, Sub-Section 3.5.1. Nevertheless it appears that its behavior in the boattailing region is the major factor, the more so as all other force coefficients of longitudinal motion were found in flight in good agreement with the predicted data. The behavior of the normal force is directly connected to that of the pitching moment, Sub-Section 3.5.3.

3.6.4 Reconsideration of the Pitching Moment Anomaly and Summary of Results

In the preceding sub-sections the inviscid blunt-cone flow field in terms of c_{p_w} along the symmetry line of the windward side of the Orbiter was studied as well as the inviscid flow expansion history of planar one-dimensional flow. The γ_{eff} approach was employed in the latter in order to account for high-temperature real-gas effects. No viscous effects, neither weak nor strong interaction, and also no low-density effects were taken into account, assuming that they play only a minor role.

The Orbiter windward side is in a large portion approximately flat, however, the dominant flow features are governed by the vehicle shape approximately up to the canopy location, which can be considered as a blunt cone shape. Where the lower side is approximately flat, the flow properties along the windward centerline are more or less representative also in lateral direction (quasi two-dimensional distribution, Sub-Section 3.2.2).

If we accept the limitations of our investigations, and flight Mach number independence in Oswatitsch's sense is given, the general results are the following:

1. Flight Mach number *and* high-temperature real-gas effects, defined as deviations from the $M_\infty = \infty$, $\gamma = 1.4$ situation, are smallest in a certain low wall Mach number interval. If at the pressure coefficient plateau, present at $0.4 \lessapprox x/L \lessapprox 0.82$, M_w lies in this presumably existing "benign" interval, Oswatitsch's Mach number independence principle is only weakly violated[35]. *Without boattailing, C_N and C_M would only be little affected, because the lower vehicle surface is predominantly exposed to this wall Mach number interval, i. e. the wall Mach number plateau lies in the "benign" wall Mach number interval.* This holds for a large angle of attack interval, as is well indicated in Figs. 3.38 and 3.39. *A RV-W-type flight vehicle with these flow properties on the lower side, and in any case without a boattailing, thus will be insensitive with regard to flight Mach number and high-temperature real-gas effects.*
2. The boattailing leads to a supersonic expansion, provided that M_w at the pressure plateau is supersonic. This is given in the case of the Orbiter at both the angles of attack $\alpha = 25°$ and $40°$, Figs. 3.38 and 3.39. *If now M_w increases sufficiently, the – yet exactly to be defined – "benign" wall Mach number interval is left. Both – dominating – high-temperature real-gas effects and – weaker – flight Mach number effects come to influence, Figs. 3.38 and 3.39. This is the cause of the pitching moment anomaly of the Orbiter, as was conjectured earlier by several authors.* Even if the effect is small, the large affected surface portion at the windward side leads to the adverse increment of the pitching moment, see the discussion at the end of Sub-Section 3.5.2. A small role plays also the forward stagnation point region.
3. The boattailing leads to a subsonic compression, if M_w at the pressure plateau is subsonic. In the case of the Orbiter at $\alpha = 40°$ the wall Mach number at the pressure plateau is slightly supersonic. There exists a higher angle of attack, where the wall Mach number will be subsonic, and then in the boattailing region c_{p_w} (and overall also C_N) will be increased, and a pitch-down increment of C_M will result. This happens for the Orbiter at $\alpha \gtrapprox 50°$, Sub-Section 3.5.2. Our results indicate that without a boattailing this pitching moment reversal would not occur. However, our analysis is not able to give an answer to what happens, if the whole lower side of a RV-W-type vehicle is in the subsonic wall Mach number regime.
4. In general it can be said, that for a RV-W-type flight vehicle boattailing should be avoided in view of the pitching moment sensitivities in the hypersonic flight domain. If boattailing is needed to improve configurational and low-speed flight properties or for other reasons, data uncertainties in high Mach number ground-facility simulation must be expected. However, with the capabilities of numerical aerothermodynamics it is now possible to identify sensitivities and uncertainties early on and to design appropriate

[35] Oswatitsch gives no criterion in [59] for the range of validity of his principle.

correction schemes for the experimental data. Besides the pitching moment sensitivity of RV-W-type flight vehicles with boattailing, other sensitivities which curtail the Mach number independence of aerodynamic properties are not known. However, control-surface deflections, winglets, and the lateral movement of the flight vehicle must be considered with care in this regard.

3.6.5 Concluding Remarks

The reader might be tempted to think the following. For the Space Shuttle Orbiter, because of the large amount of data needed for the aerodynamic data set, the ground-facility simulation had to rely on continuously running tunnels in the $M = 8$ to 10 domain, without a sufficiently accurate representation of high-temperature real-gas effects. This should be no more a problem now in view of the today available numerical simulation capabilities.

Such a thinking would be a mistake for two reasons:

1. Even if the discrete numerical methods of aerothermodynamics, together with flow-physics and thermo-chemical models, and the still increasing power of computers allow for an accurate and reliable prediction, it is always necessary in vehicle design to understand the involved physics and the data from numerical simulations. This holds not only for the pitching moment of a flight vehicle. There is no doubt, that results of well defined numerical simulations can shed light on most complicated aerodynamic/aerothermodynamic design problems.
2. Design work involves other disciplines besides aerodynamics and aerothermodynamics. These often use their own prediction methods, if they need aerodynamic coefficients or loads. These methods are not necessarily the adequate ones. Time and cost pressures in industrial design work, in addition, lead to the use of inexpensive and fast, but not necessarily accurate enough computation methods. These problems in future possibly can be overcome with more holistic approaches in design work: keyword "Virtual Product", see Section 1.2 and the prologue of Chapter 8.

Performance and cost risks must be taken serious in design and development work of hypersonic flight vehicles. Tools and knowledge need to be adequate.

3.7 Problems

Problem 3.1 We consider six re-entry trajectory points of the Space Shuttle Orbiter mission TST-2, Fig. 2.2 with approximated flight Mach numbers: 1) $H = 80.0\ km$, $M_\infty = 26.4$, 2) $H = 70.0\ km$, $M_\infty = 22.1$, 3) $H = 60.0\ km$, $M_\infty = 15.7$, 4) $H = 50.0\ km$, $M_\infty = 10.3$, 5) $H = 40.0\ km$, $M_\infty = 6.1$, 6) $H = 30.0\ km$, $M_\infty = 3.4$.

How large are the dynamic pressures q_∞ in these points? How do they compare to the constraint given in Sub-Section 2.1.2?

Problem 3.2 Compute for the six re-entry trajectory points of Problem 3.1 the total enthalpies. Add for each point the kinetic part in per cent.

Problem 3.3 The Space Shuttle Orbiter flies during re-entry $M_\infty = 26.4$ at the altitude $H = 80.0\ km$. The angle of attack is $\alpha = 40°$. The lift coefficient is assumed to be $C_L = 0.884$, [20], the reference area is $A_{ref} = 249.91\ m^2$. We assume further a very small and and time-independent flight path angle γ.

How large is a) the re-entry mass of the Orbiter, b) the g-reduction due to the curved flight path in per cent, c) the re-entry weight of the Orbiter at sea level.

Problem 3.4 The X-38 flies at very small and time-independent flight path angle γ at 80.0 km altitude. We assume its mass to be $m = 8{,}618.0\ kg$, the reference area is $A_{ref} = 21.67\ m^2$, the lift coefficient is, Mach number independent, $C_L = 0.52$. How large is the flight Mach number M_∞?

Problem 3.5 Assume that the aerodynamic data base contains for the boattailing region the mean pressure coefficient $c_{p_{w,2.a}} = 0.63$ for the angle of attack $\alpha = 40°$, Table 3.5. In flight the actual mean pressure coefficient is $c_{p_{w,2.c}} = 0.55$, Table 3.5. How large is the delta force, if flight Mach number and altitude are $M_\infty = 24$, $H_\infty = 70.0\ km$? How large are $\triangle M$ and $\triangle C_M$?

The boattailing region of the Orbiter has an area of approximately $A_{bt} = 5.9$ m x 23.0 m = 135.7 m^2. Its mean location is at $x/L_{ref} = 0.91$. The center of gravity is assumed to be located axially at $x/L_{ref}|_{cg} = 0.65$, $L_{ref} = 32.7736\ m$, [14]. Neglect the z location. The reference area is $A_{ref} = 249.91\ m^2$. The total plan area, Fig. 3.40, is $A_{plan} = 361.3\ m^2$. The reference length for the pitching moment is the mean wing chord $\bar{c} = 12.06\ m$. Neglect contributions of the axial force to the pitching moment. Assume that the delta force acts orthogonal to the x axis of the flight vehicle.

Problem 3.6 Lift and drag coefficient of the Space Shuttle Orbiter at $\alpha =$ 40° and above $M_\infty = 20$ are Mach number independent $C_L = 0.884$ and $C_D =$ 0.821, [20].

a) How large are the normal and the axial force coefficient C_N and C_X? b) Compare the normal force coefficient to very simple estimates made with the help of the data in Tables 3.4 and 3.5, for the cases 2.a and 2.c. c) Discuss the results.

Problem 3.7 Around the forward stagnation point of the Space Shuttle Orbiter we have above $M_\infty = 20$ Mach number independent for perfect gas $c_{p_{w,perfect}} \approx c_{p_{w,perfect,stag}}$ and for real gas $c_{p_{w,real,stag}} \approx 1.94$ (point No. 7 in Fig. 3.5).

The real-gas pressure coefficient leads to a pitching moment increment. How large is at $M_\infty = 22.1$ and $H = 70.0\ km$ a) $c_{p_{w,perfect,stag}}$ (what is the assumption?), b) the increment of the normal force coefficient and c) of the pitching moment coefficient. Assume that the pressures act in normal force direction (no axial force contribution to the pitching moment) on a surface portion $\triangle A_{nose}$ $= 1.0\ m^2$ with a lever arm to the center-of-gravity of $x_l = 21.3\ m$.

Problem 3.8 At $M_\infty = 22.1$ and $H = 70.0\ km$ the Mach number independent moment coefficient is $C_M =$ - 0.041 and the normal force coefficient $C_N = 1.205$. The x-location of the center-of-gravity is $x_{cog} = 0.65\ L_{ref}$, $L_{ref} = 32.7736\ m$. The plan area is $A_{plan} = 361.3\ m^2$, the boattailing area $A_{bt} = 135.7\ m^2$, and the remaining main surface area $A_{ms} = 225.6\ m^2$. The reference area is $A_{ref} = 249.91\ m^2$, the moment reference length $\overline{c} = 12.06\ m$, and the dynamic pressure $q_\infty = 1.79\ kPa$. The boattailing force N_{bt} is assumed to act at $x_{bt}/L_{ref} = 0.91$, Table 3.5.

Assume that the axial force does not contribute to the pitching moment and that the forces on the main surface and on the boattailing surface act in the direction of the normal force. In the nomenclature of Fig. 3.42 then $N \equiv F_{res}$ et cetera. Assume $F_{bt} = A_{bt}\ q_\infty\ c_{p_{w,2.a}}$ with $c_{p_{w,2.a}} = 0.63$, Table 3.5. How large are a) F_{bt}, F_{ms}, x_{cp}, x_{ms}, b) F'_{bt}, x'_{cp}, if $c_{p_{w,2.c}} = 0.55$, Table 3.5, c) the forward shift of the center of pressure $\triangle\, x_{cp}$?

References

1. Hirschel, E.H.: Basics of Aerothermodynamics. Progress in Astronautics and Aeronautics, AIAA, Reston, Va, vol. 204. Springer, Heidelberg (2004)
2. Hirschel, E.H.: Hypersonic Aerodynamics. In: Space Course 1993, vol. 1, pp. 2-1–2-17. Technical University München, Germany (1993)
3. Sacher, P.W., Kunz, R., Staudacher, W.: The German Hypersonic Experimental Aircraft Concept. In: 2nd AIAA International Aerospaceplane Conference, Orlando, Florida (1990)
4. Hirschel, E.H., Grallert, H., Lafon, J., Rapuc, M.: Acquisition of an Aerodynamic Data Base by Means of a Winged Experimental Reentry Vehicle. Zeitschrift für Flugwissenschaften und Weltraumforschung (ZFW) 16(1), 15–27 (1992)
5. Abgrall, R., Desideri, J.-A., Glowinski, R., Mallet, M., Periaux, J. (eds.): Hypersonic Flows for Reentry Problems, vol. III. Springer, Berlin (1992)
6. Lyubimov, A.N., Rusanow, V.V.: Gas Flows Past Blunt Bodies. NASA TT-F-714, Part II (1973)
7. Weilmuenster, K.J., Gnoffo, P.A., Greene, F.A.: Navier-Stokes Simulations of Orbiter Aerodynamic Characteristics. In: Throckmorton, D.A. (ed.) Orbiter Experiments (OEX) Aerothermodynamics Symposium. NASA CP-3248, Part 1, pp. 447–494 (1995)
8. Edney, B.: Anomalous Heat Transfer and Pressure Distributions on Blunt Bodies at Hypersonic Speeds in the Presence of an Impinging Shock. FFA Rep. 115 (1968)
9. Brück, S., Radespiel, R.: Navier-Stokes Solutions for the Flow Around the HALIS Configuration – F4 Wind Tunnel Conditions – DLR Internal Report, IB 129 - 96/2 (1996)
10. Hartmann, G., Menne, S.: Winged Aerothermodynamic Activities. MSTP Report, H-TN-E33.3-004-DASA, Dasa, München/Ottobrunn, Germany (1996)

11. Riedelbauch, S.: Aerothermodynamische Eigenschaften von Hyperschallströmungen über strahlungsadiabate Oberflächen (Aerothermodynamic Properties of Hypersonic Flows past Radiation-Cooled Surfaces). Doctoral Thesis, Technische Universität München, Germany (1991) (Also DLR-FB 91-42, 1991)
12. Weiland, C.: Synthesis of Results for Problem VII, Delta Wing. In: Abgrall, R., Desideri, J.-A., Glowinski, R., Mallet, M., Periaux, J. (eds.) Hypersonic Flows for Reentry Problems, vol. III, pp. 1014–1027. Springer, Heidelberg (1992)
13. Liepmann, H.W., Roshko, A.: Elements of Gasdynamics. Dover Publications, Mineola (2002)
14. Griffith, B.F., Maus, J.R., Best, J.T.: Explanation of the Hypersonic Longitudinal Stability Problem – Lessons Learned. In: Arrington, J.P., Jones, J.J. (eds.) Shuttle Performance: Lessons Learned. NASA CP-2283, Part 1, pp. 347–380 (1983)
15. Becker, J.V.: The X-15 Project. Astronautics and Aeronautics 2, 52–61 (1964)
16. McLellan, C.H.: A Method for Increasing the Effectiveness of Stabilizing Surfaces at High Supersonic Mach Numbers. NACA RM L54F21 (1954)
17. Arrington, J.P., Jones, J.J. (eds.): Shuttle Performance: Lessons Learned. NASA CP-2283 (1983)
18. Throckmorton, D.A. (ed.): Orbiter Experiments (OEX) Aerothermodynamics Symposium. NASA CP-3248 (1995)
19. Martin, L.: Aerodynamic Coefficient Measurements, Body Flap Efficiency and Oil Flow Visualization on an Orbiter Model at Mach 10 in the ONERA S4 Wind Tunnel. ONERA Report No. 8907 GY 400G (1995)
20. N.N.: Aerodynamic Design Data Book – Orbiter Vehicle STS-1. Rockwell International, USA (1980)
21. Iliff, K.W., Shafer, M.F.: Extraction of Stability and Control Derivatives from Orbiter Flight Data. In: Throckmorton, D.A. (ed.) Orbiter Experiments (OEX) Aerothermodynamics Symposium. NASA CP-3248, Part 1, pp. 299–344 (1995)
22. Hartmann, G., Menne, S., Schröder, W.: Uncertainties Analysis/Critical Points. HERMES Report, H-NT-1-0329-DASA, Dasa, Mňchen/Ottobrunn, Germany (1994)
23. Courty, J.C., Rapuc, M., Vancamberg, P.: Aerodynamic and Thermal Data Bases. HERMES Report, H-NT-1-1206-AMD, Aviation M. Dassault, St. Cloud, France (1991)
24. N.N.: HOPE-X Data Base. Industrial communication, European Aeronautic Defence and Space Company, EADS/Japanese Space Agency, NASDA (1998)
25. Tsujimoto, T., Sakamoto, Y., Akimoto, T., Kouchiyama, J., Ishimoto, S., Aoki, T.: Aerodynamic Characteristics of HOPE-X Configuration with Twin Tails. AIAA-Paper 2001-1827 (2001)
26. Pamadi, B.N., Brauckmann, G.J., Ruth, M.J., Furhmann, H.D.: Aerodynamic Characteristics, Database Development and Flight Simulation of the X-34 Vehicle. AIAA-Paper 2000-0900 (2000)
27. `http://www.dfrc.nasa.gov/gallery`
28. Pamadi, B.N., Brauckmann, G.J.: Aerodynamic Characteristics and Development of the Aerodynamic Database of the X-34 Reusable Launch Vehicle. In: Proc. 1st Int. Symp. on Atmospheric Re-entry Vehicles and Systems, Arcachon, France (1999)
29. Brauckmann, G.J.: X-34 Vehicle Aerodynamic Characteristics. J. of Spacecraft and Rockets 36(2), 229–239 (1999)

30. Tarfeld, F.: Measurement of Direct Dynamic Derivatives with the Forced-Oscillation Technique on the Reentry Vehicle X-38 in Supersonic Flow. Deutsches Zentrum für Luft- und Raumfahrt DLR, TETRA Programme, TET-DLR-21-TN-3104 (2001)
31. Labbe, S.G., Perez, L.F., Fitzgerald, S., Longo, J.M.A., Molina, R., Rapuc, M.: X-38 Integrated Aero- and Aerothermodynamic Activities. Aerospace Science and Technology 3, 485–493 (1999)
32. N.N.: X-38 Data Base. Industrial communication, Dassault Aviation – NASA – European Aeronautic Defence and Space Company, EADS (1999)
33. Behr, R., Weber, C.: Aerothermodynamics – Euler Computations. X-CRV Rep., HT-TN-002/2000-DASA, Dasa, München/Ottobrunn, Germany (2000)
34. Aiello, M., Stojanowski, M.: X-38 CRV, S3MA Windtunnel Test Results. X-CRV Rep., DGT No. 73442, Aviation M. Dassault, St. Cloud, France (1998)
35. Weiland, C.: X-38 CRV, S4MA Windtunnel Test Results. Dasa, X-CRV Report, HT-TN-001/99-DASA, Dasa, München/Ottobrunn, Germany (1999)
36. Görgen, J.: CFD Analysis of X-38 Free Flight. TETRA Programme, TET-DASA-21-TN-2401, Dasa, München/Ottobrunn, Germany (1999)
37. Daimler-Benz Aerospace Space Infrastructure, FESTIP System Study Proceedings. FFSC-15 Suborbital HTO-HL System Concept Family. EADS, München/Ottobrunn, Germany (1999)
38. N.N.: PHOENIX Data Base. Internal industrial communication, EADS, München/Ottobrunn, Germany (2004)
39. Häberle, J.: Einfluss heisser Oberflächen auf aerothermodynamische Flugeigenschaften von HOPPER/PHOENIX (Influence of Hot Surfaces on Aerothermodynamic Flight Properties of HOPPER/PHOENIX). Diploma Thesis, Institut für Aerodynamik und Gasdynamik, Universität Stuttgart, Germany (2004)
40. Borrelli, S., Marini, M.: The Technology Program in Aerothermodynamics for PRORA-USV. ESA-SP-487, pp. 37–48 (2002)
41. Rufolo, G.C., Roncioni, P., Marini, M., Votta, R., Palazzo, S.: Experimental and Numerical Aerodynamic Integration and Aerodatabase Development for the PRORA-USV-FTB-1 Reusable Vehicle. AIAA-Paper 2006-8031 (2006)
42. Behr, R.: CFD Computations. Private communications, EADS, München/ Ottobrunn, Germany (2007)
43. Trella, M.: Introduction to the Hypersonic Phenomena of HERMES. In: Bertin, J.J., Glowinski, R., Periaux, J. (eds.) Hypersonics. Defining the Hypersonic Environment, vol. 1, pp. 67–91. Birkhäuser, Boston (1989)
44. Etkin, B., Reid, L.D.: Dynamics of Flight Mechanics: Performance, Stability and Control. John Wiley & Sons, New York (2000)
45. Hoey, R.G.: AFFTC Overview of Orbiter-Reentry Flight-Test Results. In: Arrington, J.P., Jones, J.J. (eds.) Shuttle Performance: Lessons Learned. NASA CP-2283, Part 2, pp. 1303–1334 (1983)
46. Woods, W.C., Watson, R.D.: Shuttle Orbiter Aerodynamics – Comparison Between Hypersonic Ground-Facility Results and STS-1 Flight-Derived Results. In: Throckmorton, D.A. (ed.) Orbiter Experiments (OEX) Aerothermodynamics Symposium. NASA CP-3248, Part 1, pp. 371–409 (1995)
47. Williams, S.D.: Columbia, the First Five Flights Entry Heating Data Series, an Overview, vol. 1, NASA CR-171 820 (1984)

48. Romere, P.O.: Orbiter (Pre STS-1) Aerodynamic Design Data Book Development and Methodology. In: Throckmorton, D.A. (ed.) Orbiter Experiments (OEX) Aerothermodynamics Symposium. NASA CP-3248, Part 1, pp. 249–280 (1995)
49. Woods, W.C., Arrington, J.P., Hamilton II, H.H.: A Review of Preflight Estimates of Real-Gas Effects on Space Shuttle Aerodynamic Characteristics. In: Arrington, J.P., Jones, J.J. (eds.) Shuttle Performance: Lessons Learned. NASA CP-2283, Part 1, pp. 309–346 (1983)
50. Neyland, V.Y.: Air Dissociation Effects on Aerodynamic Characteristics of an Aerospace Plane. Journal of Aircraft 30(4), 547–549 (1993)
51. Weilmuenster, K.J., Gnoffo, P.A., Greene, F.A.: Navier-Stokes Simulations of Orbiter Aerodynamic Characteristics Including Pitch Trim and Bodyflap. Journal of Spacecraft and Rockets 31(5), 355–366 (1994)
52. Paulson Jr., J.W., Brauckmann, G.J.: Recent Ground-Facility Simulations of Space Shuttle Orbiter Aerodynamics. In: Throckmorton, D.A. (ed.) Orbiter Experiments (OEX) Aerothermodynamics Symposium. NASA CP-3248, Part 1, pp. 411–445 (1995)
53. Prabhu, D.K., Papadopoulos, P.E., Davies, C.B., Wright, M.J.B., McDaniel, R.D., Venkatapathy, E., Wercinski, P.F.: Shuttle Orbiter Contingency Abort Aerodynamics, II: Real-Gas Effects and High Angles of Attack. AIAA-Paper 2003-1248 (2003)
54. Koppenwallner, G.: Low Reynolds Number Influence on Aerodynamic Performance of Hypersonic Vehicles. AGARD-CP-428, pp. 11-1–11-14 (1987)
55. Brauckmann, G.J., Paulson Jr., J.W., Weilmuenster, K.J.: Experimental and Computational Analysis of Shuttle Orbiter Hypersonic Trim Anomaly. Journal of Spacecraft and Rockets 32(5), 758–764 (1995)
56. Goodrich, W.D., Derry, S.M., Bertin, J.J.: Shuttle Orbiter Boundary-Layer Transition: A Comparison of Flight and Wind-Tunnel Data. AIAA-Paper 83-0485 (1983)
57. Boeing, Human Space Flight and Exploration, Huntington Beach, CA. Operational Aerodynamic Data Book. Boeing Document STS85-0118 CHG 9 (2000)
58. Hayes, W.D., Probstein, R.F.: Hypersonic Flow Theory. Inviscid Flows, vol. 1. Academic Press, New York (1966)
59. Oswatitsch, K.: Ähnlichkeitsgesetze für Hyperschallströmung. ZAMP II, 249–264 (1951); Similarity Laws for Hypersonic Flow. Royal Institute of Technology, Stockholm, Sweden, KTH-AERO TN 16 (1950)
60. Wüthrich, S., Sawley, M.L., Perruchoud, G.: The Coupled Euler/ Boundary-Layer Method as a Design Tool for Hypersonic Re-Entry Vehicles. Zeitschrift für Flugwissenschaften und Weltraumforschung (ZFW) 20(3), 137–144 (1996)
61. Adams, J.C., Martindale, W.R., Mayne, A.W., Marchand, E.O.: Real Gas Scale Effects on Hypersonic Laminar Boundary-Layer Parameters Including Effects of Entropy-Layer Swallowing. AIAA-Paper 76-358 (1976)
62. Weiland, C.: Zwei- und dreidimensionale Umströmungen stumpfer Körper unter Berücksichtigung schallnaher Überschallströmungen. Zeitschrift für Flugwissenschaften und Weltraumforschung (ZFW) 24(3), 237–245 (1976); Translation Two and Three-Dimensional Flows Past Blunt Bodies with Special Regard of Supersonic Flow Close to M = 1. NASA TTF-17406 (1977)
63. Pfitzner, M., Weiland, C.: 3-D Euler Solutions for Hypersonic Mach Numbers. AGARD-CP-428, pp. 22-1–22-14 (1987)
64. Kliche, D.: Personal communication. München, Germany (2008)
65. Ames Research Staff. Equations, Tables, and Charts for Compressible Flow. NACA R-1135 (1953)

4 Aerothermodynamic Design Problems of Winged Airbreathing Vehicles

To date a fully reusable space transportation system does not exist with the capability to take-off horizontally or vertically and to land horizontally. As we have learned from Chapter 3, the orbiter vehicle of the Space Shuttle system is launched vertically like a rocket with the support of solid rocket boosters attached to the expendable tank. The re-entry process in Earth's atmosphere consists of a gliding un-powered flight and a horizontal landing on a conventional runway.

All the capsules, Chapter 5, which have transported men to and from space, are single-use vehicles, launched vertically on top of rockets. The landing, either on sea or on ground, is usually performed with the aid of a parachute system.

Although the mentioned systems represent reliable means of space transportation, the cost of delivering payloads into space remains much too high and a launch on demand is not possible. Consequently, at the end of the 1980s and during the 1990s numerous activities all over the world did aim for a development of fully reusable space transportation systems, conducting horizontal take-off and landing, allowing also for a launch on demand.

Conceptual design studies have covered single-stage-to-orbit (SSTO) and two-stage-to-orbit (TSTO) systems. SSTO concepts were considered in the United States with the National Aerospace Plane (NASP), in Great Britain with the Horizontal Take-Off and Landing System (HOTOL) and in Japan with the Aerospace Plane of the National Aerospace Laboratory.

In Europe TSTO systems were favored. The most advanced concepts considered a lower stage belonging to a new generation of hypersonic aircraft, with conventional take-off and landing capabilities. The propulsion systems, that enable the lower stages to operate with hypersonic speeds at flight levels high above the maximum ceiling of turbo jet engines, are either integrated turbojet/ramjet or turbojet/scramjet systems. One of these concepts was the German SÄNGER system with turbojet/ramjet propulsion, another one the French STAR-H system with turbojet/scramjet propulsion. All these concept and system studies were terminated in the mid 1990s.

In order to identify the important aerothermodynamic phenomena, design problems, and ground-simulation issues, we don't distinguish between TSTO and SSTO space transportation systems, but between flight vehicle classes. Hypersonic winged airbreathing vehicles in our sense are cruise and acceleration vehicles (CAV) as well as ascent and re-entry vehicles (ARV), these in view of their ascent flight mission. A CAV-type vehicle can be the lower stage of a TSTO system, or a hypersonic aircraft. In contrast to that, a ARV-type vehicle would be a SSTO-system, or a hypersonic aircraft, too. Actually, almost all of the vehicles belonging to these two classes are concepts, studied and designed to different depth. Flight experience and data are not available.

In this chapter therefore we cannot make use of flight knowledge and data like in case of RV-W- and RV-NW-type flight vehicles in Chapters 3 and 5. First we give an overview of aerothermodynamic phenomena, look at a major simulation problem, viz. that of viscous effects, and note particular aerothermodynamic trends. It all concerns CAV-type vehicles, with not strongly defined boundaries to ARV-type vehicles. This is followed by a discussion of selected flight parameters and aerodynamic coefficients of the TSTO-system SÄNGER, the reference concept of a recent German hypersonics technology programme. Since hypersonic airbreathing flight vehicles are highly coupled lift and propulsion systems, two bookkeeping schemes of aerothermodynamic and propulsion forces are then treated and general configuration aspects are discussed, with emphasis on forebody aerothermodynamics. In this context aerothermodynamic inlet ramp-flow issues are discussed, too. Finally we provide a detailed overview of waverider configurations and the related aerothermodynamic issues.

4.1 Overview of Aerothermodynamic Issues of Cruise and Acceleration Vehicles

4.1.1 Aerothermodynamic Phenomena

Regarding CAV-type flight vehicles, we often find in the literature the acceleration vehicle discussed as a vehicle for orbital access, e.g., [1]. In our diction such a vehicle would be an airbreathing ARV-type vehicle, with the contradicting design goals of small vehicle drag for ascent, and a large one for re-entry. Here we understand the acceleration vehicle mainly as lower stage of a TSTO system, which performs an ascent to upper stage separation, and then returns to Earth. However, many of the problems we address here are the same or similar for CAV- and airbreathing ARV-type vehicles, if we consider only the ascent mission of the latter. The CAV-type vehicle may employ ramjet and/or scramjet propulsion. The ARV-type vehicle operates beyond the ramjet propulsion domain and could employ scramjet propulsion up to $M_\infty \approx 12$ and then rocket propulsion. Of course turbojet propulsion is needed for the low-speed part of flight.

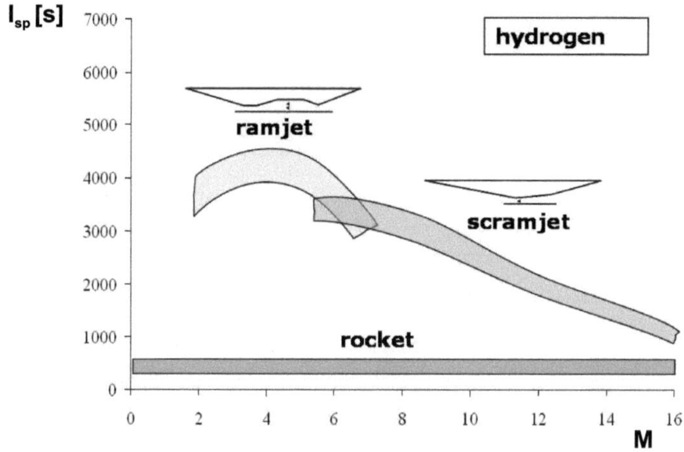

Fig. 4.1. Specific impulse I_{sp} of different (hydrogen-fuel) propulsion systems as function of the flight Mach number M, [2].

The use of the different propulsion modes for an ARV-type vehicle demands a closer look, [2]. When discussing the issue airbreathing propulsion versus rocket propulsion, usually the specific impulse I_{sp}, i. e. the impulse gained per unit mass of propellant, is considered. The larger it is, the less propellant is needed to obtain a given amount of momentum. The advantage of airbreathing propulsion over rocket propulsion is that it uses the atmospheric oxygen for the combustion process. This results in a considerably larger I_{sp}, because the oxygen does not enter the fuel balance. In Fig. 4.1 we see[1] the typical large specific impulse of the ramjet and the smaller one of the scramjet. Both impulses are decreasing with increasing flight Mach number. Both are much larger than that of the rocket engine which, however, does not depend on the Mach number.

Of course, the performance of a propulsion system in a flight vehicle depends not on the fuel consumption per time unit, but on the integrated consumption over the respective mission duration. On the other hand, the acceleration capability a of a vehicle with weight W_v is directly proportional to the thrust T and inversely proportional to the mass m_v of the vehicle:

$$a \sim \frac{T - B}{m_v}, \tag{4.1}$$

where B is a function of vehicle drag, weight, and the flight path angle.

The engine weight W_e (or mass) of course is a part of the vehicle weight (or mass). The ideal propulsion system would be a small light engine with

[1] The data underlying the curves in Figs. 4.1 and 4.2 are the result of a literature search made in the frame of the thesis [2]. The widths of the curves represent the data spread of airbreathing and rocket engines all with hydrogen as fuel.

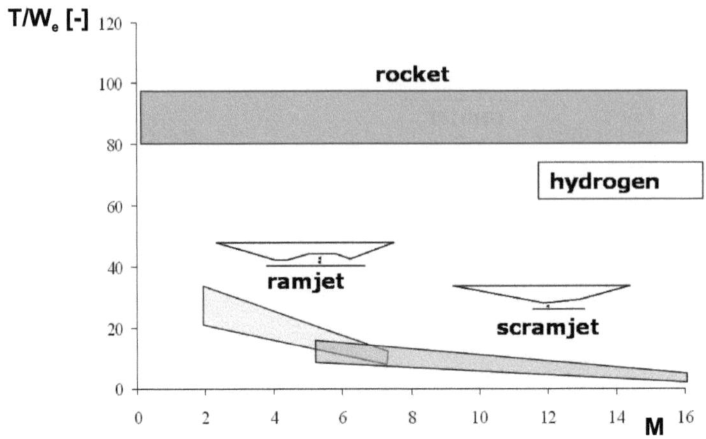

Fig. 4.2. Thrust-to-weight ratio T/W_e of different (hydrogen-fuel) propulsion systems as function of the flight Mach number M, [2].

large thrust and little fuel consumption. Hence a characteristic measure for a propulsion system is also the thrust-to-weight ratio T/W_e, Fig. 4.2. Here we see a large advantage of rocket propulsion over ramjet and scramjet propulsion. Rocket propulsion in contrast to airbreathing propulsion requires a smaller integration effort into the vehicle structure and, in particular, needs no inlet (in a broader sense no airframe/propulson integration), hence builds specifically much lighter.

However, it is the product of specific impulse and thrust-to-weight ratio, the specific performance $T\,I_{sp}/W_e$ of a propulsion system, which shows its relevance for a flight vehicle with a pure acceleration mission, viz. a ARV-type vehicle, Fig. 4.3.

In Fig. 4.3 – keep in mind that it is based on hydrogen as fuel – we see that airbreathing propulsion has, due to its large specific impulse, an advantage over rocket propulsion, but only in a limited flight Mach number range. Where to switch from airbreathing to rocket propulsion cannot be determined from this figure, because the characteristics of the whole vehicle have strong influence on the performance.

We conclude that a higher acceleration capability of a propulsion system – in this case rocket propulsion – at very large flight Mach numbers can more than compensate its smaller specific impulse. The achievement of a sufficiently positive thrust-minus-drag balance with the highly integrated lift and airbreathing propulsion system of ARV-type flight vehicles is a major factor in this consideration.

The situation is different, if we consider CAV-type vehicles as hypersonic passenger transport aircraft. Such vehicles have potentially – for long-distance flight – a high productivity (seat-miles per day). In the frame of the former German Hypersonics Technology Programme a $M_\infty = 4.4$ vehicle for 256

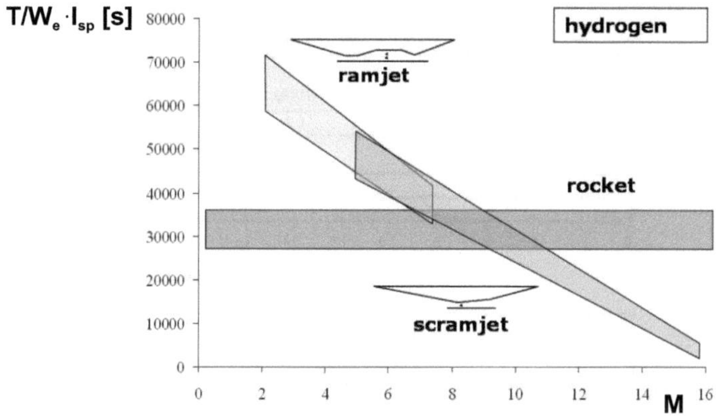

Fig. 4.3. Acceleration missions: specific performance $T/W_e{\cdot}I_{sp}$ of different (hydrogen-fuel) propulsion systems as function of the flight Mach number M, [2].

passengers (HST-230) was derived from the lower stage of the TSTO-system SÄNGER, [3]. In the recent EU-sponsored LAPCAT study, coordinated by ESA-ESTEC, propulsion concepts for $M_\infty = 4$ to 8 passenger transport aircraft were studied, [4]. For vehicles of this kind a high acceleration capability is not necessarily the driving factor, but a problem can be the degradation of passenger comfort due to the acceleration.

We show this with data from a forerunner study of the German Hypersonics Technology Programme, [5], which were published in [6]. Four reference concepts of would-be supersonic and hypersonic aircraft, LK1 to LK4, were studied in order to define key technologies. The nominal flight Mach numbers were $M_\infty = 2.2$, 3, 5, and 12.

In Fig. 4.4 for the reference concepts estimates of acceleration/deceleration distances and times of $M_\infty = 5$, 8, 12 flight, and the g-reduction with speed are shown. With assumed limits of sustained, medium to long-time acceleration ($0.5g$) and deceleration (- $0.125g$ to - $0.2g$) values thought to be acceptable for ordinary untrained passengers, we see that $M_\infty = 5$ flight possibly is the upper limit. For $M_\infty = 12$ we find rather short real nominal flight Mach number distances and times as well as a large g-reduction.

Back now to the aerothermodynamic phenomena on CAV- and ARV-type vehicles. Allowing for not strongly defined boundaries between the two vehicle classes, we give an overview of the aerothermodynamic phenomena encountered on them. We consider only the outer (external) flow path around the vehicle, flying usually at small angle of attack, Fig. 4.5, although the inner flow path is very important as well in order to achieve a positive thrust balance.

Thermal loads call for a blunt vehicle nose. Drag minimization reasons call for a small nose radius. If the latter is too small for an effective surface-radiation cooling, then active cooling becomes necessary at the vehicle nose. This might be the case for ARV-type vehicles because of their higher velocities.

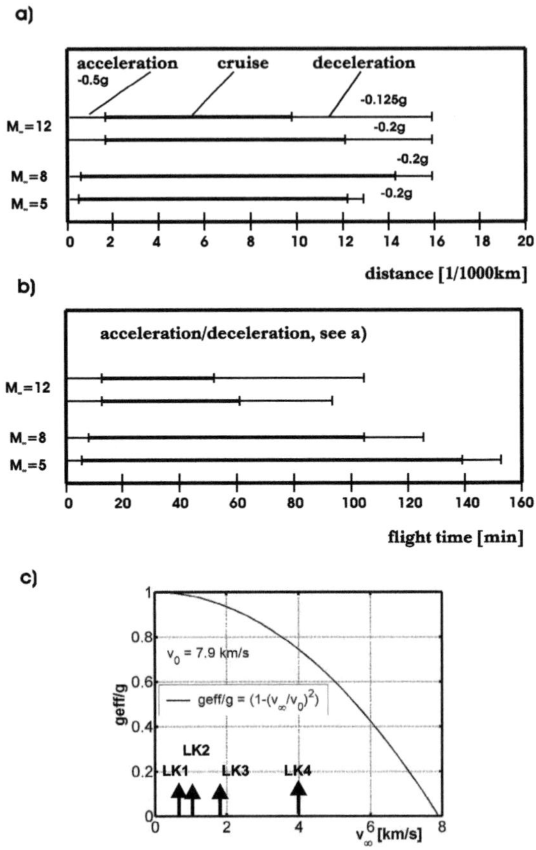

Fig. 4.4. CAV-type vehicle as hypersonic passenger transport aircraft: degradation of passenger comfort, a) acceleration/deceleration distances, acceleration/deceleration times, c) g-reduction with speed, [6].

The blunt nose gives rise to a detached bow shock and consequently also to an entropy layer on both the lower and the upper side of the airframe. The occurrence of high-temperature real-gas effects depends on the flight speed. At lower hypersonic flight Mach numbers ($M_\infty \approx 5$ to 6) stagnation point region, inlet, and boundary layer flow will be stronger affected than other flow domains. At larger Mach numbers high-temperature real-gas effects are non-negligible everywhere.

The lower side of the vehicle usually is a pre-compression surface for the inlet, Sub-Section 4.5.1. This side therefore has a larger inclination against the free-stream than the upper side, hence here in general the unit Reynolds number[2] is larger, Fig. 6.18. This means different boundary layer properties,

[2] The unit Reynolds number, composed of the boundary layer edge flow properties, $R_e^u = \rho_e u_e/\mu_e$, is an important parameter, which governs boundary layer

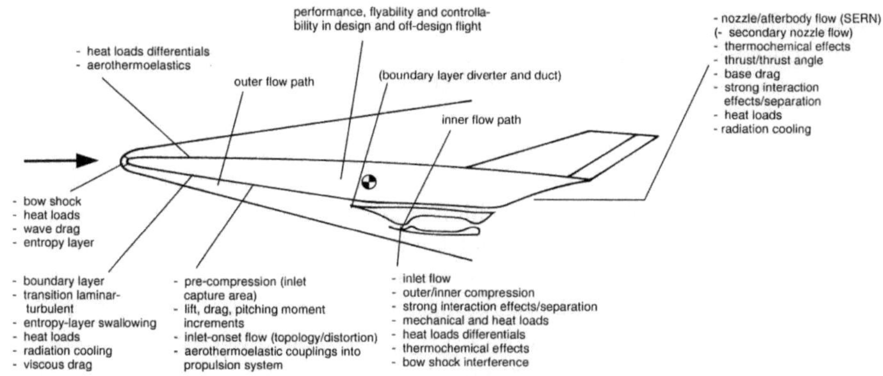

Fig. 4.5. CAV and ARV-type flight vehicles: configurational aspects and aerothermodynamic phenomena, [7].

and on the lower side smaller characteristic thicknesses of both the laminar and the turbulent boundary layer are observed, Section 9.2. Hence the thermal state is different on the lower and the upper side of the airframe, notably the surface temperatures are larger on the lower side. As an example see Fig. 4.6 on page 134. This situation may, at high flight speeds, require active cooling also on part of the lower side of the forebody. Of course thermal surface effects are different and will differently affect laminar-turbulent transition (at flight altitudes below 40.0 to 60.0 km).

While the lower side of the forebody has a positive inclination angle $\overline{\alpha}$ against the free-stream in order to allow for pre-compression, the upper side should have an inclination angle close to zero[3] in order to keep wave drag increments small. A small negative angle of the upper surface would reduce the unit Reynolds number further, increasing the characteristic boundary layer thicknesses and hence decreasing surface temperatures further. However, it is a matter of the internal layout and general trade-offs during the vehicle design, which will govern the contour of the upper side. We may well see a positive angle of attack of the upper side. This all holds also for the side contour of the airframe.

The lower side of the forebody should have a layout such that the inlet onset flow has a proper topology, Sub-Section 4.5.2. The degree of pre-compression depends on the incidence of the lower side against the free-stream. It, of course, influences directly not only the size of the capture area of the inlet, but also lift, drag, and pitching moment of the whole vehicle.

The propulsion system package (inlet, actively cooled turbojet/ramjet/scramjet engine, nozzle throat) must lie inside the bow shock for operational reasons already of the inlet, Sub-Section 4.5.5, but also in order to avoid

thicknesses, the thermal state of the surface, and the wall shear stress, Chapter 10, for more details see [8].

[3] In inviscid flow a surface with $\overline{\alpha} = 0°$ is called a free-stream surface.

wave drag increments. This in general asks for an extreme aft location of the propulsion system, and hence a disproportional long forebody. For the aerothermoelastic problems associated with a physically large forebody see Sub-Section 4.5.4.

The external inlet flow as ramp compression flow is characterized by strong interaction phenomena (shock/boundary layer interaction and corner flow), Sub-Section 4.5.5, see also Chapter 6, high-temperature real-gas effects and large mechanical (pressure) and thermal loads in the presence of radiation cooling. A boundary layer diverter is necessary for the turbojet propulsion mode, probably also for the ramjet/scramjet modes, although a recent study, [9], indicates a performance increase of a scramjet engine due to inlet boundary layer (not yet forebody boundary layer) ingestion[4].

The situation is similar at the vehicle's base where the single expansion ramp nozzle (SERN) is situated. Here thermo-chemical rate effects have to be regarded in any case. Secondary nozzle flow is present if a boundary layer diverter is employed. In case of an ARV-type vehicle, rocket engines will have to be accommodated in that area, too.

CAV- and ARV-type vehicles, the latter only in the ascent mode, are drag sensitive. Aerodynamic and propulsion forces are strongly coupled. The net thrust is the – usually small – vectorial difference of large values, viz. the difference between the inlet drag, i. e. the flow momentum entering the inlet of the propulsion system and the momentum of the engine exhaust flow leaving the nozzle, plus forces on several external surfaces, see Sub-Sections 2.2.3 and 4.4.2. Of particular importance for CAV- and ARV-type vehicles is laminar-turbulent boundary layer transition, occurring at altitudes lower than 40.0 to 60.0 km, see Sub-Section, 4.1.2, and for the latter vehicles hypersonic viscous interaction and low-density effects at large altitudes.

Summarizing we note that the external flow path of CAV- and ARV-type flight vehicles:

- is viscous effect dominated,
- is governed especially by viscous thermal surface effects,
- is strongly affected by laminar-turbulent transition because the flight altitudes are in general lower than 40.0 to 60.0 km,
- has Mach number and surface properties effects on laminar-turbulent transition and on turbulent flow phenomena,
- has radiation cooled (CAV, ARV), and in case of ARV-type vehicles possibly also actively cooled vehicle surfaces,
- includes weak (CAV) and also strong (ARV) high-temperature real-gas effects,

[4] If this is not a principal problem, it is a trade-off matter: propulsion performance gain (net thrust increment) vs. boundary layer diverter drag, mass and volume of the diverter duct (hot boundary layer flow material), momentum recovery (possible secondary combustion chamber and nozzle integrated into the SERN).

- contains strong interaction phenomena (shock/shock and shock/boundary layer interaction, corner flow, boundary layer separation),
- has gap flow at trim and control surfaces,
- includes hypersonic viscous interaction and low-density effects (ARV).

4.1.2 Major Simulation Problem: Viscous Effects

In view of the drag sensitivity of the two vehicle types, surface-radiation cooling and state of the boundary layer (laminar or turbulent) are extremely important, however, hard to treat in ground-simulation, Section 9.3, see also the discussions in [8].

We demonstrate their impact on wall temperatures and skin friction with computed data for the forebody of the SÄNGER lower stage, [8], Figs. 4.6 and 4.7. The flight parameters are given in Table 4.1.

Table 4.1. Parameters of the SÄNGER forebody computation cases, [8]. L is the length of the forebody, see Fig. 4.7, H2K is a DLR hypersonic blow-down tunnel.

Case	M_∞	H $[km]$	T_∞ $[K]$	T_w [K]	Re^u_∞ $[m^{-1}]$	L $[m]$	α [°]
Flight situation	6.8	33.0	231.5	*variable*	$1.5 \cdot 10^6$	55.0	6.0
Wind tunnel (H2K) sit.	6.8	-	61.0	300.0	$8.7 \cdot 10^6$	0.344	6.0

Fig. 4.6 shows the influence of radiation cooling, boundary layer state (laminar or turbulent) and high-temperature real-gas effects on the wall temperature[5]. The recovery temperature is only weakly affected by the boundary layer state. Real-gas effects already play a stronger role. A very strong influence is seen, when radiation cooling is present. The radiation-adiabatic temperature of the surface is not constant, much below the recovery temperature, and strongly affected by the boundary layer state. Note also the considerable temperature difference between the lower and the upper side of the forebody. For laminar flow it is about 100.0 K, for turbulent flow about 200.0 to 250.0 K.

The skin friction behavior, for the lower side of the forebody only, is shown in Fig. 4.7. The major result is that it depends on the wall temperature, weakly for laminar, strongly for turbulent flow, [8]. It follows that, if the flow is turbulent, the surface should be flown as hot as possible, i. e., as the structure and material concept permits[6]. However, not much can be said today, how this would affect laminar-turbulent transition. The influence of the thermal state of the surface in this regard is not well understood.

We show in the following two figures, for further illustration, some viscous thermal surface effects as function of the wall temperature. With the help of

[5] For the detailed phenomenological discussion see [8].

[6] This can be achieved by a tailoring of the surface emissivity coefficient ε, [8].

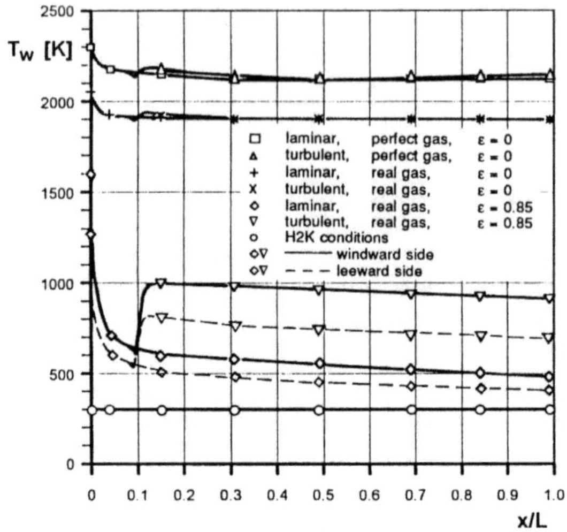

Fig. 4.6. SÄNGER lower stage, [8]: wall temperatures T_w at the lower and the upper symmetry line of the forebody in the flight and the wind-tunnel (H2K) situation (configuration see Fig. 4.7). Influence of boundary layer state, gas model, and radiation cooling.

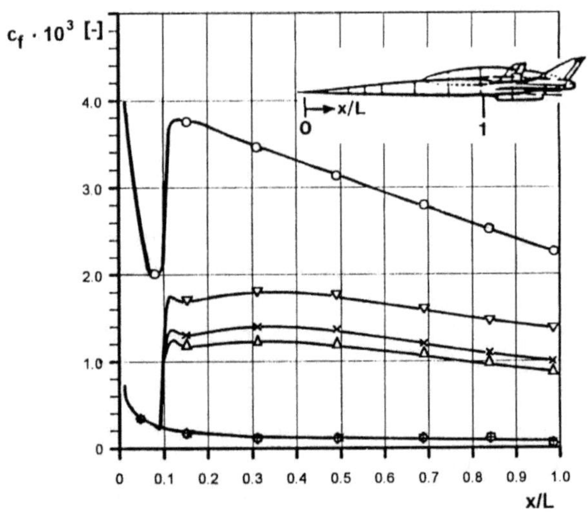

Fig. 4.7. SÄNGER lower stage, [8]: Skin-friction coefficients c_f at the lower symmetry line of the forebody in the flight and the wind-tunnel (H2K) situation (symbols see Fig. 4.6). Influence of boundary layer state, gas model, and radiation cooling.

the reference-temperature concept, see, e.g., [8], the skin-friction coefficient c_f, the heat flux in the gas at the wall q_{gw}, the boundary layer thickness δ, and the displacement thickness δ_1 are determined for both laminar and fully turbulent flow, in addition also the thickness of the viscous sub-layer of turbulent flow δ_{vs}, Section 10.

We have chosen the location $x/L = 1$, Fig. 4.7, at the symmetry line of the lower side of the forebody of the TSTO-system SÄNGER at $\alpha = 6°$. The forebody is approximated by a flat plate, the RHPM flyer. Its angle of attack is equal to the inclination angle of the lower side of the forebody, $\overline{\alpha} = 9.4$ °, Sub-Section 4.5.3. At this angle we get for perfect gas and $M_\infty = 6.8$ the shock angle $\theta = 16.006°$. In Table 4.2 the flow parameters are collected for the flight, '∞', and the RHPM boundary layer edge, 'e', condition. Note that the flight conditions differ somewhat from those given in Table 4.1, there the data of the original computations for Figs. 4.6 and 4.7 are quoted.

Table 4.2. Flow parameters of the SÄNGER forebody approximation at $\overline{\alpha} = 9.4$ and for $\gamma = 1.4$. '∞' denotes the flight, 'e' the boundary layer edge condition. R_e^u was determined with the Sutherland equation for the viscosity, perfect gas, the shock angle is $\theta = 16.0057°$.

Case	M	H [km]	T [K]	ρ [kg/m^3]	v [m/s]	Re^u [m^{-1}]	$q = 0.5\,\rho v^2$ [Pa]
∞	6.8	33.0	231.12	$1.172 \cdot 10^{-2}$	2,072.41	$1.619 \cdot 10^6$	25,167.92
e	5.22	33.0	367.13	$2.900 \cdot 10^{-2}$	2,005.40	$2.711 \cdot 10^6$	25,167.92

The results shown in Figs. 4.8 and 4.9 are for the (inlet ramp) location $x = L = 55.0\ m$. The Prandtl number chosen was $Pr = 0.737$, the recovery factors are $r_{lam} = \sqrt{Pr}$ and $r_{turb} = \sqrt[3]{Pr}$. The wall-temperature interval is $0 < T_w \leqq 2{,}500.0\ K$. In both figures, indicated by broken vertical lines, given are the total temperature T_t, the recovery temperatures $T_{r,lam}$ and $T_{r,turb}$, and the radiation-adiabatic temperatures $T_{ra,lam}$ and $T_{ra,turb}$. A look at Fig. 4.6 shows that the recovery temperatures $T_{r,lam}$ and $T_{r,turb}$ (adiabatic wall) as well as the radiation-adiabatic temperatures $T_{ra,lam}$ and $T_{ra,turb}$ (emissivity coefficient $\epsilon = 0.85$) are in good agreement with the values given there for $x/L = 1$.

In Fig. 4.8 we see in particular the strong dependence of the skin-friction coefficient c_f for turbulent flow on the wall temperature. The agreement is quite good with the values given for $x/L = 1$ in Fig. 4.7. The same holds for the coefficient for laminar flow, which, however depends only very weakly on T_w. We see the same general behavior[7] for q_{gw}: a strong dependence for turbulent, a weak one for laminar flow. Note the switches to negative signs for $T_w > T_r$.

The boundary layer thicknesses δ presented in Fig. 4.9 show also for turbulent flow a stronger dependence on T_w than for laminar flow. Really strong,

[7] This behavior in the frame of the reference-temperature approximation is discussed in detail in [8].

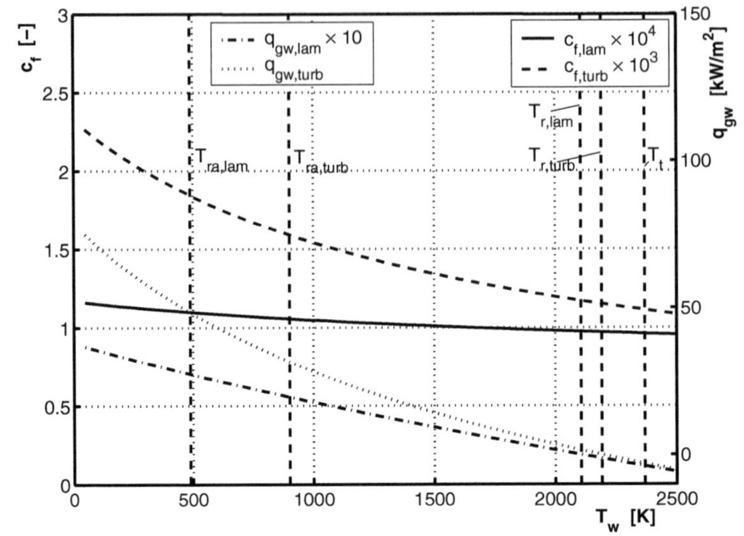

Fig. 4.8. SÄNGER lower stage, RHPM approximation of the lower side of the forebody at $\overline{\alpha} = 9.4°$, perfect gas: skin-friction coefficients $c_f = \tau_w/q_\infty$ and heat flux in the gas at the wall q_{gw} for laminar and fully turbulent flow as function of the wall temperature T_w. The location is $x = 55.0\ m$ at the lower symmetry line of the forebody.

however, is the dependence of the thickness of the viscous sub-layer, δ_{vs}. Indeed, it governs wall shear stress and heat flux in turbulent attached flow, because it is the characteristic thickness there, Section 9.2.

Important is the observation, that the boundary layer thickness in the turbulent case at the location of the first ramp of the inlet is $\delta \approx 0.68\ m$ for the radiation-adiabatic temperature. It is a measure of the height of the boundary layer diverter, if needed for ramjet operation[8]. The large value of δ, the gradient $d\delta/dT \approx 10^{-4}\ m/K$ and the impossibility to determine accurately and reliably the location of the laminar-turbulent transition location, which influences δ too, are the problem. Even if the boundary layer diverter is not needed (what amount of flow distortion is tolerated by a ramjet or scramjet engine?), the inlet ramps will operate in a thick boundary layer with unknown influence on their effectiveness.

Figures 4.6 to 4.9 show on the one hand the influence of surface-radiation cooling and the boundary layer state (laminar or turbulent) on thermal loads, viscous drag and boundary layer thicknesses, on the other hand they illustrate the simulation problem, which we have. The result of a computation for a cold-wall wind-tunnel model situation in Fig. 4.7 underlines the problem. Not forgotten should be the importance of the influence of surface properties

[8] For the boundary layer diverter height for turbojet operation at $M_\infty = 3.5$ see Problem 4.4.

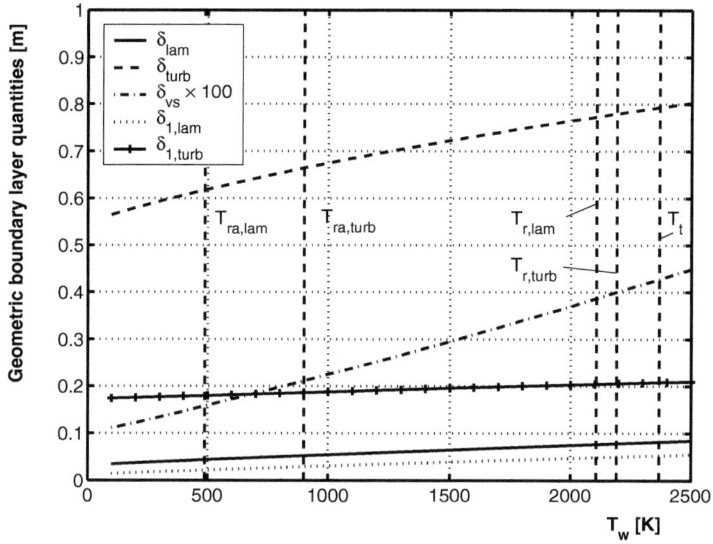

Fig. 4.9. SÄNGER lower stage, RHPM approximation of the lower side of the forebody at $\overline{\alpha} = 9.4°$, perfect gas: boundary layer thickness δ, displacement thickness δ_1 for laminar and fully turbulent flow and thickness of the viscous sub-layer δ_{vs} as function of the wall temperature T_w. The location is $x = 55.0\ m$ at the lower symmetry line of the forebody.

(necessary and permissible surface properties, Chapter 9). In view of transition and turbulent flow very smooth vehicle surfaces are needed in order to avoid increments of viscous drag and thermal loads.

We underline the importance of an accurate knowledge of the location and extent of the transition zone for CAV- and ARV-type flight vehicles, see also [8]. For the US National Aerospace Plane (NASP/X-30), [10], an ARV-type SSTO vehicle, it was reported that the uncertainty of the location of laminar-turbulent transition affects the take-off mass of the vehicle by a factor of two or more [11]. Strongly influenced by laminar-turbulent transition are mainly thermal loads (thus directly structure mass), viscous drag, and the engine inlet onset flow (height of the boundary layer diverter). For CAV-type vehicles the sensitivities may be smaller, but in any case of very large importance.

4.2 Particular Trends in CAV/ARV Aerothermodynamics

4.2.1 Stagnation Pressure

The behavior of the stagnation pressure coefficient of course is the same here as for RV-W-type vehicles. In the ramjet propulsion domain, up to $M_\infty \approx 6$, we have still a notable dependence on the flight Mach number, Fig. 3.4, but no more in the scramjet domain, up to $M_\infty \approx 12$.

4.2.2 Topology of the Forebody Windward Side Velocity Field

Other than the windward side of a RV-W-type vehicle the windward side of a CAV-type vehicle not necessarily must be predominantly flat. Fuselage and wing can, but must not be combined to have a plane lower side.

However, the lower side of the vehicles forebody must be flat in order to achieve an optimal onset flow geometry for the ramps of the inlet, Section 4.5. For a conical/semi-conical forebody we have an analogous demand.

Regardless whether we have a "pointed" or a "broad" blunt nose of the flat forebody, Sub-Section 4.5.2, we have basically the same topology of the velocity field as we have it at the flat windward side of a RV-W-type vehicle.

Downstream of the nose area, we find on the flat part of the forebody a two-dimensional (plane) flow between two primary attachment lines, see the schematic in Fig. 4.40. This assures optimum performance of the inlet and the engine, in the case of the flat forebody for a larger Mach number and angle of attack range than for a conical/semi-conical forebody.

Regarding the onset flow of trim and control surfaces located at the wing trailing edges, we note that is should be fully or at least approximately two-dimensional (optimal onset flow situation, Sub-Section 6.1.2). On the upper wing side we may have a free-stream surface.

4.2.3 Lift Generation

CAV/ARV-type vehicles fly at small angles of attack. In the ramjet domain it is possible, that the upper side of the vehicles wing/fuselage surfaces can contribute to lift, because the vacuum pressure coefficient there is still finite, Fig. 3.9. Necessary is a proper shaping, viz. a suitable negative inclination against the free stream in the respective angle of attack domain. A smaller effect is to be expected in the scramjet domain.

4.2.4 Base Pressure and Drag

CAV-type and ARV-type flight vehicles are drag sensitive. Base surfaces may contribute not much (base drag) to the total drag at high flight Mach numbers, but for these vehicles especially the transonic regime is of importance. We illustrate this in Fig. 4.10, [12]. The two-dimensional base forms, typically blunt wing trailing edges or base surfaces of some of the waverider configurations discussed in Section 4.6, show very large (negative) base pressure coefficients in the transonic regime compared to the low hypersonic regime. Three-dimensional fuselage forms appear to be less critical, but the coefficients in the transonic regime are still much larger than in the hypersonic regime.

If one takes into account the size of the respective base surface of the flight vehicle, one can imagine that the base drag can dominate the total drag in the transonic regime. This is very important, because in that regime the difference between engine thrust, see, e.g., Fig. 2.14, Sub-Section 2.2.3, and drag is anyway small, like in the hypersonic regime. In any case in all

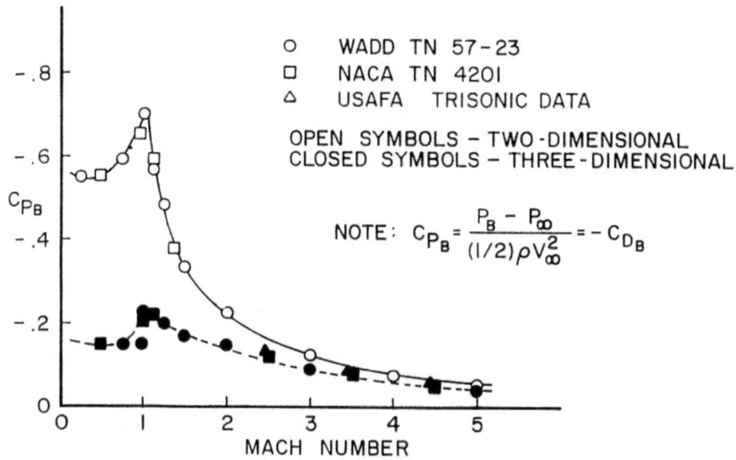

Fig. 4.10. Measured base pressure coefficient c_{p_B} of two- and three-dimensional configurations as function of the flight Mach number, [12].

Mach number regimes blunt base surfaces should be avoided. Trailing edges of wings, stabilization and control surfaces should have bluntness as small as possible from a structural point of view.

Of particular importance for TSTO space transportation systems is the base drag of the upper stage in the transonic regime. The large base surface of the HORUS orbiter of the SÄNGER system, for instance, made it necessary to devise an afterbody fairing, Fig. 4.11, similar to the tail cone which we have mentioned in Sub-Section 3.2.4 for the Space Shuttle Orbiter for ferry flights. Of course the HORUS afterbody fairing must disappear once the orbiter is launched from the lower stage. However, it was demanded that no hardware is shed in flight. No solution to the problem was found.

4.2.5 Aerodynamic Performance

A seemingly unavoidable trend of the optimum (maximum) lift-to-drag ratio of CAV-type flight vehicles, even if it seems to be contradicted by waverider results, Section 4.6, is shown in Fig. 4.12, [14].

In the lower part of Fig. 4.12 the hatched band represents the trimmed $(L/D)_{opt}$ of a number of dedicated vehicle designs, each with its layout for the given design Mach number and according to the then available technology. The lower edge of the band represents approximately the 1980 technology level, the upper that of the possible level of the year 2010 and later. Explicitly shown for orientation is the original CONCORDE design (square) and also a redraft based on the technology level of the year 2000 (black triangle).

The lower broken line finally gives $(L/D)_{opt}$ as function of the flight Mach number for a TSTO design of SÄNGER type with a design Mach number $M_{design} = 7$. The curve shows that for $M < M_{design}$ the tendency of an

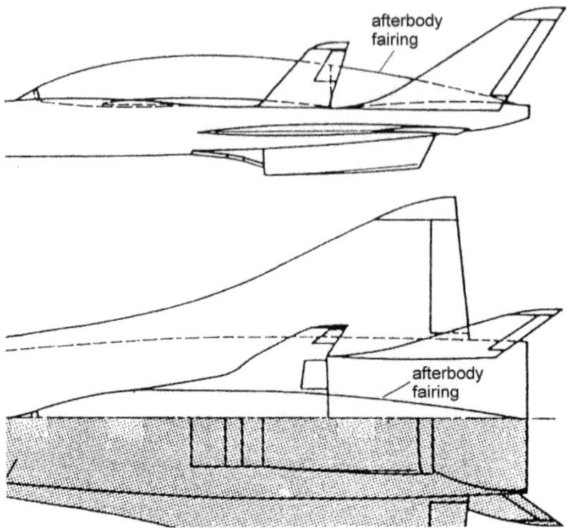

Fig. 4.11. Afterbody fairing of the HORUS Orbiter, [13].

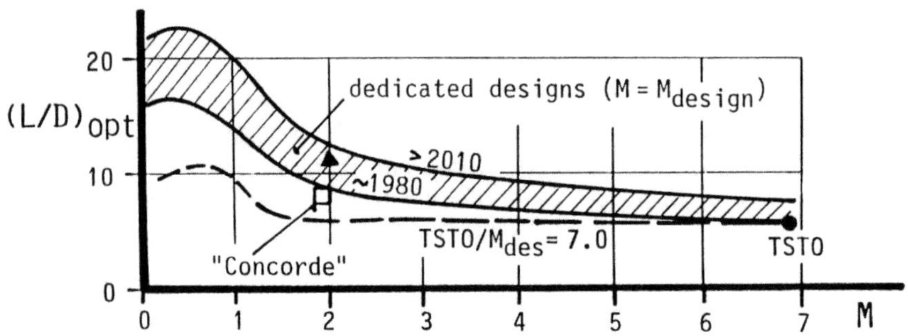

3

Fig. 4.12. Lift-to-drag ratios $(L/D)_{opt}$ of CAV-type vehicle configurations as function of the flight Mach number M, [14].

appreciable L/D increase with decreasing Mach number, as observed for waverider configurations, cannot be verified. This effect usually is seen for clean waverider configurations and inviscid aerodynamics. We show in Sub-Section 4.6.4, that viscous effects and base drag indeed change that picture and in Sub-Section 4.6.7, that an operable configuration with integrated propulsion system and aerodynamic stabilization and control surfaces behaves more or less like a conventional design.

Summarizing we can say, that the shape design of CAV- and ARV-type flight vehicles must aim for high aerodynamic quality and optimal aerodynamics/propulsion integration. This holds not only for the design Mach number. Other critical (off-) design points must be regarded as well, especially

Fig. 4.13. SÄNGER configuration with HORUS Orbiter. Synthetic image (left), [17], panel grid (right), [18].

in the transonic regime and in the Mach number regime, where the change from the turbojet mode to the ramjet/scramjet mode is made.

We observe also, that in general it might be somewhat too optimistic to work in conceptual design with correlations of the aerodynamic performance L/D like that from [15] (inviscid designs):

$$\frac{L}{D}|_{max} = \frac{4\,(M_\infty + 3)}{M_\infty}, \tag{4.2}$$

or that from [16] (waverider designs with viscous effects taken into account):

$$\frac{L}{D}|_{max} = \frac{6\,(M_\infty + 2)}{M_\infty}, \tag{4.3}$$

although they fit quite well into the hatched band of Fig. 4.12.

The particular simulation problems mentioned in Sub-Section 4.1.2, viz. laminar-turbulent transition and thermal surface effects due to radiation cooling, aggravate strongly the design task. They do not only concern the aerodynamic performance, but, since they govern strongly the thermal loads, too, also structural weight, aerothermoelastic behavior and propulsion performance, Sub-Section 4.5.

4.3 Example: Flight Parameters and Aerodynamic Coefficients of the Reference Concept SÄNGER

The lower stage of the TSTO space transportation system SÄNGER was the reference concept of the German Hypersonics Technology Programme in the 1980s/1990s, [17]. The major objective of this programme was to identify key technologies for the possible future development especially of the lower stage but also of the whole system. Both the lower and the upper stage, Fig. 4.13, were defined to a degree, [13], that a technology development and verification concept could be devised, [19]. Neither SÄNGER, nor the experimental/demonstrator vehicle HYTEX were built or did fly.

The space transportation system SÄNGER was envisaged as a fully reusable TSTO system with an airbreathing lower stage and a rocket-propelled upper

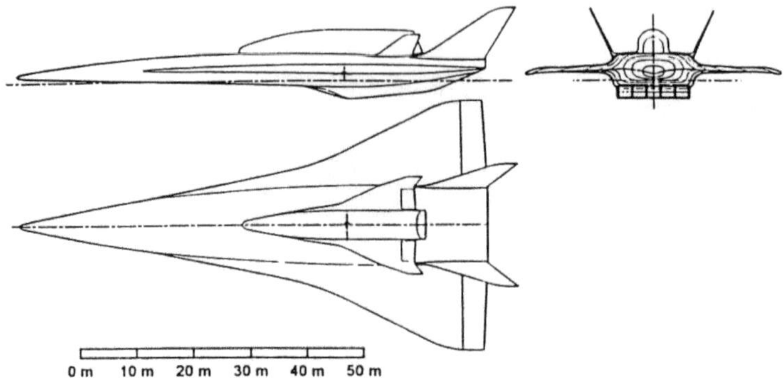

Fig. 4.14. Three-view drawing of the TSTO space transportation system SÄNGER, [13], see also [19].

stage. Both stages take off – the upper on top of the lower stage – and land horizontally like airplanes. The launch and landing sites requirements – launch from and landing in Europe – are met by a cruise capability of 3,100.0 km of the lower stage, see also Sub-Section 2.2.4, and a cross range capability of 2,500.0 km of the upper stage. The three-view drawing in Fig. 4.14 gives an impression of the system, major characteristic data of the system are listed in Table 4.3.

We make use of data from the technology programme, see, e.g., [19], also [6], in order to discuss in the following flight parameters and aerodynamic coefficients. We want to show with this some characteristic aerodynamic/ aerothermodynamic properties of CAV-type flight vehicles. However, we don't go into details of the aerodynamic shape definition and issues of flyability and controllability of such flight systems, nor do we discuss issues of upper stage integration and separation (launch). For the latter see, e.g., [20].

Consider first the flight parameters on a typical trajectory of the SÄNGER system. Fig. 4.15 shows left the flight altitude over the flight Mach number. The highest Mach number, $M_\infty = 6.8$, occurs shortly before the upper stage release. We note that this graph does not show the transonic bunt manoeuvre[9], which was a result of the study [21]. The subsequent descent of the lower stage alone is made with a higher trajectory than the ascent. The flight time, Fig. 4.15 right, is about 45 min for the ascent, and about 40 min for the descent.

The angle of attack as function of the flight Mach number is shown in Fig. 4.16 left. It drops from an initially large value at $M_\infty < 1$, to values between 4° and 7° during the cruise/climb part of the trajectory. At upper stage release, more than 14° are reached. Due to the mass shedding the angle of attack after release is very small. The yaw angle is an assumed one, dropping from 14° during take-off[10] to 1° later. For the bank angle μ_a see Sub-Section 2.2.4 and [21].

---

[9] Part of an inverted loop, Sub-Sections 2.2.3, 2.2.4.

[10] This value appears to be very large. A verification was not possible.

Table 4.3. Major data of the TSTO space transportation system SÄNGER, [13], see also [19].

	Lower stage	Upper stage
Fuselage length	82.5 m	32.45 m
Fuselage height	4.5 m	5.4 m
Fuselage width	14.4 m	5.2 m
Span width	45.1 m	17.7 m
Engines	turbo/ram (LH_2/air)	rocket (LH_2/LOX)
Number of engines	5	1
Thrust	max 500.0 kN each	1,500.0 kN
Payload	115.0 to	unmanned 7.0 to
Gross take-off weight	410.0 to (with upper stage)	115.0 to

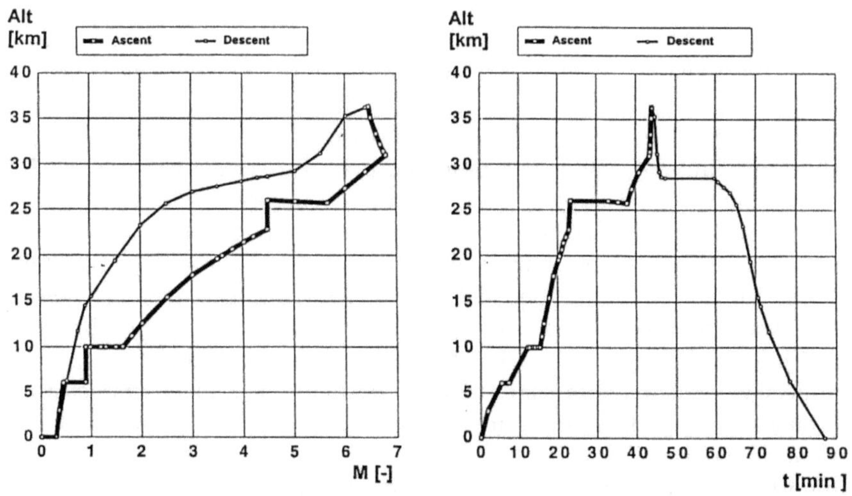

Fig. 4.15. Flight Mach number M vs. altitude Alt of a SÄNGER flight up to upper stage release (left) and time history (right), [19].

The flight velocity in Fig. 4.16 right and the data in the following Figs. 4.17 and 4.18 are given for a larger Mach number domain than SÄNGER has. These data stem from beyond-the-reference-concept considerations in [19], where two vehicle systems with the same length, $L = 82.5\ m$, at maximum speed of 3.0 and 4.0 km/s, were studied. The velocity increases almost linearly with the Mach number. The Reynolds number, based on the vehicle length, first increases to $O(10^9)$ and then drops to $O(10^8)$, showing that the attached

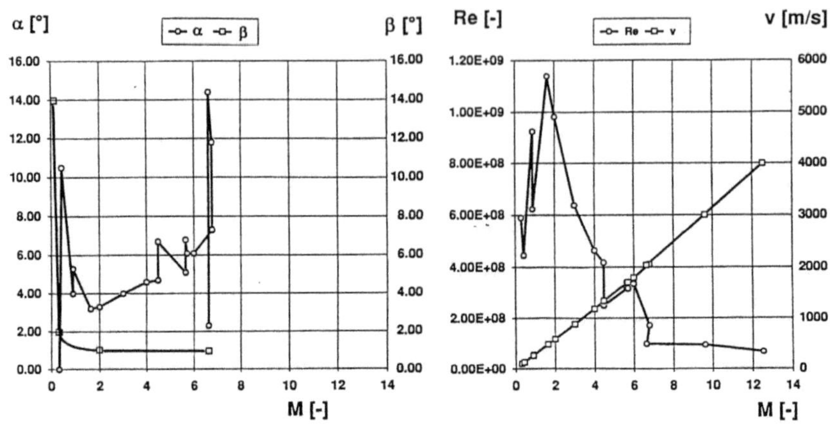

Fig. 4.16. Angle of attack α and of yaw β (left), as well as Reynolds number Re and flight speed v (right) as function of the flight Mach number M of the lower stage of a SÄNGER flight up to upper stage release, [19]. Reynolds number and flight speed are given for SÄNGER and two beyond-the-reference-concept flight systems.

viscous flow on SÄNGER will be turbulent on a large part of the vehicle surface, which certainly holds also for the two other concepts.

Static free-stream temperature T_∞ and pressure p_∞ in Fig. 4.17 (left) reflect the approximately 50.0 kPa dynamic pressure trajectory. Fig. 4.17 (right) shows the total enthalpy H_0, and the total temperature for both perfect, T_{0p}, and equilibrium real gas, T_{0r} (forebody nose tip, Euler calculation). At $M_\infty = 6.8$ the high-temperature real-gas effects are not yet strong ($\triangle T_0 \approx 200.0\ K$), but not negligible, see also Fig. 4.6. Beyond the reference concept design Mach number they become very large, and likely non-equilibrium effects will have to be regarded in the external and in the internal flow path of the respective vehicle. Note that the total enthalpy at a given trajectory point is constant everywhere in the inviscid flow portions of the external and the internal flow path.

An important parameter is the Knudsen number Kn, see, e.g., [8], which tells us whether we are in the continuum flow domain or beyond in the rarefied flow regime. For values $Kn \gtrsim 0.01$ rarefaction effects must be regarded in the aerodynamic design of the vehicle and/or its components. Since the Knudsen number is based, besides on the free-stream mean-free path, also on a characteristic length scale, it is provided in Fig. 4.18 (left) as function of the flight Mach number with several lengths L_{ref} as parameter.

The result is that up to the largest Mach number the Knudsen number is not critical for the whole vehicle, i. e. the flow past it is a continuum flow. For the two smallest length scales $L_{ref} = 1.0\ cm$ and $1.0\ mm$, which may represent the inlet's cowl-lip diameter, measurement orifices, et cetera, rarefaction effects must be regarded for flight Mach numbers $M_\infty \gtrsim 6$.

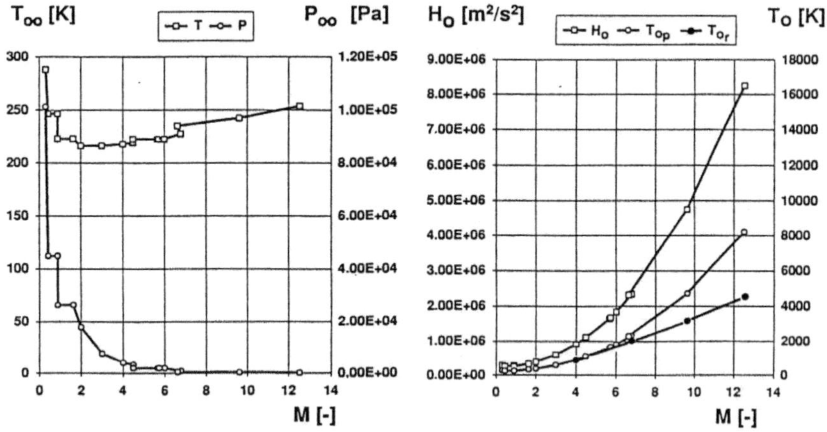

Fig. 4.17. Static free-stream temperature T_∞ and pressure p_∞ (left), and total enthalpy H_0, total temperature perfect gas T_{0p}, and equilibrium real gas T_{0r} (right) as function of the flight Mach number M of a SÄNGER flight up to upper stage release, and two beyond-the-reference-concept flight systems, [19].

Similarly, hypersonic viscous interaction effects, [8], which locally can lead to large pressure and temperature rises and may affect also global forces and moments, must be regarded at flight Mach numbers $M_\infty \gtrsim 7$, Fig. 4.18 (right). The viscous interaction parameter V_2' takes into account, other than V_1', the hot vehicle surface via the Chapman-Rubesin parameter[11].

We note with regard to rarefaction and viscous interaction effects, that the designer must be aware of these phenomena. They may affect performance, mechanical and thermal loads of the vehicle and its components/subsystems. Ground-facility simulation may be restricted in the quantification of these effects, whereas computational simulation, as long as laminar-turbulent transition and turbulence are not involved, in principle has no restrictions.

In the following we discuss extracts from the aerodynamic data sets of longitudinal motion of the SÄNGER system with and without the upper stage HORUS. The aerodynamic data bases, from which we present the extracts in Figs. 4.21 and 4.22, are composed of data from wind tunnel experiments, results from approximate engineering methods and some from numerical simulations, Fig. 4.19.

The wind-tunnel and computational data, which went into the aerodynamic data base of SÄNGER, were obtained with a configuration without propulsion system, Fig. 4.20.

This is the customary approach, because an active or even inactive propulsion system usually cannot be integrated in a ground-simulation model[12].

[11] In [8] the viscous interaction parameter is denoted $\overline{V}$.

[12] For passenger aircraft, for instance, wind-tunnel tests especially for the low-speed phases (take-off, approach and landing) are customarily made today with so called propulsion simulators.

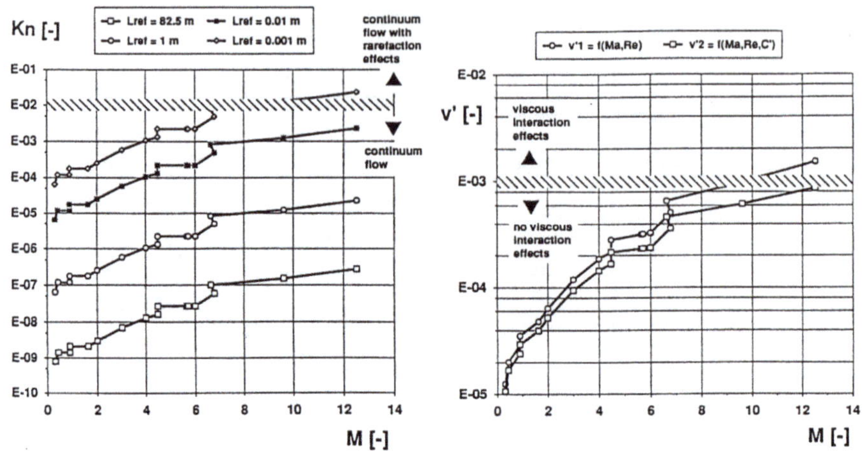

Fig. 4.18. Knudsen number Kn for different characteristic lengths L_{ref} (left), and viscous interaction parameters $V'1$ and $V'2$ (right) as function of the flight Mach number M of a SÄNGER flight up to upper stage release, and two beyond-the-reference-concept flight systems, [19].

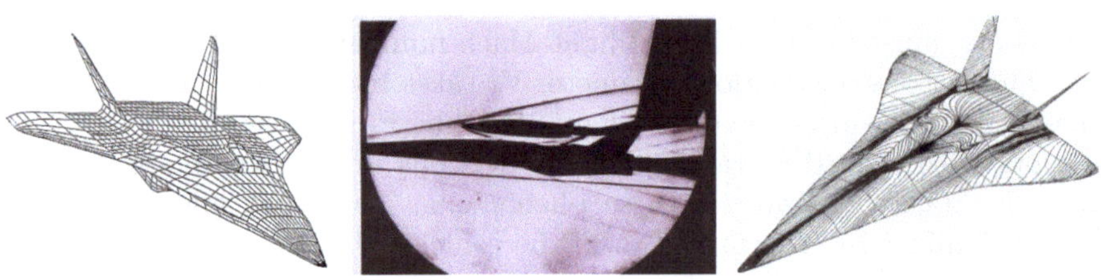

Fig. 4.19. SÄNGER configuration without HORUS Orbiter: panel grid (left), [18], and skin-friction lines evaluated from a Navier-Stokes solution (right). The figure in the middle shows a wind-tunnel schlieren picture of the stage separation process, [22].

Note the side fences on the lower side of the configuration. They represent the side walls of the propulsion system and help to gather aerodynamic coefficients for the lateral motion. For the final data base of the flight vehicle aerodynamic and propulsion data are added with the help of a book-keeping method, Section 4.4.

First we look at aerodynamic coefficients of the longitudinal motion of the SÄNGER configuration with the HORUS orbiter, Fig. 4.21. The HORUS orbiter has the afterbody fairing mentioned above. Lift and drag coefficients have the lowest values at $\alpha \geqq 0°$ for the highest Mach number ($M_\infty = 7$), increase with decreasing Mach number to a maximum around transonic speed ($0.9 \leqq M_\infty \leqq 1.1$) and drop again for subsonic velocities. For Mach numbers below $M \lessapprox 1.3$ we observe non-linear lift behavior for $\alpha \gtrapprox 8°$. The

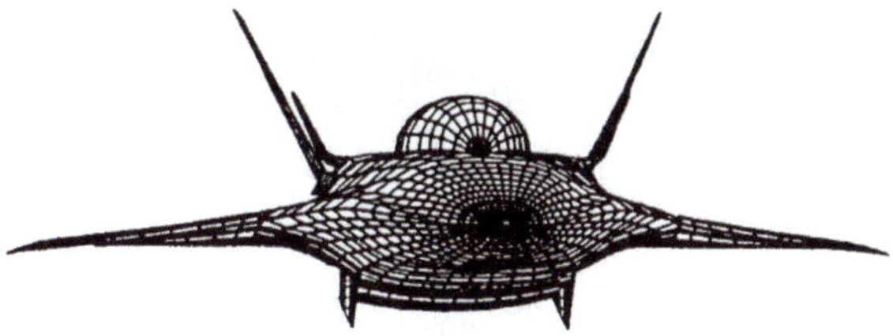

Fig. 4.20. Panel model of the SÄNGER with HORUS wind-tunnel configuration without propulsion system, [13].

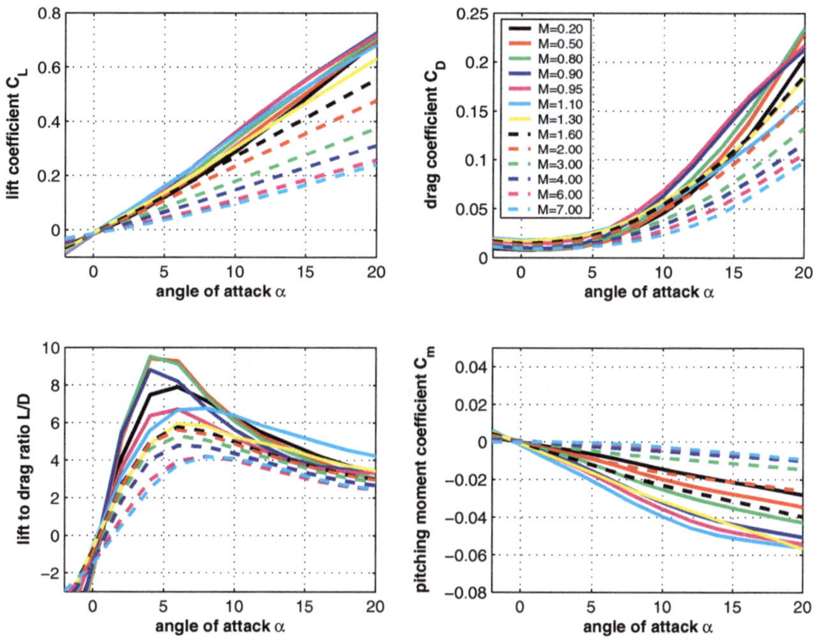

Fig. 4.21. Longitudinal motion of the SÄNGER configuration with HORUS orbiter: extract from the aerodynamic data base for the whole Mach number range. Moment reference point $x_{ref} = 0.65\, L_{ref}$. Data source: [18].

aerodynamic performance L/D, Fig. 4.21 (lower left), reaches the maximum values ($L/D \approx 9.5$, $\alpha \approx 4°$) for $0.5 \leqq M_\infty \leqq 0.8$. This behavior is different from the one of RV-W-type vehicles, Section 3.3, where the maximum L/D values are always achieved for low subsonic Mach numbers.

Mach number independence, at least for the drag coefficient, seems not yet to have been reached for $M = 6$ to 7. This holds also for SÄNGER without HORUS orbiter, Fig. 4.22. We find a similar result for the Scram 5 configuration, [2], Figs. 4.33 to 4.36. This is in contrast to what we find for RV-W-type, Section 3.3, and RV-NW-type flight vehicles, Sub-Section

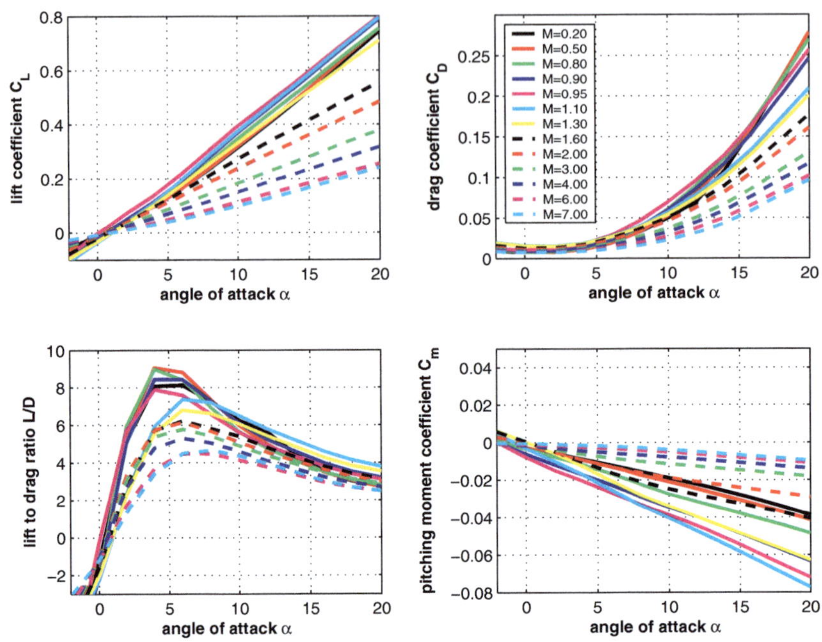

Fig. 4.22. Longitudinal motion of the SÄNGER configuration without HORUS orbiter: extract from the aerodynamic data base for the whole Mach number range. Moment reference point $x_{ref} = 0.65\, L_{ref}$. Data source: [18].

5.3.2, where we see Mach number independence for $M_\infty \gtrapprox 5$ to 6. The conclusion is, that slender vehicle shapes reach Mach number independence at larger Mach numbers than blunt ones, which appears to be plausible, see also [8].

In the whole Mach number range static stability can be observed with the strongest gradient $dC_m/d\alpha$ in the transonic flight regime. For the Mach numbers $M_\infty > 3$ this gradient is small. Actually the aerodynamic shape layout was made with an intentionally small $dC_m/d\alpha$ for the cruise Mach number – originally $M_\infty = 4.5$ – in order to keep the trim drag small. Trim for $\alpha > 0$ seems to be critical for the center-of-gravity position $x_{cog} = 0.65\, L_{ref}$.

There is also a need to know the aerodynamic data of the lower stage alone, since after upper stage separation the lower stage of a TSTO system has to fly back to its home base. In our case, Fig. 4.22, the aerodynamic coefficients of the lower stage behave in general very similar to the ones with the HORUS orbiter, Fig. 4.21.

Nevertheless we observe some minor differences in particular for $\alpha > 10°$. There the vehicle is statically more stable, especially in the subsonic and transonic range (compare the pitching moment diagrams), but the trim problem remains. Also lift and drag are somewhat larger. Finally, the maximum lift-to-drag ratio is a bit lower with $L/D \approx 9.0$ than in the other case, whereas

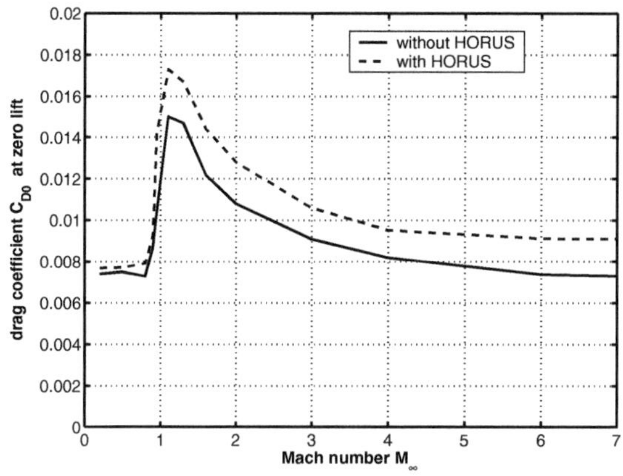

Fig. 4.23. Longitudinal motion of the SÄNGER configuration with and without HORUS: zero-lift drag coefficient C_{D_0} as function of the flight Mach number M_∞.

for the highest Mach number $M_\infty = 7.0$, L/D is higher compared to the SÄNGER with HORUS orbiter case.

The drag coefficient at zero lift of the SÄNGER configuration with and without HORUS is shown in Fig. 4.23. Note the large values in the transonic regime. Mach number independence is reached earlier for the configuration with HORUS, as is to be expected.

For the higher Mach numbers the data in Figs. 4.21 and 4.22, there especially the drag coefficients, and the data in Fig. 4.23 must be considered with reservations. This is due to the inaccurate quantification/simulation of two phenomena discussed in Section 4.1, in particular the location of the laminar-turbulent transition zone[13] and the viscous thermal surface effects due to radiation cooling of the airframe's surface. In addition we must observe, that the ground-facility measurements at the large Mach numbers have been made at too low Reynolds numbers and with constant cold-wall temperature model surfaces.

4.4 General Configurational Aspects: Highly Coupled Lift and Propulsion System

4.4.1 Some Vehicle Shape Considerations

Airbreathing hypersonic flight vehicles are drag sensitive akin to all aircraft-like vehicles. The smaller the drag, the smaller the required installed engine performance, the fuel consumption and, depending on the mission, the mass

[13] At low flight Mach numbers (and altitudes), say, lower than $M_\infty \approx 2$, the attached viscous flow past the vehicle can be considered as fully turbulent, which may reduce the ambiguity of ground-simulation data.

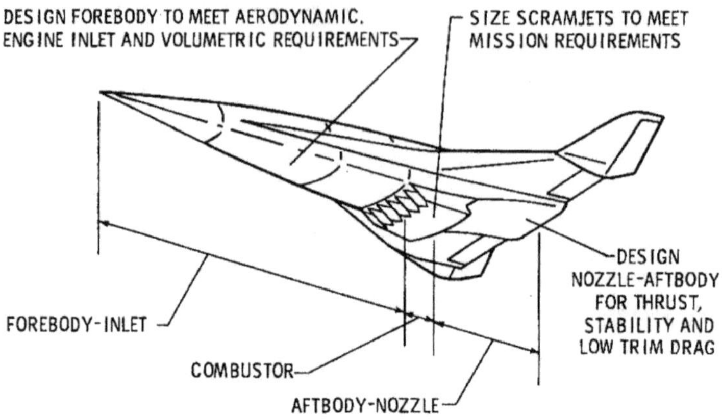

Fig. 4.24. Schematic of integrated lift and propulsion system of a CAV- or ARV-type flight vehicle, and the tasks of vehicle design, [23].

and volume of fuel, hence also of the fuel tank. The drag of an aircraft consists of the skin-friction drag, the form drag (the excrescence drag can be considered as part of it), and the induced drag. At flight speeds above the critical Mach number, the wave drag appears.

The wave drag depends on the flight Mach number, and on the shape and the angle of attack of the vehicle, [8]. In order to keep the wave drag small, CAV- and ARV-type flight vehicles, the latter for its ascent mission, must have very slender shapes and fly at small angles of attack. This is in contrast to RV-W-type vehicles. The sketch of the vehicles and the bow shock shapes in Fig. 6.2, Sub-Section 6.2.1, illustrate this.

At hypersonic flight Mach numbers the slenderness (half span/length) ratio of swept wings of CAV- and ARV-type flight vehicles should be $s/l \lessapprox 0.3$, [15]. Behind this is essentially the subsonic leading edge philosophy in order to reduce the wing's wave drag. Smaller values should not be pursued, because then the low-speed properties become insufficient (too low $dC_L/d\alpha$, critical dutch-roll behavior at high angles of attack, divergent motions (lateral/directional stability)). Low-speed properties typically are improved by flared wings (double delta, ogee wing, et cetera), see, e.g., Fig. 3.2.

The propulsion system package plays a large role with regard to the slenderness of the whole vehicle, too. The larger it is, the more it must be aft-located in order to avoid interference with the bow shock and the resulting wave drag increments, Sub-Section 4.5.1. Fig. 4.24 shows the configuration of a scramjet propelled vehicle, [23], which serves here as general schematic for CAV- or ARV-type flight vehicles.

The size of the propulsion system package is first of all governed by the necessary capture area of the inlet(s). This can be effectively reduced, if forebody pre-compression is employed, which indeed is the rule for airbreathing hypersonic vehicles. It reduces also the required engine size and weight, compared to the "free-stream" inlet of an isolated or podded engine. The pre-compression is achieved by an inclination of the forebody surface ahead of the inlet against the free-stream. Details regarding the forebody shape, flow and flow-topology issues are discussed in Sub-Section 4.5.2.

The size of the propulsion system package is also governed by that of the exhaust nozzles. They need a large area ratio because of the large pressure ratio, especially due to the low ambient pressure at high flight altitudes. The classical rocket engine bell nozzle cannot be employed because its exit plane (diameter) would be too large. This would mean, besides the matter of bow shock interference, a very high landing gear in order to permit vehicle rotation at take-off as well as the necessary large angle of attack at landing approach just before touch-down.

The way-out is the single expansion ramp nozzle (SERN), which essentially is an asymmetric external nozzle. It permits an effective exhaust expansion, has a natural radiation cooling potential (question of the influence of water vapor if hydrogen is used as propellant), and reduces the size of the propulsion system package. The drawback of such a nozzle is the change of the thrust vector in magnitude and direction along the flight trajectory, Fig. 2.15, Sub-Section 2.2.3. For the particularly critical transonic regime a possible remedy is secondary air injection into the engine nozzle flow, see, e.g., [24].

These considerations show us the major characteristics of CAV- and ARV-type flight vehicles, specifically, that they are highly coupled lift and propulsion systems. The deciding part of the external and the internal flow paths is at the lower side of the vehicle[14], Fig. 4.24. It is also noted, that the size of the external part of the propulsion system (forebody, inlet) and the external nozzle area of airbreathing flight vehicles grow with increasing cruise Mach number M_∞, Fig. 4.25, [23].

The tasks of vehicle design, Fig. 4.24, are the proper shaping of the forebody in view of the internal vehicle layout, the integrated forebody/inlet design, the housing of the core propulsion engine(s), and the SERN shaping. The latter is not only a matter of propulsion effectiveness, but in view of the highly coupled lift and propulsion system – for a typical point-mass force polygon see Fig. 2.14 – a matter of vehicle trim and stability. Thrust vector orientation and associated trim-drag penalties along the flight trajectory are of utmost importance. Of equal importance is the problem of engine flame-out. The vehicle has to be trimable even if the thrust vector disappears.

In the following section we discuss issues of aerothermodynamic airframe/propulsion integration, Section 4.5. Before we do that, we address force-accounting or bookkeeping procedures.

---

[14] Note that at the lower side of the vehicle, forebody and inlet are considered here as combined element.

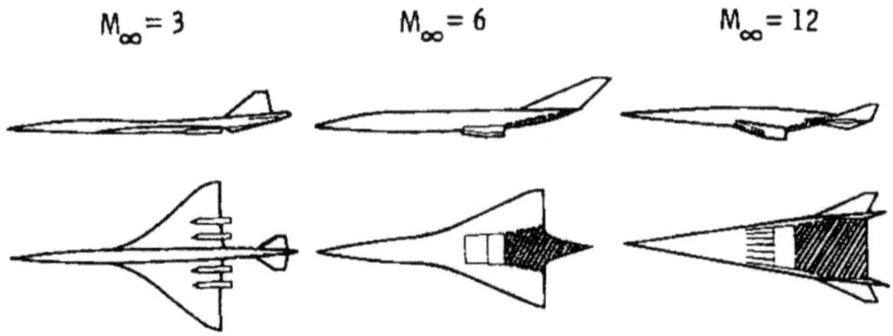

Fig. 4.25. Growth of the size of the external part of the propulsion system (forebody, inlet) and the external nozzle area (hatched) of airbreathing flight vehicles with cruise Mach number M_∞, [23].

4.4.2 Bookkeeping of Aerothermodynamic and Propulsion System Forces

In Section 4.3 we presented longitudinal force and moment coefficients of the TSTO system SÄNGER, which were found with the vehicle configuration without propulsion system. This data must be combined with the data of the propulsion system in order to find the overall – in our case – longitudinal forces and moments acting on the vehicle.

Several bookkeeping methods are imaginable. In the following we sketch first that used for the longitudinal motion of the SÄNGER system, [25]. It is governed by the fact, that the aerodynamic data are created predominantly by ground-facility simulation. Hence aerodynamic forces found in the wind-tunnel and propulsion system forces must be combined to obtain the overall forces and moments.

In Fig. 4.26 we show the principle split of the contributing surfaces of the airframe and the propulsion system. The propulsion-relevant stream-tube A_∞ at infinity is contracted, see Sub-Section 4.5.3, to A_0 at the beginning of the first ramp of the inlet. The propulsion-relevant stream-tube of the jet at the end of the SERN is denoted A_9. The problem is now that the measurements of the aerodynamic forces were made with a model without propulsion system package, Section 4.3 and Fig. 4.20.

Based on [25], an approach to account for the different forces was developed in [26], which we sketch in the following. In Fig. 4.27 the full line denotes the actual vehicle contour, the broken line that of the wind-tunnel model, and finally the dotted lines the propulsion system surfaces.

At the flat lower side of the wind-tunnel model the surface F_1 is that in the region of the inlet ramps, F_2 that in the region of the lower side of the propulsion system, F_3 that in the region of the SERN, and finally $F_2^{'}$ that

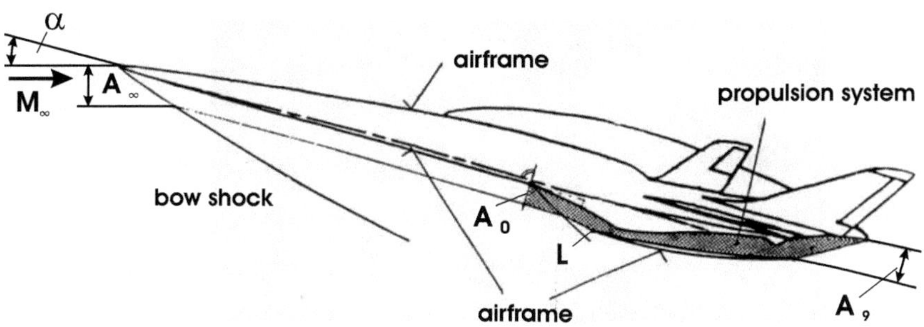

Fig. 4.26. Schematic of the bookkeeping for the longitudinal motion of the SÄNGER system: surface parts with aerodynamic forces (airframe) and with forces of the propulsion system (hatched), [25]. Stream-tube areas: A_∞ at infinity, A_0 at the beginning of the first ramp of the inlet, A_9 of the jet at the end of the SERN. The location of the cowl lip is denoted by L.

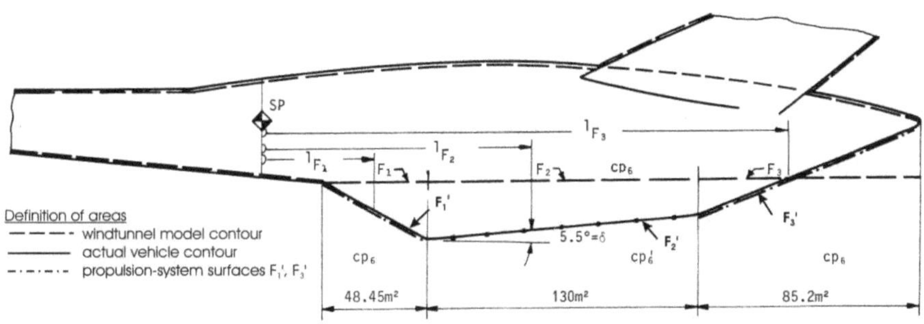

Fig. 4.27. Schematic of the force accounting for the SÄNGER system, [26]. *SP* denotes the center of gravity.

of the flat lower side of the cowl of the propulsion system[15]. The forces on these areas are accounted for by the aerodynamic data set. The areas $F_1^{'}$ and $F_3^{'}$ are the propulsion system inlet and exit surfaces, the forces/momentum fluxes acting on them are accounted for by the propulsion system data set.

The wind-tunnel model was designed to meet the bookkeeping procedure defined in [25]. Where the propulsion system is located, the lower side of the model has the flat surfaces F_1, F_2, and F_3, which are instrumented with pressure measurement gauges. The measurements of the forces and moments as well as of the pressure distribution of the surfaces F_1, F_2, and F_3 were made at the nominal angle of attack α. The pressure field measured on F_2 is projected on the cowl surface $F_2^{'}$. Because the cowl has an inclination angle δ = - 5.5°, the related pressure field of F_2 was measured with the effective angle of attack $\alpha_{eff} = \alpha - \delta$.

---

[15] This surface corresponds to the surface F_2 as we will see below.

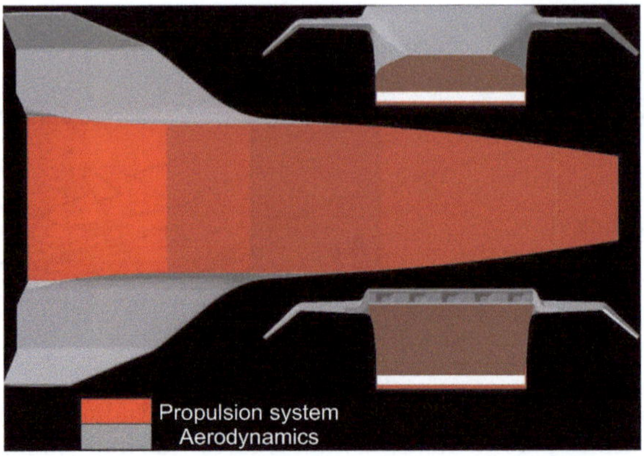

Fig. 4.28. Airbreathing orbiter study: configuration and bookkeeping surfaces at the lower side of the scramjet propelled flight vehicle Scram 5, [2]. Dark marked is the propulsion system surface, and grey the aerodynamic surface. The white surface in the front view (above) is the inlet opening, and in the rear view (below) the nozzle opening.

The aerodynamic data set, called the "truncated" data set is then found by accounting the contributions in the following way:

$$truncated\ data\ set = experimental\ data\ set -$$
$$- contributions\ of\ F_1,\ F_2,\ and\ F_3 + contribution\ of\ F_2^{'},$$

which holds for lift, drag, and the pitching moment. The contributions of the four surfaces F_1 to F_3, and $F_2^{'}$ are only normal forces, because only pressure fields are measured, not the skin friction. This of course in principle could be made, but then, as mentioned at the end of Section 4.3, for the larger Mach numbers of the truncated data set inaccuracies would exist regarding the surface $F_2^{'}$. These are due to insufficient knowledge of the location of the laminar-turbulent transition zone (location of tripping devices) and the inability to take into account the viscous thermal surface effects due to radiation cooling of the airframe's surface.

The final combined aerodynamic and propulsion system data set is found accordingly:

$$combined\ data\ set = truncated\ data\ set + propulsion\ system\ data\ set.$$

Another bookkeeping possibility is to consider the forces on the whole lower side as belonging to the propulsion system. We illustrate this in more detail than in the SÄNGER case with data from [2], see also [27], where the airbreathing upper stage (orbiter) of a TSTO space transportation system, Fig. 4.28, was studied and the system globally optimized.

The orbiter vehicle, called Scram 5, is a ARV-type vehicle in our terminology. It has twelve separate scramjet modules (not shown in the figure). For

Table 4.4. Airbreathing orbiter study: geometric data of the scramjet propelled flight vehicle Scram 5, [2].

Length	44.9 m
Span	27.0 m
Reference wing area	389.7 m^2
Reference wing length	15.8 m
x-location of cog	23.8 m
z-location of cog	+ 0.2 m

Table 4.5. Airbreathing orbiter study: flight parameters of the two considered trajectory points, [2].

Trajectory point	M_∞	H [km]	p_∞ [kPa]	q_∞ [kPa]	α [°]
1	6	27.3	$1.8{\cdot}10^3$	45.0	5.0
2	12	36.8	$4.5{\cdot}10^2$	45.0	5.0

flight Mach numbers larger $M_\infty = 12$ it is propelled by five rocket modules, located above the SERN (rear view in the figure). The relevant geometric data of the airbreathing orbiter are given in Table 4.4.

The reference trajectory of Scram 5 after separation from the lower stage up to switch-over to rocket propulsion at $M_\infty = 12$, is a $q_\infty = 45.0\ kPa$ trajectory. The flight parameters of the two trajectory points, which we consider, are given in Table 4.5.

The scramjet control volume – dark marked surface in Fig. 4.28 – encompasses the whole lower side of the flight vehicle without the wings. We show it in the side view in Fig. 4.29. Note that the lower side of the forebody is counted as Ramp 1. It has a length of 15.88 m. At zero angle of attack it is inclined against the free-stream by $\overline{\alpha}_0 = 3.5°$ (configuration angle, page 165). Ramp 2 is fixed with 4.82 m length and 8° inclination. Ramp 3 is variable, with 5.3 m length and 11° to 12.8° inclination. The variable inlet cowl lip finally has a length of 1.25 m and an inclination from - 7° to - 4.5°.

With the single surface elements of the scramjet control volume and the forces acting on them – depending on the flight attitude and the engine setting – the resulting force and pitching moment due to the propulsion system in view of the flight vehicle can be found.

We discuss now as example the contributions of the single vehicle-surface elements of the scramjet control volume to the pitching moment for the two trajectory points given in Table 4.5. In Fig. 4.30 the locations of the force acting lines (center-of-pressure of each surface element) in relation to the

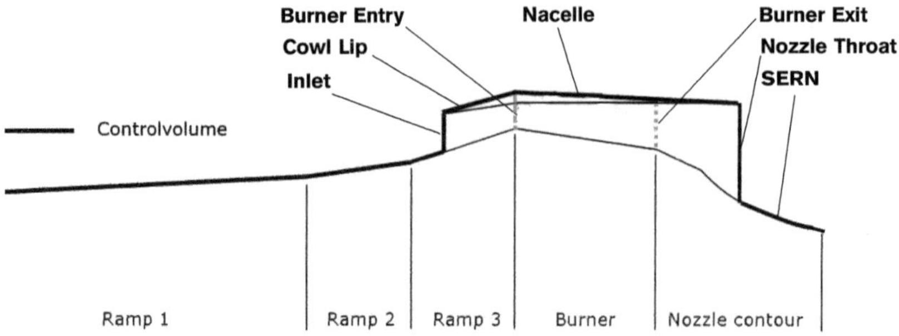

Fig. 4.29. Airbreathing orbiter study: side view of the lower side of Scram 5 (shown upside down) with the schematic of the single vehicle-surface elements of the scramjet control volume, [2].

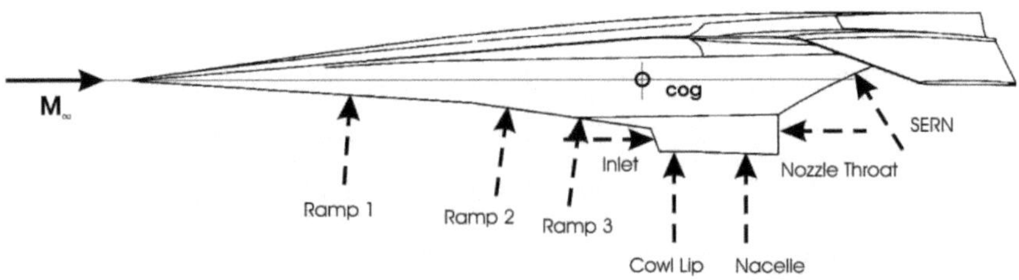

Fig. 4.30. Airbreathing orbiter study: schematic of locations of the force acting lines on the single vehicle-surface elements of the scramjet control volume.

center-of-gravity are shown. Intuitively we can see whether a surface element will give a pitch-up or pitch-down contribution.

In Figs. 4.31 and 4.32 the pitching moment contributions of the single vehicle-surface elements of the scramjet control volume are shown. Qualitatively they are the same in both trajectory points. Surface element 1 alone due to its location gives a large pitch-up contribution, that of element 2 is much smaller, and those of element 3 and the cowl lip are quite small. The lower surface of the nacelle due to its small inclination to the free-stream gives a small pitch-down contribution. The flow entering the inlet gives a large pitch-down contribution, whereas the flow exiting the nozzle gives a large – the largest – pitch-up contribution. Finally the SERN due to its location adds a large pitch-down increment.

The contributed moments are quantitatively different for the two trajectory points. The dynamic pressure q_∞, is the same for both points, but in trajectory point 1 we have a larger pressure p_∞ than in point 2, Table 4.5.

Consider first the ramp-surface elements. The pressure increases across the oblique ramp shocks, as well as the ambient pressure p_∞, are larger for trajectory point 1. This results in larger moment contributions in point 1 than in point 2.

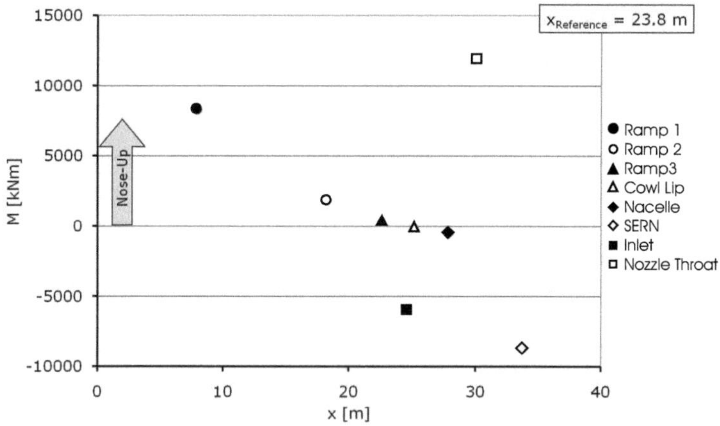

Fig. 4.31. Airbreathing orbiter study: pitching moment contributions of the single vehicle-surface elements of the scramjet control volume of Scram 5, Fig. 4.29, at $M_\infty = 6$, $H = 27.3\ km$ and $\alpha = 5°$ (trajectory point 1 in Table 4.5), [2].

The contributions of the inlet is larger at trajectory point 2 than at point 1. This is due to the fact that we have the shock-on-lip situation at point 2 ($M_{design} = 12$). This means for point 1 – note that dynamic pressure and angle of attack are the same for both points – that mass and momentum flux entering the inlet are smaller there. Hence the contributed moment is smaller. This is reflected by the contributions of the nozzle which is also larger at point 2 than at point 1. The SERN contribution, finally, is smaller at trajectory point 2.

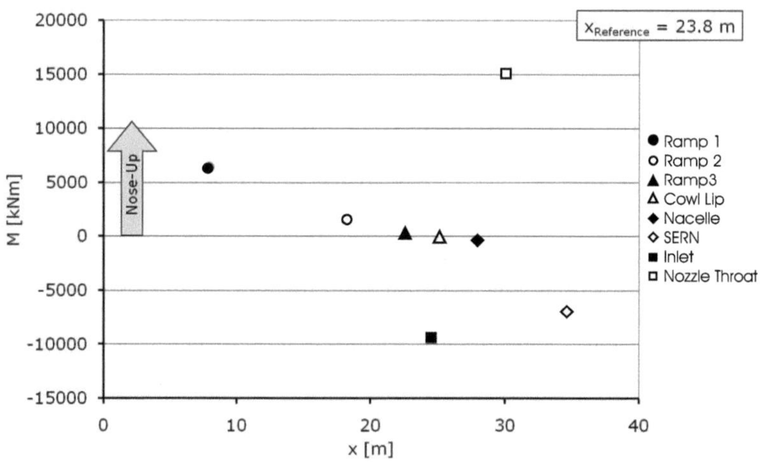

Fig. 4.32. Airbreathing orbiter study: pitching moment contributions of the single vehicle-surface elements of the scramjet control volume of Scram 5, Fig. 4.29, at $M_\infty = 12$, $H = 36.8\ km$ and $\alpha = 5°$ (trajectory point 2 in Table 4.5), [2].

In the following we discuss the aerodynamic data set[16] of Scram 5 for the longitudinal motion, [2]. Taking into account the different propulsion modes of the orbiter vehicle, three flight domains are distinguished:

1. $4 \leqq M_\infty \leqq 12$ (active scramjet domain): ascent phase, after separation from the lower stage, the scramjet is active. The contributions of the vehicle-surface elements of the scramjet control volume (red surfaces in Fig. 4.28) are not included in the aerodynamic data set. They are covered by the propulsion data set.
2. $12 \leqq M_\infty \leqq 30$ (closed scramjet domain): rocket propelled ascent phase, the scramjet is closed, the contributions of all vehicle surfaces are taken into account in the aerodynamic data set. Re-entry flight is made with the vehicle upside-down and the ramjet, too, closed.
3. $4.0 \geqq M_\infty \geqq 0.3$ (cold flow-through domain): the scramjet is open, but not active (cold flow-through mode). This is in order to "fill" the SERN and to reduce the base drag. Also here the contributions of the vehicle-surface elements of the scramjet control volume are not included in the aerodynamic data set.

In domain 1 the angle of attack range is $1° \lessapprox \alpha \lessapprox 5°$. Note that at the domain boundaries Mach numbers appear twice as parameter because the different domains were defined as overlapping.

The lift coefficient C_L as function of the angle of attack α is given in Fig. 4.33. The influence of the different propulsion modes is the following: in domain 3 $dC_L/d\alpha$ is largest (note the very small non-linearity for $M_\infty = 0.3$ and 1.2), with $C_L = 0$ lying around $\alpha = 1°$ to 1.5°. In domain 1, $dC_L/d\alpha$ in general is smaller, the location of $C_L = 0$ varies from approximately $\alpha = 0°$ to 2.6°. For domain 2 we observe Mach number independence, the smallest $dC_L/d\alpha$, and $C_L = 0$ at approximately $\alpha =$ - 0.6°.

The drag coefficient C_D as function of lift coefficient C_L is given in Fig. 4.34. The lowest zero-lift drag is found for domain 1, since there the viscous and pressure drag contributions of the surface elements of the scramjet volume are not accounted for in the aerodynamic data set[17]. The lowest drag levels are rather close to each other, however at different positive values of C_L. We see further, at positive C_L, an increase of the gradients dC_D/dC_L with increasing Mach number. For domain 2 we observe again Mach number independence, with zero-lift drag up to a factor two larger than in the scramjet domain. The largest zero-lift drag is found in domain 3 for $M_\infty = 1.2$, which is still a transonic Mach number. For $M_\infty = 0.3$ and 4 we have a zero-lift drag slightly below that for the high Mach number domain.

---

[16] All data in [2] were found by means of computational methods.

[17] Like noted for the SÄNGER data, we have also here principle data uncertainties with regard to the location of the laminar-turbulent transition zone and the viscous thermal surface effects due to radiation cooling of the airframe's surface. These concern both the aerodynamic and the scramjet data set.

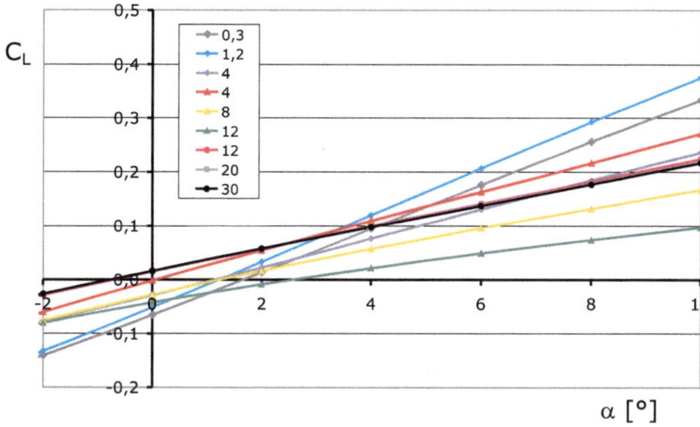

Fig. 4.33. Airbreathing orbiter study: lift coefficient C_L of Scram 5 as function of the angle of attack α in the flight Mach number range $0.3 \leqq M_\infty \leqq 30$, [2].

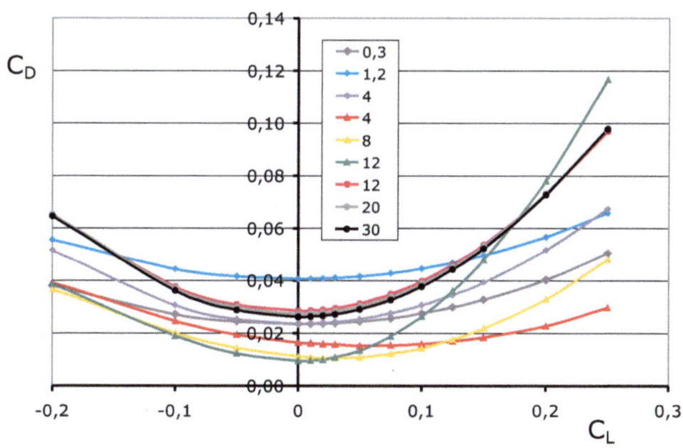

Fig. 4.34. Airbreathing orbiter study: drag coefficient C_D of Scram 5 as function of the lift coefficient C_L in the flight Mach number range $0.3 \leqq M_\infty \leqq 30$, [2].

The zero-lift drag behavior is reflected by the lift-to-drag ratio L/D as function of the lift coefficient C_L, Fig. 4.35. L/D is lowest in domain 2, with Mach number independence and the flat maximum slightly below $L/D = 3$ at $C_L \approx 0.15$. In domain 1 we have the largest value, $L/D \approx 9$, at $C_L \approx 0.2$. This decreases with increasing Mach number down to $L/D \approx 4$, at $C_L \approx 0.075$. In domain 3 we have the lowest values for $M_\infty = 1.2$ again, and the largest for $M_\infty = 4$, all lying at large C_L values.

Finally we look at the characteristics of the pitching moment as function of the angle of attack, Fig. 4.36. It is presented only for domains 1 and 2, because the data for the low Mach number domain (domain 3) are questionable, [2]. In these two domains the moment is unstable up to neutrally stable. For

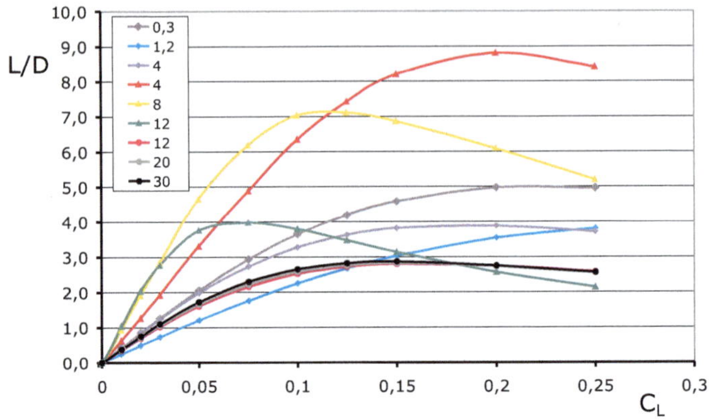

Fig. 4.35. Airbreathing orbiter study: lift-to-drag ratio L/D of Scram 5 as function of the lift coefficient C_L in the flight Mach number range $0.3 \leqq M_\infty \leqq 30$, [2].

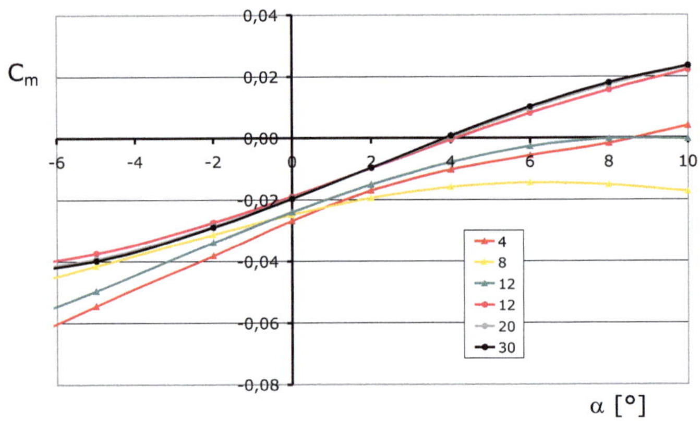

Fig. 4.36. Airbreathing orbiter study: pitching moment coefficient C_M of Scram 5 as function of the angle of attack α in the flight Mach number range $4 \leqq M_\infty \leqq 30$, $x_{ref} = x_{cog} = 23.8\,m$, $z_{ref} = z_{cog} = 0.2\,m$, [2].

the high Mach number domain we observe Mach number independence. In any case the vehicle configuration needs further refinement in order to assure trimmability and stability.

We note at last some of the conclusions in [2] and [27], see also Sub-Section 4.1.1: The scramjet lacks thrust potential, at least for the time being. The uncertainties concerning its performance and the high sensitivities of this propulsion system do not make it for some time to come a viable candidate for an upper stage of a TSTO space transportation system. For this, rocket propulsion is still the mode with the best prospect of success.

4.5 Issues of Aerothermodynamic Airframe/Propulsion Integration

In this section we address issues of aerothermodynamic airframe/propulsion integration. We treat only the highly integrated case, i. e. not the so called podded case, as we find it at the SR 71. The highly integrated case probably is the only viable approach for hypersonic flight. The advantages are the possible forebody pre-compression and the potential of flow-alignment ahead of the inlet(s). Disadvantages are that a boundary layer diverter is needed, at least for the turbojet propulsion mode, probably also for the ramjet/scramjet modes, which leads to the diverter drag, and also the possible mutual influence of the engines in case of malfunction.

The main topics of the following sub-sections are the lower side forebody aerothermodynamics in view of pre-compression – the forebody as zeroth or first stage of the ramp inlet – and optimum inlet/engine onset flow. Because forebody and inlet ramp flow are closely connected, we sketch also issues of inlet ramp flow, Sub-Section 4.5.5, the latter in view of the trim- and control-surface flow problems which we address in Chapter 6.

4.5.1 Forebody Effect

The shape of the forebody is influenced not only by aerodynamic demands, but strongly also by overall vehicle shape demands like inboard volume and center-of-gravity location, and also by mechanical and thermal loads. The lower side of the forebody is of special interest in view of inlet pre-compression and inlet/engine onset flow for both CAV- and ARV-type flight vehicles.

A look at Fig. 4.37 reveals the major forebody flow effect. The forebody causes a bow shock which at the design – the largest – flight Mach number should just envelope the propulsion system package in longitudinal and lateral direction (shock-on-lip condition, page 173).

If the bow shock would lie closer to the forebody, it would, firstly, interact with the external inlet shock system with detrimental influence on inlet and engine performance. Secondly, the bow shock would be deflected outwards with a large undesirable wave drag increment as a consequence. If the bow shock would lie further away, for no other (systems) reasons, the vehicle forebody simply must be considered as being too long.

We can formulate the description of the flow situation shown in Fig. 4.37 inversely in two ways:

- With given location of the propulsion system package: the larger the cross-section (height) of it, the larger the necessary bow shock opening angle ($\Rightarrow$ restriction of flight Mach number),
- or, with given bow shock opening angle: the larger the cross-section of the propulsion system package, the farther back the necessary propulsion system package location ($\Rightarrow$ increase of vehicle length and weight).

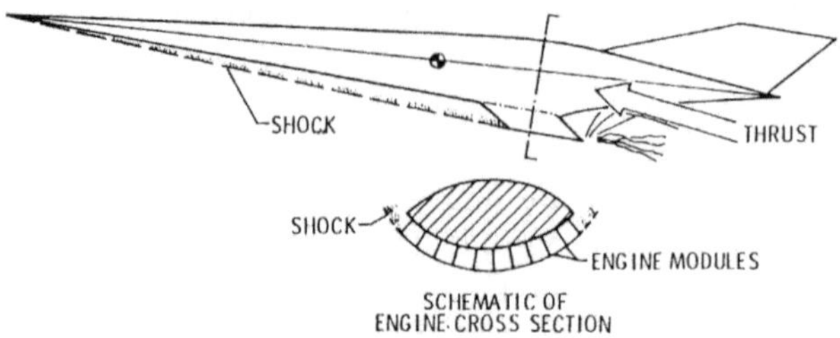

Fig. 4.37. Schematic of a scramjet vehicle with semi-conical forebody and semi-circular inlet/engine arrangement, [1].

These issues can be influenced by the degree of inclination, $\overline{\alpha}$, eq. (4.4), of the forebody's lower surface against the free-stream[18]. The inclination leads to the so called pre-compression of the inlet/engine flow, which in turn leads to a reduction of the capture area of the inlet and the cross-section of the propulsion system package. The angle of inclination of the lower side of the forebody governs the degree of pre-compression and hence the size of the cross-section of the propulsion system package. In the following several aspects of this effect are studied.

4.5.2 Flat Versus Conical Lower Side of Forebody

Before we study the pre-compression topic in detail, we note that the shape of the lower side of the forebody governs the topology of the inlet/engine onset flow field. Already the onset flow of the ramp inlet – we don't count in the following the lower side of the forebody as ramp – must be as uniform as possible in lateral direction on the relevant parts of the flight trajectory in order to assure optimum performance of the inlet and the engine without complex operation schedule.

A semi-conical (semi-circular) forebody, Fig. 4.37, or a fully conical forebody, Fig. 4.38, the latter with circular inlet/engine arrangement, is therefore a viable solution, but also a forebody with flat lower side, which has a two-dimensional inlet onset flow and inlet/engine arrangement.

Semi-conical and fully conical forebodies at a given Mach number at zero angle of attack (cone axis aligned with the free-stream direction) yield fixed pre-compression and a well defined conical inlet onset flow. As additional benefit we have the Mangler effect which keeps the boundary layer thin, see, e.g., [8].

However, at angle of attack, also aeroelastic (bending) forebody effects (see below), would change the onset flow pattern in a detrimental way. Fig. 4.39 shows this schematically for a positive α (a not too large negative α would

[18] We use the term inclination regardless of whether the lower side of the forebody is flat or conical.

Fig. 4.38. Schematic of a (zero angle of attack flight) scramjet accelerator vehicle with conical forebody and circular inlet/engine arrangement, [1].

result in a convergent flow pattern with a detachment line). The onset flow of the first inlet ramp would have lateral gradients, reduced pre-compression, and, very important, also the effective engine mass flow rate would be reduced. Of course one has to ask, how much of this can be tolerated, i. e. how sensitive engine performance is in this regard.

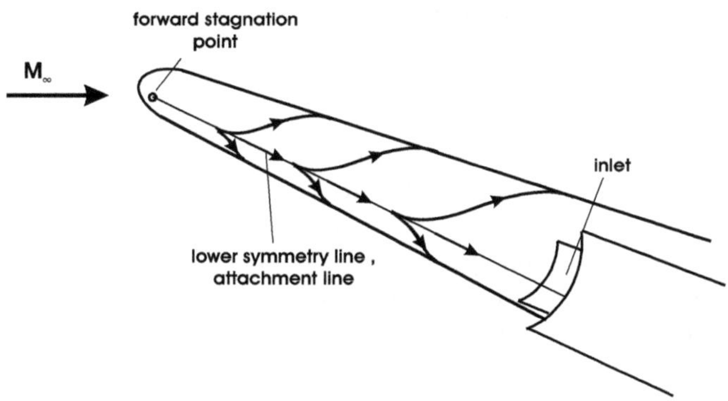

Fig. 4.39. Schematic of divergent streamline/skin-friction line pattern (features exaggerated) at the lower side of a blunt conical forebody at positive angle of attack.

We state hence, that a conical or semi-conical forebody will have the desired flow topology for a large flight Mach number range[19], but that it will become more and more non-axisymmetric and so-to-speak non-uniform for $\alpha \neq 0$. In contrast to this, a forebody with a flat lower side has the desired topology for both a large Mach number *and* angle of attack range.

The flat lower side of a forebody yields a two-dimensional onset flow of the first ramp of the inlet. Since this holds only between the primary attachment lines, Fig. 4.40, this means a certain restriction of the usable flow portion. However, a change of angle of attack at a given flight Mach number would change the degree of pre-compression, but would preserve the two-dimensional onset flow pattern.

[19] We keep in mind, that the design Mach number – shock-on-lip with margin – is the upper bound of the Mach number range.

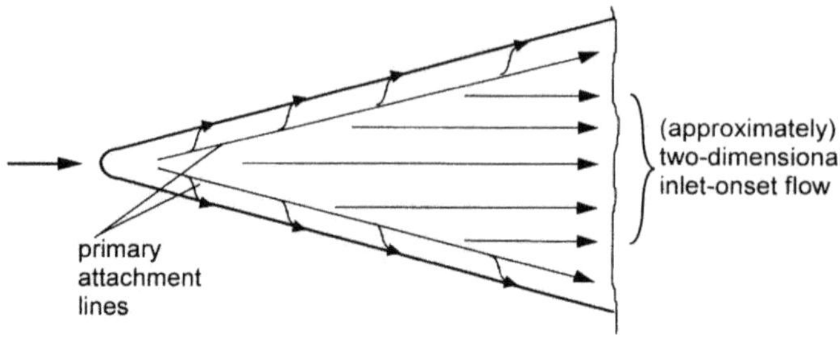

Fig. 4.40. Schematic of streamline/skin-friction line pattern at the lower side of a flat forebody (lateral extension exaggerated) at angle of attack.

In this regard we note, too, that a forebody with a "broad" blunt nose like shown in Fig. 4.28 yields a beneficial lateral bow shock uniformity across the whole forebody width compared to a "pointed" blunt nose forebody, Fig. 4.14. This is important for the shock-on-lip situation at maximum flight Mach number[20].

Indeed so far the majority of studies of airbreathing hypersonic flight vehicles show a forebody with flat lower surface and a broad blunt nose, note in this respect also the configuration of NASA's X-43A.

Can the lower side of the forebody deviate from the flat form? The demands from the side of the propulsion system are – depending on whether ramjet or scramjet propulsion is considered – flow uniformity, high static and total pressure, low Mach number, large mass flow rate, small distortion, and small boundary layer thickness (boundary layer diverter?) of the inlet onset flow on the relevant trajectory parts. Further required is the shock-on-lip condition at maximum (inlet design, page 174) flight Mach number. From the side of vehicle aerodynamics, large lift, low drag, and a suitable pitching moment are demanded.

Regarding the lower side's longitudinal shape, obviously it should be straight all over. A convex shape would lead to a flow expansion, which would reduce the pre-compression effect, however would reduce the growth of the boundary layer thickness. A concave shape, on the other hand would decelerate the flow further, thus increasing the pre-compression effect, but it would enhance strongly the growth of the boundary layer thickness.

In the frame of the German Hypersonics Technology Programme systematic studies of forebody shapes were performed, see, e.g., [28], [29]. Pointed blunt (narrow) forebodies with a slightly concave lower side cross-section

[20] We note, however, that a broad blunt nose is not necessary for a two-dimensional onset flow. The pointed blunt nose gives as well such a flow, as indicated in Fig. 4.40. This is a typical flow feature of delta wings or fuselages with flat lower side, Sub-Section 3.2.2. The forebody with pointed blunt nose, however, leads to smaller drag, lift, pitching moment, and aerodynamic load than the broad-nose shape.

were found to meet best inlet and vehicle aerodynamic (longitudinal motion) performance demands.

4.5.3 Principle of Forebody Pre-compression

The principle of pre-compression effectuated by a forebody with flat lower side can simply be explained, regardless whether the forebody has a broad or pointed blunt nose.

The lower side of the forebody has an inclination, angle $\overline{\alpha}$, against the free-stream. This angle is the sum of the configuration angle $\overline{\alpha}_0$, i. e., the inclination angle of the lower side of the forebody against the vehicle's x-axis at zero angle of attack, and the angle of attack α:

$$\overline{\alpha} = \overline{\alpha}_0 + \alpha. \tag{4.4}$$

Because the flow at the lower side is two-dimensional between the primary attachment lines, we can consider a flat plate – the RHPM flyer – at angle of attack α. In this case $\alpha \equiv \overline{\alpha}$. Fig. 4.41 shows that the resulting stream-tube compression, which governs the size of the cross-section of the propulsion system package, is the relevant effect.

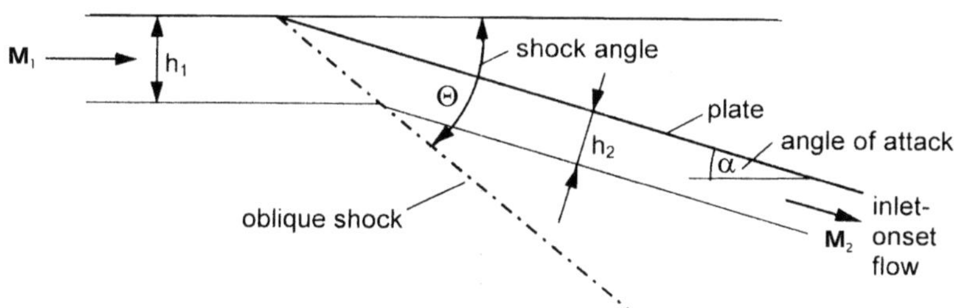

Fig. 4.41. Schematic of stream-tube compression at the RHPM flyer at angle of attack, inviscid flow. h_1 is the stream-tube height ahead of the oblique shock wave, h_2 downstream of it.

We consider the inlet capture area A_0 as measure of the cross-section of the propulsion system package, and look at its change due to pre-compression. If the mass flow rate $\dot{m}_2$ at the inlet capture area is, Fig. 4.41:

$$\dot{m}_2 = h_2 w \rho_2 v_2, \tag{4.5}$$

with w being the width of that area ($A_0 = w\, h_2$), we find the connection to the free-stream mass flow rate $\dot{m}_1$ with:

$$\dot{m}_1 = \dot{m}_2 = h_1 w \rho_1 v_1 = h_2 w \rho_2 v_2. \tag{4.6}$$

It follows, if the stream tube is compressed from h_1 to h_2, the fraction:

$$\frac{h_2}{h_1} = \frac{\rho_1 v_1}{\rho_2 v_2}. \tag{4.7}$$

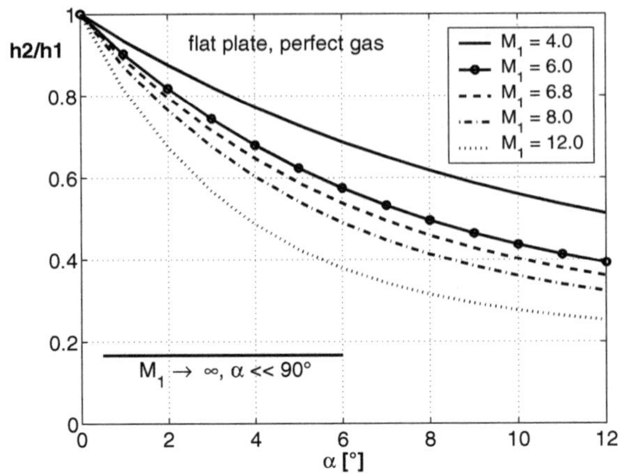

Fig. 4.42. RHPM flyer at different Mach numbers: compression of the stream-tube h_2/h_1 as function of the angle of attack α ($\equiv \overline{\alpha}$), inviscid flow, $\gamma = 1.4$.

This fraction is now determined from a geometrical consideration. From Fig. 10.1 in Chapter 10 we find, if we replace the ramp angle δ by the inclination angle $\overline{\alpha}$:

$$\frac{h_2}{h_1} = \frac{\sin(\theta - \overline{\alpha})}{\sin\theta}. \tag{4.8}$$

The shock angle θ is a function of the Mach number M_1, Fig. 4.41, and the inclination angle $\overline{\alpha}$, which for the RHPM flyer, [8], is the angle of attack α.

In Fig. 4.42 it is shown, that the pre-compression effect in terms of h_2/h_1 is already strong at small inclination angles. The ratio h_2/h_1 decreases fast with increasing α, and faster so at higher Mach number. The faster a vehicle flies, the smaller is the angle which is needed to achieve a given degree of pre-compression. High-temperature real-gas effects change the figure only a little. The theoretically attainable limit for $M_1 \rightarrow \infty$, $\alpha \ll 90°$, is $h_2/h_1 = 0.167$, see Problem 4.5.

The change of other interesting flow properties, e.g., pressure, total pressure loss et cetera can be found with the oblique shock relations, see, e.g., [8].

We consider now some data for the propulsion system of the SÄNGER lower stage[21], Fig. 4.43, [13]. At $M_\infty = 6.8$ and $H = 31.0\ km$ altitude the mass-flow rate to the five engines is $\dot{m}_{res} = 5 \cdot 255.0\ kg/s = 1{,}275.0\ kg/s$ (the data put into [13] were preliminary data, subject to changes). With the density $\rho_\infty = 1.60{\cdot}10^{-2}\ kg/m^3$ and the flight speed $v_\infty = 2{,}056.13\ m/s$ this means a capture area $A_\infty = \dot{m}_{res}/\rho_\infty v_\infty = 38.80\ m^2$.

[21] Note that the lower side of the forebody ahead of the inlets is not entirely flat in lateral direction.

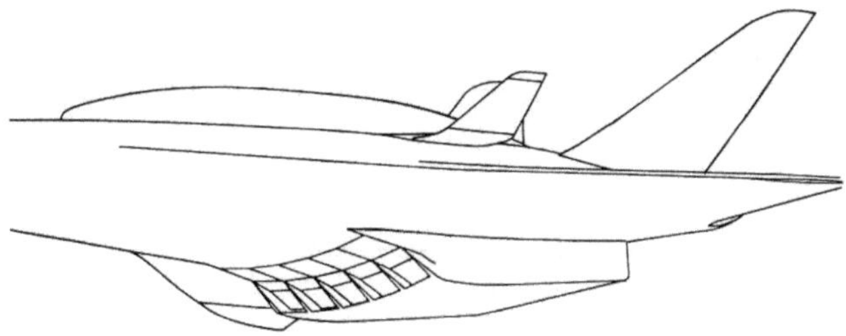

Fig. 4.43. The propulsion system package of the lower stage of SÄNGER with the five engine inlets, [13].

The actual capture area per inlet is $A_0 = 5.28\,m^2$, hence $A_{0,res} = 26.4\,m^2$ is the resulting inlet capture area[22]. The area ratio then is $A_{0,res}/A_\infty = 0.68$. Taking into account boundary layer influence et cetera, the real area ratio amounts to ≈ 0.5.

The configuration angle of the lower side of the forebody is $\overline{\alpha}_0 = 3.4°$. At the angle of attack $\alpha = 6°$ of the vehicle, the inclination of the lower side against the free-stream is $\overline{\alpha} = 9.4°$. For the RHPM-flyer at $\alpha = 9.4°$ a ratio $h_2/h_1 \approx 0.42$, Fig. 4.42, ensues.

The area ratio $A_{0,res}/A_\infty \approx 0.5$ for SÄNGER means that the capture area at infinity ahead of the flight vehicle is reduced by a factor of two at the inlets. This underlines the importance of forebody pre-compression.

Finally we have a look at the principle changes of flow parameters, net thrust, and aerodynamic parameters as function of incremental changes of the inclination $\Delta\overline{\alpha}$ of the lower side of the forebody against the free-stream, i. e. changes of pre-compression, Table 4.6. The $\overline{\alpha}$ changes are considered to be small enough, that the parameter changes are in the linear domain.

With positive $\Delta\overline{\alpha}$ we get an increase of static pressure and temperature, total pressure loss, net thrust, lift, drag, pitching moment, aerodynamic load. The inviscid Mach number decreases, whereas in a wide range of $\overline{\alpha}$ the unit Reynolds number increases, Fig. 6.18. This means a general[23] decrease of the boundary layer thickness, but an increase of thermal loads (the radiation-adiabatic temperature rises) and the wall shear stress. With negative $\Delta\overline{\alpha}$ the trends reverse.

This qualitative consideration underlines the strong coupling of the lift and the propulsion function of airbreathing hypersonic flight vehicles. If, for instance, the net thrust is increased by increasing the mass flux via the pre-compression, also the drag increases, as well as lift and pitching moment (trim

[22] The diameter of one combined turbojet/ramjet-engine is $\approx 2.0\ m$, the SERN area per engine is $\approx 10.0\ m^2$.

[23] For an approximate quantification of the trends one must take into account the boundary layer edge Mach number M_e, the unit Reynolds number Re_e^u, together with the reference temperature or enthalpy, which includes the influence of the wall temperature, [8].

Table 4.6. Principle changes of nominal flow parameters, net thrust, and aerodynamic parameters as function of incremental changes of the inclination $\Delta\overline{\alpha}$ of the lower side of the forebody against the free-stream, hence of pre-compression.

Item	$+\Delta\overline{\alpha}$	$-\Delta\overline{\alpha}$
Static pressure	⇑	⇓
Total pressure loss	⇑	⇓
Static temperature	⇑	⇓
Net thrust	⇑	⇓
Lift	⇑	⇓
Drag	⇑	⇓
Pitching moment	⇑	⇓
Aerodynamic load	⇑	⇓
M_e	⇓	⇑
Re_e^u	⇑	⇓

drag). In addition also thermal loads are increased, as well as skin friction as part of the vehicle drag, but also in view of possible erosion processes of the vehicle surface.

However, there is one performance change with $\Delta\overline{\alpha}$, that of the net thrust with forebody pre-compression, which rises grave concerns regarding the feasibility of at least large airbreathing hypersonic flight vehicles with the present available airframe technologies. This sensitivity of the net thrust is addressed next.

4.5.4 Net Thrust Sensitivity on Pre-compression

The net thrust is the small difference of large forces, which, as additional complication, act in different directions[24], Fig. 2.14. This leads to a large sensitivity of the net thrust. In the following example from the SÄNGER studies the installed thrust is defined as the actual thrust produced by the nozzle minus the flow momentum entering the inlet, the drag due to forebody boundary layer diverting, spill drag, bypass drag, and bleed drag, as experienced on the trajectory, [30]. We consider now its sensitivity on changes of the lower surface inclination angle $\overline{\alpha}$, Fig. 4.44, which is a measure of the forebody pre-compression.

The figure shows, that at $M_\infty = 6.8$ for instance, only one degree less than the nominal $\overline{\alpha}$ reduces the net installed thrust by approximately seven per

[24] All forces, hence also the net thrust, of course change, if speed, attitude, and engine settings change on the trajectory.

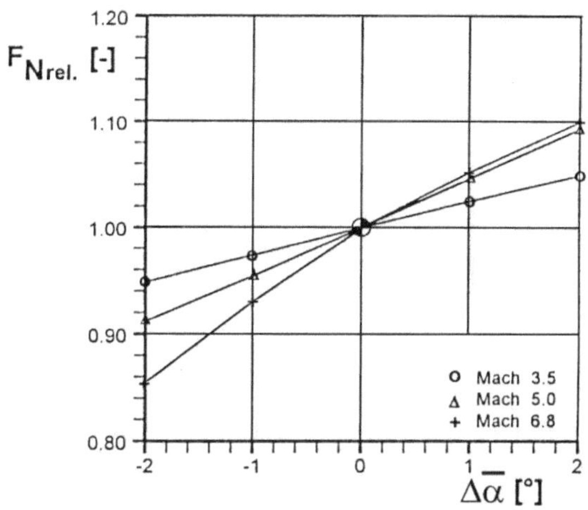

Fig. 4.44. Influence of inclination angle changes $\Delta\overline{\alpha}$ of the lower side of the SÄNGER forebody on the net installed thrust $F_{N rel.}$ for three flight Mach numbers, [30], see also [19], [13].

cent. One degree more would give about four percent more thrust. At two degree less or more the changes are approximately two times larger.

Of course, as indicated in Table 4.6, the $\overline{\alpha}$ changes alter also the vehicle drag, lift et cetera (Problem 4.6), which is not taken into account in Fig. 4.44. The important result is, that the inclination angle $\overline{\alpha}$ of the lower side of the forebody must be controlled very delicately. To lay out the necessary control means (air data and attitude sensors, propulsion control system, which includes inlet, combustor and nozzle throat control, flight control system, aerodynamic control surfaces and actuators) for minute $\overline{\alpha}$ changes in real time is a tremendous challenge, even if the airframe could be considered as being rigid.

In reality static and dynamic deformations of the forebody, in this case about 52.0 m long, lead to very large additional complications. We contrast this problem with the approach which is taken today, prologue to Chapter 8, viz. the more or less exact determination of the static and dynamic aeroelastic properties of an airframe only after the first specimen of it has been assembled. The conclusion is, that in view of the small payload fraction and the large development cost of the system, this is no more the approach of choice.

This holds all the more, because a further big problem arises, when the airframe structure and materials concept is basically that of a hot primary structure. This can for instance be a concept with integral cryogenic tanks, internal insulation, and a hot thin-sheet load-carrying structure without thermal protection system, [31]. An alternative concept would be a cold load-carrying structure with a thermal protection system though with no weight

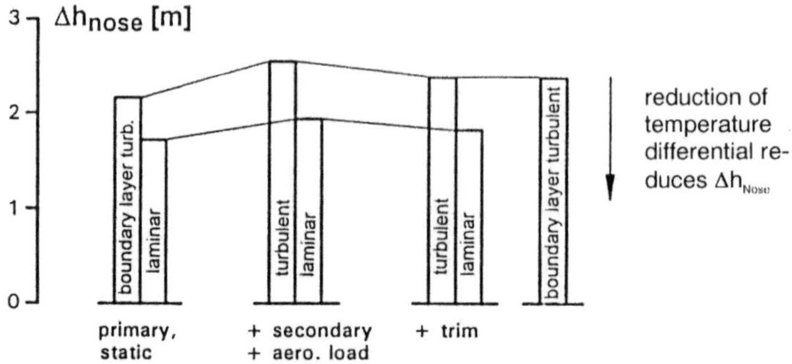

Fig. 4.45. Idealized effect of the wall-temperature difference between the lower and the upper side of the forebody (L = 55.0 m) of the SÄNGER lower stage on its static aeroelastic behavior, [32], see also [19]. M_∞ = 6.8 at 31.0 km altitude, α = 6°, $\triangle h_{nose}$: nose-up displacement.

penalty. However, in order to meet the demand of vehicle drag to remain as small as possible, such a thermal protection system must have a surface with sub-critical roughness, steps, et cetera[25].

We have seen in Fig. 4.6, that, regarding the radiation-adiabatic surface temperature, at maximum flight speed for turbulent flow a temperature difference of about 200.0 to 250.0 K exists between the lower and the upper side of the forebody. If one assumes a hot primary structure with such a design, and that heating results in deformations (statically determinate construction) rather than in additional stresses, this temperature difference, which is of realistic order of magnitude, would bend up ("bananization" effect) the forebody as shown in Fig. 4.45, [32].

Compared to the cold forebody, we get a nose-up displacement of the assumed hot structure of about 2.0 m. Secondary effects, and the aerodynamic load and trim increase this to approximately 2.5 m. A proper insulation at the lower side, or an alternate cold primary structure and materials concept, however, could reduce this effect almost completely. For the hot primary structure and materials concept a tailoring (reduction) of the emissivity coefficient of the upper side of the forebody would reduce the temperature difference and hence reduce the effect, too. In addition it would reduce the turbulent viscous drag there.

Nothing is known about the influence of thermal effects on the dynamic aeroelastic properties of the forebody and the airframe in general. In the frame of the German Hypersonics Technology Programme feasibility studies, see [31], have shown, that static structural ground tests for the large airframe of the lower stage of SÄNGER with a hot primary structure at $T \approx 1000.0$ K appear to be possible, but not so dynamic tests. As a way out, the numerical

[25] For the concept of necessary and permissible surface properties, Chapter 9.

simulation of the aerothermoelastic properties of the airframe in the frame of the "transfer-model concept", [33], [31], [19], was proposed. This concept has at its core the multidisciplinary numerical simulation of the real-elastic airframe, Sub-Section 8.2.2. It should permit to determine the relevant static and dynamic aerothermoelastic properties of the airframe with sufficient accuracy and reliability much earlier than with today's approach.

We conclude, that forebody pre-compression is an effective means to reduce the cross-section of the propulsion system package. However, it leads to a sensitivity of the net thrust on the degree of pre-compression which technologically is very challenging, especially in view of the aerothermoelastic properties of the airframe[26]. We have discussed this on the basis of data from the German Hypersonics Technology Programme, but the effect has been shown to exist also in other studies, see, e.g., [34].

In view of the small payload fractions of, e.g., space transportation systems and the large development cost, a new approach to the design and development of such systems appears to be necessary. However, the problem needs to be analyzed in more depth. Topics to be treated in detail are: impact of effective disturbances of the inclination angle of the lower side of the vehicle's forebody on net thrust, drag, lift, pitching moment, pressure and thermal loads; receptivity and amplification of disturbances (atmospherical, and vehicle/propulsion system control-induced disturbances); role of airframe aerothermoelasticity; optimization of the degree of pre-compression in the overall vehicle design process; optimization of the forebody design in view of pre-compression, et cetera.

4.5.5 Inlet Ramp Flow

We have seen above, that the forebody sometimes is considered to be the first ramp of a multi-ramp inlet with external compression. In any case, the external inlet flow is closely connected to the forebody flow. In order to round the picture of the flow past the windward side of a CAV- or ARV-type flight vehicle, we consider now some aerothermodynamic issues of inlet ramp flow.

Required pre-compressor Mach numbers of turbojet engines, like combustion chamber Mach numbers of ramjet engines, are in the range $M \approx 0.4$ to 0.6, whereas the combustion-chamber Mach numbers of scramjet engines are of the order $M \approx 2$ to 3. The deceleration of the flight-speed air stream to these Mach numbers is either fully or at least partly attained with the help of shock waves, see, e.g., [8]. Because the total pressure loss associated with this deceleration mode should be as small as possible, usually a sequence of oblique shock waves is employed, i. e. multiple inlet ramps. In praxis the outer part of the inlet (external compression) is a two- or three-ramp inlet.

If we look at CAV-type and ARV-type vehicles, we distinguish between cruise and acceleration vehicles, Section 4.1. For the first vehicle type

[26] This concerns medium to large scale flight vehicles, but probably not small vehicles with a highly rigid airframe.

generally flat lower shapes are envisaged, for the latter also conical shapes, see, e.g., [1]. We have, in any case, at the ramps of the inlet – either flat or conical – flow phenomena and hence design problems, see, e.g., [35], [36], which are basically the same as those we have at aerodynamic control surfaces with optimal onset flow geometry, Chapter 6.

A major feature of the flight vehicles in the background of our considerations is the forebody pre-compression, which was discussed in Sub-Section 4.5.1 and the following sub-sections. It is a major element in the design of an inlet, [37], but, because of the interdependency of inlet, combustion chamber and nozzle, and also of the whole flight vehicle. Because we intend to discuss here only issues of inlet ramp aerothermodynamics and to show the relation between the flow phenomena found on inlet ramps and on control surfaces, we don't consider forebody pre-compression anymore explicitly.

In the following we concentrate on flat two-dimensional inlet-ramp configurations. Two-dimensional means that we don't consider edge and side-wall flow phenomena. For a discussion of the flow and simulation problems of the complex strong interaction phenomena in a realistic three-ramp inlet configuration, which include longitudinal corner flow, see, e.g., [38], [39].

We show such a situation in Fig. 4.46, [40]. The three ramps of the inlet have the angles 5°, 5°, and 10°. In the upper figure the oil traces of the glancing ramp shocks at the side wall are clearly discernible. The longitudinal corner-flow field is caused by the ramp shocks interacting with the side-wall boundary layer flow. It shows a considerable total pressure loss, which is illustrated in the lower part of the figure. There p_p is the pitot pressure and p_0 the free-stream total pressure.

Because we have no forebody flow, and a rather low Reynolds number in this case, boundary layer transition happens on the model surface, with Görtler vortices obviously induced already by the first ramp. They become sufficiently strong on the third ramp, to lead to the typical wall-heating pattern (striations), middle image in Fig. 4.46, see also Sub-Section 6.3.3. The Stanton number is defined by $St = q_{gw}/[\rho_\infty v_\infty c_p(T_r - T_w)]$, with the recovery temperature approximated by $T_r = 0.9\ T_t$.

In Fig. 4.47 we give the schematic of a three-ramp inlet, which is typical for combined turbojet/ramjet propulsion. The inlet onset flow a) has been pre-compressed by the lower side of the forebody, which is not indicated in the figure. The forebody boundary layer at least for the turbojet mode must be diverted (boundary layer diverter) in order to avoid unwanted distortion of the inlet flow, and eventually of the flow entering the engine[27]. In the external compression regime b) the three oblique ramp shocks are shown to be focussed on the lip of the inlet cowl. The bow shock of the forebody is not shown.

The interesting flow phenomena around the junction – equivalent to the hinge line of control surfaces – of the ramps and asymptotically on the ramp

[27] However, if the forebody boundary layer is diverted, the "new" ramp boundary layer will initially be laminar and may have detrimental flow separation effects at the ramp junctions.

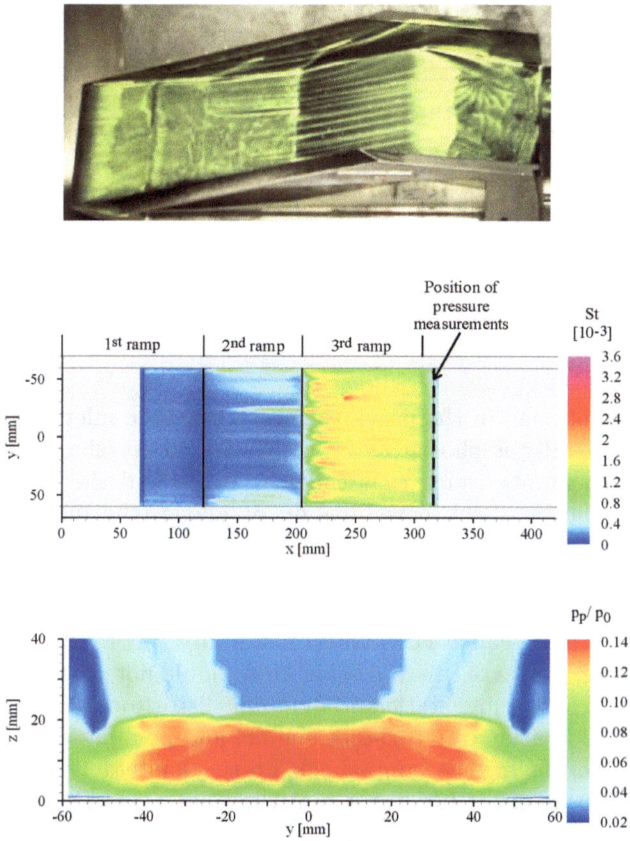

Fig. 4.46. Three-ramp inlet-flow experiment, [40]. From above: oil-flow visualization, heat-flux distribution (infrared image), and pitot-pressure distribution, $M_{onset} = 6$, $Re_{l,onset} = 2.1 \cdot 10^6$, $l = 0.42\ m$. The flow is from the left to the right.

surfaces are the same as those at aerodynamic control surfaces. In addition the effective ramp-shock angles, depending on the ramp angles and on the displacement properties of the ramp boundary layers are an important issue.

The layout of the ramp inlet with a given forebody pre-compression is made at the design flight Mach number for the shock-on-lip situation, where the forebody bow shock and all ramp-shock surfaces intersect the lip bow shock surface at the same location at the lip, Fig. 4.48. In this situation the drag of the inlet is minimized and the smallest cumulative total pressure loss ensues which, however, basically is a function of the number and inclination of the bow shock and the ramp shocks.

We emphasize that the shock-on-lip situation is defined by the design flight Mach number M_∞ *and* the inclination $\overline{\alpha}$ of the (flat) lower side against the free stream. With a given forebody shape hence the pair 'design flight Mach number' and 'design angle of attack' defines the shock-on-lip situation.

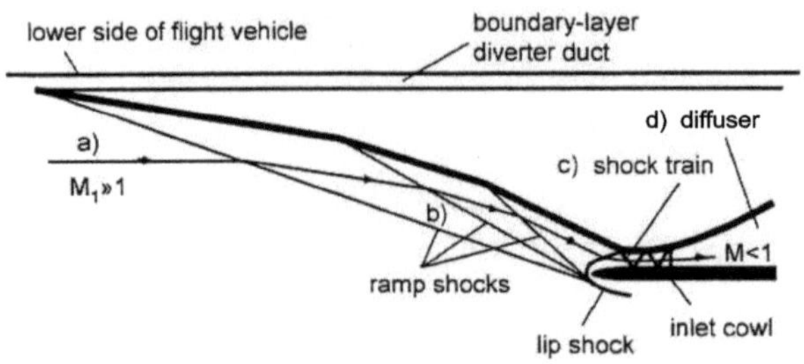

Fig. 4.47. Schematic of a three-ramp inlet without forebody bow shock, [8]. The flow is from the left to the right.

However, as mentioned in the previous sub-section, the inlet design flight Mach number generally is chosen to be somewhat larger than the vehicle design flight Mach number such that the bow shock and the ramp shocks interaction with the lip shock never lies on or above (the inlet is located below the fuselage) and – somewhat behind – the cowl lip. This margin is to assure, that the slip surface(s), which result from the interaction with the lip shock, never can come into contact or close to the inner cowl boundary layer, which can lead to the so called Ferri-Nucci instability (inlet buzz), [41], [38].

In any case disturbances of this and other kind should not enter the internal engine flow path. The ensuing static and dynamic distortions are detrimental especially for the operation of a turbojet engine, and possibly also of a ram-jet/scramjet engine. Having the ramp-shock/lip-shock interaction at all flight Mach numbers outside of the lip means that the Edney type IV shock/shock interaction, [42], see, e.g., also [8], with its large thermal and mechanical loads on the lip structure, is avoided, too.

The choice of the ramp angles is dictated by the needed flow deceleration (pressure recovery) together with the desired – with margin – shock-on-lip situation. The optimal ramp-angle domain for each ramp is that leading to the case b) in Fig. 6.9: viscous flat-plate/ramp flow with non-separating boundary layer. In this way undue total pressure loss and thickening of the ramp boundary layer do not occur, which would influence the effective ramp and shock angles. Avoided are then also the large thermal loads associated with case c), Fig. 6.9.

The properties of the onset flow boundary layer of each ramp depend, like in the case of aerodynamic control surfaces, on the initial conditions, entropy-layer swallowing and location of laminar-turbulent transition on the forebody, the boundary conditions, especially the temperature of the radiation cooled wall, and the running length, usually in presence of constant external inviscid flow along the ramp surface.

The initial conditions for the first ramp are either given, with boundary layer diversion, by the new start of it at the blunt nose of the ramp, or without diversion, by the forebody boundary layer, which can be very thick according to its initial and boundary conditions and the forebody length, see above. In

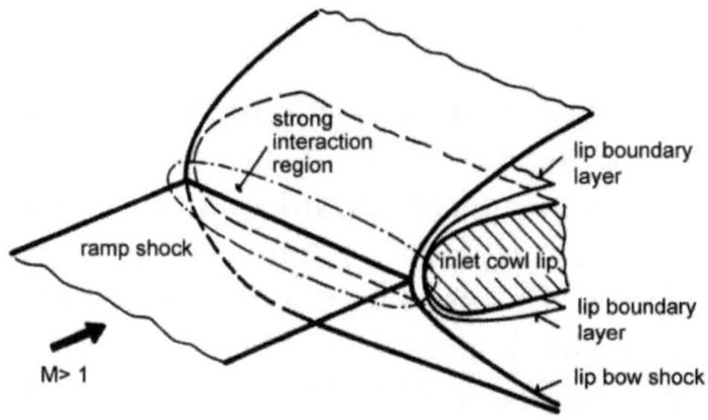

Fig. 4.48. Schematic of the shock-on-lip situation (only one ramp shock shown) without margin, [8].

the latter case the ramp(s) may be buried more or less in the boundary layer, and choking phenomena may occur in the inlet, [43]. At any successive ramp the initial conditions are governed by the virtual origin of the boundary layer, with the jump of the unit Reynolds number from the preceding flat plate to the ramp as important parameter, Fig. 6.18, Sub-Chapter 6.3.3.

Bleed, i. e., removal of boundary layer material through the ramp surface in order to avoid or reduce separation[28], case c) in Fig. 6.9, should be minimized or better avoided completely[29]. Any bleed leads to the so-called bleed drag, because flow momentum is lost, which in general cannot be recovered because of the large choking in the bleed passage. The consequence is a reduced net thrust of the propulsion system, not to mention the configurative measures to remove and finally dump the high-temperature bleed air. Bleed in the throat region in order to start the inlet is another issue, which we don't discuss here. If bleed is necessary it can be applied at a ramps junction, or, to be preferred, ahead of it, in order to remove low-momentum boundary layer material already before the interaction occurs.

In view of the flight Mach number and altitude range in which the inlet must operate, and regarding the systems aspects of the flight vehicle, anyway a trade-off is necessary, taking into account forebody pre-compression – for the particular problems associated with that see the preceding sub-sections – number of ramps, cowl lip shape, external and internal compression, et cetera.

4.6 Waverider Configurations

In this section we consider waverider shapes of possible CAV-type flight vehicles. These shapes offer – in their pure form – very good aerodynamic properties, e.g., regarding the lift-to-drag ratio. We discuss the basic principles of their design, give examples, look at the influence of viscous effects and thermal loads issues, and finally at issues of a waverider as an operational flight vehicle.

---

[28] For bleed as shock wave/boundary layer interaction control means see, e.g., [44].

[29] This concerns also sidewall bleed at realistic inlet configurations, applied in order to control the viscous interaction phenomena there.

4.6.1 Introduction

In the beginning of the 1990s there was a renewed interest in the space exploration community to transport payload into space by advanced space transportation systems. These systems should be able to take off and land horizontally like conventional airplanes. Compared to expandable rocket launched non-winged vehicles the aerodynamic lift should be used to reduce transport costs for large payload masses significantly. Further they should have a higher mission and landing site flexibility.

At that time, Single-Stage-To-Orbit (SSTO) and Two-Stage-To-Orbit (TSTO) vehicles (see also the introduction to this Chapter) were considered as very promising candidates for such transportation systems, where for example, the lower stage of the TSTO system was foreseen to be a hypersonic spacecraft propelled by an advanced airbreathing propulsion system. The efficiency of such a space transportation system is directly related to the aerodynamic performance, the lift-to-drag ratio L/D. Normally, hypersonic flight vehicles exhibit $L/D|_{max}$ values ranging from 4 to 6, [45].

Of course, there are several possible shapes of a hypersonic spacecraft which in general should be of slender type, [46], [27]. In particular the waverider shape seems to have a very promising aerodynamic potential. This idea was firstly proposed and published by T. R. F. Nonweiler in 1959, [47].

In the following Sub-Sections we deal with aspects of the design of hypersonic spacecraft on the basis of the waverider idea. Here, we focus on:

- waverider design methods based on wedge flow, cone flow, osculating cone flow,
- influence of viscous effects on the aerodynamic performance,
- off-design behavior particularly in the subsonic and the transonic regimes,
- the leading edge sharpness constraint,
- trim and control surface integration,
- upper surface design for L/D amendment,
- volume and volume efficiency constraints,
- the integration of a propulsion system.

4.6.2 Design Methods

The design of a waverider is an inverse process which means that for a given flow field the optimum vehicle shape is constructed. This shape fulfills the conditions of the prescribed flow field. The design process itself (based on an inviscid flow field) requires as input:

- the definition of a generating shock surface,
- the definition of the vehicle's leading edge(s) on the generating shock surface,

- the free-stream Mach number $M_\infty = M|_{design}$,
- the shock angle of the generating shock surface $\sigma|_{design}$[30].

The lower side of a waverider is determined by the stream surface emanating from the prescribed leading edge. Therefore, the shock surface is attached to this leading edge. Since, as well known, the pressure downstream of shock waves is increased, this stream surface generates the lift of the shape and the corresponding wave drag. So the maximum aerodynamic performance L/D for a given Mach number is one main objective of an optimization process.

The upper side of the waverider is often a free-stream surface. Due to the nominally sharp leading edge and the fact that on the upper side the deflection angle is zero the application of the oblique shock relations exhibits in the inviscid limit no contribution to the aerodynamic coefficients. The lift is produced exclusively by the lower surface and it exists no interference between the flow of the lower and the upper side. The consequence is, that no lift is lost owing to flow-spilling effects from the lower to the upper side. For rounded leading edges this situation changes, which will be discussed below in more detail. Nevertheless improved waverider designs generate upper surfaces with expanding flow, thus producing lift, too, approximately of the order of 10 per cent of the total lift, [45].

Design by wedge flow

The simplest way to construct the lower surface of a waverider is to apply a wedge flow, where for a given design Mach number $M|_{design}$ and a wedge angle δ, the shock angle $\sigma|_{design}$ is settled. The flow conditions are exactly determined by the oblique shock relations, see, e.g., [8]. Defining the trace of the leading edge on the shock surface and constructing the stream surface by the stream lines emanating straight from the leading edge, the result is a shape, which looks like the "caret" ascii symbol ($\wedge$), and was therefore called the "caret wing", Fig. 4.49, [46], [47]. The compression surface is built by two plane surfaces which intersect at a ridge line lying on the lower surface of the original two-dimensional wedge. Another solution obtained by this procedure is shown in Fig. 4.50, where two caret shapes are combined.

Design by singular cone flow

Whereas wedges generate plane shock surfaces with constant flow conditions behind them, a more general extension of the above mentioned design principle can be attained by employing circular cone flow with zero angle of attack. As is well known, the shock is curved in cross-sectional planes and the flow variables are constant along rays emanating from the vertex of the cone. Introducing conical coordinates and assuming inviscid flows, the profiles of

[30] In the literature the shock angle is often denoted by θ, e.g., [8], and also the present book.

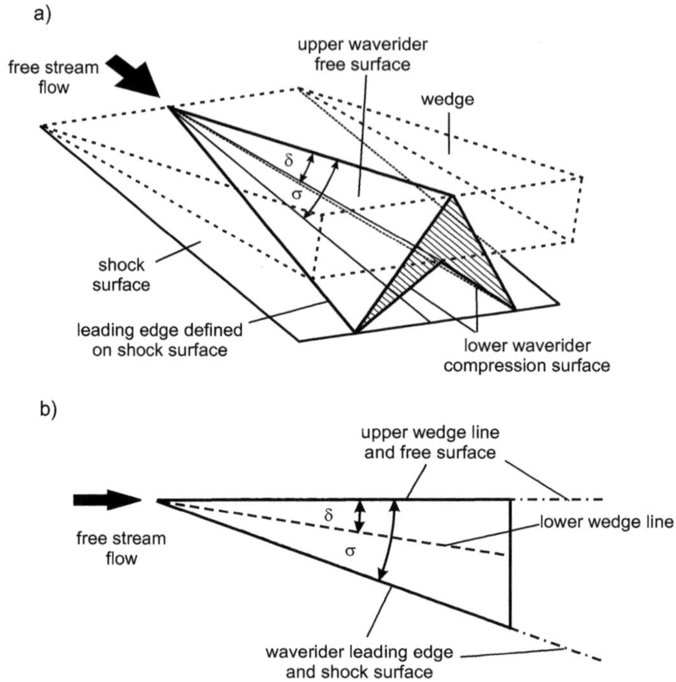

Fig. 4.49. Waverider shape based on wedge flow, classical "caret wing", δ wedge angle, σ shock angle, a) three dimensional view, b) side view, [46], [47].

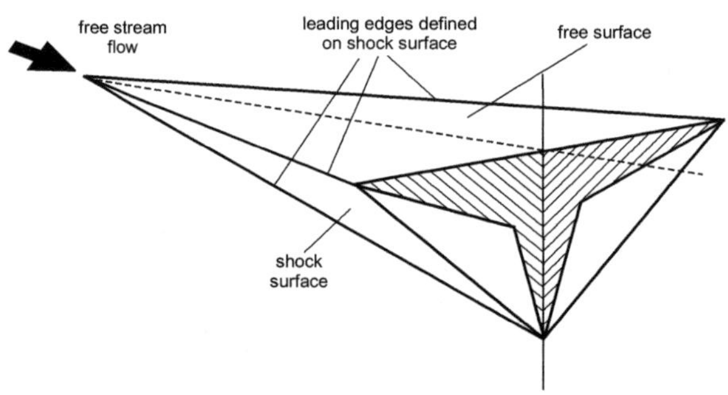

Fig. 4.50. Non-conventional waverider shape based on a combination of two slanting wedge/caret geometries, [46].

the flow variables between the shock and the cone surface can be exactly calculated by solving the Taylor-Maccoll[31] equation, [48], [49]:

[31] By definition of the free-stream Mach number M_∞ and the shock angle σ the velocity components v_r and v_θ behind the shock wave are known. Then with solving eq. (4.9) every cone flow can be determined, where the corresponding cone angle δ is obtained by the condition $v_\theta = 0$ for $\theta = \delta$.

$$\frac{\gamma-1}{2}\left[1-v_r^2-\left(\frac{dv_r}{d\theta}\right)^2\right]\left(2v_r+\frac{dv_r}{d\theta}\cot\theta+\frac{d^2v_r}{d\theta^2}\right)- \\ -\frac{dv_r}{d\theta}\left(v_r\frac{dv_r}{d\theta}+\frac{dv_r}{d\theta}\frac{d^2v_r}{d\theta^2}\right)=0, \qquad (4.9)$$

where v_r is the velocity component along a conical ray, θ is the angle of the ray referred to the cone axis (see Fig. 4.51 b), γ is the ratio of specific heats. The velocity component normal to the conical ray, assuming irrotational flow, is defined by[32] $v_\theta = dv_r/d\theta$. With the definition of the leading edge trace on the shock surface and the construction of the stream lines emanating from the leading edge by evaluating the flow field calculated by eq. (4.9), the stream surface representing the lower side of the waverider is designed. Fig. 4.51 exhibits the geometrical relations of the procedure.

For completion it should be mentioned that waveriders were also designed on the basis of inclined circular cones (angle of attack $\alpha \neq 0$) as well as on elliptic cones, either aligned or inclined. In these cases the shock strength is not constant and the generating shock produces entropy profiles. The flow field itself is three dimensional and analytical solutions are only available for small deviations from the circular cone flow by applying the small perturbation theory, [50].

Design by osculating cone flow

The waverider design by using singular cone flow is in principle restricted by the fact that the generating shock surface is all the time a piece of a conical shock surface. This reduces the design flexibility, in particular with respect to the compliance with other mission requirements on the waverider than aerodynamics, for example the volume and the volume efficiency or the off-design behavior.

Therefore in the beginning of the 1990s the "osculating cone concept" was developed [51], [52][33]. The basic idea is, that for a prescribed shock profile in the exit plane an envelope of a shock surface is constructed by the conical shocks of span-wise osculating cones, Fig. 4.52.

There exists at every point normal to the shock profile a conical flow plane, with a constant shock angle σ, which is considered as independent of the other flow planes in its neighborhood, Fig. 4.53. The curvature at the shock profile determines the local cone radius and with that, whereby the shock angle is held constant, the length of the cone is defined, Fig.4.54.

Due to the fact that the shock angle in span-wise direction is constant, only one single solution of the Taylor-Maccoll equation, eq. (4.9), is required to define the entire flow field. Once this quasi three-dimensional flow field exists,

[32] This is the result of $rot\, v = 0$ formulated in spherical coordinates for axisymmetric conical flow conditions.

[33] For a detailed description of this procedure see also [53] and [54].

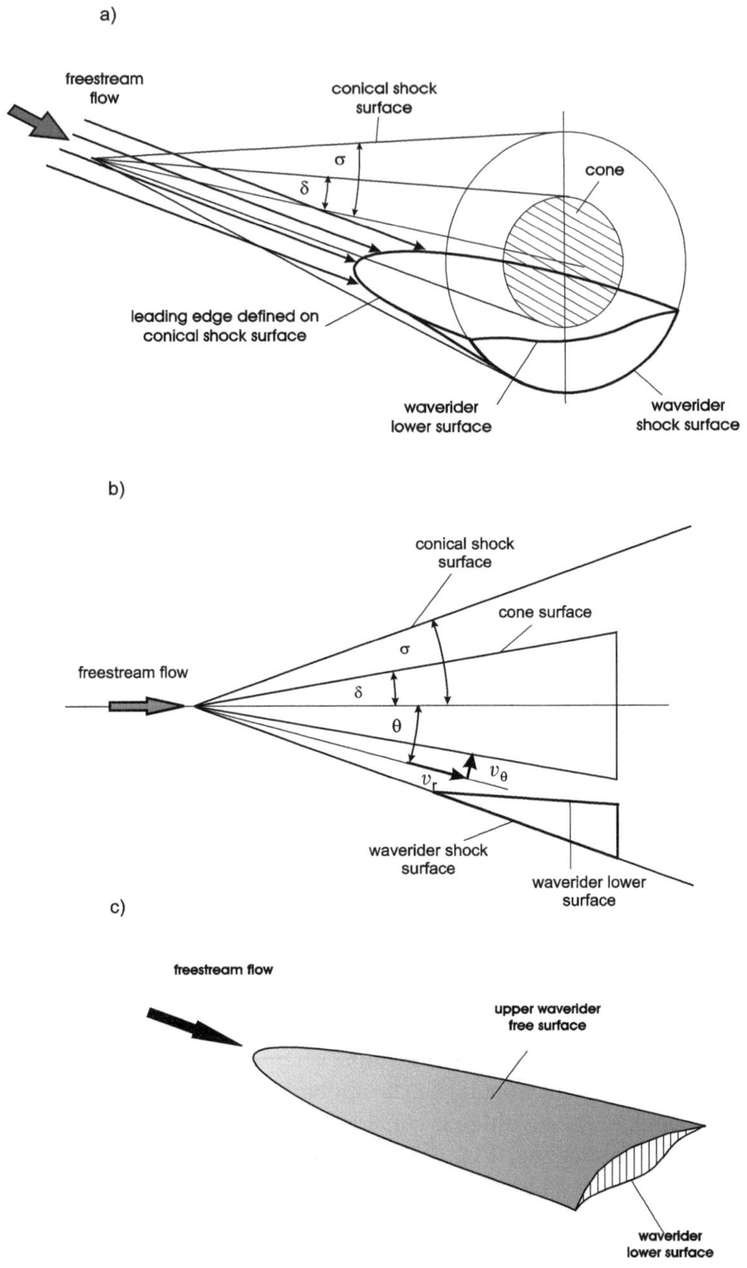

Fig. 4.51. Waverider design based on singular cone flow with zero angle of attack, δ cone angle, σ shock angle, a) three dimensional view of the design strategy, b) side view of the design strategy, c) typical design result. Data sources: [45], [46], [52].

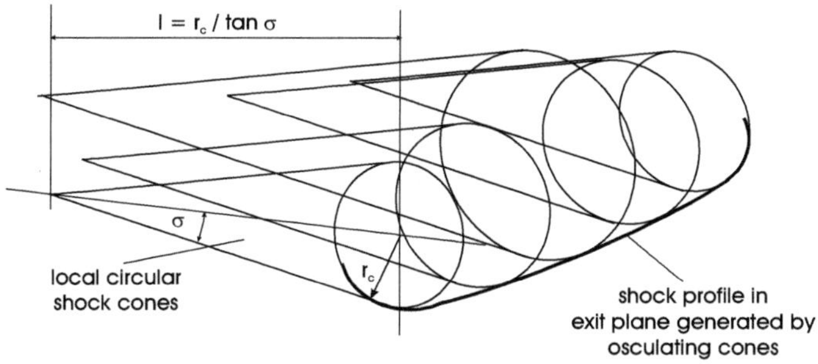

Fig. 4.52. Waverider design based on osculating cone flow, definition of local cone length l with local cone radius r_c, [51], [52], [54].

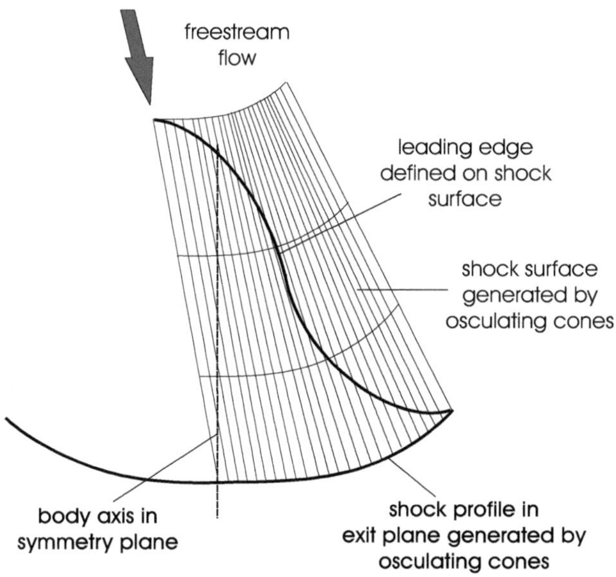

Fig. 4.53. Definition of the leading edge on the shock surface, which is generated by osculating cones, [51], [53], [54].

the stream surface defining the lower side of the waverider is calculated by tracking the stream lines starting at the leading edge.

A more general method, where not only the shock shape is varied, but also the shock strength (which means for a given Mach number $M|_{design}$ to vary the shock angle $\sigma|_{design}$), is proposed in [55]. Since in that case the flow field can no longer be composed by analytical solutions, this requires the development of an appropriate numerical method. For this reason a cross-stream marching method based on the Euler equations was derived, [55].

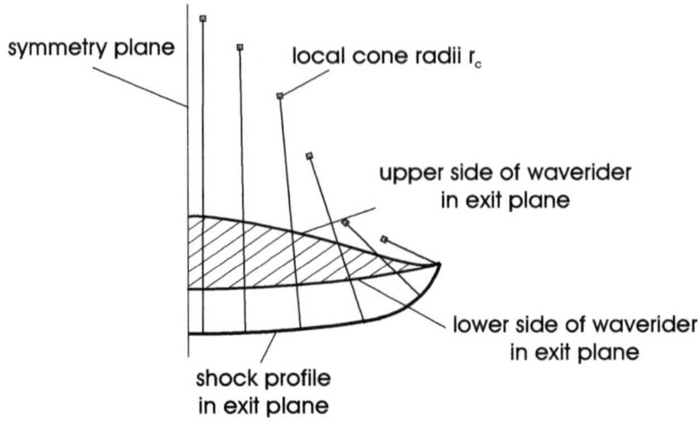

Fig. 4.54. Definition of the local radius r_c of osculating cones, [51], [53], [54].

4.6.3 Exemplary Waverider Designs

On the basis of the singular-cone flow method some authors have designed families of optimized waveriders where viscous effects, Sub-Section 4.6.4, are integrated in the optimization process, [45], [56], [57]. Mostly, the boundary layers were determined by integral methods which provide fast solutions, necessary during optimizations. An example, taken from [45], is exhibited in Fig 4.55. For a design Mach number $M|_{design} = 6$ and shock angles $\sigma|_{design} = 11°, 12°, 13°, 14°$ optimized waverider shapes are calculated[34]. The state of the boundary layers was transferred from laminar to turbulent by the application of a transition model, which was based on semi-empirical relations. The wall temperature was $T_w = 1{,}100.0\ K$ and the Reynolds number $Re = 122.4 \cdot 10^6$. The highest L/D value[35] ($L/D = 8.07$) is obtained for $\sigma|_{design} = 12°$. Further, Fig. 4.55 shows the perspective and front view of the shape with $\sigma|_{design} = 12°$.

An example of a waverider, designed by the osculating cone method, is given in Fig. 4.56. With the design code WIPAR[36], [58], operated in the inviscid mode, the shape of the waverider was determined for the design conditions $M|_{design} = 12$ and $\sigma|_{design} = 9.1°$. Generally, the aerodynamic performance L/D of inviscid waverider designs depends on the design Mach number, the shock angle and the shock contour. When using the osculating cone method different shock shapes can be chosen. Only waverider shapes designed for constant conditions, described above, have the same L/D value.

Despite the fact that the data base in this regard is very rare, we have constructed a three-dimensional surface, which contains all the known

[34] Note, that for every prescribed shock angle $\sigma|_{design}$ the optimization process creates distinct waverider shapes.

[35] All the L/D values given here do not encompass the base drag.

[36] WIPAR stands for "Waverider Interactive Parameter Adjustment Routine" and was developed by K. B. Center. Details can be found in [58].

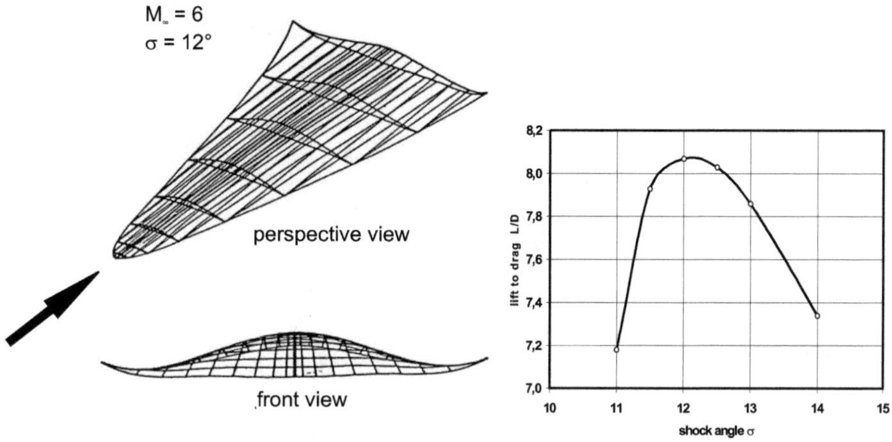

Fig. 4.55. L/D optimized waverider designs for various shock angles $\sigma|_{design}$, based on singular cone flow including viscous effects. The design Mach number is $M|_{design} = 6$, [45].

performance data[37], for inviscid waverider design as function of the design Mach number M_∞ and the design shock angle $\sigma|_{design}$, Fig. 4.57. There are two general trends. The first one is the slight increase of the performance L/D with decreasing design Mach number (down to $M_\infty \approx 4.0$), and the second one concerns the increase of the performance with decreasing design shock angle, which behaves exponentially for lower σ values.

The fitted surface is described by the following equation

$$z(x, y) = f_1(x) f_2(y), \tag{4.10}$$

with

$$\begin{aligned} f_1(x) &= \cos^2((\varphi_0 + (x - m1)\, a_1)) \cdot \\ &\quad \cdot (a_2 \cos^2 \varphi_0 - 1)/\cos^2 \varphi_0 + a_3, \\ f_2(y) &= a_4 + a_5 e^{-a_6(y - a_7)}, \end{aligned}$$

and $a_1 = 12.00$, $a_2 = 1.50$, $a_3 = 0.992$, $a_4 = 7.30$, $a_5 = 59.00$, $a_6 = 0.80$, $a_7 = 5.00$, $\varphi_0 = 10.0$, $m_1 = 4.50$, $z \equiv L/D$, $y \equiv \sigma$, $x \equiv M_\infty$.

It is difficult to compare aerodynamic performance data of existing results due to the specialized nature of waverider design analysis. Nevertheless, the L/D quantities in Table 4.7 give a hint of the range of the aerodynamic performance.

The performance values which pertain to [45] are relatively high for design Mach numbers up to $M|_{design} \approx 10$, but unfortunately the singular

[37] The data are based on [53], [54]. The shock surfaces for the osculating cone method were held constant.

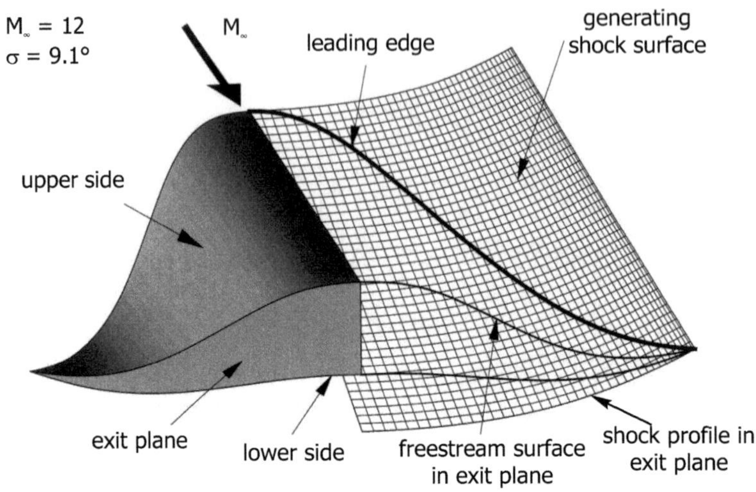

Fig. 4.56. Waverider design based on osculating cone flow for the shock angle $\sigma|_{design} = 9.1°$ and the design Mach number $M|_{design} = 12$, [54].

cone method generates shapes which hardly fulfill the requirements for a practical space transportation vehicle. Such a vehicle needs at least an adequately good low speed capability, a sufficient volume and volume distribution for the integration of tanks, propulsion system, payload, and equipment, as well as the ability to integrate aerodynamic stabilizer, trim and control surfaces. It seems that all this is hard to realize with a shape like the one of Fig. 4.55.

As already mentioned, the potential of the waverider shapes to meet the above requirements for an operational vehicle, designed by using the osculating cone methods, is much more promising, Table 4.7.

The data for the waverider of Fig. 4.56, based on an inviscid design, is the outcome of numerical flow solutions including viscous effects, where the boundary layer was assumed to be fully turbulent, [54]. The wall temperature was $T_w = 1{,}000.0\ K$ and the Reynolds number $Re = 200 \cdot 10^6$. Whereas the investigation of [52] with the goal to examine the influence of the design Mach number provides high L/D values, the outcome of the work, reported in [59] with the goal to amend the low speed performance, exhibits considerably lower L/D values, Table 4.7.

These examples provide some insight into the design problem of vehicles based on the waverider concept. We state here that only the aerodynamic performance of a vehicle, which is able to successfully operate along the whole flight trajectory, is of practical need.

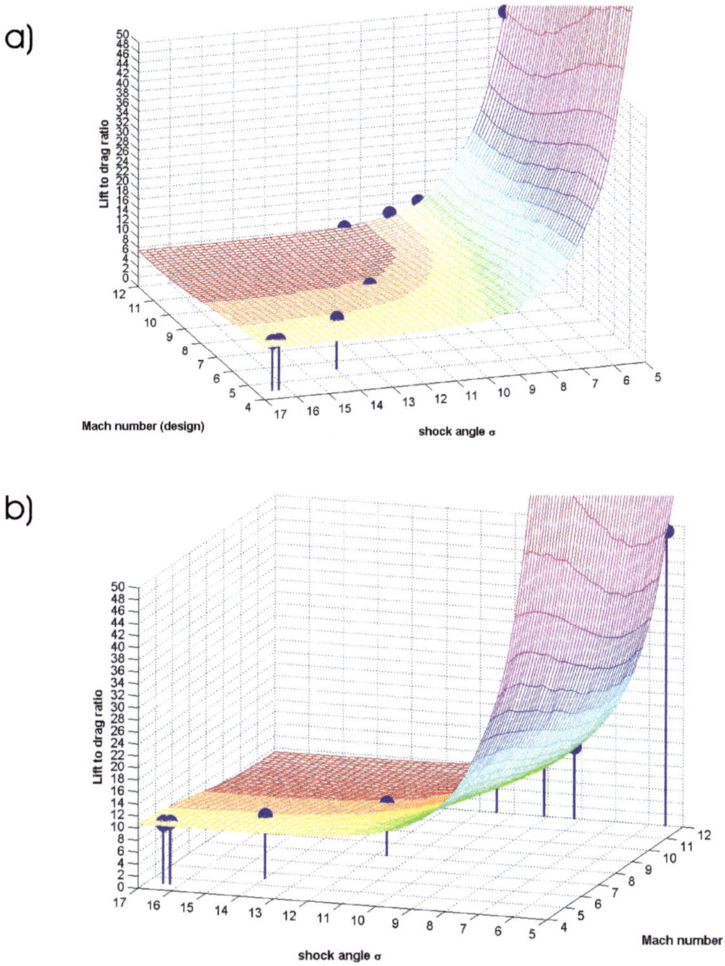

Fig. 4.57. Inviscid performance data L/D as function of the design Mach number and the design shock angle. View from left a), view from right b). Data (blue dots) sources: [53], [54].

4.6.4 Influence of Viscous Effects on the Aerodynamic Performance

The classical waverider design process as suggested by Nonweiler, [47], is a procedure employing inviscid flow fields, which are described by analytical equations. A realistic design, therefore, requires to take into account viscous effects already during the design process.

Viscous effects in this context are wall-shear stress and weak or strong (hypersonic viscous interaction) displacement of the inviscid flow by the boundary layer, as well as at high altitudes slip-flow effects. At altitudes below 40.0

Table 4.7. Aerodynamic performance data of various waverider designs (* values extrapolated).

$M_\infty\|_{design}$	$L/D\|_{designpoint}$	$Re\|_{design}$	$\sigma\|_{design}$	Conditions and Remarks	Ref.
4	8.91	−−	−−	singular cone method	[45]
6	8.07	$1.22 \cdot 10^8$	12°	design goal:	
10	7.00	$0.63 \cdot 10^8$	−−	max. L/D	
25	4.27	$1.4 \cdot 10^6$	9°		
4.5	6.55	$2.0 \cdot 10^8$	16.3°	osculating cone method	[54]
5.5	6.52	$2.0 \cdot 10^8$	14°	design goal: max.L/D	
12	6.15	$2.0 \cdot 10^8$	9.1°	for shock profile, thickness	
				distribution, planform unchanged	
4	5.55	$2.5 \cdot 10^8$	−−	osculating cone method	[59]
6	5.20	$2.5 \cdot 10^8$	−−	design goal: enhancing	
				low speed performance	
4.5	7.43	$1.0 \cdot 10^8$	−−	osculating cone method	[52]
8	7.28	$1.0 \cdot 10^8$	−−	design goal:	
12	7.30*	$1.0 \cdot 10^8$	−−	influence of design	
20	7.30*	$1.0 \cdot 10^8$	−−	Mach number on L/D	

to 60.0 km, boundary layers are initially laminar and become then turbulent. However, today the location of laminar-turbulent transition cannot be predicted accurately and reliably, [8]. Therefore either parametric studies of the influence of the transition location or computations with the assumption of fully turbulent flow should be performed.

The wall-shear stress of a turbulent high speed boundary layer depends much stronger on the wall temperature than that of a laminar one. For that reason design work in general should employ the radiation-adiabatic wall temperature, [8]. The choice of a computation method for the quantification of viscous effects must take into account these phenomena. Since boundary layer methods in general are not able to describe strong interaction phenomena, like the mutual effects of vortices, shocks and boundary layers in global flow separation regimes, Navier-Stokes methods are needed if such phenomena are present.

Viscous effects were taken into account for the first time by K. G. Bowcutt et al., [45]. Unfortunately, the aerodynamic performance values for the

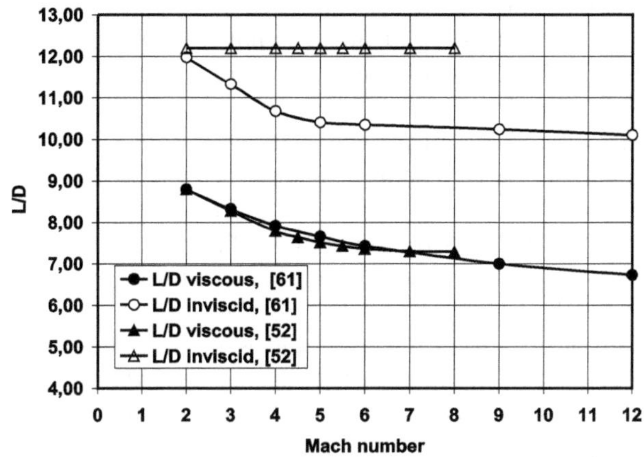

Fig. 4.58. Influence of viscous effects on the aerodynamic performance of two waveriders designed by the osculating cone method, [52], [61]. $M|_{design} = 12$ in both cases.

waveriders given in this reference are only the viscous ones and no information about the inviscid values of these shapes is available. Therefore, we refer to later investigations where on the one hand for the design of the waveriders viscous effects are included, but on the other hand the performance of the design was checked by several viscous and inviscid methods, [52] to [54], [60], [61].

Fig. 4.58 shows the aerodynamic performance as function of the free-stream Mach number for two waverider designs of the type displayed in Fig. 4.56. The first one, [52], has the design Mach number $M|_{design} = 12$ and the shock angle $\sigma|_{design} = 8.17°$. The flow fields of the off-design Mach numbers were determined by a shock expansion method for the inviscid part, and the viscous effects were modelled by adding the skin-friction of a flat plate with a fully turbulent boundary layer ($T_w = 1{,}000.0\ K$, $Re = 10^8$). The second one, [61], has again the design Mach number $M|_{design} = 12$, but a shock angle of $\sigma|_{design} = 9.1°$. In that case the inviscid flow field was determined by a numerical simulation method solving the Euler equations whereas the viscous effects were treated as before ($T_w = 1{,}000.0\ K$, $Re = 10^8$ to 10^9).

Generally, it seems that for high supersonic and hypersonic Mach numbers the inviscid L/D values are nearly Mach number independent. This is in contrast to the viscous L/D values which decrease with increasing Mach number. Considering the case of [61], we observe that viscous effects reduce the aerodynamic performance roughly between 25 and 30 per cent in the Mach number range considered.

For a better assessment of the quality of these results it is worthwhile to have the data of more sophisticated numerical aerothermodynamic simulation methods, which take into account the phenomena mentioned at the beginning of this Sub-Section. In Table 4.8 we have listed L/D values calculated with the Euler and the Navier-Stokes equations. In that case the waverider shape investigated was designed for a wind tunnel situation with $M|_{design} = 8$ and $\sigma|_{design} = 11.7°$, [54], [62]. The Navier-Stokes solution with the wind tunnel Reynolds number $Re = 1.1 \cdot 10^6$ and the wall temperature $T_w = 295.0\,K$ shows for angles of attack, which generate nearly the highest L/D values ($\alpha = 0°, 3°$), a dramatic reduction of the L/D quantities compared to Euler solutions. For $\alpha = 3°$, for example, this amounts to 46 per cent. This is attributed to hypersonic viscous interaction due to the low Reynolds number.

However, for the typical flight Reynolds number $Re = 180 \cdot 10^6$ the reduction is more moderate with 23 per cent and is in the range of the values given above. It is a typical behavior that the viscous drag diminishes with increasing Reynolds number (but rises with the dynamic pressure q_∞, see, e.g., [8]). For completeness we refer to the fact that the flow for the low Reynolds number case was considered to be laminar, whereas for the high Reynolds number case turbulent flow was assumed with transition at $x/L = 0.03$. The wall temperature in the latter case was assumed to be constant with $T_w = 1{,}000.0\,K$.

Finally, we note here the observation, that waveriders, which are designed by an optimization process where viscous effects are included, have approximately the same magnitude of wave drag and friction drag, [45].

4.6.5 Off-Design and Low Speed Behavior

In the early days, when the waverider idea was coming up, there was the prevailing opinion, that the high L/D values are only attainable at the design point. It was argued that the maximum lift is exclusively achievable if the creating shock is attached to the leading edge, preventing flow spilling from the lower side to the upper side of the vehicle.

Later investigations have changed this view, in particular with methods able to analyze the details of a flow field, i. e., the methods of numerical

Table 4.8. Comparison of Euler and Navier-Stokes results for demonstrating the aerodynamic performance losses owing to viscous effects. Data taken from a waverider study with $M|_{design} = 8$ and $\sigma|_{design} = 11.7°$, [54], [62].

α	$L/D\|_{Euler}$	$L/D\|_{Navier-Stokes}$ *laminar* $Re = 1.1 \cdot 10^6$	$L/D\|_{Navier-Stokes}$ *turbulent* $Re = 180 \cdot 10^6$
0°	8.73	3.97	−−
3°	6.36	4.35	5.17
10°	3.49	3.13	−−

aerodynamics. We show in Fig. 4.59, that indeed the aerodynamic performance increases, when the flight Mach number M_∞ is decreased relative to $M|_{design}$.

Two of the curves ($\diamond$ [52], $\bullet$ [61]) in Fig. 4.59 have already been shown in Fig. 4.58 and pertain to waveriders designed by the osculating cone method. The third one ($\circ$) belongs to the modified shape [54], [61], Fig. 4.64 (left), and the fourth one ($*$) to a waverider designed by the singular cone method for $M|_{design} = 4$, [56]. All curves show the same tendency, but the last one provides lower L/D values and a stronger negative gradient with the Mach number. This can partly be explained on the one hand by the Reynolds number, which is one and a half order of magnitude smaller than the Reynolds number of the other cases (see Sub-Section 4.6.4), and on the other hand by the fact that on the upper side of the shape a smooth ogive-cylinder was integrated.

What is the reason for the generally observed increase of aerodynamic performance for flight Mach numbers smaller than the design Mach number? The numerical studies have revealed that for supersonic and hypersonic off-design Mach numbers, lower than the design Mach number, two effects come into play:

1. If we decrease continuously the free-stream Mach number M_∞, we observe that from a certain M_∞ value the shock will detach from the leading edge depending on the local sweep angle. In the vicinity of the highest sweep angle the shock detaches first. There, the flow expands around the leading edge, generating on the upper side a pressure decrease with the consequence of growing lift.

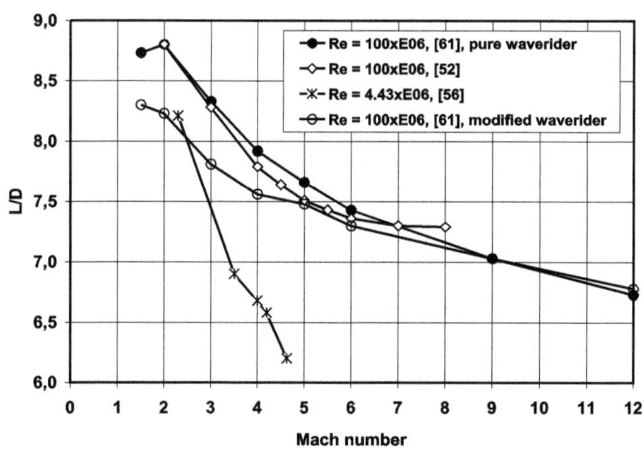

Fig. 4.59. Off-design behavior in the supersonic and the hypersonic regime for various waverider designs. Aerodynamic performance L/D including viscous effects, $M|_{design} = 12$, [52], $M|_{design} = 4$, [56], $M|_{design} = 12$, [61].

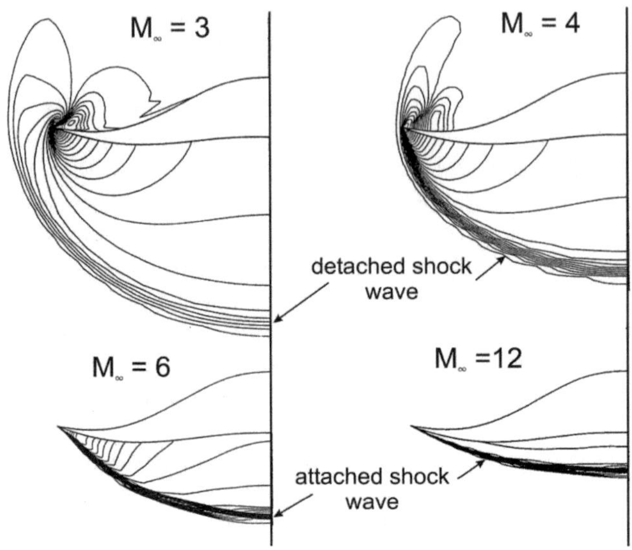

Fig. 4.60. Shock waves in the exit plane at off-design Mach numbers for a waverider with $M|_{design} = 12$ and $\sigma|_{design} = 9.1°$, Euler solution, [61].

We demonstrate the effect with Fig. 4.60, [61], where the shock waves in the exit plane found with Euler solutions are plotted for various free-stream Mach numbers. For the design Mach number $M_\infty = 12$ we have all along the leading edge an attached shock wave. This is nearly maintained down to $M_\infty \approx 6$, where the shock detachment begins at the leading edge position with the largest sweep angle, [54], but in the exit plane the shock is still attached, Fig. 4.60 lower left. However, for $M_\infty = 4$ in the regime where the shock is detached from the leading edge an expansion around this leading edge is discernible, which is intensified for smaller Mach numbers. This expansion leads to a small supersonic region on the upper side of the wing, which is terminated in inboard direction by an embedded cross-flow shock. The resulting low pressure increases the lift L. The wave drag is not substantially increased by the weak embedded cross flow shocks, because the upper surface is aligned with the free-stream. Therefore the aerodynamic performance L/D will grow.

2. By the further decrease of M_∞ the flow separates along the subsonic leading edge (primary separation) and generates a vortex sheet, which curls up and forms a leading edge vortex above the upper side of the vehicle[38].

[38] The vortex sheet in reality consists of the two boundary layers from below and above the leading edge. The mechanism by which in Euler solutions vorticity is created is mainly due to flow kinematics, thermodynamics (entropy gradient) and artificial viscosity of the numerical method. For more information see, e.g., [63].

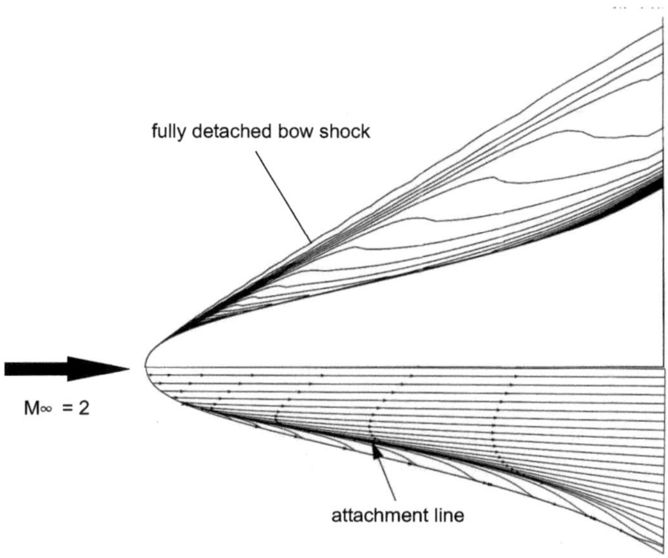

Fig. 4.61. Detached bow shock (upper half) and streamlines on the upper surface (lower half) for the off-design Mach number $M_\infty = 2$ of a waverider with $M|_{design} = 12$ and $\sigma|_{design} = 9.1°$, Euler solution, [54].

This again leads to an substantial increase of the lift (non-linear lift). Once more, if the upper side is a free-stream surface, no wave drag is produced by this effect and the aerodynamic performance will grow.

In Fig. 4.61, [54], the surface streamlines and especially an attachment line can be seen on the upper side of the shape (free-stream Mach number $M_\infty = 2$). The leading edge is subsonic in the interval $0.04 \lessapprox x/L \lessapprox 0.97$. We can speak here of a pseudo spilling, because the attaching flow comes from the free-stream of the lower side of the shape, which, though, is the flow behind the bow shock.

We consider now how these flow phenomena effect the magnitude of the lift coefficient C_L. In Fig. 4.62 the shares of the lower side and the upper side to the total lift coefficient as function of the free-stream Mach number are displayed. For free-stream Mach numbers $M_\infty \lessapprox 6$ the upper side produces a contribution to the lift which grows with decreasing M_∞ and reaches for $M_\infty = 2$ a share of approximately 17 per cent of the total C_L value.

The sketch of the topology of the flow field past the waverider shape in a cross section at $x/L \approx 0.5$, Fig. 4.63, helps to understand the streamline pattern and the attachment line shown in Fig. 4.61. In a cross section attachment locations, separation locations et cetera show up as singular points.

On the lower side of the shape two attachment lines exist. Because the flow between them is more or less parallel, we count them as quarter saddle points S_1'' and S_2''.

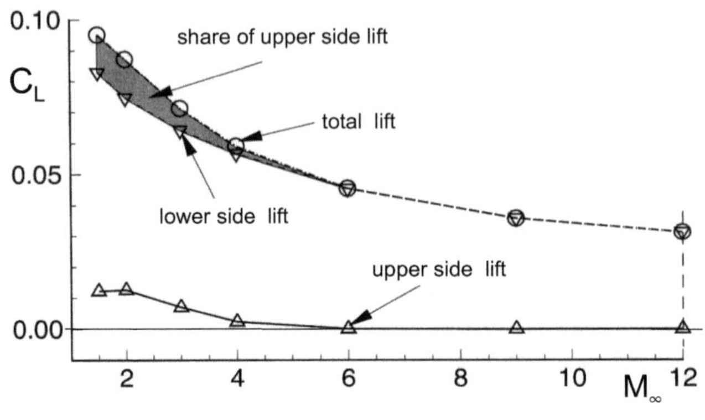

Fig. 4.62. Influence of the upper side of a waverider on the lift coefficient, $M|_{design} = 12$, $\sigma|_{design} = 9.1°$, Euler solution, [54].

At the leading edges the flow separates (flow-off separation, [64]) and we count two half saddle points S_1' and S_2'. The resulting vortex sheets curl up to form vortices (focus points F_1, F_2). Below and inboard of the vortices, flow attaches (the attachment line in Fig. 4.61) on the upper surface (half saddle points S_3' and S_4'). This flow has not left the lower vehicle shape in a separation process, instead it is original free-stream flow, which has crossed, with a slight deflection, the bow shock[39].

The situation, finally, between the attachment lines on the upper side is not so clear, but we can postulate a detachment line in the symmetry plane, which we count as half saddle point S_5'.

In closing this discussion we note that the flow pattern in the cross section obeys the topological rule which connects the numbers of singular points: saddle points S, half saddle points S', quarter saddle points S'', nodal and focal points ($N = F$), and half nodal and focal points ($N' = F'$), [65]:

$$\left(\Sigma N + \frac{1}{2}\Sigma N'\right) - \left(\sum S + \frac{1}{2}\sum S' + \frac{1}{4}\sum S''\right) = -1. \qquad (4.11)$$

Realistic viscous flow exhibits second-order and sometimes higher-order separation phenomena on the upper side of the configuration. Separation on the upper side can happen also due to the cross-flow shocks. If the leading edge is rounded, the primary separation line will lie on the upper side, again with second-order et cetera separation phenomena present[40].

[39] This flow carries the original total enthalpy. The diverging flow pattern at S_3' and S_4', Fig. 4.61, which is present also at S_1'' and S_2'', together with this original total enthalpy, causes the typical hot-spot situation at attachment lines during hypersonic flight, [8].

[40] The reader will find details regarding such flow cases (Blunt Delta Wing, BDW) in [8].

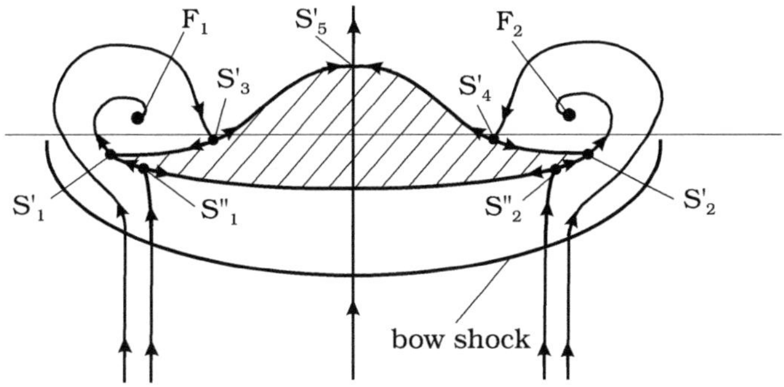

Fig. 4.63. Sketch of the topology of the flow field past the waverider shape of Fig. 4.61 in a cross section at $x/L \approx 0.5$ (exaggerated shape and flow features).

The literature for the analysis of the flow behavior in the subsonic and the transonic regime of waveriders is very rare. In [66] information is given about experiments in NASA Langley's Low Speed Tunnel using the so called LoFLYTE^{TM} model[41]. This configuration is based on a waverider approach, but with contour changes necessary for the integration of stabilization, trim and control surfaces, Sub-Section 6.6.1, as well as engine nacelle and canopy.

So far no performance data for subsonic and transonic Mach numbers with respect to a pure waverider shape are available. We therefore refer to a configuration which is a modification of the shape shown in Fig. 4.56. The reasons for the modifications are to design a vehicle with enough volume for tanks and equipment and with the capability to integrate elevons and ailerons, [54], [60], [61], [67] to [69].

To achieve that, the upper surface, originally a free-stream surface, was changed. In order to maintain the necessary volume, the part around the symmetry plane was not modified, but the outer part was replaced by an expansion surface, so that a sharp trailing edge was generated, which has also the effect to reduce the base drag. Furthermore, between the wing tip and the nearly unchanged central part, two plane surfaces were defined for the integration of elevons and ailerons which demand straight hinge lines. This has required some slight modifications also of the lower surface[42], Fig. 4.64 left.

In Fig. 4.59 the L/D values of the modified waverider are also plotted ($\circ$ [61]), showing how the configurational changes alter the aerodynamic performance. Down to $M_\infty \approx 8$ the L/D losses are very small, then they grow, reaching a maximum at $M_\infty = 2$ with approximately 7 per cent.

[41] LoFLYTE^{TM} is the trademark of Accurate Automation Corporation, Chattanooga, Tennessee.

[42] This configuration was designed at the German Aerospace Establishment DLR and named the "DLR-F8" wing.

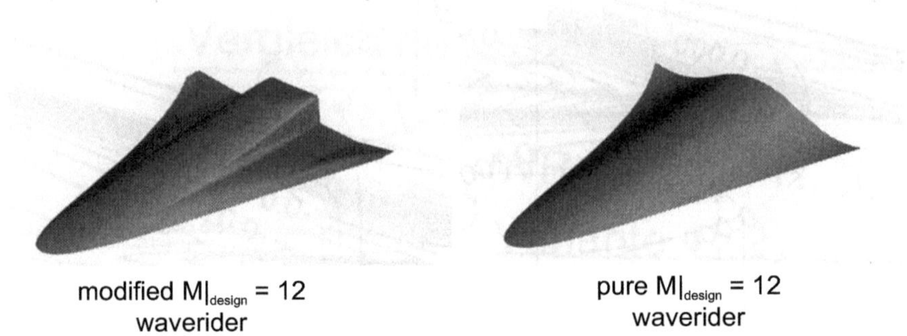

Fig. 4.64. Comparison of waverider shapes with $M|_{design} = 12$, [54], [61]. Left modified, right original shape (see Fig. 4.56).

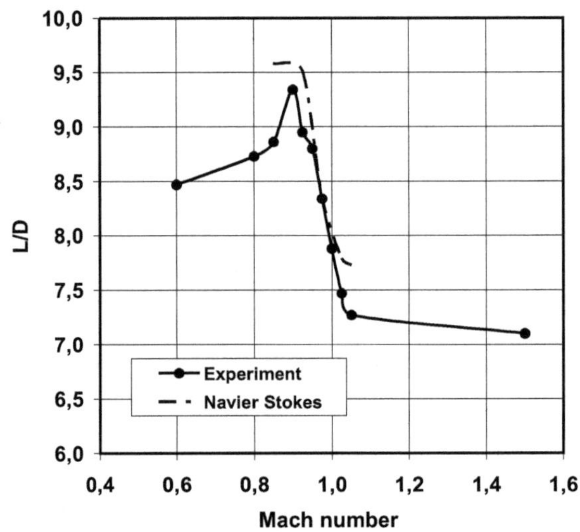

Fig. 4.65. Aerodynamic performance of the modified waverider (Fig. 4.64 left), in the subsonic and the transonic regime, [60]. Comparison of experimental data and Navier-Stokes results. Angle of attack $\alpha \approx$ - 1°, $Re_\infty \approx 6.5 \cdot 10^6$.

The behavior of the aerodynamic efficiency in the subsonic and the transonic regime is shown in Fig. 4.65. In the low supersonic regime, when the flight Mach number decreases, the suction along the leading edge region (see Fig. 4.61) increases, which produces higher (non-linear) lift and therefore an increase of L/D.

The supersonic flow field contains two major shock waves. The first one is the front shock wave (that is the remaining part of the generating shock surface, which brings about the lift in hypersonic flight), and the second one is the terminating shock wave at the trailing edge due to the expanding flow on the upper surface. When the flight Mach number is further diminished

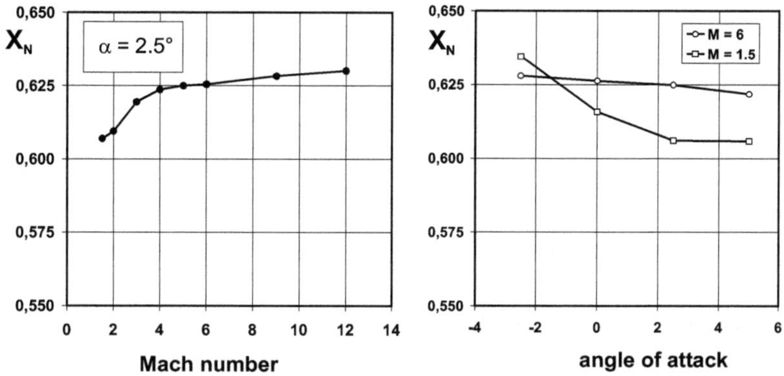

Fig. 4.66. Neutral point location X_N as function of Mach number (left) and angle of attack (right) for a waverider with $M|_{design} = 12$ and $\sigma|_{design} = 9.1°$, Fig. 4.64, [54].

into the transonic regime, the front shock disappears and the terminating shock moves in upstream direction while it weakens. This leads to a drastic reduction of the wave drag. Although the lift is also reduced, but due to the distinct leading edge vortex less than the wave drag, the aerodynamic performance jumps from approximately 7.2 ($M_\infty = 1.05$) to 9.4 ($M_\infty = 0.9$) and has still a value of 8.5 at $M_\infty = 0.6$, Fig. 4.65. The reliability of this outcome is endorsed by the good agreement between the experimental data and the Navier-Stokes results in the transonic regime.

4.6.6 Static Stability Considerations

It is not possible to determine beforehand for a waverider shape the longitudinal stability, since this needs at least a clear definition of the center-of-gravity of the fully integrated flight vehicle and in case of a powered plane the impact of the propulsion system on aerodynamics and flight mechanics. But what one can do, is to exhibit the characteristics of the neutral point, Chapter 7, as function of the Mach number and the angle of attack. This enables one to assess the impact of necessary shape modifications on the longitudinal stability, like the ones for volume enhancement or improved volume distribution, integration of propulsion system and aerodynamic controls as well as for base drag reduction.

For the waverider of Fig. 4.64 (left) the location of the neutral point x_N is down to a Mach number $M_\infty \approx 4$ only slightly Mach number dependent, which is due to the fact that down to this Mach number the leading edge is supersonic. For lower Mach numbers this situation changes and x_N decreases more rapidly which leads to a potential reduction of stability, Fig. 4.66 (left), [54]. The angle of attack dependency for a flight Mach number $M_\infty = 6$ is rather low, whereas for $M_\infty = 1.5$ the neutral point moves upstream with increasing α. This is due to the increased influence of the leading edge vortex on the upper surface, Fig. 4.66 (right).

In any case we observe that in this example the neutral point lies rather far forward. To achieve a corresponding location of the center-of-gravity is problematic. On the other hand, if the static margin is too large, trim drag will become large, if it is small, all data uncertainties need to be small, too, in order to reduce design risks.

For completeness we point to some literature where work on lateral stability issues for several waverider shapes is discussed: [52], [53], [66], [70], [71].

4.6.7 Waverider – Operable Vehicle?

To make a waverider an operable flight vehicle, specific modifications of the shape have to be made. That is what we already have learned from the modified waverider shape (DLR-F8) discussed in the last Sub-Section. There, for the integration of aerodynamic trim and control surfaces (straight hinge lines) and the reduction of the base area (which reduces the base drag in particular in the subsonic and the transonic flow regime), the upper surface was modified such that a straight sharp trailing edge is achieved, Fig. 4.64 (left). In addition in the central part of the configuration the volume necessary for integration of the tank, the engine nozzle and the equipment is established. As could be shown, all that has not reduced very much the L/D value in the high supersonic and hypersonic regime and the maximum reduction of 7 per cent (at $M_\infty = 2$) was indeed moderate, Fig. 4.59.

Another critical point is the sharpness of the leading edge. Thermal loads are proportional to the geometry of the leading edge: $\sim (\cos\varphi)^{0.5} R^{-0.5}$, [8], where φ is the local sweep angle and R the local nose radius of the leading edge[43]. Materials and structure limitations hence require a finite bluntness of the leading edge, which must be larger at low-sweep portions than at high-sweep portions. On the other hand such bluntness should be as small as possible since it generates additional wave drag.

In order to get an impression about the sensitivity of the aerodynamic efficiency with respect to leading edge bluntness, we have listed in Table 4.9 some data originating from [54], [60], [67], [70]. Euler solutions were conducted for the pure waverider shape, Figs. 4.56 and 4.64 (right), with a leading edge radius of $R/L_{ref} = 0.0008$, regardless of the local sweep angle. This means that for a real vehicle with $L_{ref} = 70.0\ m$ the radius is $56.0\ mm$. The table shows a substantial influence, in particular, for high Mach numbers, which is not surprising, since the wave drag increases with increasing Mach numbers.

System studies and multidisciplinary research on fluid-structure interactions done in [72], [73] have revealed that the thermal load problem at the

[43] This is the proportionality for laminar flow, that for turbulent flow is similar, but with different exponents. The proportionality for the non-swept blunt leading edge for laminar flow is simply $\sim R^{-0.5}$, since $\varphi = 0°$.

Table 4.9. Influence of leading edge bluntness on the aerodynamic performance of a waverider with $M|_{design} = 12$ and $\sigma|_{design} = 9.1°$, Figs. 4.56 and 4.64. Comparison based on Euler solutions, data taken from [54], [60], [67], ($|^*$ values linearly extrapolated).

M_∞	$L/D|^{(1)}$ sharp leading edge	$L/D|^{(2)}$ R/Lref = 0.0008	$L/D|^{(3)|^*}$ R/Lref = 0.0001	$\frac{(1)-(2)}{(1)}$ in per cent	$\frac{(1)-(3)|^*}{(1)}$ in per cent
12	10.20	7.97	9.92	21.8	2.73
9	10.21	8.35	9.98	18.2	2.27
6	10.24	8.88	10.07	13.3	1.66
3	11.35	10.55	11.25	7.0	0.88

leading edge can be solved even with leading edge radii of order $R/L_{ref} \approx 0.0001$. By assuming this value and having in mind that there is a linear dependency of the wave drag on the leading edge radius R, which is justified by investigations reported in [54], we get the values in the 4th and the 6th column of Table 4.9. These values should be verified by data from more detailed investigations (which are actually not available) or from free-flight experiments. Nevertheless, they point to the conclusion, that possibly thermal loads related leading edge bluntness does not play a substantial role for the aerodynamic efficiency of waveriders.

In order to design waverider shapes, which have practical relevance, it is worthwhile to obey two general rules.

The first one says, that with increasing angle of the generating shock surface the thickness of the vehicle shape increases (and therefore the available volume), at the price of a diminished aerodynamic performance. The second one deals with the curvature of the contour in cross sections, which is much smaller for high design Mach numbers compared to low design Mach numbers. Small or moderate curvatures ease the integration of aerodynamic trim and control surfaces, tanks and other equipment. Since, as already shown in Fig. 4.58, the aerodynamic performance L/D is only marginally dependent on the Mach number in the hypersonic regime, it could be a possible practice, to use a high Mach number for the shape design, say $M|_{design} = 12$, and to have a cruise Mach number of $M|_{cruise} = 6$, see also Table 4.9.

Of course, the general goal of the waverider idea is to develop an efficient hypersonic aircraft. This implies, as was several times mentioned above, the integration of various devices and volumes. In this regard we look now at a waverider which was designed by the singular cone method for $M|_{design} = 4$, [56]. Since wind-tunnel experiments were made with this shape, already the baseline shape did possess on the upper side a volume enhancement for the installation of the balance and the sting adapter, which looks like a smooth canopy, Fig. 4.67.

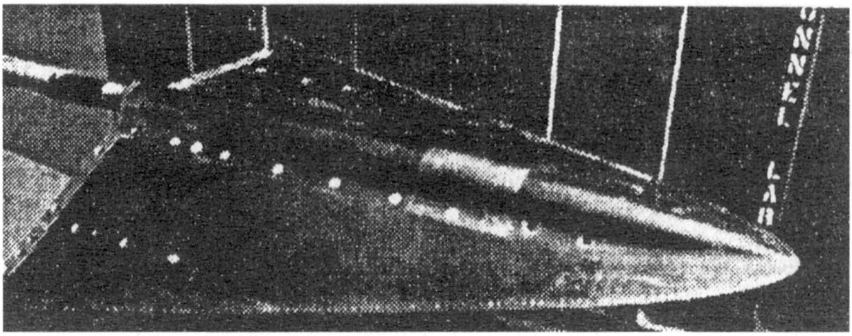

Fig. 4.67. Photograph of baseline waverider designed with the singular cone method for $M|_{design} = 4$ including volume enhancement for providing the balance housing, [56].

A propulsion system with an inlet and a nozzle/expansion ramp, elevons and ailerons – but no vertical stabilizer – and a faceted canopy[44] were integrated into this shape[45], [56]. In Table 4.10 the aerodynamic efficiency for the fully integrated and the clean configuration is compared. It shows a drastic decrease of the L/D values. At the design Mach number the propulsion system contributes approximately 20 per cent to this efficiency loss and the elevons and ailerons approximately 14 per cent, [56].

Again, we remind the reader, that it is difficult to assess and compare such values, since there is no common basis for the design changes of the various investigations. Nevertheless, these data give a preliminary hint in which range aerodynamic performance losses are to be expected.

Regarding the question of operability there are further issues, which need to be considered.

We had a look at longitudinal (static) stability, Sub-Section 4.6.6. It is a general problem of CAV-type flight vehicles to place the propulsion system and the fuel tanks, which will be emptied during flight, such that trim and stable flight can be achieved along the entire trajectory. An additional problem is the aerothermoelastic behavior of the airframe, and that the axial thrust coefficient and the thrust vector angle of propulsion are Mach number dependent, Sub-Section 2.2.3. To fly a large CAV-type vehicle statically unstable, is a dangerous and uncertain option, that would demand a tremendous research and development effort in the area of active control on top of the basis effort to create such a vehicle.

The integration of airframe/airbreathing propulsion on the lower side of the waverider can be accomplished. The principle problems are those discussed in Section 4.5. Also for the waverider the sensitivity of the net thrust on changes of the angle of attack, Sub-Section 4.5.4, exists.

---

[44] The faceted canopy produces additional wave drag compared to the prior mentioned smooth canopy.

[45] Only the geometrical impact is considered.

Table 4.10. Aerodynamic efficiency loss due to the integration of vehicle components into a clean waverider configuration. Shape designed by the singular cone method, $M|_{design} = 4$, [56].

M_∞	$L/D\|_{max}$ clean configuration	$L/D\|_{max}$ fully integrated	$\Delta(L/D)\|_{max}$ performance loss in per cent
2.30	8.21	4.93	39.95
4.00	6.68	4.69	29.79
4.63	6.20	4.58	26.13

Besides the longitudinal stability the lateral or directional stability is of importance. In a natural way it is possibly given only by the configuration shown in Fig. 4.50. This means that in general one or two vertical stabilizers with large wetted surfaces are necessary. These put more mass to the aft of the vehicle (cog location!) and deteriorate further the ratio of volume to surface of the vehicle, with the associated increase of viscous drag.

Another problem is the low roll damping property of waverider shapes. Together with the intrinsic roll-couplings found on such slender configurations it puts high demands on vehicle control.

Further, the low-speed behavior pertains not only the L/D at low speed, Sub-Section 4.6.5, but strongly the stability and control properties at large angles of attack particularly during approach and landing. Side wind sensitivity there is strong, because vortex dynamics, especially vortex breakup, becomes critical for strongly swept leading edges.

All these issues – some of them, however, also given with other vehicle shapes – are probably the reason, why the interest in pure waverider shapes is dwindling. If their operational L/D is not much better than those of other vehicles like given in Fig. 4.12, the incentives to use them are few. However, the question is, how much elements of pure waverider shapes can be combined with blended body and other configuration concepts in order to optimize their aerodynamic performance.

4.7 Problems

Problem 4.1 Consider the lower side of the forebody of the lower stage of the TSTO system SÄNGER. Estimate for the location $x = 55.0\ m$ a) the skin-friction coefficient and b) the heat flux in the gas at the wall for laminar and fully turbulent flow. Assume perfect gas, a Prandtl number $Pr = 0.74$, and further for the laminar case $T_w =$ const. $= 490.0\ K$, and for the turbulent case $T_w =$ const. $= 880.0\ K$. Employ simply at a flat plate the external inviscid flow data given in Table 4.2, case e. Compare the results with the respective data given in Fig. 4.8.

Problem 4.2. Compute with the data of Problem 4.1 for the location $x = 55.0\ m$ the boundary layer thickness δ and the displacement thickness δ_1 for laminar and fully turbulent flow as well as the scaling thickness of the viscous sub-layer δ_{vs} for fully turbulent flow. Compare the results with the respective data given in Fig. 4.9.

Problem 4.3. Repeat the computations from Problem 4.1 and 4.2 with the most simple representation of the lower side of the forebody, viz. a flat plate at zero angle of attack. Employ the external inviscid flow data given in Table 4.2, case ∞. Take the wall temperatures from Problem 4.1. Compare the results with those from Problem 4.1 and 4.2.

Problem 4.4. Compute with the approach of Problem 4.3 the flow data at the lower forebody location $x = 55.0\ m$ of the lower stage of SÄNGER flying with $M_\infty = 3.5$ at $H = 20.0\ km$. Take all free-stream parameters from Table 12.1. Assume perfect gas, fully turbulent flow and $T_w = 690.0\ K$. a) Why is the heat flux in the gas at the wall, q_{gw}, so small? b) The boundary layer thickness δ_{turb} is a measure for the height of the boundary layer diverter. Compare it with the ones found in the Problems 4.2 and 4.3. c) What happens to the skin-friction, if the wall temperature is decreased?

Problem 4.5. Derive from eq. (4.8) for large Mach numbers M_1 and small angles $\overline{\alpha}$ the limit $h_2/h_1 = 0.167$ in Fig. 4.42. For large Mach numbers M_1 and small angles $\overline{\alpha}$ we have, see, e.g., [8], the relation:

$$\theta = \frac{\gamma + 1}{2} \overline{\alpha}.$$

Problem 4.6. The lower side of the forebody of SÄNGER flies at $M_\infty = 6.8$ with an inclination angle $\overline{\alpha} = 9.4°$. If $\overline{\alpha}$ is changed by $\Delta\overline{\alpha} = -\ 1°$ and $+\ 1°$, how large is in each case the change in per cent of the lift and the drag coefficient. Use Newton's theory, see, e.g., [8]. Discuss the results in view of what Fig. 4.44 says.

Problem 4.7. The transformation of spherical coordinates r, θ, φ to cartesian coordinates is given by, see the figure below:

$$x = r \sin\theta \cos\varphi,$$
$$y = r \sin\theta \sin\varphi,$$
$$z = r \cos\theta.$$

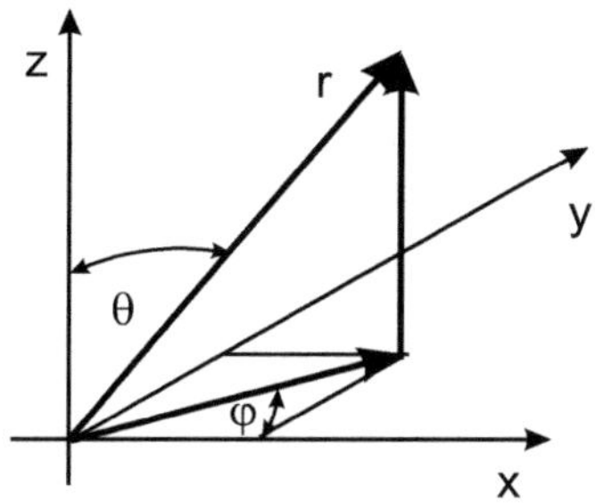

Definition of spherical coordinates.

Eq. (4.9) is valid for conical, axisymmetric and irrotational flow with the condition $v_\theta = \dfrac{\partial v_r}{\partial \theta}$.

Prove that $v_\theta = \dfrac{\partial v_r}{\partial \theta}$ by using the irrotationality condition rot $\underline{v} = 0$ in spherical coordinates.

Problem 4.8. The neutral point concept, Sub-Section 7.4, assumes that the first right hand side term of the upper part of eq. (7.18) vanishes, which means that there exists no offset in the z-direction of the center-of-gravity ($z_{cog} = 0$).

The problem defined here will show how accurate the center-of-gravity has to be embodied to meet the above assumption. For waveriders flying with hypersonic Mach numbers the following statements are true:

- the location of the neutral point is nearly independent of Mach number and angle of attack,
- the difference between the locations of the neutral point x_N and the center-of-gravity x_{cog} is typically of the order of 0.5 per cent.

For a waverider of the type shown in Fig. 4.64 (right) we extract from [54] for the flight Mach number $M_\infty = 8$ the aerodynamic coefficients given in the table:

α	C_L	C_D
0°	0.0470	0.0117
3°	0.0902	0.0203
10°	0.2120	0.0683

Assume the length of the waverider to be $L_{ref} = 80.0\ m$. Determine for the three angles of attack the permissible z offset, if the influence of this offset is not to exceed 1 per cent of the pitching moment coefficient C_m.

References

1. Hunt, J.L.: Hypersonic Airbreathing Vehicle Design (Focus on Aero-Space Plane). In: Bertin, J.J., Glowinski, R., Periaux, J. (eds.) Hypersonics. Defining the Hypersonic Environment, vol. 1, pp. 205–262. Birkhäuser, Boston (1989)
2. Hornung, M.: Entwurf einer luftatmenden Oberstufe und Gesamtoptimierung eines transatmosphärischen Raumtransportsystems (Design of an Airbreathing Upper Stage and Global Optimization of a Transatmospheric Space Transportation System). Doctoral thesis, Universität der Bundeswehr, Neubiberg/München, Germany (2003)
3. Koelle, D.E., Sacher, P.W., Grallert, H.: Deutsche Raketenflugzeuge und Raumtransporter-Projekte. Bernard und Graefe, Bonn, Germany (2007)
4. Steelant, J.: A European Project on Advanced Propulsion Concepts for Sustained Hypersonic Flight. In: 2nd European Conference for Aerospace Sciences (EUCASS), Brussels, Belgium, VKI (2007) ISBN 978-2-930389-27-3
5. Hirschel, E.H., Hornung, H.G., Oertel, H., Schmidt, W.: Aerothermodynamik von Überschallflugzeugen (Aerothermodynamics of Supersonic Aircraft). Technical Report: MBB/LKE112/HYPAC/1/A, MBB, München/Ottobrunn, Germany (1987)
6. Hirschel, E.H.: Aerothermodynamic Phenomena and the Design of Atmospheric Hypersonic Airplanes. In: Bertin, J.J., Periaux, J., Ballmann, J. (eds.) Advances in Hypersonics. Defining the Hypersonic Environment, vol. 1, pp. 1–39. Birkhäuser, Boston (1992)
7. Perrier, P.C., Hirschel, E.H.: Vehicle Configurations and Aerothermodynamic Challenges. AGARD-CP-600 3, C2-1–C2-16 (1997)
8. Hirschel, E.H.: Basics of Aerothermodynamics. Progress in Astronautics and Aeronautics, AIAA, Reston, Va. Springer, Heidelberg (2004)
9. Mack, A., Steelant, J., Togiti, V., Longo, J.M.A.: Impact of Intake Boundary Layer Turbulence on the Combustion Behaviour in a Sramjet. In: 2nd European Conference for Aerospace Sciences (EUCASS), Brussels, Belgium, VKI (2007) ISBN 978-2-930389-27-3
10. Williams, R.M.: National Aerospace Plane: Technology for America's Future. Aerospace America 24(11), 18–22 (1986)
11. Shea, J.F.: Report of the Defense Science Board Task Force on the National Aerospace Plane (NASP). Office of the Under Secretary of Defense for Acquisition, Washington, D.C (1988)
12. Nicolai, L.M.: Fundamentals of Aircraft Design. METS, Inc, San Jose, Cal. (1975)
13. Hauck, H.: Leitkonzept SÄNGER – Referenz-Daten-Buch. Issue 1, Revision 2, Dasa, München/Ottobrunn, Germany (1993)
14. Staudacher, W.: Entwurfsprobleme luftatmender Raumtransportsysteme. Space Course 1995, Universität Stuttgart, Germany (1995)
15. Küchemann, D.: The Aerodynamic Design of Aircraft. Pergamon Press, Oxford (1978)
16. Anderson Jr., J.D.: Introduction to Flight, 4th edn. McGraw Hill, New York (2000)
17. Kuczera, H., Krammer, P., Sacher, P.W.: SÄNGER and the German Hypersonic Technology Programme. IAF Congress Montreal, Paper No. IAF-91-198 (1991)

18. Kraus, M.: Aerodynamische Datensätze für die Konfiguration SÄNGER 4-92 (Aerodynamic Data Set for the SÄNGER Configuration 4-92). Technical Report: DASA-LME211-TN-HYPAC-290, Dasa, München/Ottobrunn, Germany (1992)
19. Hirschel, E.H.: The Technology Development and Verification Concept of the German Hypersonics Technology Programme. AGARD-R-813, pp. 12-1–12-15 (1996)
20. Weiland, C.: Stage Separation Aerothermodynamics. AGARD-R-813, pp. 11-1–11-28 (1996)
21. Bayer, R., Sachs, G.: Optimal Return-to-Base Cruise of Hypersonic Carrier Vehicles. Zeitschrift für Flugwissenschaften und Weltraumforschung (ZFW) 19(1), 47–54 (1992)
22. Esch, H.: Kraftmessungen zur Stufentrennung am MBB-SÄNGER Konzept bei Ma = 6 im Hyperschallkanal H2K. DLR Internal Report: IB - 39113 - 90C18 (1990)
23. Edwards, C.L.W., Small, W.J., Weidner, J.P.: Studies of Scramjet/Air-frame Integration Techniques for Hypersonic Aircraft. AIAA-Paper 75-58 (1975)
24. Berens, T.M.: Thrust Vector Optimization for Hypersonic Vehicles in the Transonic Mach Number Regime. AIAA-Paper 93-5060 (1993)
25. Herrmann, O., Schmitz, D.: Vorläufiges "Bookkeeping"-Verfahren für Hyperschall-Flugzeuge. MBB-FE125/124-TN-HYP-31, MBB, München/Ottobrunn, Germany (1988)
26. Lifka, H.: Bookkeeping-Korrekturen des Aerodynamischen Datensatzes der Längsbewegung für SÄNGER bezüglich Triebwerkseinbau. MBB-FE121-TN-HYPAC-0133, MBB, München/Ottobrunn, Germany (1990)
27. Lentz, S., Hornung, M., Staudacher, W.: Conceptual Design of Winged Reusable Two-Stage-To-Orbit Space Transport Systems. In: Jacob, D., Sachs, G., Wagner, S. (eds.) Basic Research and Technologies for Two-Stage-To-Orbit Vehicles, pp. 9–37. WILEY-VCH, Weinheim (2005)
28. Berens, T.M., Bissinger, N.C.: Study on Forebody Precompression Effects and Inlet Entry Conditions for Hypersonic Vehicles. AIAA-Paper 96-4531 (1996)
29. Berens, T.M., Bissinger, N.C.: Forebody Precompression Performance of Hypersonic Flight Test Vehicles. AIAA-Paper 98-1574 (1998)
30. Schaber, R.: Einfluss entscheidender Triebwerksparameter auf das Leistungsverhalten eines Hyperschall-Antriebs. MTU-N94-EP-0001, MTU, München/Karlsfeld, Germany (1994)
31. Hirschel, E.H.: The Technology Development and Verification Concept of the German Hypersonics Technology Programme. Dasa-LME12-HYPAC-STY-0017-A, Dasa, München/Ottobrunn, Germany (1995)
32. Staudacher, W., Wimbauer, J.: Design Sensitivities of Airbreathing Hypersonic Vehicles. AIAA-Paper 93-5099 (1993)
33. Hirschel, E.H.: The Hypersonics Technology Development and Verification Strategy of the German Hypersonics Technology Programme. AIAA-Paper 93-5072 (1993)
34. Rick, H., Bauer, A., Esch, T., Hollmeier, S., Kau, H.-P., Kopp, S., Kreiner, A.: Hypersonic Highly Integrated Propulsion Systems – Design and Off-Design Simulation. In: Jacob, D., Sachs, G., Wagner, S. (eds.) Basic Research and Technologies for Two-Stage-To-Orbit Vehicles, pp. 327–346. WILEY-VCH, Weinheim (2005)

35. Heiser, W.H., Pratt, D.T.: Hypersonic Airbreathing Propulsion. AIAA Education Series, Washington, D.C (1994)
36. Seddon, J., Goldsmith, E.L.: Intake Aerodynamics, 2nd edn. AIAA Education Series, Washington, D.C. (1999)
37. Bissinger, N.C., Schmitz, D.: Design and Wind Tunnel Testing of Intakes for Hypersonic Vehicles. AIAA-Paper 93-5042 (1993)
38. Fisher, S.A., Neale, M.C., Brooks, A.J.: On the Sub-Critical Stability of Variable Ramp Intakes at Mach Numbers Around 2. Aeronautical Research Council, R. and M. No. 3711, London, U.K (1972)
39. Reddy, D.R., Benson, T.J., Weir, L.J.: Comparison of 3-D Viscous Flow Computations of Mach 5 Inlet with Experimental Data. AIAA-Paper 90-0600 (1990)
40. Henckels, A., Gruhn, P.: Experimental Studies of Viscous Interaction Effects in Hypersonic Inlets and Nozzle Flow Fields. In: Jacob, D., Sachs, G., Wagner, S. (eds.) Basic Research and Technologies for Two-Stage-To-Orbit Vehicles, pp. 383–403. WILEY-VCH, Weinheim (2005)
41. Ferri, A., Nucci, L.M.: The Origin of Aerodynamic Instability of Supersonic Inlets at Subcritical Conditions. NACA RM L50K30 (1951)
42. Edney, B.: Anomalous Heat Transfer and Pressure Distributions on Blunt Bodies at Hypersonic Speeds in the Presence of an Impinging Shock. FFA Rep. 115 (1968)
43. Dailey, C.L.: Supersonic Diffuser Instability. J. Aeronautical Sciences 22(11) (1955)
44. Delery, J.M.: Shock Wave/Turbulent Boundary Layer Interaction and its Control. Progress in Aerospace Sciences 22, 209–280 (1985)
45. Bowcutt, K.G., Anderson Jr., J.D., Capriotti, D.: Viscous Optimized Hypersonic Waveriders. AIAA-Paper 87-0272 (1987)
46. Küchemann, D.: Hypersonic Aircraft and Their Aerodynamic Problems. Roy. Air. Establ., Technical Memorandum No. Aero 849 (1964)
47. Nonweiler, T.R.F.: Aerodynamic Problems of Manned Space Vehicles. J. Roy. Aeron. Soc. 63, 521–528 (1959)
48. Taylor, G.I., Maccoll, J.W.: The Air Pressure on a Cone Moving at High Speeds. Proc. Roy. Soc. (A) 139, 278–311 (1933)
49. Anderson Jr., J.D.: Modern Compressible Flow, 3rd edn. McGraw Hill, New York (2003)
50. Rasmussen, M.L.: Waverider Configurations Derived from Inclined Circular and Elliptic Cones. J. of Spacecraft 17(6), 537–545 (1980)
51. Sobieczky, H., Dougherty, F.C., Jones, K.D.: Hypersonic Waverider Design from Given Shock Waves. In: Proceedings of 1st Int. Hypersonic Waverider Symposium, University of Maryland (1990)
52. Eggers, T., Sobieczky, H., Center, K.B.: Design of Advanced Waveriders with High Aerodynamic Efficiency. AIAA-Paper 93-5141 (1993)
53. Eggers, T., Radespiel, R.: Design of Waveriders. Proceedings: Space Course, Paper No. 6, Technische Universität München, Germany (1993)
54. Eggers, T.: Aerodynamischer Entwurf von Wellenreiter Konfigurationen für Hyperschallflugzeuge (Aerodynamic Design of Waverider Configurations for Hypersonic Aircraft). Doctoral Thesis, Technische Universität Braunschweig, Germany, DLR Forschungsbericht 1999 - 10 (1999)
55. Jones, K.D., Dougherty, F.C., Seebass, A.R., Sobieczky, H.: Waverider Design for Generalized Shock Geometries. AIAA-Paper 93-0774 (1993)

56. Cockrell, C.E., Huebner, L.D., Finley, D.B.: Aerodynamic Performance and Flow-Field Characteristics of Two Waverider-Derived Hypersonic Cruise Configurations. AIAA-Paper 95-0736 (1995)
57. Cockrell, C.E.: Interpretation of Waverider Performance Data Using Computational Fluid Dynamics. AIAA-Paper 93-2921 (1993)
58. Center, K.B.: Interactive Hypersonic Waverider design and Optimization. Ph.D. Thesis, University of Colorado, Boulder (1993)
59. Miller, R.W., Argrow, B.M., Center, K.B., Brauckmann, G.J., Rhode, M.N.: Experimental Verification of the Osculating Cones Method for Two Waverider Forebodies at Mach 4 and 6. AIAA-Paper 98-0682 (1998)
60. Hummel, D., Blaschke, R.C., Eggers, T., Strohmeyer, D.: Experimental and Numerical Investigations on Waveriders in Different Flight Regimes. In: Proceedings of 21st ICAS Congress, ICAS-98-2,1,4, Melbourne (1998)
61. Eggers, T., Strohmeyer, D., Nickel, H., Radespiel, R.: Aerodynamic Off-Design Behavior of Integrated Waveriders from Take-off up to Hypersonic Flight. ESA SP-367 (1995)
62. Waibel, M.: Theoretische Untersuchungen über Stoss-Grenzschicht Wechselwirkungen an einem $M_\infty = 8$ Wellenreiter (Theoretical Investigations of Shock/ Boundary Layer Interactions at a $M_\infty = 8$ Waverider). Diploma Thesis, DLR Fotrschungsbericht 94-12, Germany (1994)
63. Hirschel, E.H.: Vortex Flows: Some General Properties, and Modelling, Configurational and Manipulation Aspects. AIAA-Paper 96-2514 (1996)
64. Hirschel, E.H.: Evaluation of Results of Boundary-Layer Calculations with Regard to Design Aerodynamics. AGARD-R-741, pp. 6-1–6-29 (1986)
65. Peake, D.J., Tobak, M.: Three-Dimensional Interaction and Vortical Flows with Emphasis on High Speeds. AGARDograph 252 (1980)
66. Hahne, D.E.: Evaluation of the Low-Speed Stability and Control Characteristics of a Mach 5.5 Waverider Concept. NASA TM-4756 (1997)
67. Eggers, T., Strohmeyer, D.: Design of High L/D Vehicles Based on Hypersonic Waveriders. Future Aerospace Technology in the Service of the Alliance, AGARD-CP-600, Vol. 3, Paper C18 (1997)
68. Strohmeyer, D., Eggers, T., Haupt, M.: Waverider Aerodynamics and Preliminary Design for Two-Stage-to-Orbit Missions, Part 1. J. of Spacecraft 35(4), 450–458 (1998)
69. Strohmeyer, D., Eggers, T., Heinze, W., Bardenhagen, A.: Planform Effects on the Aerodynamics of Waveriders for TSTO Missions. AIAA-Paper 96-4544 (1996)
70. Eggers, T., Radespiel, R., Waibel, M., Hummel, D.: Flow Phenomena of Hypersonic Waveriders and Validation of Design Methods. AIAA-Paper 93-5045 (1993)
71. Strohmeyer, D.: Lateral Stability Derivatives for Osculating Cones Waveriders in Sub- and Transonic Flow. AIAA-Paper 98-1618 (1998)
72. Heinze, W., Bardenhagen, A.: Waverider Aerodynamics and Preliminary Design for Two-Stage-to-Orbit Missions, Part 2. J. of Spacecraft 35(4), 459–466 (1998)
73. Haupt, M., Kossira, H., Radespiel, R.: Analyse von aerodynamisch belasteten Flügelvorderkanten mit der Methode der Fluid-Struktur-Kopplung. DGLR Jahrestagung, Paper DGLR-JT95 - 115, Annual Book No. 3 (1995)

5 Aerothermodynamic Design Problems of Non-Winged Re-Entry Vehicles

The transport of payload into space, either suborbital or orbital, and its return to the Earth's surface, is known to require the development and construction of suitable vehicles which are able to withstand the very severe thermal and mechanical (pressure and shear stress) loads encountered during such a mission. In the early days of space transport, the designers had the feeling that the vehicle shapes should be as simple and compact as possible. So, capsules and probes as the most important types of non-winged re-entry (RV-NW-type) vehicles were born.

In this chapter we deal with a few major aerothermodynamic design problems of RV-NW-type flight vehicles. Aerothermodynamic phenomena, high Mach number and total enthalpy effects as well as particular trends in aerothermodynamics of RV-NW-type vehicles are mostly similar to those of RV-W-type vehicles, Sections 3.1 and 3.2. However, we consider also vehicles for re-entry from higher altitudes than treated in Chapter 3, and also vehicles operating in non-Earth atmospheres.

A Lunar return of APOLLO takes place with a velocity much higher than those typical of RV-W-type vehicles, therefore we find in this case much more severe thermo-chemical phenomena. Also the flight in others than Earth's atmosphere leads to particular thermo-chemical description problems. We abstain from giving an overview of the special aerothermodynamic issues of such vehicles.

First, a general overview of the topics treated is given, strategies for atmospheric entry and orbital transfers are sketched, and configurational aspects are discussed. Because RV-NW-type flight vehicles as a rule have no aerodynamic stabilization, trim and control surfaces, we concentrate our considerations in two of the three main sections on issues of static and dynamic stability. A general treatment of static stability also of these vehicles is given in Chapter 7. The last section is dedicated to a discussion of thermal loads.

5.1 Introduction and Entry Strategies

The class of non-winged re-entry vehicles considered here comprises the ballistic entry probes (Sub-Class 1), the traditional capsules like APOLLO and SOYUZ (Sub-Class 2) as well as blunted cones, bicones and bent bicones (Sub-Class 3). While, normally, the capsules do not have aerodynamic control surfaces, the sub-class of cones may have some, in particular body flaps for longitudinal trim and, in case of multi-functional control surfaces with inclinations with respect to the lateral axis, also for roll control and lateral stability, Fig. 5.1 c).

For the capsule sub-class the lift-to-drag ratio during atmospheric re-entry is mostly in the range of $0.3 \lessapprox L/D \lessapprox 0.4$. Already here it should be mentioned, in order to avoid confusion, that for capsules a positive lift will only be obtained for negative angles of attack (if classical aerodynamic definitions are used, Fig. 7.3), because the aerodynamic lift force is caused predominantly by the front part (heat shield) of the vehicle, Fig. 5.1 a), whereas in case of a bicone the lift force is brought about by the whole body, Fig. 5.1 b). The explanation of this behavior is given in Section 5.3.

The aerodynamic efficiency of the blunted cone sub-class is somewhat higher and lies between $0.7 \lessapprox L/D \lessapprox 1.4$. In that case the contribution to the aerodynamic forces and moments is distributed over the whole body.

It is the intention of this chapter to provide the reader with detailed information about:

- the shape of some typical non-winged vehicles of the above mentioned three sub-classes,
- the requirements on their aerodynamic performance due to mission definition,
- some aspects of their aerodynamic data bases,
- static and dynamic stability,
- the role of the center-of-gravity regarding flyability and controllability,
- the influence of some geometrical shape variations on the aerodynamic coefficients,
- aerodynamic trim including parasite trim states,
- the influence of high-temperature real-gas effects on aerodynamic forces and moments as well as on aerodynamic trim,
- thermal loads.

To understand what kind of aerodynamic performance space vehicles must have and which thermal loads the configurations have to withstand, some terms describing the various strategies for atmospheric entry and transfer between orbits are now explained.

5.1.1 Aerobraking

Direct entry of probes and capsules into the atmosphere of any planets (Earth, Mars, Venus, Titan, et cetera), where the entry velocity is strongly reduced to

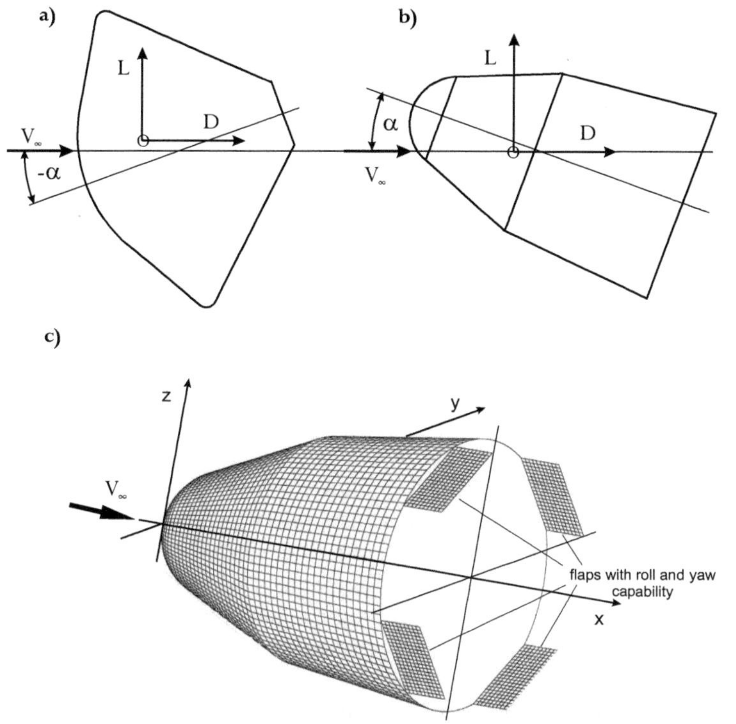

Fig. 5.1. Angle of attack for entry probe or capsule, a), and bicone, b), required for positive lift L. Sketch, illustrating the role of multi-functional control surfaces on bicones for pitch, yaw and roll, c).

a low descent speed, is called aerobraking entry. In principle it should always be possible to conduct an aerobraking entry if the following requirements can be satisfied (see also Chapter 2):

- resistance against thermal loads,
- minimization of vehicle mass ⇒ thermal protection system weight,
- restricted g-loads (n_t, n_n) ⇒ deceleration limit depending on payload,
- tolerable landing distortion (deviation from nominal landing position, recovery on ground or in water),
- minimum influence of atmospheric uncertainties[1] due to not well explored planets.

In reality, the entry strategy of a RV-NW-type vehicle has to be adjusted to the specific mission (entry velocity, entry angle, density of atmosphere, endurable g-loads, et cetera) which results in the decision to use either a ballistic vehicle, or a low L/D, or a moderate L/D lifting vehicle. Additionally,

[1] This holds even for the Earth atmosphere, Chapter 2. For properties of the Earth atmosphere see Chapter 12.

it could be necessary to decrease the orbital velocity by a boost of a retro-rocket system or by an aerocapturing maneuver, e.g., in case of large entry velocity. The entry corridor is bounded by a certain low entry angle (shallow entry) beyond which the vehicle leaves the atmosphere again, and by a certain high entry angle (steep entry), above which the g-loads reach too high values or the aerothermal loads can not be mastered.

Further, aerobraking is employed in order to support the initial propulsion boost during orbital transfer. Ballistic vehicles having no lift need usually a lot of passes through the atmosphere (e.g., for elliptic orbits in the periapsis regime of interplanetary missions) in order to reduce the speed for the target orbit, since the reduction per pass is low due to the limited energy reduction by aerodynamic drag in rarefied gas regimes, [1].

5.1.2 Aerocapturing

The main problem for orbital transfer, planetary (Mars, Venus, Moon, et cetera), and Earth return missions with high entry velocities is to diminish properly the energy of the vehicle, that is to reduce the velocity relative to the surface of the planet to be approached. Since the 1960s studies were undertaken to develop the physical and technological basis for reducing velocities by aeroassisted orbital maneuvers, [2]. A typical example of such a maneuver is as follows.

For an orbital transfer, say from geostationary (GEO) to low Earth orbit (LEO), the vehicle dips into the atmosphere, conducts an approximately constant drag flight, controlled by the lifting capability, until the velocity increment (Δv_∞) for a stable motion in the target orbit is reached, and skips back out of the atmosphere into just this target orbit. This process is called an aerocapturing mission.

In principle the flight control of the maneuver (lift control) can be carried out either by banking operations (see Chapter 2) or by angle of attack variations. Since the technology for lift control by pitch movement is rather complex and expensive in terms of system construction, in reality only bank-angle control systems are considered. The process described above is the same for a planetary mission (e.g. in the joint CNES–NASA Programme for Mars Sample Return, [3]), where again the velocity decrement is achieved by a single sufficiently deep atmospheric pass to transfer the vehicle from its hyperbolic trajectory to the target orbit about the planet.

During the 1970s and the early 1980s researchers had the opinion that aerocapturing maneuvers requires vehicles with lift-to-drag values larger than unity, which can only be provided by slender or bent bicones. Further investigations have shown that the aerocapturing capability can also be achieved with vehicles having a $L/D \approx 0.3$, [1], [4], but the ballistic factor has to be low. In order to broaden the physical basis for this space-mission concept in the United States a research programme was initiated, named Aeroassist Flight Experiment AFE, [5].

5.1.3 Ballistic Flight – Ballistic Factor

Ballistic flight is flight without lift: $L = 0$. In the flight mechanical equations for space applications the factor:

$$\beta_m = \frac{m}{A_{ref} C_D} , \tag{5.1}$$

called the ballistic factor or parameter, Sub-Section 2.1.1, with m being the mass, A_{ref} the reference area, and C_D the aerodynamic drag coefficient, plays a particular role, [6], [7]. This quantity is a measure for the manner how probes perform a ballistic entry in any atmosphere (Earth, Mars, Venus, Titan moon, et cetera) with a specified landing distortion. Generally, the system concept manager of a space mission has to decide, considering budget, costs, mission and/or vehicle reliability, tolerable landing distortion and so on, which kind of atmospheric entry the capsule or probe should conduct: either ballistic or lifting.

Ballistic probes have the advantage that they do not require guidance and control precautions. Therefore these concepts are less costly than the lifting ones, but they need low ballistic factors for direct entry. Low ballistic factor means: large reference area, high drag coefficient and low mass. Normally, for all missions known, an appropriate mass reduction is critical. The magnitude of drag is limited by the semi-apertural cone angle $\phi = \pi/2 - \Theta_1$, Figs. 5.3 and 5.4, which however can generate static stability problems, since the center-of-pressure is moved forward. A good compromise is a cone angle of $50° \lessapprox \phi \lessapprox 70°$ (HUYGENS and BEAGLE2: $\phi = 60°$, OREX: $\phi = 50°$).

A proper means for reducing the ballistic factor is the increase of the frontal area (A_{ref}), but one should have in mind that this could increase the thermal heat-shield mass. On the other hand, low ballistic factors provide low thermal loads. Finally, the nose radius R_1, Fig. 5.3, does not affect very much the drag, but a large nose radius can contribute to a reduction of the mass of the Thermal Protection System (TPS) due to a decrease in the magnitude of the surface, [1], [7]. An upper limit for the ballistic factor of ballistic vehicles seems to be $\beta_m \approx 60.0\ kg/m^2$. Lifting capsules, such as the ARD, APOLLO and VIKING, can have much larger values since their entry flight can be guided and controlled by an onboard stability and control system, Table 5.1.

Fig. 5.2 shows a comparison of flight trajectories for a ballistic re-entry (OREX) and a low L/D re-entry (ARD).

5.2 General Configurational Aspects

5.2.1 Ballistic Probes

In the past there were some scientific space exploratory missions to other planets or moons of planets of our solar system using ballistic probes. In the early days of space exploration the probes PIONEER (1978) and MERCURY

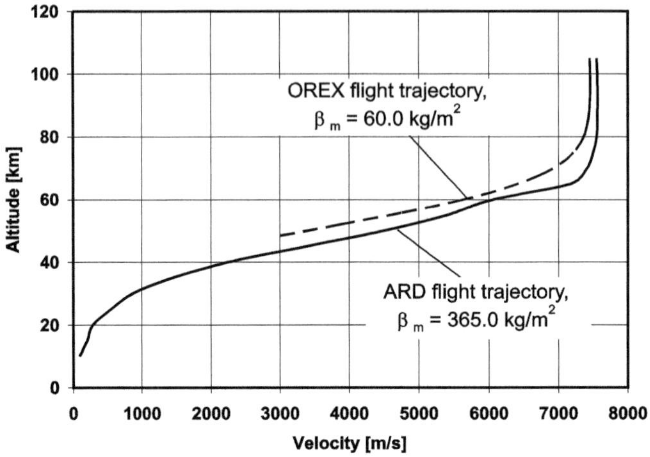

Fig. 5.2. Flight trajectories of OREX, [8] and ARD, [9], in the altitude velocity map.

orbiter (1959-1963) were flown to the planet Venus, VIKING (1975-1982) (with some lifting capability) was travelling to Mars and GALILEO (1989-1995) went to the planet Jupiter.

Since the Saturn moon Titan has an atmosphere, from which some scientists expect that extra-terrestrial life will develop in the future, the European HUYGENS probe was designed in the 1990s in order to explore the atmosphere and the ground conditions. It was launched in 1997 as a passenger

Table 5.1. Ballistic factor $\beta_m = m/(A_{ref} C_D)$ of ballistic probes and lifting capsules.

Vehicle	Mass $[kg]$	A_{ref} $[m^2]$	Drag C_D	Ballistic factor β_m $[kg/m^2]$	Ref.
OREX	761.0	9.08	$\approx 1.40^{\alpha=0^\circ}_{M=\infty}$	60.0	[10], [11]
EDV No.3	42.9	0.7854	$0.9595^{\alpha=0^\circ}_{M=\infty}$	57.0	[12]
HUYGENS	≈ 300.0	5.73	$\approx 1.52^{\alpha=0^\circ}_{M=\infty}$	34.0	[13] to [15]
BEAGLE2	60.0	0.636	$\approx 1.45^{\alpha=0^\circ}_{M=\infty}$	65.0	[16], [17]
ARD	2,800.0	6.16	$1.247^{\alpha=-22.8^\circ}_{M=10}$	365.0	[18]
APOLLO	5,470.0	12.02	$1.247^{\alpha=-22.7^\circ}_{M=10}$	365.0	[19], [20]
VIKING type	9,200.0	15.20	$1.391^{\alpha=-23.9^\circ}_{M=10}$	435.0	[21]

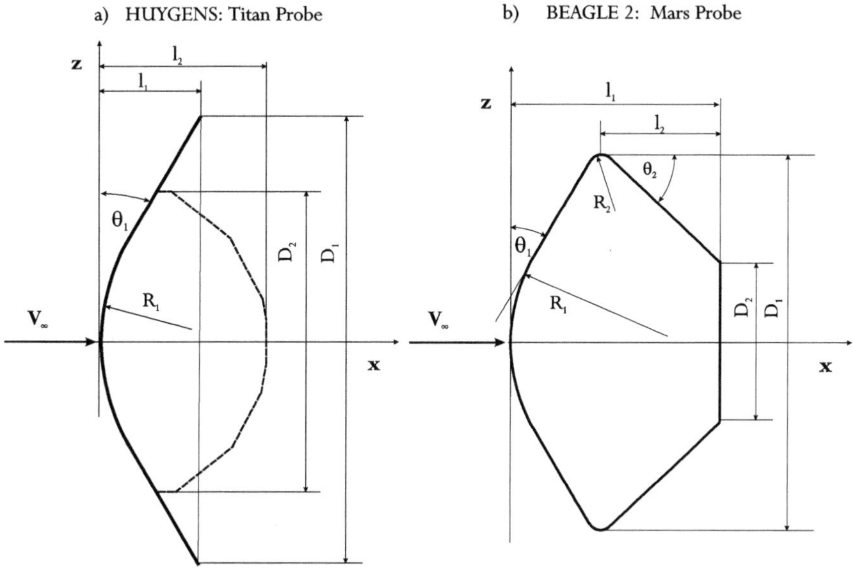

Fig. 5.3. Shape definition of the ballistic probes HUYGENS, [1] to [14], [15], and BEAGLE2, [16], [17]. HUYGENS has a rugged back contour, therefore it is idealized with a dashed line.

on-board of the American Mariner-Mark II CASSINI orbiter. After a flight of roughly seven years HUYGENS has conducted a very successful entry into the Titan atmosphere in January 2005.

To ensure a stable ballistic flight and to master the thermal loads during entry in a not well known atmosphere was the aerothermodynamic challenge of this mission. The composition of Titan's atmosphere consists approximately of 87 per cent N_2, 10 per cent Ar and 3 per cent CH_4 (in molar fractions). The geometrical definition of the HUYGENS probe is given in Fig. 5.3 a) and Table 5.2.

In the frame of a recent space mission to the planet Mars the British small and low-cost probe BEAGLE2 was ejected from ESA's "Mars Express" (launched in June 2003) in order to conduct a ballistic entry into Mars' atmosphere. The capsule had a mass of 60.0 kg with a payload of 30.0 kg. Once having arrived at the Martian surface, a six month scientific mission was planned to follow. The Martian atmosphere consists essentially of 97 per cent CO_2, 3 per cent N_2 (in molar fractions) and some traces of Ar. Besides the provision of a reliable aerodynamic data base for a safe landing on the surface, the determination of the thermal loads was the main task of the planned mission [16], [17]. Unfortunately, BEAGLE2 was lost without knowing the exact reasons. Fig. 5.3 b) shows the shape of BEAGLE2.

The space programme in Japan had the objective to develop an unmanned winged orbiter called HOPE. To reach this goal several demonstrators were

Table 5.2. Geometrical data and mission information of ballistic capsules.

Vehicle	Mission	v_e [km/s]	l_1 [mm]	l_2 [mm]	D_1 [mm]	D_2 [mm]	R_1 [mm]	R_2 [mm]	θ_1 [°]	θ_2 [°]
HUYGENS [14], [15]	Titan	6.0	620.5	985.0	2,700.0	1,790.0	1,250.0		30	
BEAGLE2 [16], [17]	Mars	5.63	499.5	212.0	900.0	371.8	417.0	29.0	30	43.75
OREX [10], [11]	Earth LEO	7.4	1,060.0		3,400.0	1,735.0	1,350.0	100.0	40	40

designed and developed for getting aerodynamic and aerothermal data (and data for other disciplines like flight mechanics and vehicle control) in real free-flight environments. The Orbital Re-entry Experiment OREX was one of these demonstrators. It had a successful flight in Earth orbit and a subsequent ballistic re-entry in February 1994. The main tasks of this flight were to test the reliability of the TPS system (which was that one developed for HOPE) and to collect data of the hypersonic and supersonic aerodynamic and aerothermal behavior. We give, without further discussion, the data of OREX in Table 5.2 and Fig. 5.4.

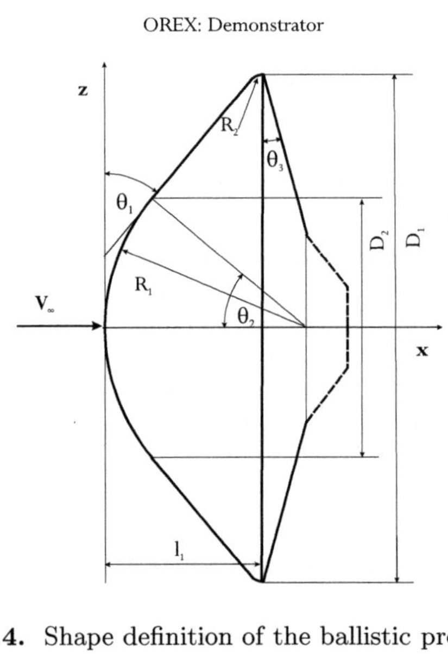

Fig. 5.4. Shape definition of the ballistic probe OREX, [10], [11].

5.2.2 Lifting Capsules

Capsules flying with an $L/D > 0$ while entering an atmosphere are called lifting capsules. If they have an axisymmetric shape, their angle of attack necessarily must be negative in order to achieve positive lift, Section 5.3. Mission information about such RV-NW-type vehicles, the American APOLLO and the Russian SOYUZ vehicle being the most prominent ones, is given in Table 5.3.

There is no doubt that the aerothermodynamics of the APOLLO capsule are one of the best known. Due to the large number of flights in the 1960s and 1970s either in Earth orbit or of Lunar return the free-flight data base is remarkable. During the design phase of APOLLO most of the aerothermodynamic data was obtained from ground-simulation facility experiments, [19], [20], [22], [28] to [30]. The heat transfer measurements in the hypersonic flow regime were conducted in "cold" hypersonic tunnels since no real high-enthalpy facilities were available at that time.

Some thirty years later in Europe the Atmospheric Re-entry Demonstrator (ARD) was developed. Its shape was a sub-scaled APOLLO configuration with a modified rear part. In a first iteration the aerodynamic data base for ARD was taken from APOLLO and later on improved. The advent of powerful numerical simulation methods had made it possible to strongly increase the understanding of complicated flow fields with multiple interactions of shocks, vortices and boundary layers, either attached or separated, with the influence of hot gases in thermodynamic equilibrium or non-equilibrium, with finite-rate catalytic wall conditions, and so on.

Also new high-enthalpy facilities were available in Europe with the HEG in Germany and the F4 tunnel in France. For ARD these new capacities were employed during the development phase. ARD was successfully flown in

Table 5.3. Mission information of lifting capsules.

Vehicle	Mission	V_e $[km/s]$	Ref.
APOLLO	Earth LEO	7.67	[20], [22]
	Lunar return	10.76	
ARD	Earth LEO	7.4	[18], [23]
SOYUZ	Earth LEO	7.9	[24]
VIKING_1	Earth LEO	7.9	[25]
AFE	Earth GEO	10.36	[4]
	Mars entry	5.70	
CARINA	Earth LEO	7.6	[26], [27]

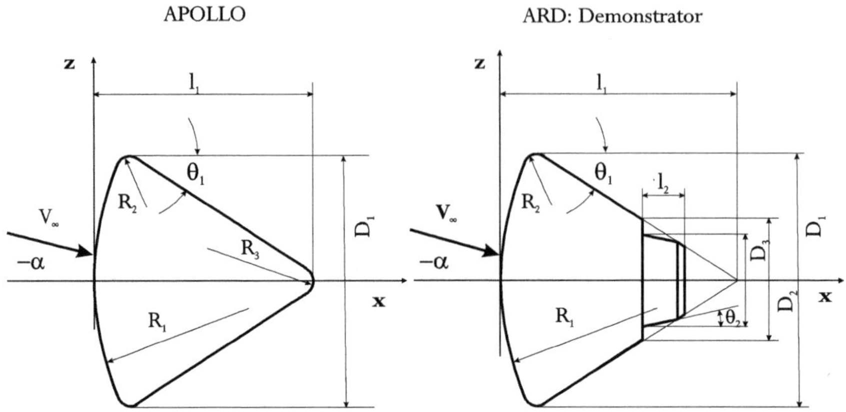

Fig. 5.5. Shape definitions of the lifting capsules APOLLO, [20], [22], and ARD, [18], [23].

October 1998 and was recovered in the Pacific Ocean, [18], [23], [31]. In Fig. 5.5 the shapes of APOLLO and ARD are plotted, while the corresponding geometrical values are listed in Table 5.4.

The Russian lifting capsule SOYUZ was the space-transportation system to the Russian space station MIR and is since the year 2001, besides the US Space Shuttle System, the transportation system guaranteeing the access to the International Space Station (ISS). Further, it acts as a rescue vehicle for the Space Station crew in case of any injury or sickness of the crew members, [24].

VIKING type shapes are interesting configurations if non-winged solutions are sought for the transport of man to and from space. In the frame of

Table 5.4. Geometrical data of lifting capsules, Figs. 5.5 to 5.7.

Vehicle	l_1 [mm]	l_2 [mm]	l_3 [mm]	D_1 [mm]	D_2 [mm]	D_3 [mm]	R_1 [mm]	R_2 [mm]	R_3 [mm]	θ_1 [°]	θ_2 [°]
APOLLO	3,529.0			3,912.0			4,694.0	196.0	232.	33	
ARD	2,594.0	460.0		2,800.0	1,317.0	1,015.	4,694.0	140.0		33	12
SOYUZ	2,142.0	1,778.	936.	2,200.0	980.0		2,235.0	978.0		11	7
VIKING1	3,740.0			4,400.0			2,200.0	88.0		80	25
AFE	376.0			4,267.0						30	17
CARINA	1,263.0	482.0		1,078.0	634.0		4,380.0	2,124.0		13	

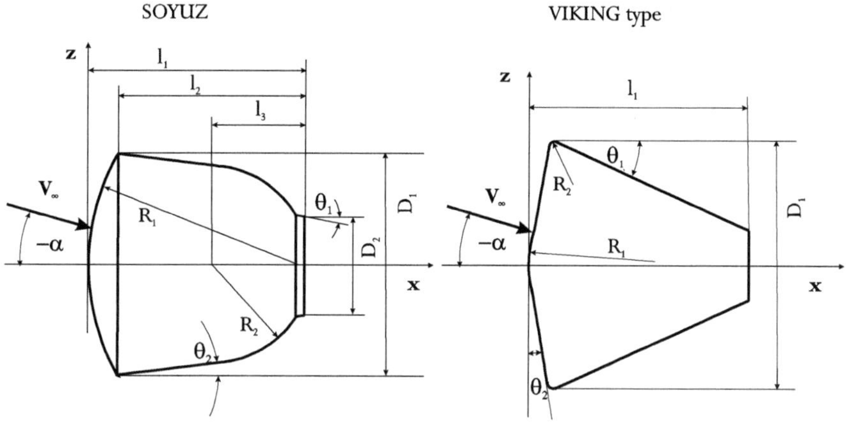

Fig. 5.6. Shape definitions of lifting capsules SOYUZ, [24] and VIKING type, [25].

ESA's post-HERMES Manned Space Transportation Programme (MSTP) and the Crew Transport Vehicle (CTV) activities, VIKING type shapes were investigated in very large detail by wind tunnel experiments, approximate engineering methods and highly sophisticated numerical simulation methods, Fig. 5.6.

Since the beginning of the space era discussions about the advantage of aeroassisted orbital transfer vehicles have taken place. To realize this technique requires a very good knowledge of the aerodynamic and aerothermal behavior of the vehicle with respect to performance and controllability as well as thermal loads. As was already mentioned, in order to improve the existing data bases in this regard, NASA had started in the 1980s a project with a generic configuration called Aeroassisted Flight Experiment (AFE), Fig. 5.7.

Due to the asymmetric shape of AFE, the aerodynamic performance with $0.25 \lessapprox L/D \lessapprox 0.3$ is reached for a trim angle of attack $\alpha_{trim} \approx 0°$ with respect to the x-coordinate.

The expected advantages of this shape are twofold. Firstly, an impingement of the shear layer on the payload, located behind the heat shield, is more unlikely for moderate angle of attack variations, or it happens farther downstream of the base compared to axisymmetric bodies. Secondly, the heating at the shoulders of the front shield, often the peak heating regime, is lower for AFE ($\alpha_{trim} \approx 0°$), due to the larger radii at the shoulders, than for axisymmetric shapes with trim angles $\alpha_{trim} \approx -20°$.

The main goal of the aeroassisted orbital transfer technique is to reduce the relative orbital speed with the help of the atmosphere if for example an orbit transfer (from geostationary to low Earth orbit) or an atmospheric re-entry with supercritical speed (Lunar return) has to be conducted. The advantage of this process, today called aerocapturing, Sub-Section 5.1.2, is the possibility to dramatically decrease (up to 50 per cent) the total orbiter mass. This

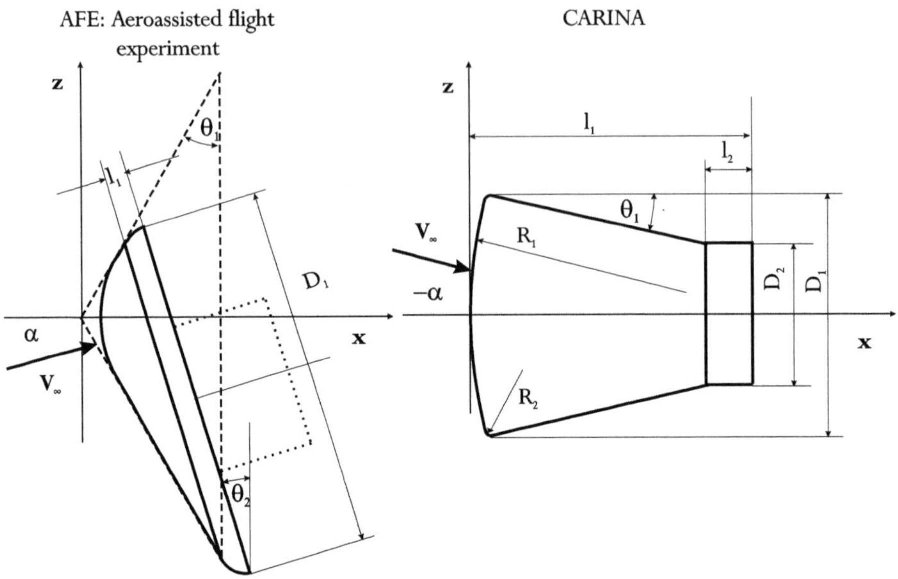

Fig. 5.7. Shape definitions of lifting capsules AFE, [3], [32], [33], and CARINA, [26].

is mainly due to the fact that no (or a reduced) chemical propulsion system including the propellant, is needed compared to conventional missions.

Since the beginning of this century there is a renewed interest in this technique in the frame of the Mars exploration activities, where a Mars Sample Return Orbiter (MSRO) was generically defined and in detail investigated, which has an AFE like shape. The realization of this project (later than the year 2013 according to ESA's exploration plan), would be the first aerocapturing mission ever performed, [3], [34].

In the 1990s the Italian Space Agency (ASI) has supported a satellite project named Capsula di Rientro Non Abitata (CARINA) for performing microgravity experiments in space. This system has the capability to conduct an atmospheric re-entry. The re-entry module of this system had a configuration based on the APOLLO/GEMINI shape and should have been able to return a payload mass of about 130.0 kg, Fig. 5.7 (right). An aerodynamic data base was established for the hypersonic, the supersonic, and the transonic Mach number range, [26], [27].

5.2.3 Bicones

Since a long time various bicones, fat (bluff) bicones, slender bicones, bent bicones, Fig. 5.8, were considered for particular space missions and some preliminary studies have been made. The advantage of these configurations is the higher lift-to-drag ratio L/D compared to simple capsules. Fat bicones

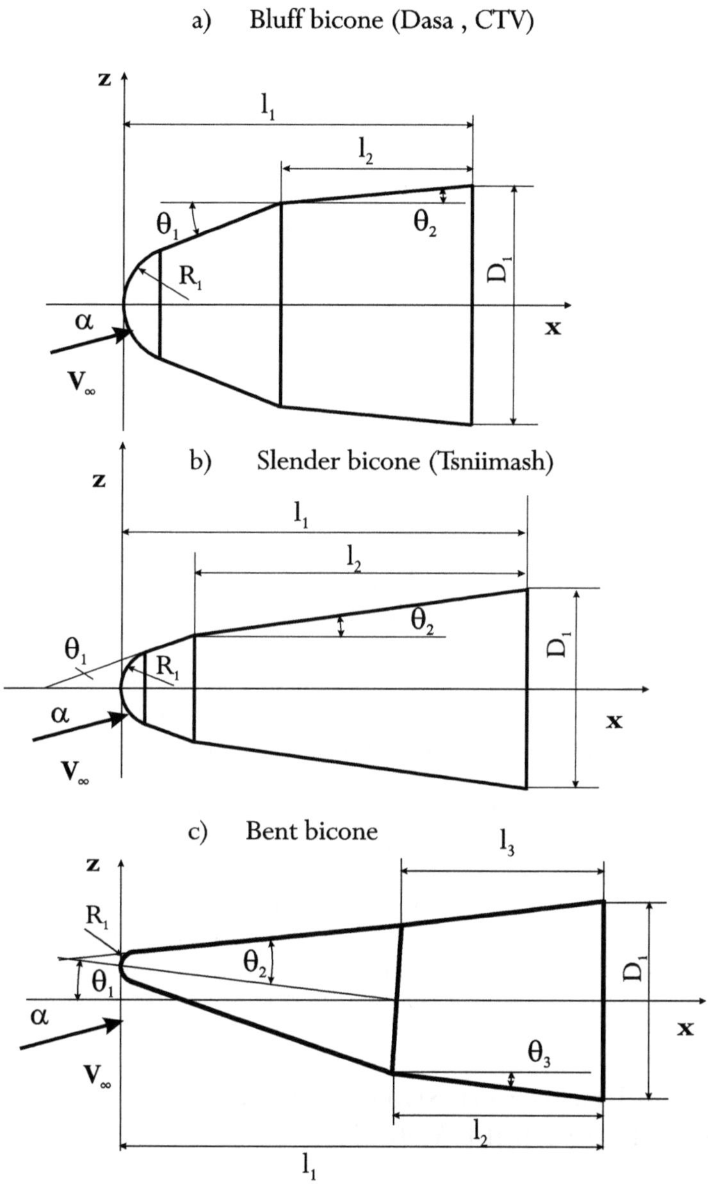

Fig. 5.8. Shape definitions of a) fat (bluff) bicone, [36], b) slender bicone, [24], and c) bent bicone, [37].

have a $L/D \approx 0.6$, slender ones a $L/D \approx 0.9$ and bent bicones will even reach values up to $L/D \approx 1.4$[2]. In contrast to the classical axisymmetric RV-NW-type vehicles, these shapes achieve lift with a positive angle of attack.

[2] The US Shuttle has a $L/D \approx O(1)$, Chapter 3.

Table 5.5. Geometrical data and mission information of bicones.

Vehicle	Mission	v_e [km/s]	l_1 [mm]	l_2 [mm]	l_3 [mm]	D_1 [mm]	R_1 [mm]	θ_1 [°]	θ_2 [°]	θ_3 [°]
CTV Dasa [36]	Earth LEO	7.6	6,300.0	3,425.0		4,400.0	1,056.0	22	5.4	
CTV ESA [24]	Earth LEO	7.6	6,830.0	3,745.0		4,398.0	882.0	20	7	
Slender Bicone [24]	Earth LEO	7.6	8,395.0	6,863.0		4,159.0	796.0	20	8	
Bent Bicone[3] [35], [37]	Earth LEO	7.6	182.52	80.85	77.32	76.20	5.79	7	12.84	7

Orbital transfer operations, where only the altitude of the orbit is changed, may be feasible with vehicles with a $L/D \approx 0.3$, but for missions with a change of the inclination of the target orbit, higher aerodynamic performance is necessary during the aerocapturing phase.

In general, bicones are appropriate for missions where a large cross-range capability, good maneuverability, low landing distortion (vehicle recovery), low entry loads are required, and for high entry velocities and thin atmospheres (low deceleration).

Up to now none of these vehicles have reached a development state for performing a free flight (neither orbital nor suborbital). In the US literature a few papers report about some investigations in this field, e.g., [35], [37].

In Europe some activities were performed in the frame of ESA's Crew Transport Vehicle (CTV) studies, [1], [36]. Also in Russia there are some preliminary studies on biconic shapes, [24]. Three of these biconic shapes can be found in Fig. 5.8 and the geometrical parameters are listed in Table 5.5.

5.3 Trim Conditions and Static Stability of RV-NW-Type Flight Vehicles

In this Section we discuss the aerodynamic capabilities and potentials of the various non-winged vehicles. For that it is necessary to define the coordinate systems applied including the aerodynamic forces and moments, and to show what trim conditions and static stability mean. General formulas and definitions describing the aerodynamic state of all kind of vehicles are found in Chapter 7.

[3] The data of the bent bicone are those of a wind tunnel model.

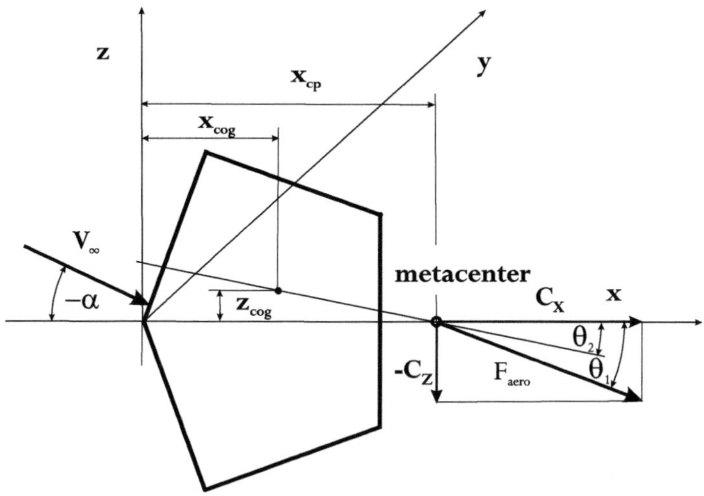

Fig. 5.9. Definition of the metacenter.

5.3.1 Park's Formula

For capsule-like shapes it is observed, at supersonic and hypersonic Mach numbers, that the acting line of the resultant aerodynamic force crosses, for different angles of attack, the axis of symmetry (namely the x-coordinate) approximately always at the same position. This intersection point is called the metacenter with the coordinate x_{cp}, Fig. 5.9.

In cases where the aerodynamic coefficients are known only for a few discrete angles of attack, this observation can be a helpful cognition for the determination of the trim angle of attack. This is often the situation, if the aerodynamic data are obtained with the help of numerical simulation methods, where non-equilibrium thermodynamics, catalytic walls, turbulent flow et cetera are taken into account, which makes the computations (still) very expensive and time consuming.

We assume that x_{cp} and the force coefficients C_X and C_Z are nearly independent of α (for $M_\infty \gtrapprox 2$) and $\partial C_m/\partial\alpha$ is approximately constant[4]. With eq. (7.2) we define:

$$L_{ref}C_m(\alpha_j)|_{cog} - C_Z(\alpha_j)(x_{cog} - x_{cp}) + C_X(\alpha_j)z_{cog} = 0, \qquad z_{cp} = 0, \quad (5.2)$$

$$C_m(\alpha_{trim})|_{cog} = 0, \quad (5.3)$$

$$C_m(\alpha_j)|_{cog} + (\alpha_{trim} - \alpha_j)\frac{\partial C_m(\alpha_j)}{\partial\alpha} = 0, \quad (5.4)$$

[4] This holds, for instance, for the VIKING_2 shape, Fig. 5.10, and also some others in the following sub-section.

$$-\left\{\frac{L_{ref}\alpha_j}{C_Z(x_{cog}-x_{cp})}\frac{\partial C_m(\alpha_j)}{\partial \alpha}\left(\frac{\alpha_{trim}}{\alpha_j}-1\right)+1\right\}\frac{C_Z(\alpha_j)}{C_X(\alpha_j)}+\frac{z_{cog}}{x_{cog}-x_{cp}}=0. \tag{5.5}$$

In these equations α_{trim} denotes the trim angle of attack and α_j the angle of attack, where the aerodynamic coefficients are known. Further we obtain from eq. (7.15):

$$-L_{ref}\alpha_j\frac{\partial C_m(\alpha_j)}{\partial \alpha}=-\alpha_j\frac{\partial C_Z(\alpha_j)}{\partial \alpha}(x_{cog}-x_{cp})+\alpha_j\frac{\partial C_X(\alpha_j)}{\partial \alpha}z_{cog}. \tag{5.6}$$

With $\alpha_j\ \partial C_Z(\alpha_j)/\partial\alpha \approx C_Z$, $\partial C_X(\alpha_j)/\partial\alpha \approx 0$, $\tan\theta_1 = C_Z/C_X$, and $\tan\theta_2 = z_{cog}/(x_{cp}-x_{cog})$, finally eq. (5.5) has the form:

$$\alpha_{trim}=-\alpha_j\frac{\tan\theta_2}{\tan\theta_1}. \tag{5.7}$$

This is Park's formula, [38], [39], which allows to find the trim angle of attack from aerodynamic coefficients given at a discrete trajectory point in a suitable vicinity of the trim angle. Later in this chapter we will demonstrate the applicability of this formula by some examples.

5.3.2 Performance Data of Lifting Capsules

In this Sub-Section we give an overview about aerodynamic coefficients[5] of some of the shapes presented in Section 5.2. Since most of the shapes are bodies of revolution (the AFE shape is considered only for the yaw angle $\beta = 0$) the coefficients C_X, C_Z, C_m, L/D, Fig. 7.3, describe the aerodynamic performance. The coefficient for the dynamic stability $C_{mq} + C_{m\dot{\alpha}}$ will be treated separately in Section 5.4.

The configurations of VIKING type shapes are characterized by the following geometrical relations: $R_1/D_1,\ R_2/D_1,\ l_1/D_1,\ \theta_1,\ \theta_2$. The values of the VIKING_1 shape, Fig. 5.6, are $R_1/D_1 = 0.5,\ R_2/D_1 = 0.02,\ l_1/D_1 = 0.85,\ \theta_1 = 80°,\ \theta_2 = 25°$. VIKING_2 has a different aft cone angle with $\theta_2 = 20°$ and the reference diameter amounts to $D_1 = L_{ref} = 4{,}400.0$ mm [21].

In Fig. 5.10, aerodynamic data of the VIKING_2 shape are given. Note, as explained above in this section, capsules have a positive lift only for negative angle of attack. Therefore all coefficients are plotted versus negative angle of attack. Further, the conventions of the signs are defined by Fig. 7.3.

The data of Fig. 5.10 are taken for $0.5 \leqq M_\infty \leqq 3.97$ from wind tunnel experiments. The $M_\infty = 10$ values are based on Euler calculations with

[5] The reader should note the custom in RV-NW-type vehicle aerodynamics, that the integral aerodynamic forces and coefficients are given in terms of the axial force X and the normal force Z (body axis system). For the transformation into the flight path system see Section 7.6.

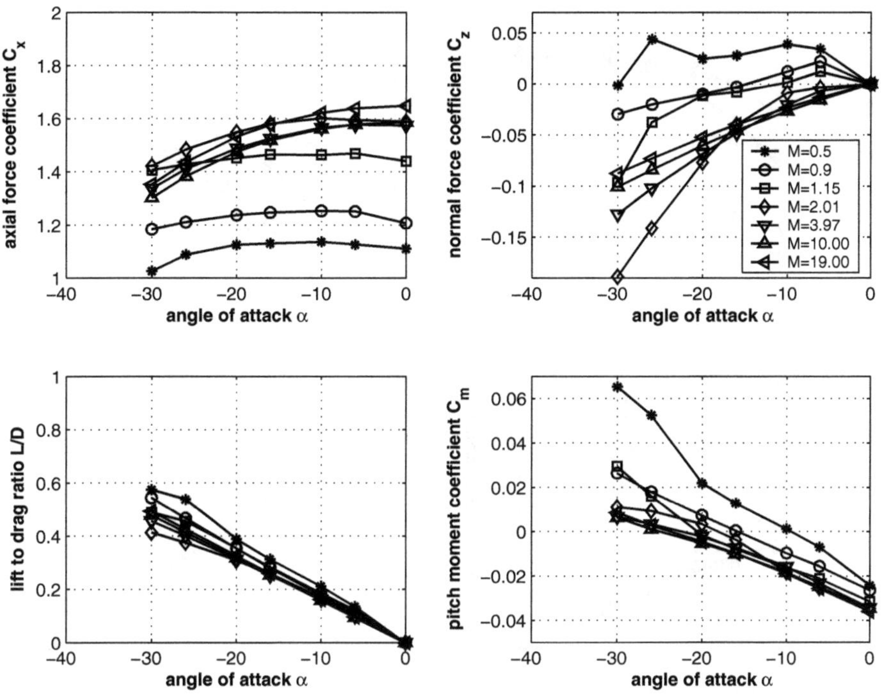

Fig. 5.10. Aerodynamic data of the VIKING_2 shape. Moment reference: $x_{ref} = 0.34\ D_1$, $z_{ref} = 0.0218\ D_1$. Data source: [21].

perfect gas assumption, while for the Mach number $M_\infty = 19$ an Euler computation is used with real gas in non-equilibrium thermodynamic state. Since the trim angle of attack varies between $\alpha_{trim} \approx -10°$ ($M_\infty = 0.5$) and $\alpha_{trim} \approx -25°$ ($M_\infty = 19$), it seems possible to fly in the whole Mach number range with L/D between 0.2 and 0.4.

All the aerodynamic coefficients, plotted in Fig. 5.10, exhibit a non-monotonic behavior with regard to the Mach number with extreme values in the transonic regime. However, for $M_\infty \geqq 3.97$ there exists nearly a Mach number independence for C_Z, C_m and L/D, which, however, is not so clear for C_X. Finally, for the reference point chosen the capsule is statically stable in the whole Mach number and angle of attack regime, since the condition $C_{m\alpha} < 0$ is met everywhere, see Fig. 7.9.

Fig. 5.11 exhibits the aerodynamics of the APOLLO capsule. This shape produces L/D values which are dependent on the Mach number in a similar way as the VIKING type shapes, and L/D's of 0.3 are realistic, [40], [41]. The variation of the trim angles for the various Mach numbers is in the same range as for the VIKING_2 shape. Again static stability is preserved in the whole Mach number regime.

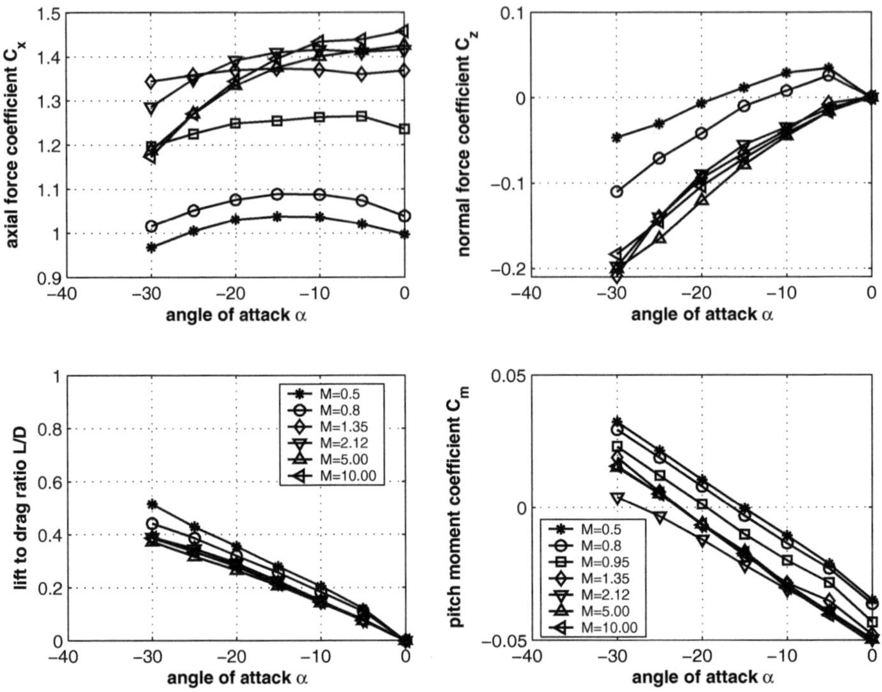

Fig. 5.11. Aerodynamic data of the APOLLO shape. Moment reference: $x_{ref} = 0.265\,D_1$, $z_{ref} = 0.035\,D_1$. C_m and C_X, as well as L/D and C_Z have the same legend. Data source: [40].

Another famous capsule, besides APOLLO, is the Russian SOYUZ. It is the vehicle which serves the ISS. It is of interest to see the relatively high spreading of the L/D values with respect to the Mach number, Fig. 5.12.

This data set is completely generated by wind tunnel results and the plotted values are taken from [24]. The highest Mach number measured is $M_\infty = 5.96$. Of course, it is not clear if apparent changes of the aerodynamic coefficients will occur for hypersonic Mach numbers up to 30, but a look at the data for VIKING_2 or APOLLO reveals that the differences are probably low. So, SOYUZ is able to fly in the hypersonic regime with $L/D \approx 0.3$ for a trim angle $\alpha_{trim} \approx -26°$. A further increase of the aerodynamic performance ($\Rightarrow$ higher L/D) seems hardly possible, whatever the z_{offset} of the center-of-gravity is, Sub-Section 5.3.3. Static stability is given for all the Mach numbers tested.

Comparing the three vehicles discussed above, the VIKING_2 shape has obviously the best potential in aerodynamic performance, but this shape was never flown as a manned space transporter.

AFE has an interesting shape, which is the only non-axisymmetric one, Fig. 5.7. About ten years after the respective NASA technology programme (see above) in the late 1980s and the beginning 1990s, a renewed interest in

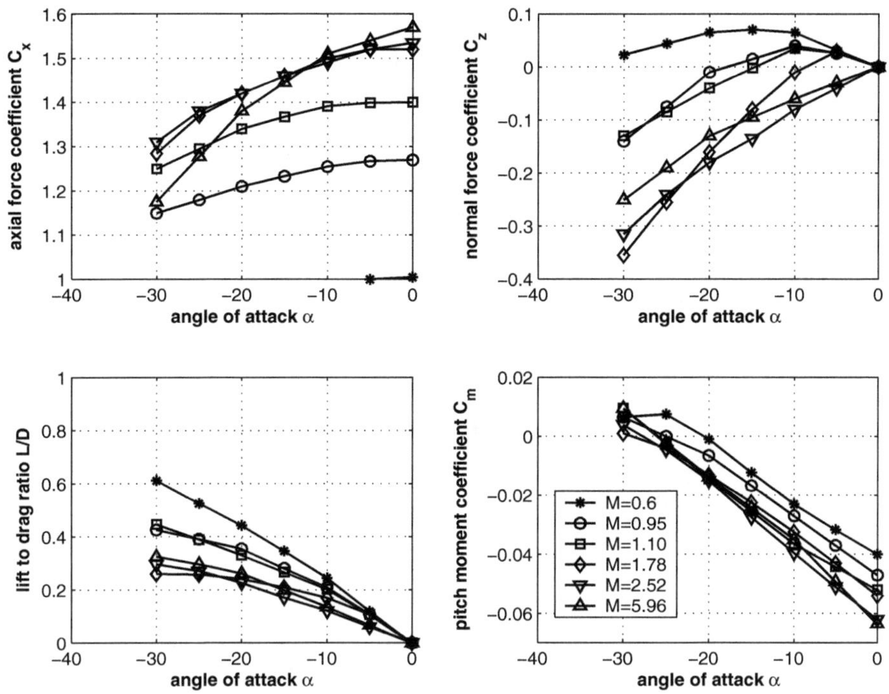

Fig. 5.12. Aerodynamic data of the SOYUZ shape. Moment reference: $x_{ref} = 0.370\, D_1$, $z_{ref} = 0.039\, D_1$. Data source: [24].

this shape arose at the European Space Agency (ESA) in the frame of the Mars Sample Return Orbiter (MSRO) activities. The original shape had a diameter of $D_1 = 4,267.0\ mm \,\widehat{=}\, 14.0\, ft$ which was reduced in the MSRO case to $D_1 = 3,657.0\, mm \widehat{=} 12.0\, ft$. Fig. 5.13 shows C_X, C_Z, C_m and L/D for some hypersonic Mach numbers.

For the Mach number $M_\infty = 11.8$ experiments were conducted in the Hypervelocity Free-Flight Aerodynamic Facility (HFFAF) at NASA Ames, [42]. From the experimental conditions it seems that this facility is able to duplicate nearly all the parameters of a real hypersonic free-flight, namely the free-stream pressure, density and temperature, as well as the velocity of the model. Therefore one could expect that the data reflect properly the influence of the real-gas behavior, if Mach number independence, Section 3.6, is given. On the other hand the data reduction requires the flight-mechanical evaluation of the trajectory of the model inside the facility, which is obviously not a simple task, [42].

Further for $M_\infty = 9.55$ and 5.94 data were measured in NASA Langley's cold hypersonic wind tunnel, [43]. The Langley data are close together for both Mach numbers. The pitching moment is larger for the Langley data compared to the HFFAF data, which was not expected, since at least for

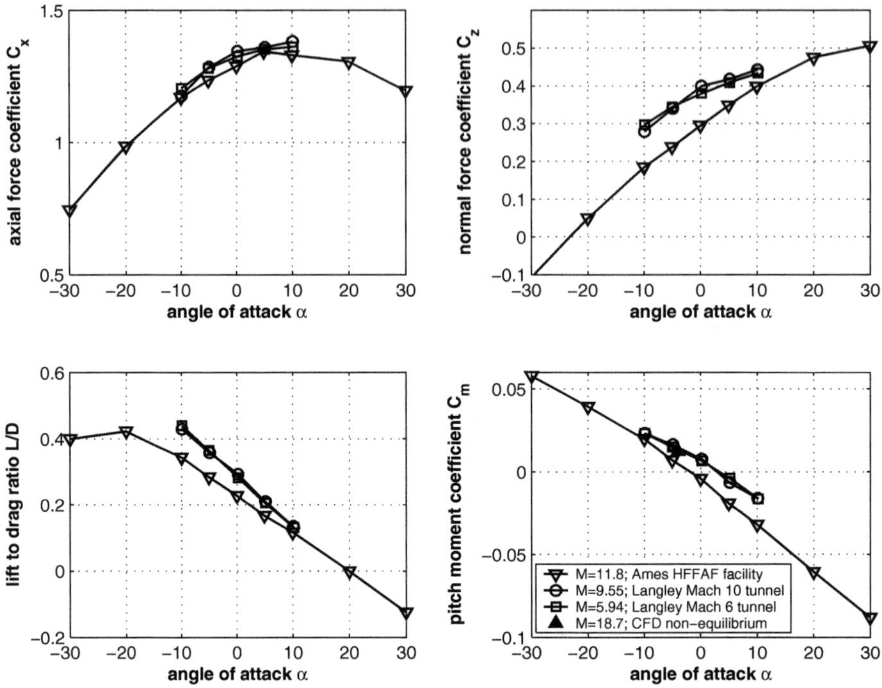

Fig. 5.13. Aerodynamic coefficients in the hypersonic flow regime for the AFE shape. Moment reference: $x_{ref} = 0.2509\, D_1$, $z_{ref} =$ - $0.2301\, D_1$, measured from the origin. Data source: [42], [43].

axisymmetric shapes the real-gas effects normally increase the pitching moment and increase the magnitude of the trim angle.

The pitching moment of a complete non-equilibrium CFD solution (for the Martian atmosphere) for $M_\infty = 18.7$ and $\alpha =$ - $4°$, is given in [44] and is plotted in Fig. 5.13, lower right. It is closer to Langley's data. Nevertheless, other numerical investigations show, [3], that in the hypersonic flight regime the trim angle would be approximately - 1°, which is in a better agreement with the HFFAF data than the Langley data.

From the above discussion and that one from Sub-Section 5.3.7 about the influence of real-gas effects, we note two points:

- First, there is a high-temperature real-gas effect on the trim angle.
- Second, the ground-facility simulation seems to be very difficult. CFD methods with the most advanced thermodynamic models, if properly validated, offer the most promising results.

5.3.3 Controlled Flight and the Role of the Center-of-Gravity

Every non-winged vehicle with an axisymmetric shape is only able to conduct a lift based, trimmed flight, if a z-offset of the center-of-gravity exists.

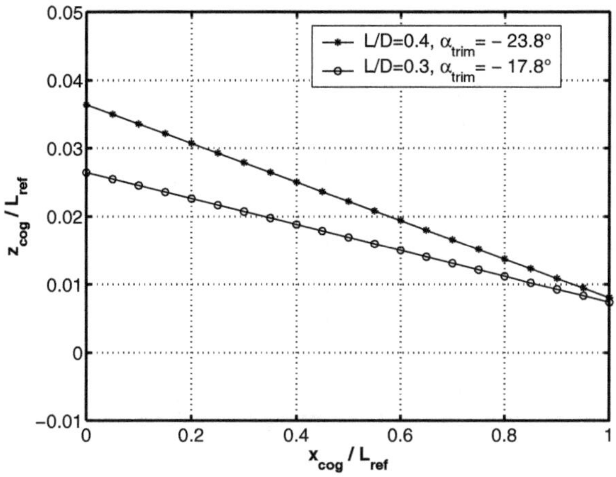

Fig. 5.14. Center-of-gravity positions for two trim conditions of the VIKING_2 shape.

Otherwise the trim angle of attack is zero leading to zero lift and lift-to-drag ratio.

To discuss this in more detail we use the aerodynamic data of the VIKING_2 shape, Fig. 5.10. From the lift-to-drag graph we can extract that a hypersonic performance of $L/D = 0.3$ can be achieved by a trim angle of $\alpha_{trim,L/D=0.3}$ = - 17.8° and $L/D = 0.4$ by $\alpha_{trim,L/D=0.4}$ = - 23.8°. With the help of eq. (7.11) we can determine all the positions of the center-of-gravity ensuring trimmed flight with a fixed L/D after the coordinates of the center-of-pressure have been determined by means of the corresponding ones of eqs. (7.5) to (7.10). In Fig. 5.14 the resulting "center-of-gravity lines" for the two design points are shown.

It should already be mentioned here, that the free-flight experience of the APOLLO capsule has exhibited that in the hypersonic flight phase the trim angle was approximately 3° lower than predicted, with the consequence of a lower L/D value. In the frame of ESA's MSTP this phenomenon was investigated in great detail on the ARD capsule, which has nearly the same aerodynamic properties as the APOLLO capsule, Fig. 5.5. The outcome was that mainly the extreme heating of the air in the shock layer in front of the vehicle is responsible for this behavior. This heating leads to the excitation of vibrational modes in the molecules, to dissociation and partly ionization (for entry speeds $v_e \gtrapprox 8.0\ km/s$). The thermodynamic state can then be in equilibrium or non-equilibrium, depending on the ambient density, which influences the surface pressure distribution. This behavior is summarized by the term "high-temperature real-gas effects" and will be treated in more detail in Sub-Section 5.3.7.

It is evident from eq. (7.11) that for $C_X/C_Z \gg 1$ the relation $z_{cog}/x_{cog} \ll 1$ holds. Therefore the z-offset of the center-of-gravity z_{cog} is the dominating quantity for ensuring flyability and controllability. Its influence on the trim

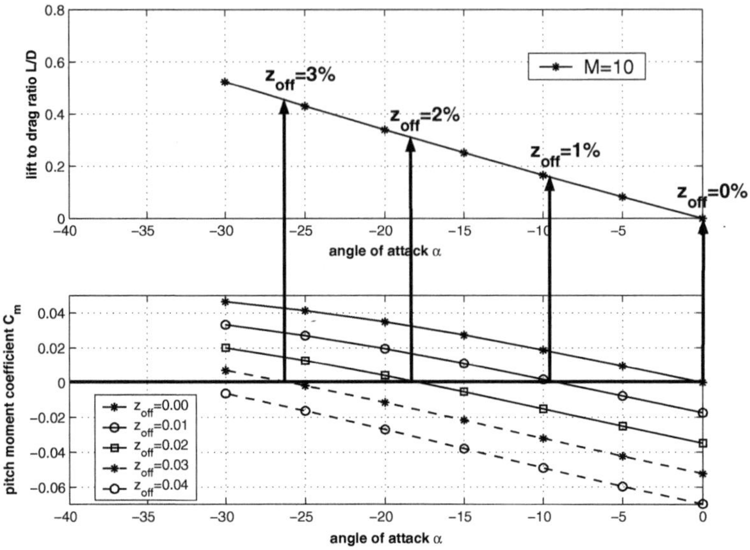

Fig. 5.15. Influence of $z_{offset} \equiv z_{cog}$ on aerodynamic trim for the VIKING_2 shape. $z_{offset} \equiv z_{off}$ is given in per cent of the reference length $L_{ref} = D_1 =$ 4,400.0 mm.

angle of attack and on L/D is very high, which is distinctly demonstrated in Fig. 5.15.

A change of 1 per cent in $z_{offset} \equiv z_{cog}$ (this is only 44.0 mm for the VIKING_2 vehicle) alters the trim angle of attack by about $\Delta\alpha_{trim} \approx 9°$. On the other hand changes in the x-coordinate of the center-of-gravity x_{cog} influence only little the trim angle, which can be concluded from Fig. 5.16. A shift of 15 per cent of the moment reference $x_{ref} \equiv x_{cog}$ (in our example 660.0 mm) leads to a trim angle change of merely $\Delta\alpha_{trim} \approx 2°$.

5.3.4 Sensitivity of Aerodynamics Against Shape Variations

It is important to get a feeling what contour changes of a capsule may cause. We demonstrate this with VIKING_3 as baseline shape. The geometrical relations in this case are $R_1/D_1 = 0.5$, $R_2/D_1 = 0.02$, $l_1/D_1 = 0.95$, $\theta_1 = 80°$, $\theta_2 = 16°$ (see Fig. 5.6 with $\theta_2 = 16°$). The investigation is valid for hypersonic flow conditions, [45]. The moment reference point is for all pitching moment diagrams given by $x_{ref} = 0.33\, L_{ref}$, $z_{ref} = 0.02\, L_{ref}$.

R_1/D_1 variation

The influence of R_1/D_1 variations on the aerodynamic performance is rather weak and always $\Delta(L/D) \lessapprox 1$ per cent. However, R_1 should have a reasonably large value, since the thermal loads are directly related to the inverse of the square root of R_1.

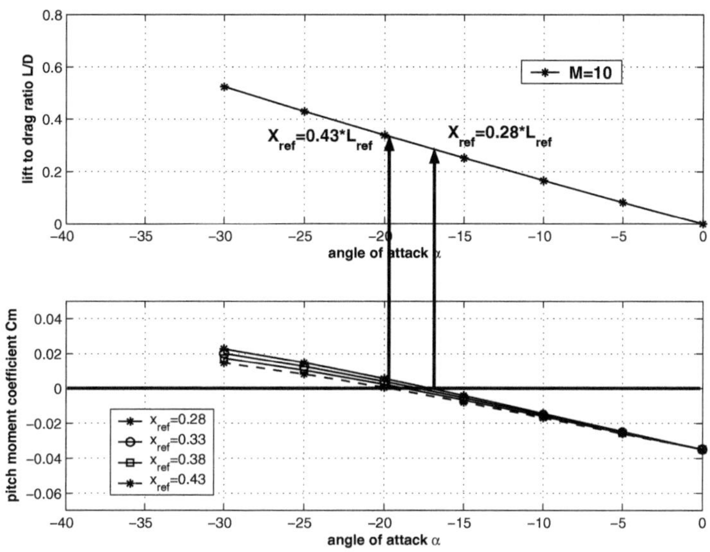

Fig. 5.16. Influence of $x_{ref} \equiv x_{cog}$ on aerodynamic trim for the VIKING_2 shape. $L_{ref} = D_1 =$ 4,400.0 mm.

R_2/D_1 variation

Let us consider what happens if R_2/D_1 is increased to 0.10, Fig. 5.17. First, a remarkable decrease in the aerodynamic performance L/D can be observed and second, the pitching moment increases, which results in a trim angle shift of roughly 3.5° ($\alpha_{trim,R_2/D_1=0.02} \approx$ - 20.0° $\Longrightarrow \alpha_{trim,R_2/D_1=0.10} \approx$ - 16.5°).

The pitching moment increase means that a pitch-up effect occurs. The static stability is seen to be slightly increased with $|\partial C_m/\partial\alpha|_{R_2/D_1=0.10} > |\partial C_m/\partial\alpha|_{R_2/D_1=0.02}$, [25]. Of course, the thermal loads at the shoulder are reduced for $R_2/D_1 = 0.10$ due to the diminished flow expansion there compared to the $R_2/D_1 = 0.02$ case.

Physical explanation

The increase of R_2/D_1 can be considered as a reduction of bluntness of the capsule. The axial force is considerably diminished due to this reduction, which reduces the lift L and thus L/D. The pressure force on the leeward side is more reduced than the one on the windward side which leads to a pitch-up effect. This result is confirmed by the work in [46], where a 70° spherical cone with different shoulder radii is investigated by employing a numerical method solving the Euler equations.

l_1/D_1 variation

The influence of changing l_1/D_1 from 0.85 to 1.05 is investigated. While L/D is nearly the same for $l_1/D_1 = 0.85$ and 1.05, the pitching moment is quite

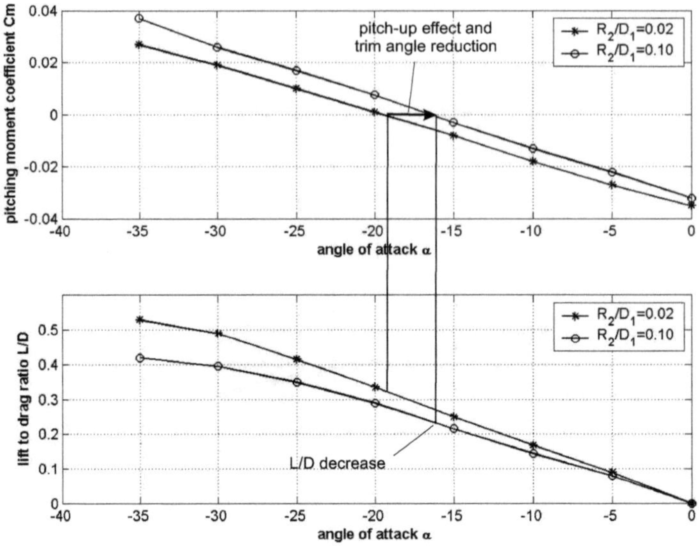

Fig. 5.17. Influence of R_2/D_1 on aerodynamic coefficients based on the VIKING_3 shape. Data source: [45].

higher for $l_1/D_1 = 0.85$ compared to $l_1/D_1 = 1.05$, which leads to a reduced trim angle α_{trim}, Fig. 5.18.

θ_1 variation

The half cone angle θ_1 determines the magnitude of L/D, which grows with increasing θ_1. To achieve a value of $L/D \approx 0.3$, a minimum of $\theta_1 = 60°$ is required, [32]. Further, the trim angle α_{trim} is decreased with increasing θ_1, which is due to the fact that the center-of-pressure moves downstream. This results in a rise of the pitching moment. In an example given in [32], θ_1 was changed from 70° to 75°, which has caused a reduction in the magnitude of the trim angle α_{trim} from - 20.8° to - 19.5°.

θ_2 variation

The variation $14° \leqq \theta_2 \leqq 18°$ is investigated, [25]. For $\alpha <$ - 20° the pitching moment decreases slightly and L/D increases when θ_2 increases. For higher values of θ_2 this effect is shifted to higher angles of attack.

Physical explanation

The increase of the aft cone angle θ_2 leads to an increase of the negative normal force coefficient C_Z, which reduces the lift. The flow past shapes with lower θ_2 values "sees" earlier the aft cone part, producing this negative C_Z increase, which on the other hand, brings an increase of the pitch-up effect, Fig. 5.19.

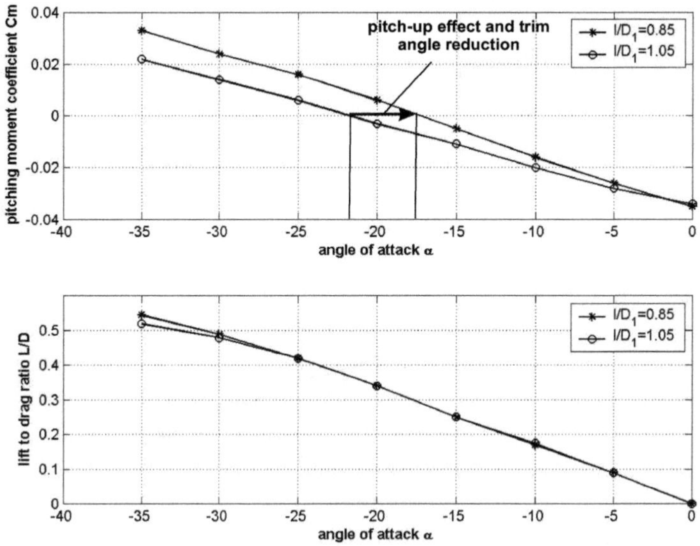

Fig. 5.18. Influence of l_1/D_1 on aerodynamic coefficients based on the VIKING_3 shape. Data source: [45].

5.3.5 Parasite Trim

During the development and testing of the classical capsules APOLLO and SOYUZ it was observed, that over an angle of attack range $0° \geqq \alpha \geqq$ - 360° the pitching moment C_m could meet the trim and stability conditions ($C_m = 0$, $\partial C_m/\partial\alpha < 0$) also at other points besides the nominal one. These points are called "parasite trim points". There are at least three reasons why it must be prevented that the vehicle turns into such non-nominal trim positions:

- the re-entry process can only be successfully conducted with the heat shield pointing forward in order to cope with the mechanical and thermal loads,
- the parachute landing system can be deployed only if the apex cover can be jettisoned properly, which requires the heat shield pointing forward,
- in the launch abort case the escape procedure requires definitely a capsule heat shield in pointing-forward attitude.

The best solution of this problem would be given by a change to a vehicle shape which prevents the existence of parasite trim points. During the APOLLO program a lot of tests were done with keels, spoilers and strakes, but obviously none of these devices did solve the problem satisfactorily, [19], [22].

A typical C_m plot showing one parasite trim point (VIKING_2 shape) is given in Fig. 5.20. The nominal trim point is at $\alpha_{trim,nominal}$ = - 7.9°, while the parasite trim point has the value $\alpha_{trim,parasite}$ = - 141°.

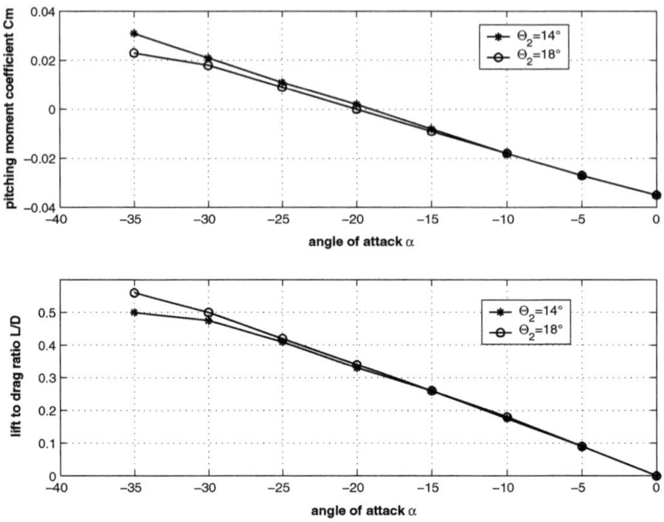

Fig. 5.19. Influence of θ_2 on aerodynamic coefficients based on the VIKING_3 shape. Data source: [25].

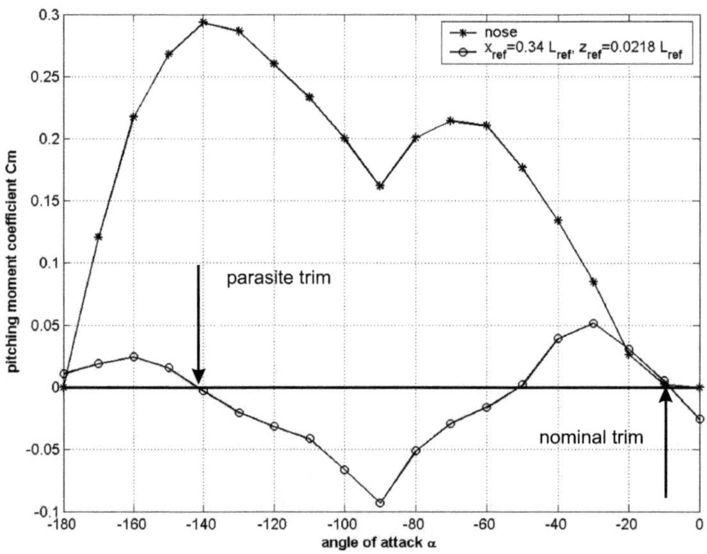

Fig. 5.20. Pitching moment of the VIKING_2 shape showing a parasite trim point for $M_\infty = 0.7$ (nose: $x_{ref} = 0, z_{ref} = 0$). Data source: [21].

The APOLLO capsule possesses over the whole Mach number range one parasite trim point, which is somewhat fluctuating in the subsonic, transonic and low supersonic regimes, while for higher Mach numbers this trim point

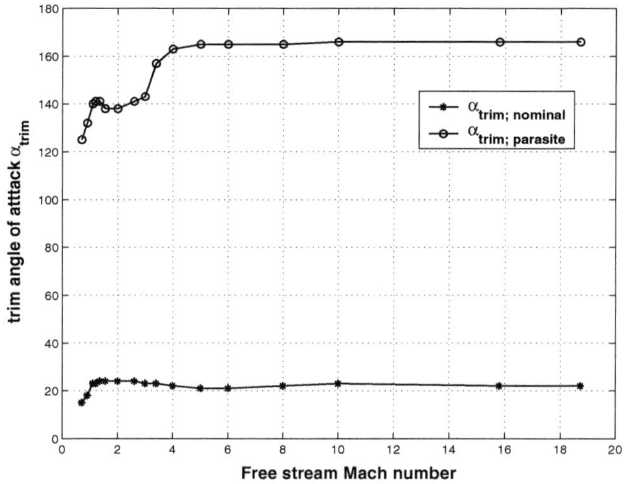

Fig. 5.21. APOLLO nominal and parasite trim points as function of Mach number. Data source: [22].

becomes independent of the Mach number, Fig. 5.21. This is valid for the center-of-gravity location $x_{cog}/D_1 = 0.657$ (measured from the apex) and $z_{cog}/D_1 = 0.035$.

The data available give hints that at least three parameters may influence the number and the location of parasite trim points. The first one is the x-component of the center-of-gravity. The larger x_{cog}, i. e., the more the center-of-gravity lies away from the apex, the more likely is the appearance of one or more parasite trim points or, the other way around, with an appropriately low x_{cog} value the appearance of these points can be avoided. Secondly, the Reynolds number has an influence. For larger Reynolds numbers one can often observe more than one parasite trim point, the reasons for this are not clear. The third parameter having an influence is the flight Mach number, again for not clear reasons. The lower the flight Mach number, the higher is the probability of the occurrence of parasite trim points. In Table 5.6 data are listed for SOYUZ and the VIKING_2 shape demonstrating this behavior.

5.3.6 Performance Data of Bicones

An alternative to the classical capsules are the bicones which can provide more than twice the lift-to-drag ratio L/D compared to those of capsules. Constraints of the internal lay-out (payload accommodation) and of the launch system (faring restrictions) may lead to a fat bicone shape like the one shown in Fig. 5.8 a). For this shape the L/D in hypersonic flight is approximately 0.65. As already mentioned, axisymmetric bodies can only be trimmed if the center-of-gravity is off the axis of symmetry. For the moment reference point applied here, Fig. 5.22, with $z_{ref} = 0.0\,L_{ref}$ the vehicle is

Table 5.6. Appearance of parasite trim points depending on center-of-gravity position and Reynolds number. x_{cog} and z_{cog} are non-dimensionalized with L_{ref}.

Vehicle	M_∞	x_{cog}	z_{cog}	$\alpha_{trim,nominal}$	$\alpha_{trim,parasite}$	Source of data
SOYUZ	1.10	0.30	0.0357	- 22°	none	wind tunnel [24]
		0.45	0.0204	- 18°	- 132°	
VIKING_2	∞	0.30	0.0230	- 23°	none	engineering
		0.375	0.0205	- 23°	- 174°	methods [47]
VIKING_2						wind tunnel [21]
$Re=0.25\cdot10^6$	0.70	0.34	0.0218	- 7.9°	- 141°	
$Re=3.70\cdot10^6$				- 13.26°	- 121° (1)	
					+ 122° (2)	
					+ 146° (3)	

stable in the supersonic and hypersonic regime and unstable in the subsonic and the transonic regime, but cannot be trimmed, because for all Mach numbers $C_m \neq 0$ at $\alpha > 0$.

This problem can be overcome either by a suitable selection of the center-of-gravity (which may be restricted by the internal lay-out of the vehicle) or by employing aerodynamic devices like flaps and brakes (which complicates the design and the control system). The data set plotted in Fig. 5.22 was established by applying approximate methods like the local inclination methods for supersonic and hypersonic Mach numbers and Panel methods for subsonic Mach numbers, [48]. For all the aerodynamic coefficients the Mach number dependency is clearly non-monotonic with extreme values in the transonic regime, as was already discussed for the capsules in Sub-Section 5.3.2. The coefficients are getting independent of the Mach number for $M_\infty \gtrapprox 5.0$.

For two points ($M_\infty = 1.5$, $\alpha = 20°$ and $25°$) Euler solutions were generated with the method reported in [49], [50]. As one can see, the agreement with the other data in Fig. 5.22 is rather good (except for C_X at $\alpha = 25°$), which proves the reliability of the engineering method used. To get a bit more insight into the general flow field, the Mach number isolines (left) and the wall pressure distribution (right) are plotted in Fig. 5.23 for $M_\infty = 1.5$, $\alpha = 20°$.

It is interesting to observe in the left part of the figure (Mach number isolines) the embedded shock on the leeward (upper) side at $x/L \approx 0.25$, which is generated due to an overexpansion of the flow (the wall pressure

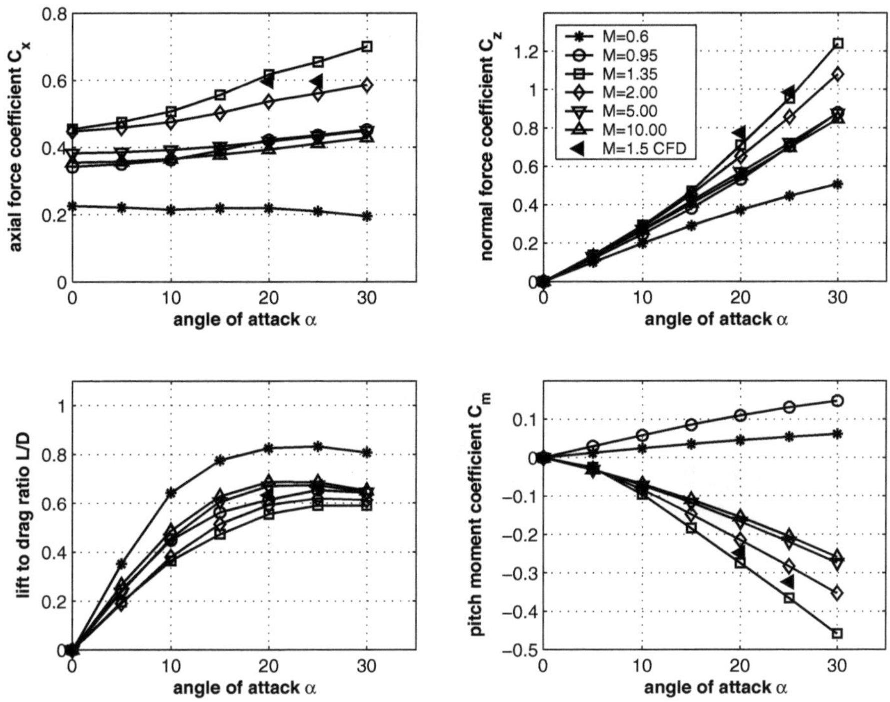

Fig. 5.22. Aerodynamic data of the bluff bicone shape, Fig. 5.8 a). Moment reference: $x_{ref} = 0.25\, L_{ref}, z_{ref} = 0.0\, L_{ref}$. Data source: [48].

does not correspond to the cone deflection!), which can only be restored in supersonic flow by a compression shock[6].

A further increase of L/D can be attained with a slender bicone, Fig. 5.8 b), where the reduced diameter D_1 leads to decreased axial and drag forces and a slight increase of the normal force, Fig. 5.24, [24].

The maximum L/D value for hypersonic flow amounts to 1.08. The data are assembled from wind tunnel results ($0.6 \leqq M_\infty \leqq 4.0$) and from engineering solutions ($M_\infty = 5.96$). On the other hand, the internal lay-out and the payload accommodation can be better realized with the bluff bicone (CTV) shape. As we can extract from the pitching moment graph in Fig. 5.24, for the selected moment reference point, the vehicle is unstable in the subsonic and transonic regime and only slightly stable in the supersonic and hypersonic area. Trim can only be achieved for $M_\infty = 2.53$.

The question arises, whether there exists a center-of-gravity location, where for all Mach numbers statically stable and trimmed flight can be secured. In Fig. 5.25 the pitching moment for such a point ($x_{ref} = 0.42\, L_{ref}$,

[6] Since the embedded shock is formed in the flow field away from the wall, it has no clear footprint in the wall pressure distribution, Fig. 5.23 right at $X \approx 1.75$.

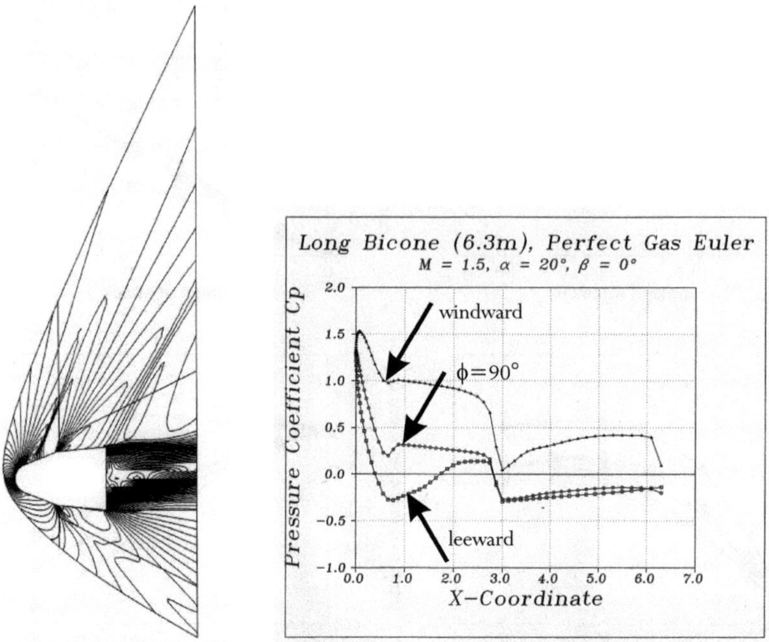

Fig. 5.23. Mach number isolines (left) and wall pressure distributions (right) in three planes of the bluff bicone shape, $M_\infty = 1.5, \alpha = 20°$. Data source: [48].

z_{ref} = -0.1467 L_{ref}, Bicone_Tsnii shape) is plotted, but it seems rather doubtful if in practice the lay-out designer can realize this *cog* location. The Bicone_Dasa shape does not have such a point, which means that for Mach numbers lower than unity other arrangements have to be made.

The hypersonic flow for the bent bicone, as shown in Fig. 5.8 c), was investigated experimentally in wind tunnels (M_∞ = 6 and 10), [37]. Bent bicones generate even for an angle of attack $\alpha = 0°$ an asymmetric flow field which supports the trim capability of the shape, as can be seen in Fig. 5.26. Hypersonic trim is possible for realistic center-of-gravity locations and angles of attack.

The figure shows, compared to that one for the symmetric bicone, an additional reduction of the axial force, in particular for low angles of attack, which lets the lift-to-drag ratio grow to a maximum of $L/D \approx 1.45$. It is still of order 1.0 at hypersonic trim ($22° \lessapprox \alpha_{trim} \lessapprox 24°$).

Despite the fact, that the aerodynamic performance of bicones is superior to that of classical capsules, none of these configurations were ever flown. This may have a historical background, because the aerodynamic data bases are much more complete for capsules and in addition for them real flight experience is available. Since this is not the case for bicones, the possible risks in this regard have obviously prevented any development of such systems.

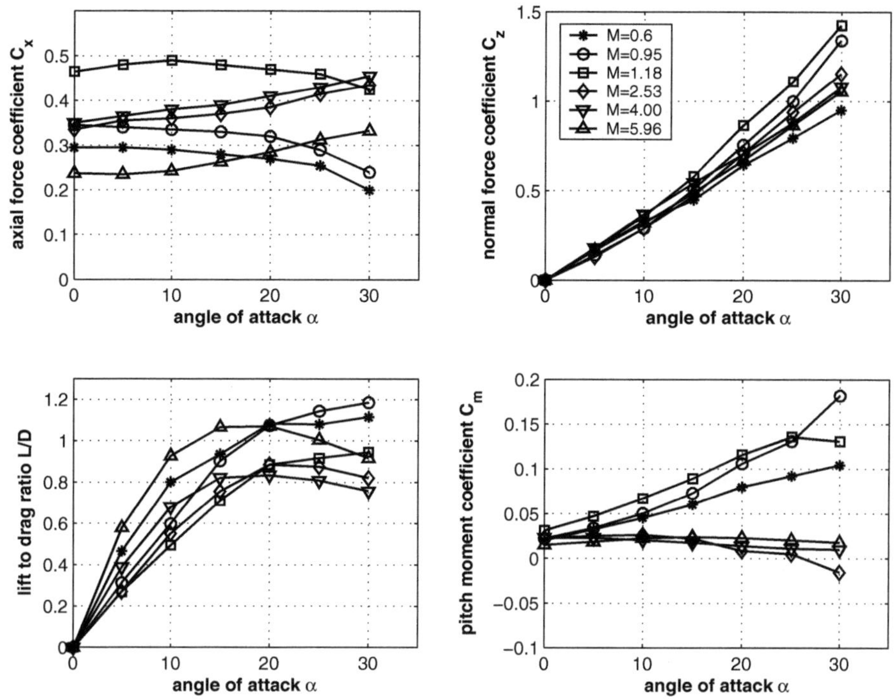

Fig. 5.24. Aerodynamic data of the slender bicone (Tsniimash) shape, Fig. 5.8 b). Moment reference: $x_{ref} = 0.57\,L_{ref}, z_{ref} =$ - $0.0667\,L_{ref}$. Data source: [24].

5.3.7 Influence of High-Temperature Real-Gas Effects on Forces and Moments

The first indications that high-temperature real-gas effects can have a considerable influence on the aerodynamic forces and moments at hypersonic speed were given by the APOLLO experience. There, the trim angle measured during flight was approximately 3° lower than predicted. Also, the observation during the first Space Shuttle Orbiter flight, Sections 3.5 and 3.6, that a non-predicted pitch-up moment was generated, which was due to a forward shift of the center-of-pressure, has likely the same physical cause. At that time it was argued for APOLLO that the following three physical phenomena could be responsible for that:

- compressibility (Mach number effects),
- hypersonic viscous interaction,
- high-temperature real-gas effects.

Today we know that the main effect is due to the thermodynamic state of the air which is strongly heated up in the hypersonic bow shock layer.

During the APOLLO project neither numerical nor experimental means were available for investigating a possible influence of real-gas effects on the

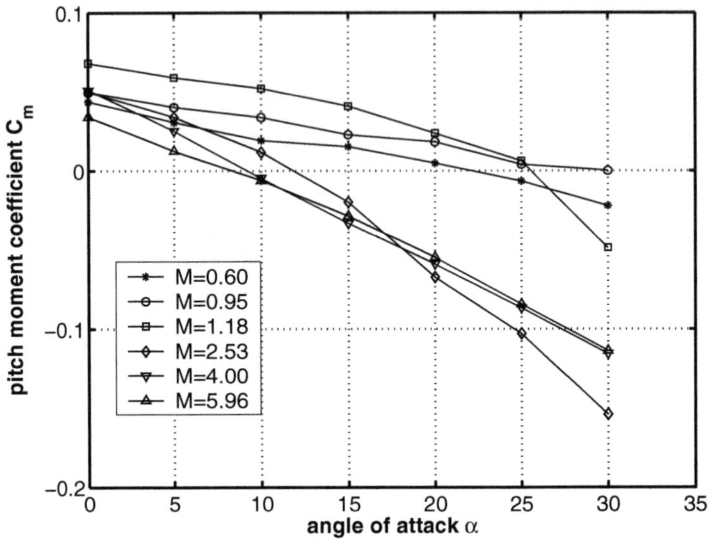

Fig. 5.25. Pitching moment for stable and trimmed flight of the slender bi-cone (Tsniimash) shape, Fig. 5.8 b). Moment reference: $x_{ref} = 0.42\,L_{ref}, z_{ref} =$ -0.1467 L_{ref}. Data source: [24].

aerodynamic behavior. With the advent of numerical simulation methods solving the Euler, Navier-Stokes or the Boltzmann equation, and new high-enthalpy ground-simulation facilities, this situation has changed. In [38] a Navier-Stokes solver with an equilibrium and non-equilibrium real-gas approach was applied to the front part of a two-dimensional APOLLO like shape.

This method contained besides the chemical reactions also the vibrational and electron excitations in a non-equilibrium state. Since non-equilibrium was assumed, the vibrational and electron excitation states were described by a second temperature T_{vibr}, which can be quite different from the rotational-translational temperature T depending on the degree of vibrational non-equilibrium[7]. The computations showed that indeed the pitching moment of the generic APOLLO-like shape was increased due to the real-gas effects, with the consequence of reduced trim angles. Further, the peak of the pressure profile along the wall was slightly shifted to the windward side which is characteristic for flows with real gases.

Since these effects are so important for reliable entry flights from space, in the European MSTP with the Atmospheric Re-entry Demonstrator ARD, strong efforts were undertaken to reveal this problem. More than 120 complete three-dimensional Euler and Navier-Stokes computations at predicted (pre-flight) and measured (post-flight) trajectory points were conducted with

[7] The governing equations for this approach can be found in detail in Chapter 11.

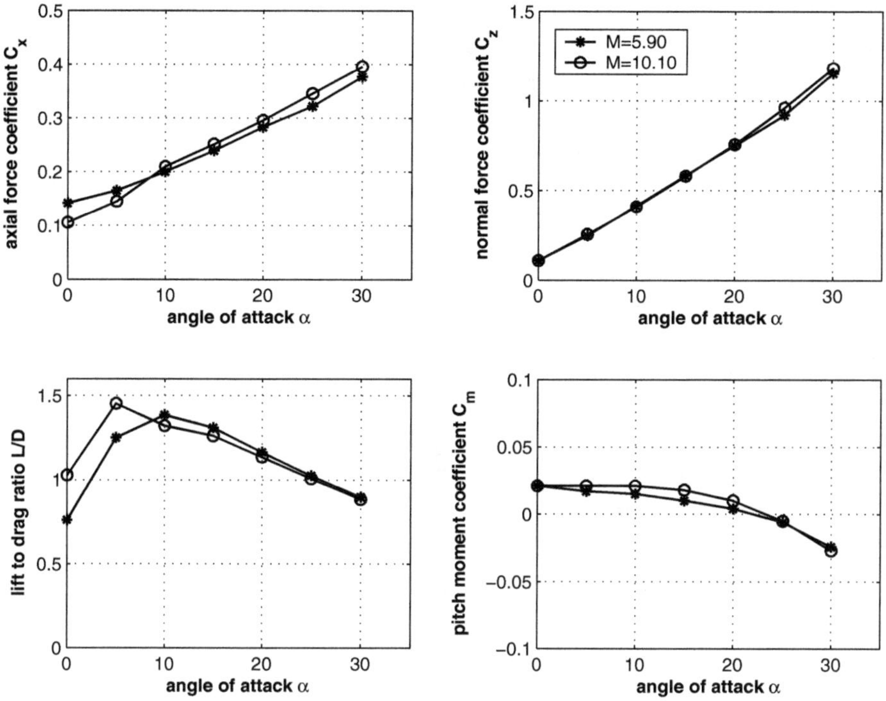

Fig. 5.26. Aerodynamic coefficients in the hypersonic regime for the bent bicone shape, Fig. 5.8 c). Moment reference: $x_{ref} = 0.554\,L_{ref}$, $z_{ref} = 0.0\,L_{ref}$. Data source: [37].

perfect, equilibrium and non-equilibrium thermochemical states of the air, [9], [51], [52]. A summary of these results is given in Fig. 5.27.

All the computations were conducted for an angle of attack α = - 20°. Mach number independence is well discernable for C_X, C_Z, and L/D, but not independence from the thermodynamic state of the gas. The axial force coefficient C_X is best represented by the equilibrium assumption (upper left) and the normal force coefficient C_Z by the non-equilibrium one (upper right). L/D is not much affected, but the non-equilibrium state seems to be the appropriate one (lower left).

Finally, the trim angle α_{trim} in particular for high Mach numbers, agrees fairly well with the non-equilibrium data (lower right). The trim angles are computed with Park's formula, eq. (5.7). Indeed, for high Mach numbers the differences of α_{trim} between perfect gas predictions and flight are more than 2°, which is in agreement with the observations during APOLLO flights. From these results it can be concluded that perfect gas simulations, either numerically or experimentally, are not appropriate for re-entry flows with high Mach numbers, say $M_\infty \gtrsim 6$.

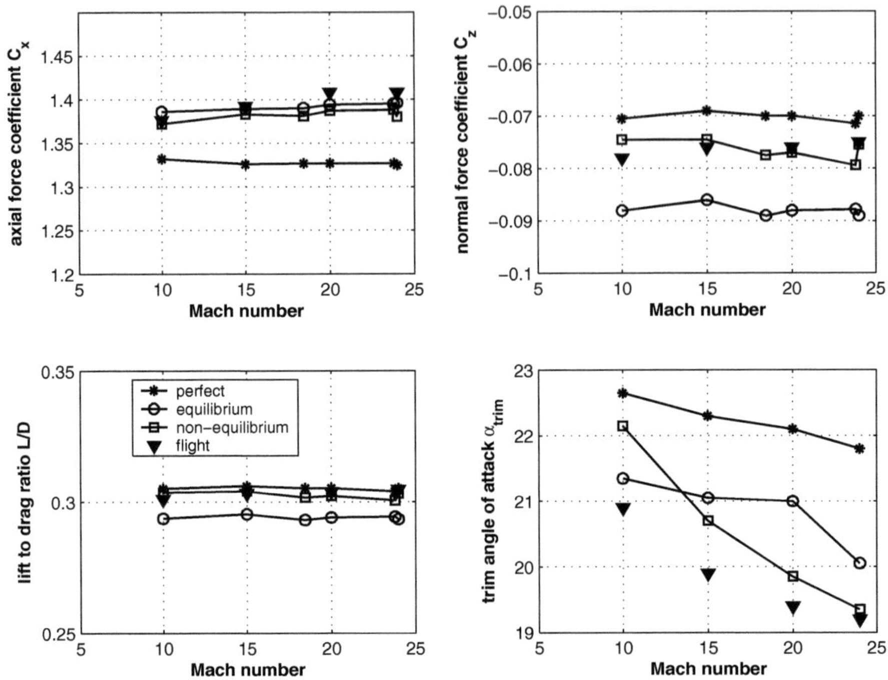

Fig. 5.27. Influence of high-temperature real-gas effects on the aerodynamics of the ARD capsule, angle of attack α = - 20°. Data source: [9], [51], [52], center-of-gravity: $x_{cog} = 0.260\,D_1$, $z_{cog} = 0.0353\,D_1$.

5.4 Dynamic Stability

When a capsule or ballistic probe enters a planetary atmosphere the incidence of this vehicle (angle of attack) evolves like an oscillator responding to the aerodynamically static and dynamic forces and moments. If the vehicle behaves statically stable the static torque always tends to restore the vehicle towards the trim position during its oscillatory motion. If the vehicle behaves dynamically stable the amplitude of the oscillatory motion then is damped[8], reaching asymptotically[9] the stable trim position.

Since the 1960s it is well-known that capsules and ballistic probes, widely used in planetary exploration missions, exhibit often dynamic instabilities during the landing phase. This may concern flight velocities in the low supersonic, the transonic and the subsonic flow regime. The final landing operation, which is often supported by an appropriate parachute system, where the elements of

[8] Long-period lightly damped oscillatory modes are called phugoid modes in aircraft flight dynamics.

[9] Due to the flight along the trajectory with decreasing altitude and increasing density.

this system (drogue chute $\Longrightarrow$ pilot chute $\Longrightarrow$ parachute, or similar) have to be extracted, mostly from a canister inside the vehicle, and successfully deployed, requires a dynamically stable flight state of this vehicle.

Because of the importance of dynamically stable flight in the different Mach number regimes, we give here a compact account of the issues of dynamic stability of NW-RV-flight vehicles, because in general this cannot be found in the literature.

Generally, the dynamic stability is defined by, [54]:

$$\xi = C_D - \frac{\partial C_L}{\partial \alpha} + \left(\frac{L_{ref}}{r}\right)(C_{mq} + C_{m\dot{\alpha}}), \tag{5.8}$$

with C_D the drag coefficient, C_L the lift coefficient, $C_{mq} + C_{m\dot{\alpha}}$ the dynamic derivative of pitch motion, the so called pitch damping coefficient, and r the radius of gyration of the vehicle around the pitch axis.

The vehicle is dynamically stable, if the dynamic stability coefficient

$$\xi < 0. \tag{5.9}$$

Condition eq. (5.9) is often used in ballistic range tests where it is assumed that the aerodynamic coefficients are constant for angle of attack variations, [53], [54].

However, our main interest consists in the determination of the dynamic derivative of pitch motion $C_{mq} + C_{m\dot{\alpha}}$. So we restrict the following discussion to the description of experimental and numerical methods regarding the damping properties of oscillatory movements of RV-NW-flight vehicles.

5.4.1 Physics of Dynamic Instability

From the beginning of the investigations of dynamic instability of blunt[10] or very blunt[11] shapes, it was argued that the near wake flow plays an important role regarding any dynamically unstable behavior, [55], [56]. Further, experimental investigations revealed that also nose-induced flow separation and reattachment make the vehicle more unstable.

Since no full and clear understanding of the related physical phenomena is available up to now, we list here some experimentally observed phenomena and corresponding interpretations. Partly they are contradicting each other, which illustrates the poor present state of the art.

- The wake influences the pitch rotation. When the vehicle and with that the aft part of the configuration comes down during the pitch movement, for example to $\alpha_3 = 0°$ at time t_3, Fig. 5.28, the aft part and, of course, the wake resides in the flow field generated by the front part of the vehicle at an earlier instant with $\Delta\alpha > 0°$ for the time $t_2 \approx t_3 - \Delta t$. The time lag

[10] Spherical cones and bicones.

[11] Capsules and ballistic probes.

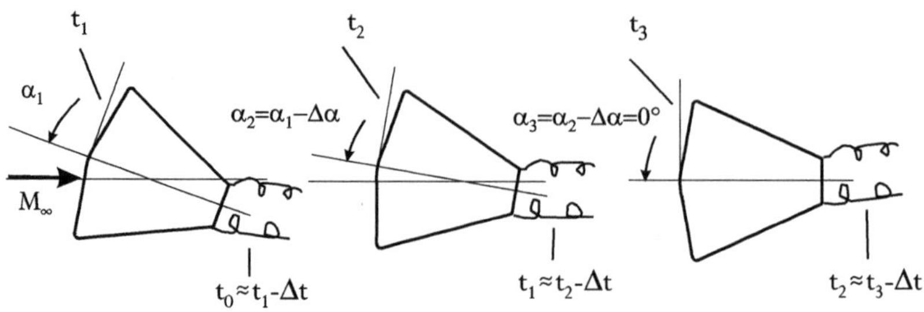

Fig. 5.28. Sketch of oscillatory pitch motion.

Δt is needed to propagate the flow information from the front part to the aft part and the wake. So, a destabilizing moment is generated, compared to the steady flow field at a fixed $\alpha_3 = 0°$ position, Fig. 5.28, [55].

- An experience from many test cases with flow separation is that opposite effects on dynamic and static stability often may exist. The flow separation induces an increase of the static stability and causes dynamic instability, see for example Fig. 5.29, [55].
- Rounded bases on spherically blunted cones make these shapes dynamically more unstable, Fig. 5.30.
- The dynamic pressure plays a role. During re-entry in a planet's atmosphere there is a rapid increase of the dynamic pressure q_∞ due to the density increase. It reaches a maximum when the decrease of v_∞^2 becomes stronger than the increase of ρ_∞. Typically, the vehicle behaves dynamically stable as long as q_∞ increases and gets unstable close to the maximum and beyond, [53]. An often used explanation is, that the vehicle experiences a weakened static restoring moment, due to the decreasing dynamic pressure, [54].
- As mentioned above the vehicles are more unstable with decreasing Mach number, which is explained with the growing influence of the pressure distribution at the aft part of the vehicle for lower Mach numbers.
- Hysteresis effects play a role, [57], [58]. The analysis of the unsteady wake structure suggests that the vortical flow in the wake moves up and down with a non-dimensional frequency, defined by the Strouhal number[12], which amounts approximately to $Sr = 0.2$. A second frequency in the flow field is associated with the free shear layer, mostly generated near the shoulder of a capsule or entry probe, where vortices are formed by the Helmholtz instability, having a Strouhal number $Sr \approx 2$. In contrast to that, the natural flight frequency of the non-winged vehicles discussed in [57], derived from the solution of the equations of angular motion, Sub-Section 5.4.2, is

[12] $Sr = fD/v_\infty$, with f the frequency, D the reference vehicle diameter and v_∞ the free-stream velocity.

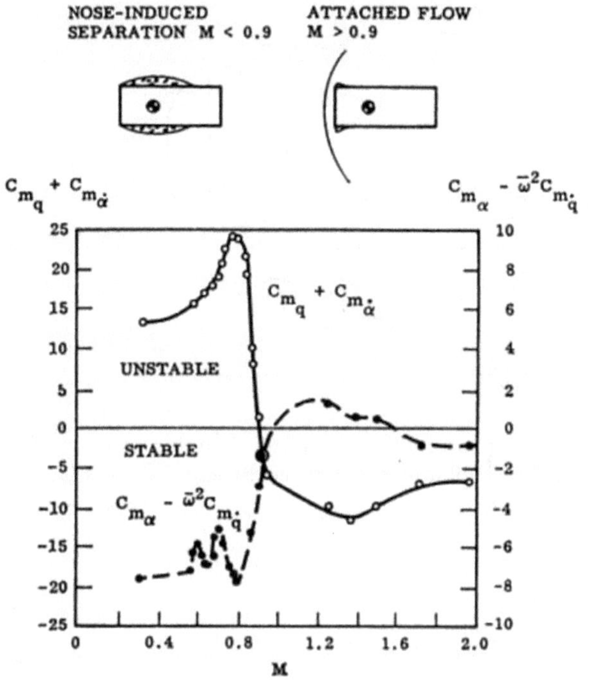

Fig. 5.29. Typical stability behavior. Flow field past a canister with separation at $\alpha_0 = 0$ and pitch angle $\Delta\Theta = 1.5°$, [55]. Coefficient of dynamic stability $C_{mq} + C_{m\dot{\alpha}}$ (left axis) and of static stability $C_{m\alpha}$ - $\overline{\omega}^2 C_{m\dot{q}}$ (right axis, this is the notation in [55] for the static stability) as function of the lower range of the flight Mach number M.

at least one order of magnitude lower than the aforementioned Strouhal numbers indicate. Therefore it was argued that there is no resonance coupling possible between the angular motion and the frequencies of the unsteady flow structure which induce the dynamic instability.

Instead, the pressure distribution along the vehicle's aft part surface indicates during the oscillation a hysteresis effect, which means that at an instantaneous angular position, say at $\Theta = \Theta_c$, the pressure distribution during upward movement is different compared to the one during downward movement. In the case that the pitching moment C_m, evaluated by the integration of the pressure distribution around the center-of-gravity, is larger during upward compared to downward movement, an additional moment is generated which acts in the direction of the angular motion. Thus the dynamic oscillations are less damped and could become unstable. Fig. 5.31 demonstrates this behavior for a two-dimensional APOLLO-like shape[13].

[13] The unstable tendency of the system is demonstrated by the direction of rotation of the moment coefficient C_m, which is clockwise in Fig. 5.31.

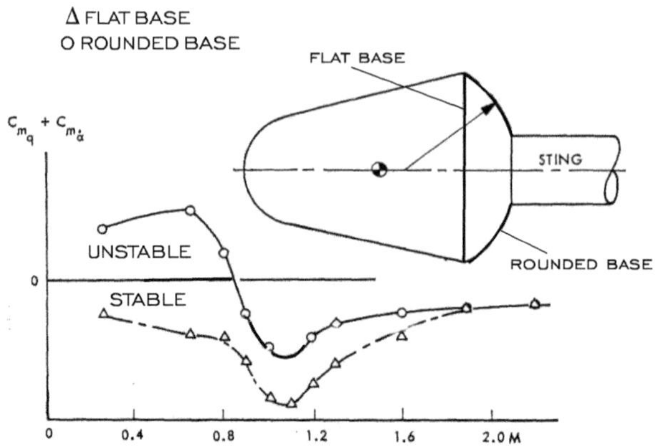

Fig. 5.30. Effect of the base contour on pitch damping at $\alpha_0 = 0°$ and angular position $\Delta\Theta = 1.5°$, [55]. Dynamic stability coefficient $C_{mq} + C_{m\dot{\alpha}}$ as function of the flight Mach number M.

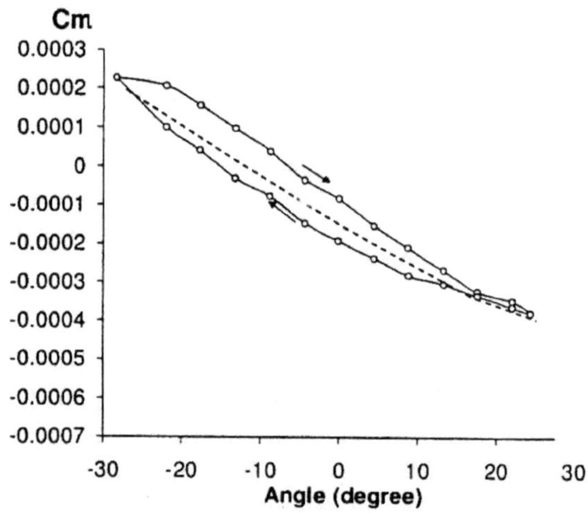

Fig. 5.31. Hysteresis due to oscillatory motion: pitching moment C_m as function of pitch angle θ, [57].

- The importance of the hysteresis effect for the dynamic stability is also described in [59]. The hysteresis effect is manifested by pressure oscillations due to the body movement at the aft part of the vehicle. A detailed analysis on the basis of reliable CFD results of flow fields with fixed angular positions[14], revealed the existence of a longitudinal vortex pair generated just behind the recirculation region of the vehicle wake. From that it was conjectured that the magnitude of the lift slope $C_{L\alpha}$ plays an important

[14] For steady free-stream conditions.

role for the dynamic stability in the sense that larger $C_{L\alpha}$ values lead to a more unstable behavior.

5.4.2 Equation of Angular Motion

The general equation of the rate of change of angular motion reads, [60]:

$$T_f \left(\frac{d\Omega}{dt}\right)_g = T_f \left(\frac{d\Omega}{dt}\right)_f + \underline{\Omega}_f \times T_f \underline{\Omega}_f = \underline{Q}^a_f, \tag{5.10}$$

$$T = \begin{pmatrix} I_{xx} & -I_{xy} & -I_{zx} \\ -I_{xy} & I_{yy} & -I_{yz} \\ -I_{zx} & -I_{yz} & I_{zz} \end{pmatrix}, \ \underline{\Omega} = \begin{pmatrix} p \\ q \\ r \end{pmatrix}, \ \underline{Q}^a = \begin{pmatrix} L \\ M \\ N \end{pmatrix}, \tag{5.11}$$

with T being the tensor of the moment of inertia, $\underline{\Omega}$ the vector of angular velocity, and $\underline{Q}^a$ the vector of aerodynamic moments. The subscript g denotes the inertial, Earth-fixed (geodetic), and f the non-inertial (vehicle fixed, i. e., body-axis) coordinate system, which are explained in Fig. 5.32. The aerodynamic entities are specified by the superscript a. L, M, N are the total roll, pitch and yaw moments, and p, q, r are the angular roll, pitch and yaw velocity components.

The three components of the vector-matrix equation, eq. (5.10), have the form:

$$I_{xx}\dot{p} - I_{xy}\dot{q} - I_{zx}\dot{r} + q(-I_{zx}p - I_{yz}q + I_{zz}r) - r(-I_{xy}p + I_{yy}q - I_{yz}r) = L,$$

$$-I_{xy}\dot{p}+I_{yy}\dot{q}-I_{yz}\dot{r}+r(I_{xx}p-I_{xy}q-I_{zx}r)-p(-I_{zx}p-I_{yz}q+I_{zz}r) = M, \tag{5.12}$$

$$-I_{zx}\dot{p} - I_{yz}\dot{q} + I_{zz}\dot{r} + p(-I_{xy}p + I_{yy}q - I_{yz}r) - q(I_{xx}p - I_{xy}q - I_{zx}r) = N.$$

For vehicles, being symmetrical in the x-z plane with $I_{xy} = I_{yz} = 0$, they reduce to:

$$I_{xx}\dot{p} - I_{zx}\dot{r} + q(-I_{zx}p + I_{zz}r) - rI_{yy}q = L,$$

$$I_{yy}\dot{q} + r(I_{xx}p - I_{zx}r) - p(-I_{zx}p + I_{zz}r) = M, \tag{5.13}$$

$$-I_{zx}\dot{p} + I_{zz}\dot{r} + p(I_{yy}q) - q(I_{xx}p - I_{zx}r) = N.$$

The angular orientation of the vehicle in the inertial space is described by the Euler angles ψ (heading angle), θ (pitch angle) and ϕ (bank angle), Fig. 5.32:

$$\underline{\Omega}_f = \begin{pmatrix} p \\ q \\ r \end{pmatrix}_f = \begin{pmatrix} 1 & 0 & -\sin\theta \\ 0 & \cos\phi & \sin\phi\cos\theta \\ 0 & -\sin\phi & \cos\phi\cos\theta \end{pmatrix} \begin{pmatrix} \dot{\phi} \\ \dot{\theta} \\ \dot{\psi} \end{pmatrix}. \tag{5.14}$$

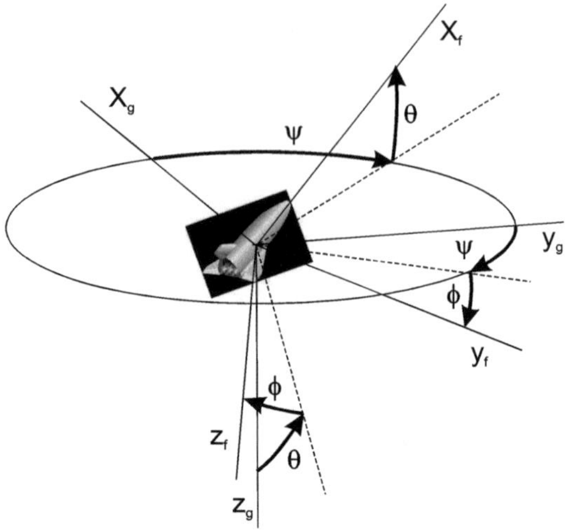

Fig. 5.32. Definition of the Euler angles ϕ, θ, ψ transforming geodetic (g) to vehicle fixed (f) coordinate systems.

The components of the vector-matrix equation, eq. (5.14), in detail read:

$$p = \dot{\phi} - \dot{\psi} \sin\theta,$$

$$q = \dot{\theta} \cos\phi + \dot{\psi} \sin\phi \cos\theta, \tag{5.15}$$

$$r = -\dot{\theta} \sin\phi + \dot{\psi} \cos\phi \cos\theta.$$

Due to the rotational symmetry of most of the RV-NW-type flight vehicles a consideration of the angular motion restricted to the x-z (pitch) plane is often sufficient (one degree of freedom of rotation) for determining the dynamic stability. Thus $\phi = \psi = p = r = 0$, and eqs. (5.13) and (5.15) reduce to:

$$I_{yy}\, \dot{q} = I_{yy}\, \ddot{\theta} = M. \tag{5.16}$$

Consider now the flight of a RV-NW-Type vehicle. For convenience we assume straight level flight, Fig. 5.33.

The vehicle, after being disturbed, may oscillate in the pitch plane initially with a small angular amplitude when the disturbance is small. The question is now whether this oscillatory movement will be damped or will be amplified. In the first case we speak about a dynamically stable vehicle, in the latter about a dynamically unstable vehicle.

For an oscillatory movement with small pitch angles $\theta(t)$ around a fixed angle of attack α, the moment M in eq. (5.16) is written as the sum of two terms:

$$M(\theta) = \frac{\partial M^a}{\partial \theta}\, \theta + \frac{\partial M^a}{\partial \dot{\theta}}\, \dot{\theta}, \tag{5.17}$$

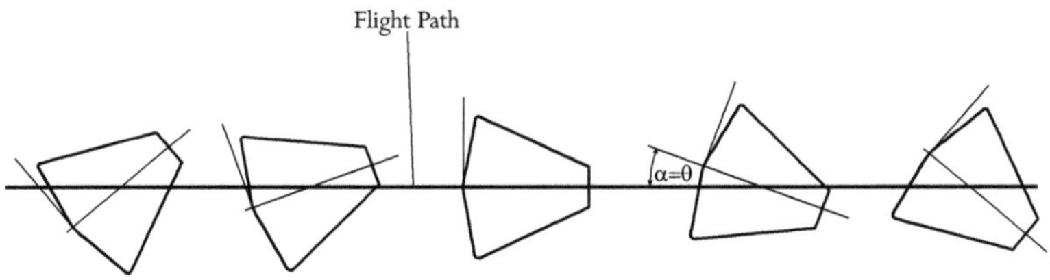

Fig. 5.33. Straight level flight of an RV-NW-Type vehicle which oscillates in the pitch plane.

where $\dot{\theta}$ is the rate of change with time t of pitch movement. The assumptions and conditions which are used in the derivation of eq. (5.17) are described in detail in [61].

The first term of this relation is called the aerodynamic restoring moment, because for $M^a_\theta < 0$ (static stability), it effectuates $M \rightarrow 0$, i. e., stable flight. The second term is called the aerodynamic damping moment, because for $M^a_{\dot{\theta}} < 0$ (dynamic stability), it reduces M as function of time t, $M(t) \rightarrow 0$, i. e., the pitch oscillation is damped with time.

The movement shown in Fig. 5.33 can be considered as the sum of two movement forms, namely a pitch oscillation with angular pitch velocity q at constant angle of attack α, and an angle of attack oscillation with constant pitch angle θ. Superimposed they result in the straight flight path shown in the figure.

In the first movement form the vehicle moves with oscillating pitch angle θ (θ is the position of the vehicle's x-axis) and constant angle of attack along a curved flight path, Fig. 5.34. The pitching moment hence is a function of the pitch velocity $q = \dot{\theta}$, and its rate of change with q is $\partial M/\partial q = M_q$.

In the second movement form the vehicle moves along a curved flight path with constant pitch angle, which acts on the vehicle as an angle of attack change with time, Fig. 5.35. The pitching moment hence is a function of the angle of attack velocity $\dot{\alpha}$, and its rate of change with $\dot{\alpha}$ is $\partial M/\partial \dot{\alpha} = M_{\dot{\alpha}}$.

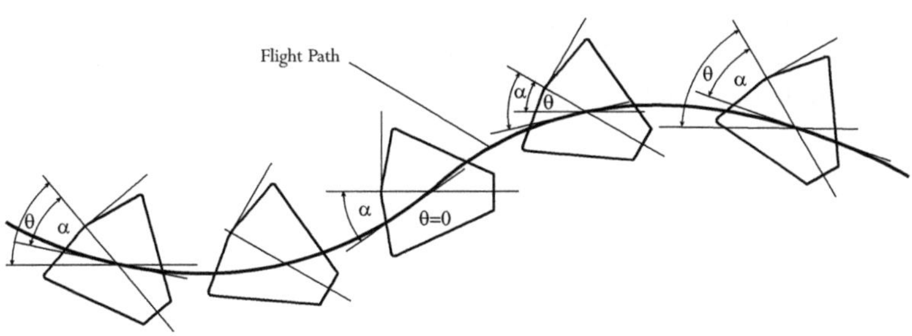

Fig. 5.34. Pitch plane: oscillation with pitch velocity q at constant angle of attack α leads to M_q.

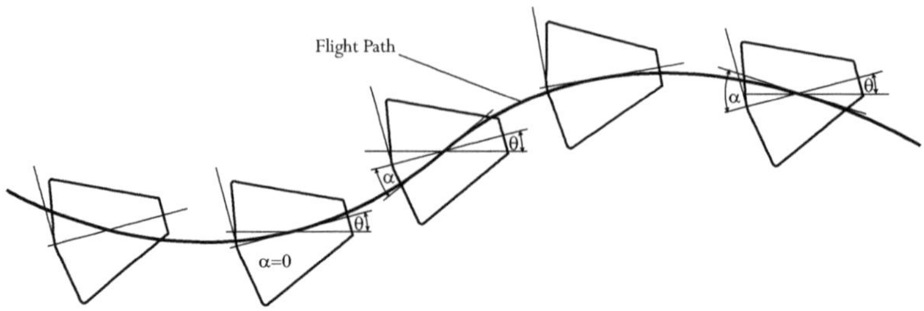

Fig. 5.35. Pitch plane: angle of attack oscillation at constant pitch angle θ leads to $M_{\dot{\alpha}}$.

This means that the aerodynamic damping moment can thought to be composed of two terms:

$$M^a_{\dot{\theta}} = M^a_q + M^a_{\dot{\alpha}}. \tag{5.18}$$

And finally we obtain

$$M = M^a_\theta \, \theta + (M^a_q + M^a_{\dot{\alpha}}) \, \dot{\theta}. \tag{5.19}$$

In case of wind tunnel experiments (free or forced oscillation) the measured total moment M in eq. (5.17) has to be expanded by the components of the elastic restraints considered as a tare moment M^t (for more details see [61] and [53]):

$$M = (M^a_{\dot{\theta}} + M^t_{\dot{\theta}}) \, \dot{\theta} + (M^a_\theta + M^t_\theta) \, \theta. \tag{5.20}$$

In the non-dimensional form we find for the aerodynamic coefficient of pitch damping, also called dynamic derivative of pitch motion[15]:

$$C_{mq} + C_{m\dot{\alpha}} = M^a_{\dot{\theta}} \frac{2 v_\infty}{q_\infty A_{ref} D^2}. \tag{5.21}$$

If its value is negative, damping occurs, the vehicle is dynamically stable.

The coefficient of the slope of the pitching moment, i. e. its derivative with respect to the angle of attack, reads:

$$C_{m\alpha} = M^a_\theta \frac{1}{q_\infty A_{ref} D}. \tag{5.22}$$

If its value is negative, the vehicle is statically stable. D denotes the diameter of a RV-NW-type vehicle.

The reader is asked to note that a vehicle can be statically stable but dynamically unstable and vice versa, see, for example, Fig. 5.29.

---

[15] $C_{mq} + C_{m\dot{\alpha}}$ is the usual notation for the dynamic derivative of pitch motion.

5.4.3 Experimental Methods

Up to now the dynamic stability of flight vehicles is determined mainly by experimental methods. The most often used wind tunnel techniques are, [61], [62], [53]:

- free oscillation technique,
- free-to-tumble technique,
- forced oscillation technique,
- free flight technique[16].

As already mentioned in Sub-Section 5.4.2, for RV-NW-type flight vehicles it is sufficient to determine the pitch damping coefficient $C_{mq} + C_{m\dot{\alpha}}$ for a proper assessment of the dynamic stability of the vehicle. In the following we treat shortly the capabilities of the first three techniques including the data reduction processes, which are most often used for capsule and entry probe investigations.

Free oscillation technique

This method is probably the earliest and obviously the simplest technique to determine the damping coefficient. The model, mounted on a sting and fixed by an elastic flexure, is deflected to an initial amplitude, then released and the decay of the oscillatory motion in the presence of flow is observed. No complicated drive or control system is required. On the other hand, this technique fails if the model is dynamically unstable, which means that the initial amplitude is amplified and no control of the angular motion is possible. Nevertheless, such behavior would indicate that the model possesses a strong dynamic instability.

For the determination of the dynamic coefficient we consider the equation of angular motion (harmonic motion with small amplitudes) in the form of eqs. (5.16) and (5.20):

$$I_{yy}\ddot{\theta} = (M^a_{\dot{\theta}} + M^t_{\dot{\theta}})\,\dot{\theta} + (M^a_{\theta} + M^t_{\theta})\,\theta. \tag{5.23}$$

For further consideration we rewrite eq. (5.23) in the form:

$$I_{yy}\ddot{\theta} + C\dot{\theta} + K\theta = 0, \tag{5.24}$$

where the damping parameter C, and the restoring parameter K are:

$$C = -(M^a_{\dot{\theta}} + M^t_{\dot{\theta}}), \quad K = -(M^a_{\theta} + M^t_{\theta}). \tag{5.25}$$

The solution of eq. (5.24), which is of interest for us, represents the conditions, where the system will oscillate and reads[17]:

$$\theta = A_1 e^{(a+bi)t} + A_2 e^{(a-bi)t} \qquad \text{or}$$
$$\theta = \theta_0 e^{-(C/2I_{yy})t} \sin(\omega_d\, t \pm \phi), \tag{5.26}$$

[16] In ballistic wind tunnels.

[17] The roots of the characteristic equation are complex.

where $a = -C/2I_{yy}$, $b = \omega_d = \sqrt{\frac{K}{I_{yy}} - \left(\frac{C}{2I_{yy}}\right)^2}$ is the damped natural frequency, and A_1, A_2, θ_0, ϕ are arbitrary constants. The oscillatory system eq. (5.24) is damped if $a < \omega_n$ or $-C/2I_{yy} < \sqrt{K/I_{yy}}$, where ω_n is the undamped natural frequency.

From the measured time history of the oscillatory motion the decrease of the amplitude θ as function of time is known. By definition of the logarithmic decrement at times t_1 and t_2 we obtain from eq. (5.26) (with $\sin(\omega_d t \pm \phi) = 1$):

$$\begin{aligned} \ln\frac{\theta_2}{\theta_1} &= -(C/2I_{yy})(t_2 - t_1), \\ C &= -\frac{2I_{yy}f}{C_{yr}}\ln\frac{\theta_2}{\theta_1}, \end{aligned} \tag{5.27}$$

where $t_2 - t_1 = C_{yr}/f$. C_{yr} is the number of cycles damped during $\Delta t = t_2 - t_1$ and f the frequency of oscillation.

The restoring (static) parameter is:

$$K = I_{yy}\omega_d^2 + I_{yy}\left(\frac{C}{2I_{yy}}\right)^2, \tag{5.28}$$

where the second right-hand side term is usually very small and can be neglected. Therefore we have

$$K = I_{yy}\omega_d^2. \tag{5.29}$$

As mentioned in Sub-Section 5.4.2, the dynamic damping and the static restoring moment consist of the aerodynamic moments and the moments generated by tares[18] in the mechanical system. The tares are given from wind-off measurements and we find, [61]:

$$\begin{aligned} C &= -(M^a_{\dot\theta} + M^t_{\dot\theta})_w, \\ K &= -(M^a_{\theta} + M^t_{\theta})_w, \\ M^a_{\dot\theta} &= (M^a_{\dot\theta} + M^t_{\dot\theta})_w - (M^t_{\dot\theta})_v, \\ M^a_{\theta} &= (M^a_{\theta} + M^t_{\theta})_w - (M^t_{\theta})_v, \end{aligned} \tag{5.30}$$

where the subscripts w and v denote wind-on and vacuum conditions[19], respectively.

Finally we obtain:

$$M^a_{\dot\theta} = -2I_{yy}\ln\frac{\theta_2}{\theta_1}\left[\left(\frac{f}{C_{yr}}\right)_w - \left(\frac{f}{C_{yr}}\right)_v\frac{f_v}{f_w}\right], \tag{5.31}$$

[18] Tare damping comprises the damping by the mechanical system, the damping capacity of the materials and the damping of still air in wind-off condition.

[19] Vacuum conditions are used in order to prevent any influence of still air damping.

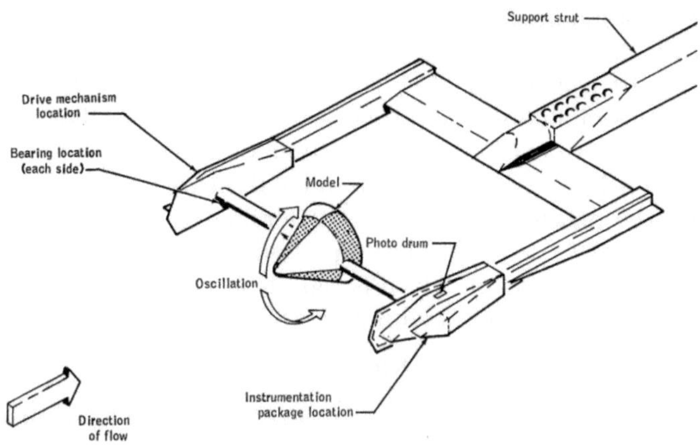

Fig. 5.36. Model installation in free-to-tumble dynamic tests, [22].

$$M_\theta^a = -I_{yy} \left[(\omega_d)_w^2 - (\omega_d)_v^2 \right] , \tag{5.32}$$

which includes a correction for the tare damping considering the different oscillation frequencies in the wind-on and wind-off case. With the relations eqs. (5.21) and (5.22) the aerodynamic pitch damping coefficient of the vehicle is known.

Free-to-tumble technique

Models mounted on a transverse rod through its center-of-gravity are statically balanced and are able to tumble freely through an α range from 0° to 360°, Fig. 5.36.

This free-to-tumble technique was extensively employed during the APOLLO programme, [19], [22], [53]. In some Mach number ranges tare damping is negligible and for Mach numbers, where tare damping is expected to be a fractional part of the total damping, corrections to the input data are applied. Consequently, in eq. (5.23) $M_{\dot\theta}^t$ and M_θ^t are deleted and we have:

$$I_{yy}\ddot\theta - \frac{q_\infty A_{ref} D^2}{2v_\infty} \left(C_{mq} + C_{m\dot\alpha} \right) \dot\theta - q_\infty A_{ref} D \, C_{m\alpha} \, \theta = 0. \tag{5.33}$$

In eq. (5.33) $C_{mq} + C_{m\dot\alpha}$ is the only unknown, because I_{yy} is measured in advance and $C_{m\alpha}$ is known from static tests. Therefore the pitch damping coefficient is found by an iteration process, changing the value of $C_{mq} + C_{m\dot\alpha}$ until the measured time history coincides with the computed one.

Forced oscillation technique

An experimental means to quantify the dynamic instability[20] is given by the forced oscillation technique. The model is fixed, for example, on a

[20] And not only the damping capability as in the free oscillation technique.

crossed-flexure pivot mounted on a sting and is forced to oscillate by a special apparatus, [61], [62], [53].

The oscillation frequency should be at or near the resonant frequency of the model, since this minimizes the torque required to sustain the oscillation and simplifies the determination of the dynamic derivatives. The resonance conditions can be achieved by tuning the oscillation frequency of the forcing torque until the phase shift between the forcing torque and the model displacement amounts to $\phi = 90°$. The equation of angular motion, which describes the motion of a model, which is rigidly suspended in an airstream and forced to oscillate by a sinusoidal function of time, has the form:

$$I_{yy}\ddot{\theta} + C\dot{\theta} + K\theta = M \cos \omega t, \tag{5.34}$$

where C and K are defined by eq. (5.25). M denotes the forcing moment and ω the oscillation frequency.

A particular solution of eq. (5.34) is given by:

$$\theta = \theta_0 \cos(\omega t - \phi). \tag{5.35}$$

By substituting the angular velocity $\dot{\theta}$ and the angular acceleration $\ddot{\theta}$ in eq. (5.34) and equating coefficients we obtain:

$$\theta_0 = \frac{M}{(K - I_{yy}\omega^2) \cos \phi + C\omega \sin \phi},$$

$$\tan \phi = \frac{C\omega}{K - I_{yy}\omega^2}. \tag{5.36}$$

In case of a constant amplitude motion and resonance conditions ($\phi = 90°$), the inertia term balances the restoring moment parameter K, and the damping moment parameter C is proportional to the forcing moment M:

$$K = I_{yy}\omega_n^2, \tag{5.37}$$

$$C = \frac{M}{\theta_0 \omega}. \tag{5.38}$$

Restoring moment parameter K and damping moment parameter C can be determined from eq. (5.37) and (5.38), if the forcing moment M, the amplitude θ_0 and the oscillation frequency ω are known from measurements[21]. With eqs. (5.30) and (5.37) we find for $M^a_{\dot{\theta}}$ and M^a_{θ}:

$$M^a_{\dot{\theta}} = -\frac{1}{\omega_w}\left[\left(\frac{M}{\theta_0}\right)_w - \left(\frac{M}{\theta_0}\right)_v\right], \tag{5.39}$$

$$M^a_{\theta} = -I_{yy}(\omega_w^2 - \omega_v^2). \tag{5.40}$$

Again, with the relations eqs. (5.21) and (5.22), the dimensionless forms of the dynamic derivative are known.

[21] For forced oscillation systems which operate off resonance, the derivation of the equations of dynamic stability can be found in [61].

5.4.4 Numerical Methods

Since the advent of powerful three-dimensional unsteady Navier-Stokes solvers in the late 1990s, which are able also to simulate turbulent flows, first attempts were undertaken to directly predict dynamic derivatives by numerical methods. In [59] the unsteady flow around a capsule shape was computed and the flow field was evaluated with the goal to reveal the physical phenomena responsible for the dynamic instability. Unfortunately, no direct determination of the damping derivatives was carried out, but a so called "Constant-Delay-Model" was developed, which enables one to derive the dynamic stability parameters from flow calculations at fixed angles of attack.

With the procedure reported in [63], [64], a forced oscillation scenario was simulated by numerical methods. The vehicle experiences, for example, an oscillatory pitch rotation $\theta(t) = \theta_0 \sin \omega t$ around a fixed angle of attack α_{fix} with a reduced frequency $k = \omega_n D / v_\infty$. The natural undamped angular velocity ω_n is determined by eqs. (5.37) and (5.22):

$$\omega_n = \sqrt{\frac{C_{m\alpha} q A_{ref} D}{I_{yy}}}. \tag{5.41}$$

With a linear approach for the pitching moment, namely

$$C_m(t) = C_m(\alpha_{fix}) + C_{m\alpha}\,\theta(t) + (C_{mq} + C_{m\dot{\alpha}})\,\dot{\theta}(t)\frac{D}{v_\infty}, \tag{5.42}$$

the dynamic damping term $C_{mq} + C_{m\dot{\alpha}}$ can be calculated by a numerical averaging process, since $C_m(t)$, $C_m(\alpha_{fix})$, and $C_{m\alpha}$ are known from the numerical solutions.

Calculations, which the authors of [63], [64] have performed for the X-24 and X-38 lifting vehicles, show very promising results and it seems that this approach has a great potential for future applications.

5.4.5 Typical Experimental Results

There are two capsules in the world which were flown more often than any other ones. The first one is the U.S. APOLLO and the second one the Russian SOYUZ, which is still in operation[22]. The dynamic stability of APOLLO was investigated with the three experimental techniques which were described in Sub-Section 5.4.3:

- free oscillation technique,
- free-to-tumble technique,
- forced oscillation technique.

[22] SOYUZ acts as crew transporter for the ISS.

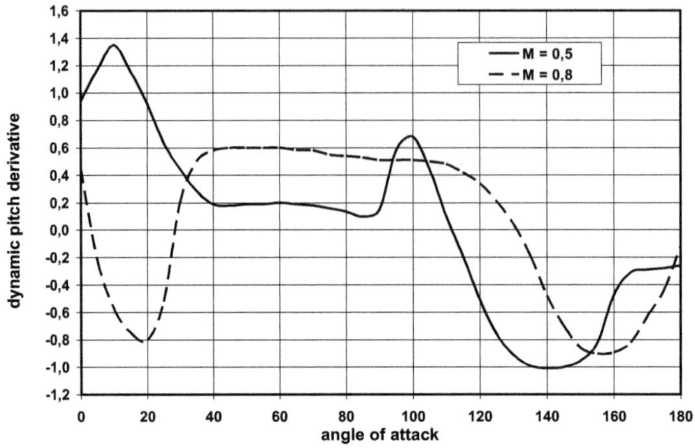

Fig. 5.37. Dynamic derivative of pitch motion $C_{mq}+C_{m\dot{\alpha}}$ as function of the angle of attack for $M_\infty = 0.5$ and 0.8. APOLLO shape, Fig. 5.5. Data source: [22].

As an example for the subsonic Mach numbers $M_\infty = 0.5$ and 0.8, Fig. 5.37 shows the dynamic pitch derivative $C_{mq} + C_{m\dot{\alpha}}$ over an angle of attack range $0° \leqq \alpha \leqq 180°$. In that case the free-to-tumble test technique was applied. With the trim angles for these Mach numbers being in the range of $\alpha_{trim} \approx 15°$, the vehicle behaves for $M_\infty = 0.8$ dynamically stable and unstable for $M_\infty = 0.5$, [22].

The SOYUZ capsule has a strong dependence of the dynamic stability[23] on the angle of attack at $M_\infty = 0.9$, Fig. 5.38. The data were received from wind tunnel experiments and verified in several free flight campaigns, [24]. Since the trim angle α_{trim} is likely of order 20°, SOYUZ behaves dynamically unstable for that Mach number. For hypersonic Mach numbers the capsule is stable and has only a weak dependence on α.

Chapman and Yates, [54], investigated the influence of non-linear aerodynamics during the data reduction process (both for experimental and numerical data) for determining the dynamic pitch derivative $C_{mq} + C_{m\dot{\alpha}}$. For the HUYGENS Titan entry probe the results are plotted in Fig. 5.39.

The values obtained with linear aerodynamics show a strong scatter with Mach number and support the impression of dynamic instability for the $\alpha < 5°$ range. The values calculated with non-linear aerodynamics are nearly constant ($\alpha = 0°, 10°$) and exhibit also for $\alpha = 0°$ a slight damping capability. Further, Fig. 5.39 shows two data of a forced oscillation experiment ($M_\infty = 2.0$) made at FFA Sweden and one data of a free oscillation experiment ($M_\infty = 2.9$) done by the Aircraft Research Association (ARA) in the U.K. These data were extracted from [53]. The two FFA Sweden data are not

[23] In the Russian literature the notation $m_z^{\bar{\omega}_z}$ is used to describe the dynamic derivative with $m_z^{\bar{\omega}_z} \sim C_{mq} + C_{m\dot{\alpha}}$.

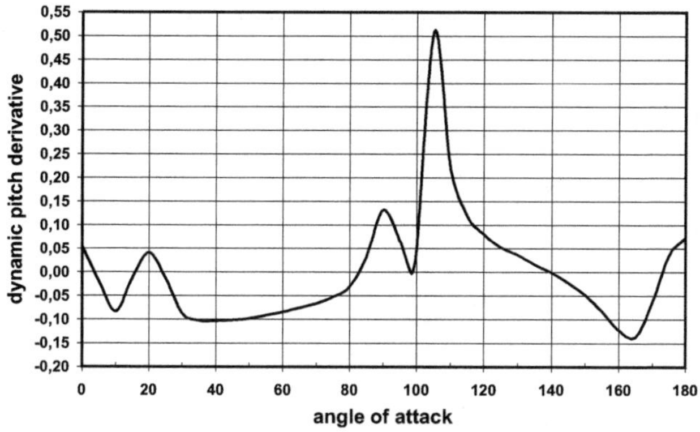

Fig. 5.38. Dynamic derivative of pitch motion $m_z^{\bar{\omega}_z}$ as function of the angle of attack for $M_\infty = 0.9$. SOYUZ shape, Fig. 5.6. Data source: [24].

in line with the general trend and it is argued in [53] that the discrepancy is due to some open questions during the execution of the experiment. The variation of the dynamic derivative with angle of attack for supersonic Mach numbers ($M_\infty = 1.8$ and 2.0) is displayed in Fig. 5.40.

Both curves follow the well-known trend that with increasing angle of attack the dynamic stability increases, but the distribution established with the non-linear evaluation procedure starts at $\alpha = 0°$ with a small damping value, while the FFA Sweden data indicates an undamped behavior (amplification) there.

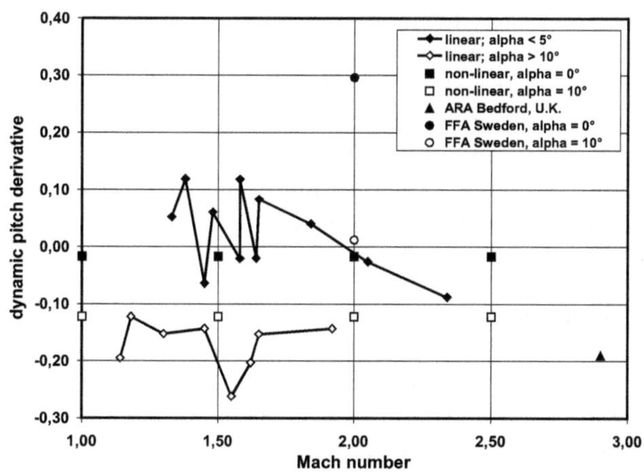

Fig. 5.39. Dynamic derivative of pitch motion $C_{mq} + C_{m\dot{\alpha}}$ as function of the Mach number. HUYGENS shape, Fig. 5.3. Data source: [53], [54].

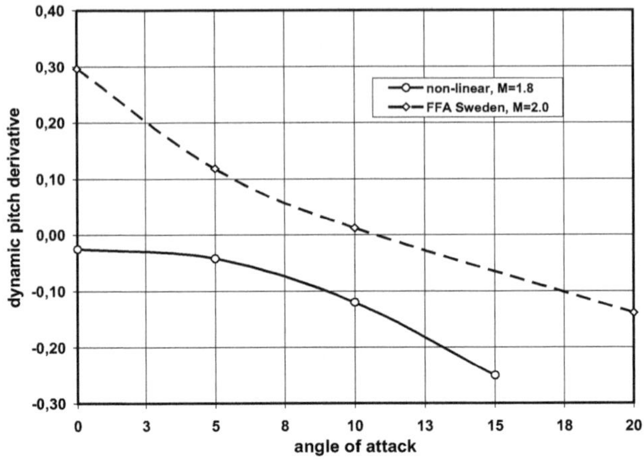

Fig. 5.40. Dynamic derivative of pitch motion $C_{mq} + C_{m\dot{\alpha}}$ as function of angle of attack. HUYGENS shape, Fig. 5.3. Data source: [53], [54].

In the 1970s the Mars entry probe VIKING for exploring the Martian atmosphere was developed in the USA. Two of these probes have entered the atmosphere of the planet Mars in 1976. At that time there were a lot of investigations regarding the dynamic stability of these entry probes [56], [53]. Since the VIKING shape is also a classical one, we present here the dynamic stability behavior, extracted from a free-oscillation experiment, for a typical flow case, namely the Mach number $M_\infty = 1.76$, the reduced frequency $k = 0.0069$ and the Reynolds number $Re = 7.5 \cdot 10^5$. Fig. 5.41 shows the results, which are taken from [56]. A strong instability in a small range around zero angle of attack ($\alpha < 1°$) can be observed, which vanishes exponentially leading subsequently to a positive and nearly constant dynamic damping.

In conclusion it seems that today the uncertainties and discrepancies in the dynamic derivative data of non-winged vehicles are still large and that more activities are necessary for amending the physical understanding. Further, the test methods and the data reduction methods, which are used also for the numerical predictions, must be improved.

5.5 Thermal Loads

One of the major concerns during the development of hypersonic vehicles is the reliable prediction of the thermal loads on the airframe. These govern the design of the Thermal Protection System (TPS), which is one of the main contributors to the vehicle mass. This strongly affects the magnitude of the payload mass, which is often the main issue for an efficient space mission. Generally, the peak wall temperature during re-entry defines the type of TPS

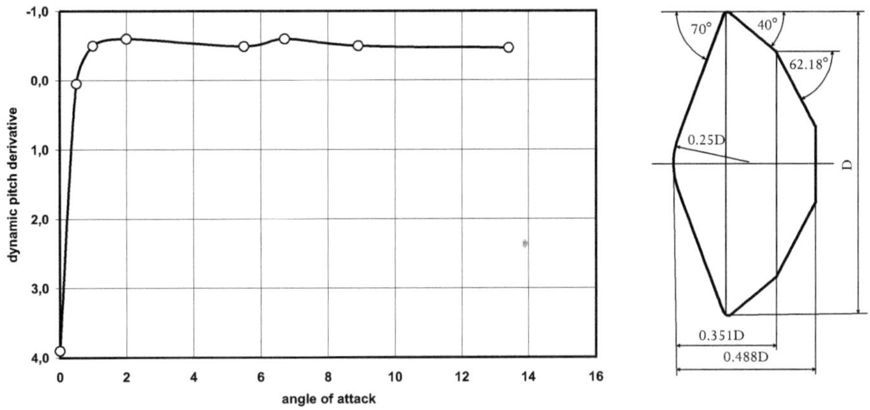

Fig. 5.41. Dynamic derivative of pitch motion $C_{mq} + C_{m\dot{\alpha}}$ as function of angle of attack, VIKING-type shape. Data source: [56].

material used, whereas the time-integrated heat-flux governs structure and thickness of the TPS.

During hypersonic flight heat is transported towards the vehicle surface by three physical transport processes, Fig. 9.2, Section 9.1, see also [65]. The first one is heat transported by diffusion in the gas towards the wall, more precisely, the heat flux in the gas at the wall[24]. The second one is non-convex radiation from other surface parts of the vehicle. At RV-NW-type vehicles this usually does not play a role. Thermal gas radiation of the vibrational excited, dissociated and ionized gas is the third one. For classical Low Earth Orbit (LEO) re-entry with entry speeds below approximately 8.0 km/s, the diffusive heat transfer q_{gw} is the dominating effect. On the other hand, for example, for Lunar return with an entry speed of approximately 10.6 km/s, the heat transfer due to thermal gas radiation $q_{rad,g}$ becomes very important and approaches values of 50 per cent of the total heat-flux. This is due to the strong increase of ionized particles which are generated during entry due to the pass of the gas through the bow shock layer.

Heat is transported away from the vehicle surface mainly by radiation cooling, but also by conduction into the structure, Section 9.1. Important can also be other cooling processes like ablation, transpiration et cetera.

In Chapter 9 we give a short sketch of the thermal state of a vehicle surface and also, in a simulation compendium, an overview of all kinds of methods for the determination of thermal loads. In Chapter 10 relations for an approximative determination of thermal loads are provided. These are useful for first guesses and, for instance, for trajectory definition purposes.

We refer the reader to Chapters 9 and 10 for details and focus our discussion now on data from three free-flight events and one technology study, namely:

[24] In the literature this is often called convective heating.

- OREX suborbital flight,
- ARD suborbital flight,
- APOLLO low Earth orbit (LEO) and Lunar return,
- VIKING-type shape technology study.

5.5.1 OREX Suborbital Flight

Since at the time of the APOLLO flight no numerical methods with adequate quality for complete 3-D flow field simulations at corresponding trajectory points were available, pre-flight predictions and post-flight comparisons were done with analytical relations mainly developed for predicting the forward stagnation point heating. In the 1990s the situation had changed. The data received by demonstrator flights of the Japanese OREX probe (1994) and the European ARD capsule (1998) were used to verify results of pre-flight numerical simulations and to formulate conditions for post-flight analysis including the comparison of the results obtained.

The OREX free-flight experiment gives the possibility to study how intensive the influence of physical phenomena, see, e.g., [65], along the re-entry trajectory on the wall heat flux is. In [10], [8] Viscous Shock Layer (VSL) and Navier Stokes (NS) solutions were generated for selected trajectory points with various degrees of modelisation, where at the wall a temperature distribution was prescribed[25]. In particular, from [8] we can learn in what regime the surface catalytic behavior, the slip conditions, the thermal non-equilibrium and the number of species in the chemical model are important. The outcome was, that in the altitude regime $105.0\,km \gtrsim H \gtrsim 84.0\,km$ slip conditions and thermal non-equilibrium must be included in the NS or VSL solutions. This provides the best results for the stagnation point heating, Fig. 5.42. The wall catalycity does not play any particular role at that altitude, since the recombination probabilities are very low.

For lower altitudes the influence of thermal non-equilibrium and slip conditions decreases and the wall catalycity becomes more important. The comparison with the OREX stagnation point free-flight data supports unambiguously this trend, Fig. 5.42.

In Fig. 5.43 the wall heat flux, q_{gw}, over the OREX surface is plotted for an altitude $H = 92.8\ km$. Again no differences between the calculation with non-catalytic (not shown) and finite-rate catalysis are present. The results with fully catalytic assumptions are far from being realistic. Further, the general trend is that for these altitudes the computations with slip conditions give the more realistic answer. Finally, the authors of [8] showed that the dependency on the number of species of the chemical model on the heat-flux is rather low, therefore they preferred a seven species model, see, e.g., [65].

[25] The wall temperature distribution was determined by a fluid-structure coupling process, where the radiation equilibrium state (radiation-adiabatic wall) was iteratively reached, [10].

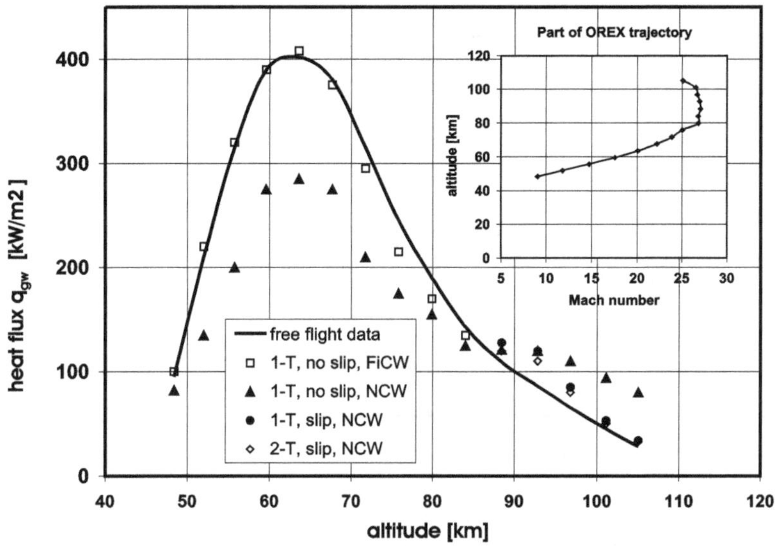

Fig. 5.42. Heat flux in the gas at the wall, q_{gw}, at the stagnation point of the OREX probe during re-entry as function of the flight altitude. Free-flight data and CFD analysis (viscous shock-layer solutions). FiCW: finite catalytic wall, NCW: non-catalytic wall, 1-T: one-temperature model, 2-T two-temperature model. Data source: [8].

5.5.2 ARD Suborbital Flight

As mentioned before, another demonstration flight was conducted by the Atmospheric Re-entry Demonstrator (ARD). For the heat-flux investigations this vehicle was subdivided into three parts, namely the front, the rear cone and the back cover part, Fig. 5.5. The magnitude of the heat-fluxes on the rear cone and back cover part is rather low compared to that on the front part, and the maximum value does not exceed 37.0 kW/m^2. Nevertheless for the proper design and sizing of the TPS system it is important to know also in this area the behavior of the flow during the flight along the trajectory. It was expected that a turbulent reattachment of the flow on the rear cone with the corresponding strong increase of the heat transfer could happen, but this was not observed. Instead, on the back cover a laminar-turbulent flow transition was detected, with a moderate increase of the heat flux, [66].

But let us focus our attention on the front part. Free-flight data are available for wall-temperatures up to $\approx$ 1,100.0 K, since beyond this value the sensors did not work properly. Depending on the location of the sensors, the measured free-flight data are spread over a Mach number range of $15 \lessapprox M\infty \lessapprox 26$ (corresponding to $51.0\, km \lessapprox H \lessapprox 77.0\, km$). The numerical simulations were carried out for $6 \lessapprox M_\infty \lessapprox 26$. Since in the thermo-chemical non-equilibrium flow regime the wall catalycity can play a major role, at the

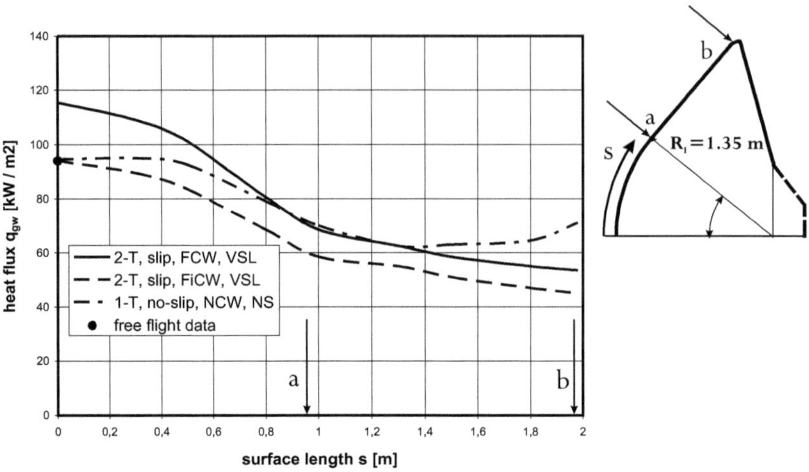

Fig. 5.43. Heat flux in the gas at the wall, q_{gw}, at the OREX probe (symmetry plane) at $v_\infty = 7{,}454.1\ m/s$, $H = 92.8\ km$, as function of the surface length s (definition in the inset). CFD analysis from viscous shock layer and Navier-Stokes solutions. FCW: fully catalytic wall, FiCW: finite catalytic wall, NCW: non-catalytic wall, 1-T: one-temperature model, 2-T two-temperature model. Free flight measurements were made only at the stagnation point ($s = 0.0m$). Data source: [8], [10].

same trajectory points computations with non-catalytic and fully catalytic walls were conducted. The wall temperature distribution was calculated with the radiation-adiabatic wall condition, [51], [66].

Let us first identify what is in agreement with the experience from OREX. It is the fact that for high altitudes $H \gtrapprox 80.0\ km$ the wall catalycity does not play any role, and that the computations with fully catalytic wall assumption give too high values there (flight time $< 4,900.0\ s$ in Fig. 5.44).

In the regime of the peak heating ($H \approx 64.5\ km$, flight time $\approx 4,950.0\ s$) fully catalytic NS computations provide lower values than the measured ones and even the data received from computations with chemical equilibrium are lower, which is in contradiction to the OREX results, Fig. 5.43. An explanation is, that this dispersion could be due to the pyrolysis effect of the front shield ablator, which probably affects the wall catalysis. The process itself is unknown.

On the windward side in flight no clear transition from the laminar to the turbulent boundary layer state could be detected, but on the leeward side (near the shoulder) at an altitude $H \approx 55.0\ km$, transition was observed, [66]. The maximum heat-flux was reached near the windward side shoulder, Fig. 5.45, with approximately $1,200.0\ kW/m^2$ and the NS solutions for laminar flow with equilibrium thermodynamics are closest to the measured data. Note that in Figs. 5.44 and 5.45 the data are interpolated between the NS solutions calculated for selected trajectory points.

5.5.3 APOLLO Low Earth Orbit (LEO) and Lunar Return

Of course, a broad database exists for the APOLLO vehicle, not only from free-flight but also from wind tunnel experiments, and several attempts were made over the time to duplicate these data by applying improved numerical simulation methods and physical modelling. We have learned from the discussion before (OREX and ARD flights), that for high altitudes ($H \gtrapprox 80.0\ km$) the wall catalysis does not play a role for the heat flux, but that with decreasing altitude this effect becomes more and more important. The work done in [30] supports this cognition, Fig. 5.46.

In [30] the heat-fluxes in the pitch plane of APOLLO (x-z plane, Fig. 5.5, 2-D approach) for a Lunar aerocapturing return are calculated by different methods, ranging from simple analytical relations, Chapter 10, up to combined Euler/boundary layer solutions including non-equilibrium real-gas effects. The flight conditions are: $H = 66.7\ km$, $M_\infty = 32.6$, $\alpha = -\ 20°$, $v_\infty = 9{,}980.0\ m/s$. The results obtained with the latter method, and shown in Fig. 5.46, exhibit a much higher heat-flux distribution for the fully catalytic wall, with a peak value of $\approx 2,000.0\ kW/m^2$, than with the finite rate catalytic model (peak heating $\approx 1,080.0\ kW/m^2$). From the general experience we conclude that the data obtained by the fully catalytic wall condition are much too high and that the data of the finite-rate catalytic model are more realistic, but further verification is necessary.

Finally, Fig. 5.47 contains from the same work the data from APOLLO wind tunnel tests, Lees' analytical equations eqs. (10.71) to (10.73), and

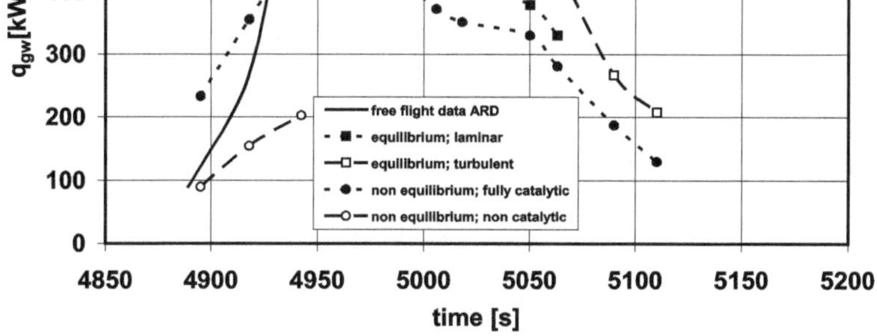

Fig. 5.44. Heat flux in the gas at the wall, q_{gw}, at the stagnation point of the ARD demonstrator along a part of the re-entry trajectory as function of the flight time. Free-flight data and Navier-Stokes data with different modelings, angle of attack $\alpha = -\ 20°$. Data source: [23], [66].

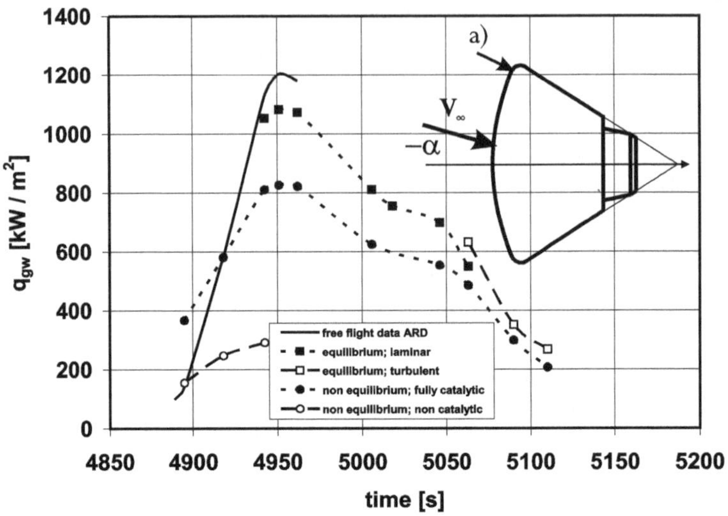

Fig. 5.45. Heat flux in the gas at the wall, q_{gw}, in point a) (peak heating regime) of the ARD demonstrator (symmetry plane) along part of the re-entry trajectory as function of the flight time (see Fig. 5.44). Free-flight data and Navier-Stokes data with different modelings, angle of attack α = - 20°. Data source: [23], [66].

an Euler/boundary layer approach (where the boundary layer equations are solved by an integral method). The wind tunnel conditions are M_∞ = 10.17, $\alpha = 33°$, $v_\infty = 1,440.0\, m/s$.

Lees' method gives a much too high peak value, whereas away from s/R = 1 the agreement with the wind tunnel data is satisfactory. Closer to the wind tunnel data are the results of the Euler/boundary layer approach, where for the peak heating also an overshoot can be observed, which, however, is considerably smaller than the one obtained with Lees formula. Nevertheless, we learn that in the early stage of a system study the application of such a simple approach like the one of Lees (but have in mind the peak heating overshoot!) can be helpful in getting a first idea of the thermal loads.

5.5.4 VIKING-Type Shape Technology Study

The flow field on the leeward side of either RV-W-type or RV-NW-type flight vehicles moving at high Mach numbers and considerably large angles of attack, contains in general vortical-type structures due to complex separation and reattachment/attachment processes of the flow.

In this context it is important to differentiate between attachment and reattachment lines. At reattachment lines the incoming flow in general is boundary layer material, which has lost already total enthalpy, for instance due to surface radiation cooling. At attachment lines, the incoming flow in

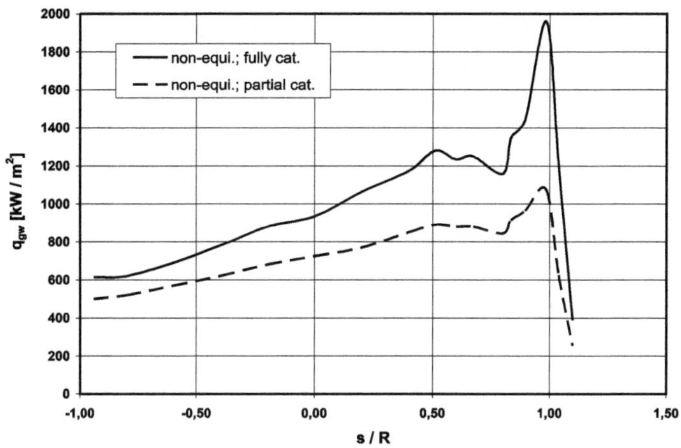

Fig. 5.46. Heat flux in the gas at the wall, q_{gw}, at the APOLLO front-shield surface (symmetry plane) as function of s/R ($s/R = 0$: center point ($z = 0$), Fig. 5.5, $s/R = 1$: stagnation point at the forward pointing shoulder). Aerocapturing Lunar return, $M_\infty = 32.6$, $H = 66.7\ km$, $\alpha = -\ 20°$. Comparison of fully and partial catalytic wall conditions. Non-equilibrium Euler/boundary layer solution. Data source: [30].

general is the outer flow[26] with the original total enthalpy, which may further enhance thermal loads. Depending on the topology of the velocity field at the lee side of a flight vehicle, both types of attachment lines may be present, [65].

The general phenomenon is called vortex/boundary layer interaction, which includes also secondary and higher-order vortex-driven flow separation and reattachment phenomena. boundary layers are becoming thick, where separation occurs (convergence of skin-friction lines) and thin in case of attachment and reattachment (divergence of skin-friction lines), [65]. In the latter case, with the attenuation of the boundary layer, the gradients of the flow variables normal to the wall grow considerably. For gases with relatively high-enthalpy this is also valid for the gas temperature with the consequence of locally high heat fluxes in the gas at the wall (q_{gw}) and large wall temperatures (hot-spot situation). At separation lines a cold-spot situation ensues.

As an example, let us consider the flow over the rear part of the VIKING_1 shape, Fig. 5.6, [45], for the moderate Mach numbers $M_\infty = 3$ at $H = 35.0$ km altitude, and $M_\infty = 5$ at $H = 41.0\ km$. The angle of attack in both cases is $\alpha = -\ 25°$. We discuss a solution of the full Navier-Stokes equations, [67], see also Chapter 11, for turbulent flow[27] with a radiation-adiabatic wall boundary condition.

In Fig. 5.48 we show the attitude of the flight vehicle for the $M_\infty = 3$ case. We have noted in Sub-Section 5.2.2 that in order to achieve, for a classical axisymmetric RV-NW-type vehicle, a $L/D > 0$, it is necessary that it is flying

[26] See in this respect Fig. 4.63 and the general discussion in [65].

[27] A two equation turbulence model is employed, [67], [68], see also Chapter 11.

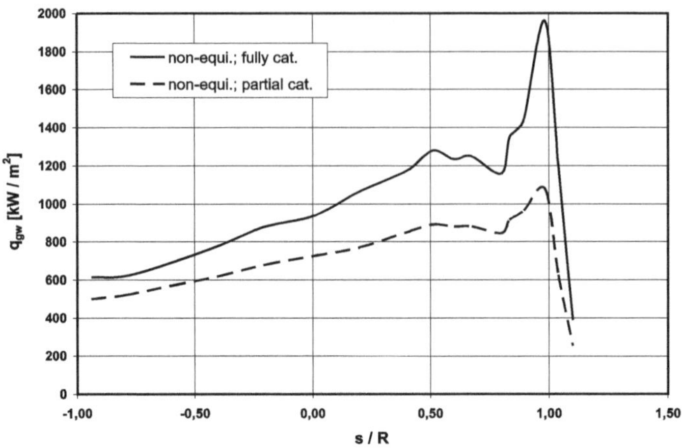

Fig. 5.47. Heat flux in the gas at the wall, q_{gw}, at the APOLLO front-shield surface (symmetry plane) as function of s/R. (s/R convention see Fig. 5.46). Comparison of different prediction methods and wind tunnel data, $M_\infty = 10.17$, $\alpha = -33°$. BLIMP/IEC3D: Euler/boundary layer integral method. Data source: [30].

with a negative angle of attack. In our case, due to the particular shape of VIKING_1, at the considered (negative) angle of attack the flow at the lower back side of the vehicle is undergoing massive separation. This is indicated by the very rugged iso-Mach lines in the figure (symmetry plane).

The separation pattern is expressed by the topology of the skin-friction lines on the vehicle surface[28]. We look first at the $M_\infty = 3$ case, Fig. 5.49. In part a) of the figure, the front view, we see that all skin-friction lines are originating in the forward stagnation point. The pattern of course is left-right symmetrical. The skin-friction lines lead over the heat shield and the shoulder to the back part and there turn downward. Immediately behind the shoulder, nearly halfway down, a convergence of skin-friction lines is discernible. It indicates squeeze-off separation, [69]. Of the two involved boundary layers one comes over the shoulder from the front part of the vehicle, the other from the back part (part b) of the figure in the neighborhood of point of 'A'.).

In part b) of Fig. 5.49 we see a very particular skin-friction line pattern on the lower side of the vehicle. We do not attempt to reconstruct the complete skin-friction line topology, because the available information is not sufficient. Since we cannot decide whether we have attachment or reattachment lines, we simply speak about attachment lines. On the upper left, coming from above, we see the separation line, which we saw already in part a) of the figure. The

[28] In general it is not possible to deduce from this surface topology fully and unambiguously the vortex and vortex sheet pattern above the surface.

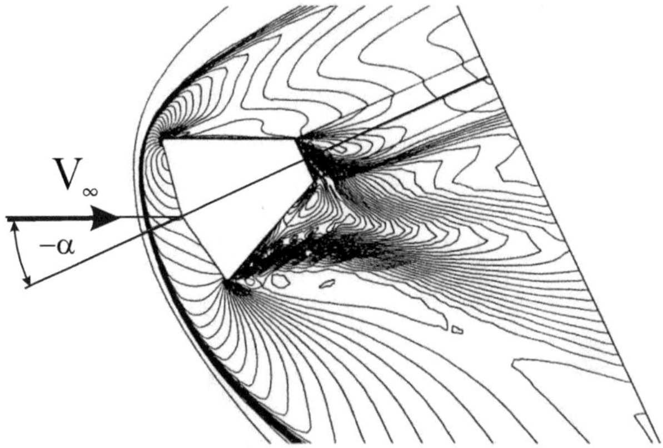

Fig. 5.48. Attitude of the VIKING_1 shape and iso-Mach lines in the symmetry plane. $M_\infty = 3$, $H = 35.0\ km$, $\alpha =$ - 25°, [45].

largest part of the lower surface then is characterized by skin-friction lines running from the back towards the front part of the vehicle[29].

We wish to identify possible attachment/reattachment and separation lines. Therefore we consider first singular points in the skin-friction line pattern, however we note only those marked with a number. The singular point '1' lies on the lower symmetry line of the body. Going upward from it, we find the points '2' to '4'. On the right-hand side of these points, the flow runs toward the back of the vehicle. On the left-hand side, as already observed, the flow runs towards the front part. The diverging pattern of the flow coming from '1' shows that we observe here an attachment line extending to the front part. In '2' it is a separation line, again an attachment line in '3' and finally a separation line in '4'. This holds, even if the patterns are partly tapering out.

In the $M_\infty = 5$ case, Fig. 5.50, although with lesser information available than for the $M_\infty = 3$ case, we see a much simpler skin-friction line pattern on the lower side of the vehicle. Again the largest part of the lower surface is characterized by skin-friction lines running from the back towards the front part of the vehicle. However, besides the attachment line on the lower symmetry line, now only one separation line is indicated, which ends in the singular point '1' in a vortex filament, which leaves the body surface. A second separation line is discernible, which originates behind the lower shoulder of the heat shield, and runs also towards '1'. All this again is present left-right symmetrical on the body surface.

[29] The numerical solution resulted in a steady flow field. This does not rule out the possibility that in reality this is only a time-averaged picture of a fluctuating flow field. Nevertheless, we assume steady flow throughout.

a)

b)

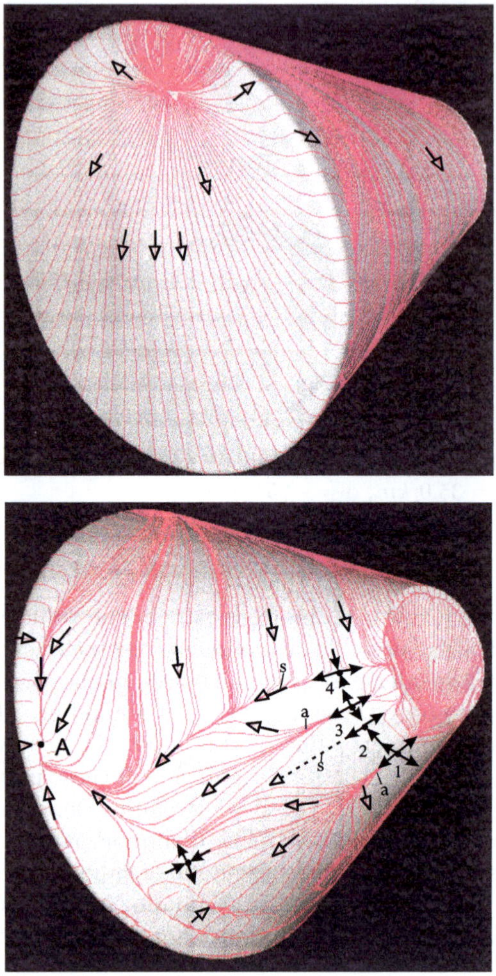

Fig. 5.49. Pattern of skin-friction lines on the surface of the VIKING_1 shape. $M_\infty = 3$, $H = 35.0\ km$, $\alpha =$ - 25°, [45]. a): front view (heat shield), b): rear view on the lower side, s: separation line, a: attachment line, broken line(s): guessed pattern.

We note that for the larger Mach number cases in [45] similarly simple skin-friction line topologies are computed. Whether in general such topologies become simpler with increasing Mach numbers is not clear.

As explained above, attachment/reattachment lines lead to hot-spot situations, while separation lines lead to cold-spot situations. In case of the radiation-adiabatic wall this holds either for the radiation-adiabatic temperature T_{ra} or for the heat flux in the gas at the wall q_{gw}. These two entities are directly related to each other, Chapter 9. In the following we consider therefore only the radiation-adiabatic temperature fields present on the lower side

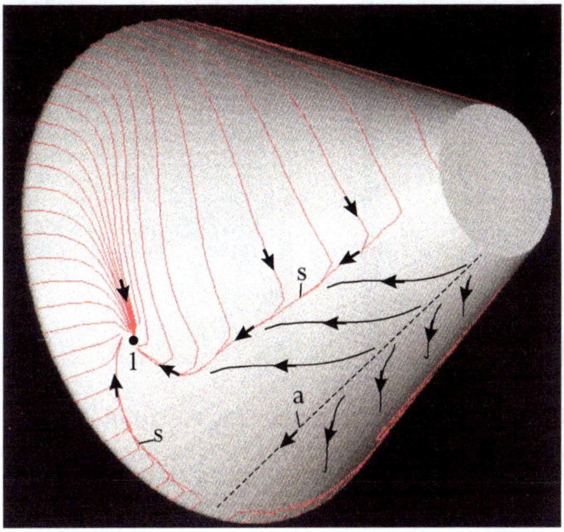

Fig. 5.50. Pattern of skin-friction lines on the surface of the VIKING_1 shape. $M_\infty = 5$, $H = 41.0\ km$, $\alpha =$ - 25°, [45]. Rear view on the lower side, s: separation line, a: attachment line, broken line(s): guessed pattern.

of the vehicle in the $M_\infty = 3$, $H = 35.0\ km$ case, Fig. 5.51 (left), and in the $M_\infty = 5$, $H = 41.0\ km$ case, Fig. 5.51 (right). Note that the figures are turned upside down, and that they have different color codes for the temperature.

In the $M_\infty = 3$ case the total temperature is around 660.0 K. Due to the surface radiation cooling we find (not shown), a radiation-adiabatic temperature of approximately 550.0 K at the stagnation point of the heat shield and at the upper shoulder, as well as at the heat-shield apex, where also a thinning of the boundary layer occurs. On the rear part, Fig. 5.51 (left), we get around point '1' (see Fig. 5.49) on the lower attachment line approximately 500.0 K, along the attachment line beginning in '3' approximately 400.0 K, which we find also on the upper side of the rear part of the vehicle. The lowest temperatures found are approximately 230.0 K to 300.0 K in the region closer to the shoulder.

In the $M_\infty = 5$ case the total temperature is around 1,360.0 K. Due to the surface radiation cooling we find (again not shown), a radiation-adiabatic temperature of approximately 950.0 K at the stagnation point of the heat shield and at the upper shoulder, and a somewhat lower temperature at the heat-shield apex. On the rear part, Fig. 5.51 (right), we get on the lower attachment line initially approximately 600.0 K, and along the attachment line approximately 300.0 K. On the upper side of the rear part of the vehicle we have temperatures around 500.0 K. The lowest temperature is found with approximately 250.0 K at the location, where the vortex filament leaves the surface, point '1' in Fig. 5.50. This is as low as the lowest temperature in the $M_\infty = 3$ case.

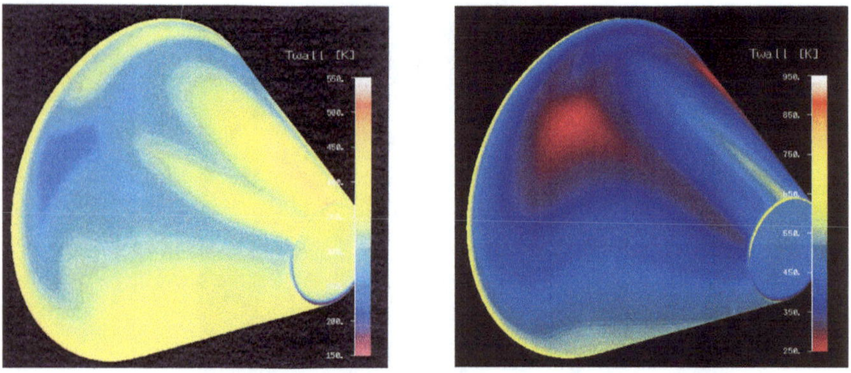

Fig. 5.51. Radiation-adiabatic wall temperature distribution, T_{ra}, on the rear part of the VIKING_1 shape for the free-stream conditions $M_\infty = 3$, $H = 35.0\ km$, $\alpha =$ - 25° (left), and $M_\infty = 5$, $H = 41.0\ km$, $\alpha =$ - 25° (right). The figures are turned upside down compared to the Figs. 5.48, 5.49, 5.50, [45].

Our results show that large surface-temperature differences can occur at the backsides of RV-NW-type vehicles because also the backsides in general are radiation cooled. The temperature differences and hot-spot situations of course become more pronounced at higher flight Mach numbers. However, since the temperature of a radiation cooled surface depends also strongly on the (unit) Reynolds number, [65], one always has to take into account flight speed and altitude, and also, of course, the attitude of the vehicle, in order to identify hot-spot and cold-spot situations. Their locations and temperature/heat-flux levels depend definitely on these parameters.

For the layout of the heat insulation of the backside of the vehicle thus the situation appears to be somewhat less clear-cut than for the layout of the frontal heat shield. Although at the backside of the vehicle the thermal loads are in general much smaller, also there the heat insulation must have a mass as low as possible.

All together we can say, that a careful analysis of the entire flow field along the relevant parts of a flight trajectory (the re-entry trajectory) is necessary for getting a reliable data set regarding thermal loads. Flow-topology dependent hot-spot and cold-spot situations cannot be predicted with simple approximate methods. Coupled Euler/3-D boundary layer solutions in general will suffice on a windward side. On a leeward side with extensive flow separation/attachment and 3D vortical patterns, only solutions of the full Navier-Stokes equations permit to describe the actual situation accurately.

5.6 Problems

Problem 5.1. Given is a wedge-like body, which flies with an angle of attack α at M_∞, Fig. 7.5. It has the length L, the width $w = 0.1\ L$, and a wedge angle θ. Assume Newtonian flow, see, e.g., [65], and determine the location of the center-of-pressure as well as the necessary location of the center-of-gravity for longitudinal trim. Assume that the center-of-gravity lies on the axis ($z = 0$) of the body. The reference area is $A_{ref} = L\,w = 0.1\,L^2$, and the reference length $L_{ref} = 0.5\,L$. Use the relations derived in Chapter 7.1.

Problem 5.2. Consider the high Mach number flow past a double wedge (Figs. 7.5, 7.10), where the Newtonian approach is valid. Verify analytically that:

a) for a wedge angle $\phi_w = 45°$ the configuration has no lift for an angle of attack $\alpha = 45°$,
b) for a wedge angle $\phi_w = 54,732°$ the lift is negative for all angles of attack $\alpha > 0$.

Problem 5.3. The moments and in particular the components of the moments are dependent on the reference point. In the beginning of a design cycle and during the system development phase the center-of-gravity may be not well determined usually due to the open questions of the internal lay-out. Therefore the aerodynamicists define a fictitious moment reference point. At the end this requires a re-evaluation of the components of the moment if the center-of-gravity is ultimately defined.

Calculate the pitching moment of the Bicone_Tsnii shape for a reference location $x_{ref} = 0.42\,L_{ref}$, $z_{ref} = -\ 0.1467\ L_{ref}$ for the flight point $M_\infty = 1.18$, $\alpha = 30°$ by using the data of Fig. 5.24, where also the pitching moment for the reference point $x_{ref} = 0.57\,L_{ref}$, $z_{ref} = -\ 0.0667\ L_{ref}$ is plotted. Compare with the data of Fig. 5.25. It is sufficient to extract approximately the data from Fig. 5.24. Use the relations derived in Chapter 7.1.

Problem 5.4. We consider a generic contour consisting of a quarter of a circle with radius r and a cylindrical part of length $2\,r$, see figure below. The angle of attack α moves between 0° and 90°.

Determine with the classical Newton method for hypersonic Mach numbers the relations for the aerodynamic coefficients C_X and C_Z and the coordinates

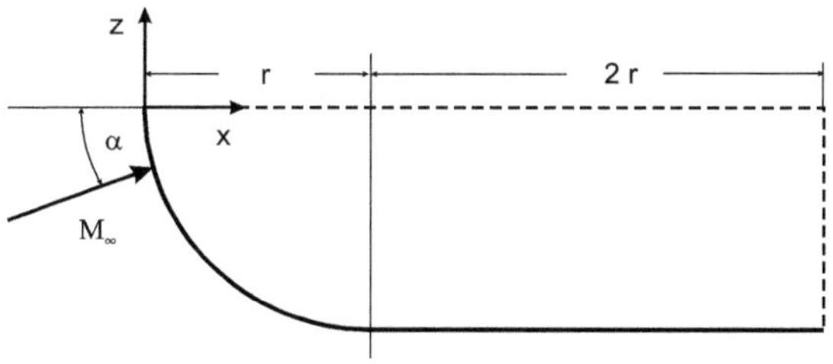

Sketch of the generic contour of problem 5.4.

of the center-of-pressure x_{cp} and z_{cp} as function of the angle of attack, first for the circle alone, and second for the complete contour. Use the three equations (7.5), (7.7), (7.9) and the information of Problems 5.1 and 5.2.

Evaluate for $\alpha = 0°$ and $\alpha = 90°$ these relations and justify them by drawing the results.

References

1. Smith, A.J.: Entry and Vehicle Design Considerations. Capsule Aerothermodynamics, AGARD-R-808, Paper 4 (1997)
2. Walberg, G.D.: A Survey of Aeroassisted Orbit Transfer. J. of Spacecraft 22(1), 3–18 (1985)
3. Charbonnier, J.M., Fraysse, H., Verant, J.L., Traineau, J.C., Pot, T., Masson, A.: Aerothermodynamics of the Mars Premier Orbiter in Aerocapture Configuration. In: Proceedings 2nd Int. Symp. on Atmospheric Re-entry Vehicles and Systems, Arcachon, France (2001)
4. Walberg, G.D., Siemers, P.M., Calloway, R.L., Jones, J.J.: The Aeroassist Flight Experiment. In: IAF Conference, Paper No. IAF-87-197, Brighton, England (1987)
5. Williams, L.J., Putnam, T.W., Morris, R.: Aeroassist Key to Returning from Space and the Case for AFE. Capsule Aerothermodynamics, AGARD-R-808, Paper 13 (1997)
6. Regan, F.J.: Re-entry Vehicle Dynamics. AIAA Education Series, New York, N.Y (1984)
7. Bertin, J.J.: Hypersonic Aerothermodynamics. AIAA Education Series, Washington, D.C. (1994)
8. Gupta, R.N., Moss, J.N., Price, J.M.: Assessment of Thermochemical Nonequilibrium and Slip Effects for Orbital Re-entry Experiment (OREX). AIAA-Paper 96-1859 (1996)
9. Paulat, J.C.: Atmospheric Re-entry Demonstrator Post Flight Analysis – Aerdynamics. In: Proceedings 2nd Int. Symp. on Atmospheric Re-entry Vehicles and Systems, Arcachon, France (2001)
10. Yamamoto, Y., Yoshioka, M.: CFD and FEM Coupling Analysis of OREX Aerothermodynamic Flight Data. AIAA-Paper 95-2087 (1995)
11. Murakami, K., Fujiwara, T.: A Hypersonic Flowfield Around a Re-entry Body Including Detailed Base Flow. ESA-SP-367 (1995)
12. Rumynski, A.N., Murzinov, I.N., Ivanov, N.M., Kazakov, M.N., Sobolevsky, V.G., Finchenko, V.S.: Comparative Multidisciplinary Analysis of Vehicles Intended for Descent on to the Earth Surface with the Mars Ground Samples. ESA-SP-487 (2002)
13. Way, D.W., Powell, R.W., Edquist, K.T., Masciarelli, J.P., Starr, B.R.: Aerocapture Simulation and Performance for the Titan Explorer Mission. AIAA-Paper 2003-4951 (2003)
14. Baillion, M., Pallegoix, J.F.: HUYGENS Probe Aerothermodynamics. AIAA-Paper 97-2476 (1997)
15. Baillion, M.: Aerothermodynamic Requirements and Design of the HUYGENS Probe. Capsule Aerothermodynamics, AGARD-R-808 (1997)

16. Burnell, S.I., Liever, P., Smith, A.J., Parnaby, G.: Prediction of the BEAGLE2 Static Aerodynamic Coefficients. In: Proceedings 2nd Int. Symp. on Atmospheric Re-entry Vehicles and Systems, Arcachon, France (2001)
17. Smith, A.J., Parnaby, G., Matthews, A.J., Jones, T.V.: Aerothermodynamic Environment of the BEAGLE2 Entry Capsule. ESA-SP-487 (2002)
18. Paulat, J.C., Baillion, M.: Atmospheric Re-entry Demonstrator Aerodynamics. In: Proceedings 1st Int. Symp. on Atmospheric Re-entry Vehicles and Systems, Arcachon, France (1999)
19. Moseley, W.C., Martino, J.C.: APOLLO Wind Tunnel Testing Programme – Historical Development of General Configurations. NASA TN D-3748 (1966)
20. Bertin, J.J.: The Effect of Protuberances, Cavities, and Angle of Attack on the Wind Tunnel Pressure and Heat Transfer Distribution for the APOLLO Command Module. NASA TM X-1243 (1966)
21. Blanchet, D., Kilian, J.M., Rives, J.: Crew Transport Vehicle-Phase B, Aerodynamic Data Base. CTV-Programme, HV-TN-8-10062-AS, Aerospatiale (1997)
22. Moseley, W.C., Moore, R.H., Hughes, J.E.: Stability Characteristics of the APOLLO Command Module. NASA TN D-3890 (1967)
23. Tran, P.: ARD Aerothermal Environment. In: Proceedings 1st Int. Symp. on Atmospheric Re-entry Vehicles and Systems, Arcachon, France (1999)
24. Ivanov, N.M.: Catalogue of Different Shapes for Unwinged Re-entry Vehicles. Final report, ESA Contract 10756/94/F/BM (1994)
25. Weiland, C.: Crew Transport Vehicle Phase A, Aerodynamics and Aerothermodynamics. CTV-Programme, HV-TN-3100-X05-DASA, Dasa, München/Ottobrunn, Germany (1995)
26. Marzano, A., Solazzo, M., Sansone, A., Capuano, A., Borriello, G.: Aerothermodynamic Development of the CARINA Re-entry Vehicel: CFD Analysis and Experimental Tests. ESA-SP-318 (1991)
27. Solazzo, M., Sansone, A., Gasbarri, P.: Aerodynamic Characterization of the CARINA Re-entry Module in the Low Supersonic Regimes. ESA-SP-367 (1995)
28. Lee, D.B., Bertin, J.J., Goodrich, W.D.: Heat Transfer Rate and Pressure Measurements Obtained During APOLLO Orbital Entries. NASA TN D-6028 (1970)
29. Lee, D.B., Goodrich, W.D.: The Aerothermodynamic Environment of the APOLLO Command Module During Superorbital Entry. NASA TN D-6792 (1972)
30. Bouslog, S.A., An, M.Y., Wang, K.C., Tam, L.T., Caram, J.M.: Two Layer Convective Heating Prediction Procedures and Sensitivities for Blunt Body Re-entry Vehicles. AIAA-Paper 93-2763 (1993)
31. Walpot, L.M.G.: Numerical Analysis of the ARD Capsule in S4 Wind Tunnel. ESA-SP-487 (2002)
32. Baillion, M., Tran, P., Caillaud, J.: Aerodynamics, Aerothermodynamics and Flight Qualities. ARCV Programme, ACRV-A-TN 1.2.3.-AS, Aerospatiale (1993)
33. Rochelle, W.C., Ting, P.C., Mueller, S.R., Colovin, J.E., Bouslog, S.A., Curry, D.M., Scott, C.D.: Aerobrake Heating Rate Sensitivity Study for the Aeroassist Flight Experiments. AIAA Paper, 89-1733 (1989)
34. Rousseau, S., Fraysse, H.: Study of the Mars Sample Return Orbiter Aerocapture Phase. In: Proceedings 2nd Int. Symp. on Atmospheric Re-entry Vehicles and Systems, Arcachon, France (2001)

35. Davies, C.B., Park, C.: Aerodynamics of Generalized Bent Biconics for Aero-Assisted, Orbital-Transfer Vehicles. J. of Spacecraft 22(2), 104–111 (1985)
36. Weiland, C., Haidinger, F.A.: Entwurf von Kapseln und deren aerothermodynamische Verifikation. In: Proceedings DGLR-Fachsymposium Strömungen mit Ablösung (not published) (November 1996)
37. Miller, C.G., Blackstock, T.A., Helms, V.T., Midden, R.E.: An Experimental Investigation of Control Surface Effectiveness and Real-Gas Simulation for Biconics. AIAA-Paper 83-0213 (1983)
38. Park, C., Yoon, S.: Calculation of Real Gas Effects on Blunt Body Trim Angles. AIAA-Paper 89-0685 (1989)
39. Paulat, J.C.: Synthesis of Critical Points. MSTP-Programme, Final Report, HT-SF-WP1130-087-AS, Aerospatiale (1997)
40. N.N.: APOLLO Data Base. Industrial communication, Aerospatiale - Deutsche Aerospace Dasa (1993)
41. Moseley, W.C., Moore, R.H., Hughes, J.E.: Stability Characteristics of the APOLLO Command Module. NASA TN D-3890 (1967)
42. Yates, L.A., Venkatapathy, E.: Trim Angle Measurement in Free-Flight Facilities. AIAA-Paper 91-1632 (1991)
43. Wells, W.L.: Measured and Predicted Aerodynamic Coefficients and Shock Shapes for Aeroassist Flight Experiment Configurations. NASA TP-2956 (1990)
44. Dieudonne, W., Spel, M.: Entry Probe Stability Analysis for the Mars Pathfinder and the Mars Premier Orbiter. ESA-SP-544 (2004)
45. Hagmeijer, R., Weiland, C.: Crew Transport Vehicle Phase A, External Shape Definition and Aerodynamic Data Set. CTV-Programme, HV-TN-2100-011-DASA, Dasa, München/Ottobrunn, Germany (1995)
46. Weilmuenster, K.J., Hamilton II, H.H.: A Comparison of Computed and Experimental Surface Pressure and Heating on 70° Sphere Cones at Angles of Attack to 20°. AIAA-Paper 86-0567 (1986)
47. N.N.: CTV Aerodynamics. Industrial communication, Aerospatiale - Deutsche Aerospace Dasa (1995)
48. N.N.: CTV Dasa Data Base. Internal industrial communication, Dasa, München/Ottobrunn, Germany (1994)
49. Weiland, C., Pfitzner, M.: 3D and 2D Solutions of the Quasi-Conservative Euler Equations. Lecture Notes in Physics, vol. 264, pp. 654–659. Springer, Heidelberg (1986)
50. Pfitzner, M., Weiland, C.: 3D Euler Solutions for Hypersonic Free-Stream Mach Numbers. AGARD-CP-428, pp. 22-1–22-14 (1987)
51. Pallegoix, J.F., Collinet, J.: Atmospheric Re-entry Demonstrator Post Flight Analysis – Flight Rebuilding with CFD Control. In: Proc. 2nd Int. Symp. on Atmospheric Re-entry Vehicles and Systems, Arcachon, France (2001)
52. Paulat, J.C.: Synthesis of Critical Points. ESA Manned Space Transportation Programme, Final Report, HT-SF-WP1130-087-AS, Aerospatiale (1997)
53. Baillion, M.: Blunt Bodies Dynamic Derivatives. Capsule Aerothermodynamics, AGARD-R-808, Paper 8 (1997)
54. Chapman, G.T., Yates, L.A.: Dynamics of Planetary Probes: Design and Testing Issues. AIAA-Paper 98-0797 (1998)
55. Ericsson, L.E., Reding, J.P.: Re-Entry Capsule Dynamics. J. of Spacecraft 8(6), 579–586 (1971)
56. Whitlock, C.H., Siemers, P.M.: Parameters Influencing Dynamic Stability Characteristics of VIKING-Type Entry Configurations at Mach 1.76. J. of Spacecraft 9(7), 558–560 (1972)

57. Karatekin, Ö., Charbonnier, J.M., Wang, F.Y., Dehant, V.: Dynamic Stability of Atmospheric Entry Probes. ESA - SP - 544 (2004)
58. Wang, F.Y., Karatekin, Ö., Charbonnier, J.M.: Low-Speed Aerodynamics of a Planetary Entry Capsule. J. of Spacecraft 36(5), 659–667 (1999)
59. Teramoto, S., Fujii, K.: Study on the Mechanism of the Instability of a Re-entry Capsule at Transonic Speeds. AIAA-Paper 2000-2603 (2000)
60. Brockhaus, R.: Flugregelung. Springer, Heidelberg (2001)
61. Schueler, C.J., Ward, L.K., Hodapp, A.E.: Techniques for Measurement of Dynamic Stability Derivatives in Ground Test Facilities. AGARDograph 121 (1967)
62. Orlik-Rueckemann, K.J.: Techniques for Dynamic Stability Testing in Wind Tunnels. AGARD-RCP-235, Paper 1 (1978)
63. Giese, P., Heinrich, R., Radespiel, R.: Numerical Prediction of Dynamic Derivatives for Lifting Bodies with a Navier-Stokes Solver. Notes on Numerical Fluid Mechanics, vol. 72, pp. 186–193. Vieweg Verlag (1999)
64. Giese, P.: Numerical Prediction of First Order Dynamic Derivatives with Help of the Flower Code. Deutsches Zentrum für Luft- und Raumfahrt, TETRA Programme, TET-DLR-21-TN-3201 (2000)
65. Hirschel, E.H.: Basics of Aerothermodynamics. Progress in Astronautics and Aeronautics, AIAA, Reston, Va, vol. 204. Springer, Heidelberg (2004)
66. Tran, P., Soler, J.: Atmospheric Re-entry Demonstrator; Post Flight Analysis: Aerothermo Environment. In: Proceedings 2nd Int. Symp. on Atmospheric Re-entry Vehicles and Systems, Arcachon, France (2001)
67. Haidinger, F.A.: Numerical Investigation of the Heat Transfer on the Front Shield of a Capsule. Deutscher Luft- und Raumfahrtkongress, DGLR - JT96 - 0137 (1996)
68. Wilcox, D.C.: Turbulence Modelling for CFD. DCW Industries, La Cañada, CAL, USA (1998)
69. Hirschel, E.H.: Evaluation of Results of Boundary-Layer Calculations with Regard to Design Aerodynamics. AGARD R-741, pp. 5-1–5-29 (1986)

6 Stabilization, Trim and Control Devices

Stabilization, trim, and control devices are the prerequisites for flyability and controllability of aerospace flight vehicles. Regarding the re-entry flight of RV-W-type vehicles, we have mentioned in Section 2.1.2, that initially at high flight altitudes the reaction control system (RCS) is the major flight-control device, whereas aerodynamic trim, stabilization and control surfaces take over further down on the trajectory. This is in contrast to CAV-type flight vehicles, where aerodynamic stabilization and control surfaces are the only devices. However, here we have the strong coupling of the thrust vector (and the aerothermoelasticity of the airframe), Sub-Section 2.2.3, into flight dynamics, trim and control of the vehicle. When ARV-type flight vehicles reach high altitudes, the situation is similar to that of RV-W-type vehicles, i. e., with decreasing control-surface effectiveness reaction control devices will have to take over. RV-NW-type vehicles (capsules) as a rule have only a RCS for flight control.

We begin this chapter with an introduction to trim and control-surface aerothermodynamics, and concentrate then on the onset flow[1] characteristics. Next treated is the asymptotic behavior of pressure, thermal state of the surface, and wall-shear stress on the control surface, approximated here as a ramp. Related issues of reaction control systems are discussed briefly. Configurational considerations are presented regarding the discussed trim and control devices. Finally the results are summarized and simulation issues are examined.

Our aim is to foster the understanding of the flow phenomena involved in the operation of stabilization, trim, and control devices and the problems related to their simulation with experimental and computational means.

6.1 Control Surface Aerothermodynamics, Introduction

In this and the following sections of the chapter we concentrate on aerothermodynamic issues of trim and control surfaces. We are concerned mainly

[1] We use the term "onset flow" for the (oncoming) flow just ahead of the respective device.

with their effectiveness. Therefore we look not only at shock/boundary layer interactions as such, which are major factors regarding especially thermal and also mechanical (pressure and shear stress) loads on such surfaces, but also at other issues, especially at the role of the control-surface onset flow. Material about shock/boundary layer interactions can be found abundantly in the literature, we point here only to [1] to [5]. Usually two-dimensional interactions are treated, because they are in general the most severe ones. In reality many three-dimensional cases can be found, see, e.g., [6] to [8]. A detailed discussion of means for control of shock wave/turbulent boundary layer interaction at transonic and supersonic Mach numbers is given in [9].

We try here do develop a somewhat wider view which emphasizes also the influence of the flow ahead of a trim or control surface, the onset flow, on the effectiveness of a surface[2]. We discuss only the two-dimensional situation, with, of course, shock/boundary layer interaction as major topic. Important for us is then the asymptotic flow behavior on the trim or control surface, i. e. the flow behavior downstream of the strong interaction domain. In general the interaction domain has a small downstream extent compared to the flap length, therefore it is the asymptotic behavior which mostly governs the effectiveness of a trim or control surface, Fig. 6.1.

6.1.1 Flap Effectiveness: Influence of the Onset Flow Field

In order to discuss effectiveness aspects of aerodynamic trim and control surfaces we look at the generic flap configuration, Fig. 6.1, which consists simply of a flat plate and a ramp. As a body flap at a RV-W-type flight vehicle, for instance, it would be located upside down at the end of the vehicle's lower side[3], see, e.g., Fig. 6.44.

Of major importance are the properties of the onset flow, which are different for different vehicle classes, depending on flight speed, altitude, vehicle shape and attitude, and on the thermal state of the surface. We consider the overall properties of the onset flow, and in addition the influence of the important similarity and geometrical parameters. Geometrical parameters are the ramp (deflection) angle, the hinge line orientation (basically we consider the optimum case of orientation: orthogonal to the onset flow), the width and length of the flap (ramp), and the width of the gap between the vehicle parent structure and the flap. Before we begin with the consideration of the overall onset flow in Section 6.2, we look at the particular flap deflection modes in hypersonic flight.

---

[2] Of course this is always a topic too, when shock/boundary layer interaction is studied. However, usually the onset flow situation is that of a laboratory experiment with immanent restrictions.

[3] The deflection angle η of horizontal surfaces in aerodynamics usually is counted positive when deflected downwards.

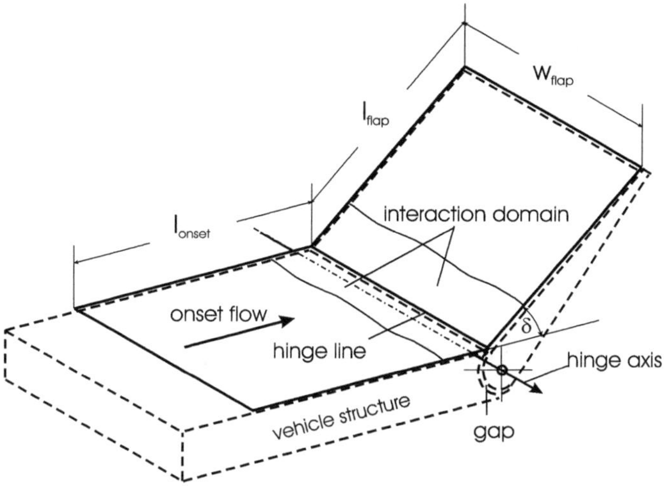

Fig. 6.1. Flat-plate/ramp configuration as generic flap configuration. We call the corner line the hinge line. Note that its location and that of the hinge axis are not exactly the same.

6.1.2 Flap Deflection Modes during Hypersonic Flight

During re-entry of a RV-W-type flight vehicle at large angle of attack, the body flap as main trim surface, as well as elevator and aileron surfaces, Fig. 6.44, usually are deflected downwards, Sub-Section 3.4.2. At the leeward sides of these surfaces anyway no or only negligible pressure forces are generated due to the hypersonic shadow effect, [10]. Upward deflection does not yield aerodynamic forces, the bodyflap is in the shadow of the fuselage, the elevons are in the wing's shadow. However, upward deflection of a surface does have an effect on the moment balance of the vehicle, because now the (upward) force acting on the initially downward deflected or undeflected lower surface is reduced or even omitted completely, Sub-Section 3.4.2.

A shadow effect, either the hypersonic one, or simply one due to massive flow separation, exists at large angle of attack for a central single vertical tail unit (stabilizer) together with the rudder, Fig. 6.44 a). On the re-entry trajectory, down to rather small angles of attack, it will first be subjected to the hypersonic shadow effect, and then lies in the separated flow domain above and behind the fuselage, Sub-Section 2.1.2. At small angles of attack, the vertical stabilizer and the – to the left or the right – deflected rudder, are effective, as long as the yaw angle of the vehicle is not too large. Otherwise also here a shadow effect is felt. Regarding vertical wing-tip stabilizers and rudders, Fig. 6.44 b), the effectiveness of the surfaces will also suffer from shadow effects at large angles of attack or yaw.

We don't see these problem at CAV-type flight vehicles which fly at small angles of attack. The effectiveness of their stabilization, trim and control surfaces, Fig. 6.46, is not reduced by shadow effects of either kind. Of course, like for low speed aircraft, angle of attack and yaw have an influence, because in principle they alter the properties of the onset flow, however, in a well understood manner.

For both RV-W-type and CAV-type flight vehicles aerothermoelasticity of the airframe and in particular of stabilization, trim and control surfaces with large mechanical and thermal loads can be a special problem. Although the effects of aerothermoelasticity are a major problem in hypersonic flight-vehicle design, we cannot treat them in the frame of this book. However, we discuss aspects of them in Chapter 8.

6.2 Onset Flow of Aerodynamic Trim and Control Surfaces

6.2.1 Overall Onset Flow Characteristics

To understand the aerothermodynamic phenomena and the thermal and mechanical loads present at a trim or control surface, the properties of the overall onset flow, together with the thermal state of the surface, need to be considered first.

Regarding the body flap and the elevator/aileron surfaces of a RV-W-type vehicle, the onset flow on a large part of the trajectory is governed by the high angle of attack, and the very blunt body shape, Figs. 6.2 b) and 3.1. The topology of the onset-flow field itself at the windward side, Sub-Section 3.2.2, plays a major role, too.

The situation is different at a CAV-type vehicle, Fig. 6.2 a), where the slender configuration flies at small angle of attack, and the onset flow of the elevator/aileron surfaces does not deviate very significantly from the free-stream flow. Also here suitable topologies of the velocity fields at the wing's lower and upper side are important, Sub-Section 4.2.2.

In order to illustrate the different flow situations at RV-W-type and CAV-type flight vehicles, we compare lower-side onset flow data for four flight cases, using the RHPM (Rankine-Hugoniot-Prandtl-Meyer) configurational

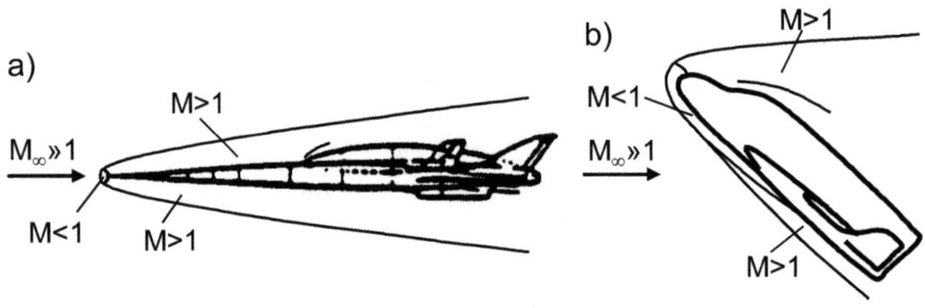

Fig. 6.2. Schematic of configurations, angles of attack, and bow shock shapes of a) a CAV-type, and b) a RV-W-type flight vehicle, [10].

Table 6.1. Flight parameters of the four cases: RV-W 1 and RV-W 2: RV-W-type vehicle (Space Shuttle Orbiter without boattailing), CAV 1: CAV-type vehicle with ramjet propulsion (SÄNGER lower stage), CAV 2: CAV-type vehicle with scramjet propulsion.

Flight vehicle:	RV-W 1	RV-W 2	CAV 1	CAV 2
M_∞ [-]	20	6	6.8	12
H $[km]$	70.0	40.0	30.0	34.2
α [°]	40.0	25.0	5.0	5.0
L_{ref} $[m]$	32.0	32.0	82.0	20.0
T_∞ $[K]$	219.69	250.33	226.51	234.27
v_∞ $[m/s]$	5,942.7	1,903.1	2,051.6	3,682.0
ρ_∞ $[kg/m^3]$	$8.75 \cdot 10^{-5}$	$4.0 \cdot 10^{-3}$	$1.84 \cdot 10^{-2}$	$9.61 \cdot 10^{-3}$
q_∞ $[kPa]$	1.54	7.24	38.72	65.0
p_∞ $[Pa]$	5.52	287.4	1,196.4	646.27
Re^u_∞ $[m^{-1}]$	$3.36 \cdot 10^4$	$4.52 \cdot 10^5$	$2.39 \cdot 10^6$	$2.19 \cdot 10^6$

approximation, Section 10.1. The first two are those of a body flap surface of a typical RV-W-type vehicle (Space Shuttle Orbiter, Sub-Section 3.3.1, however without boattailing) at $M_\infty = 24$ and $H = 70.0\ km$ with laminar flow (RV-W 1), and at $M_\infty = 6$ and $H = 40.0\ km$ with turbulent flow (RV-W 2). The other two are those of elevator/aileron surfaces of a typical CAV-type vehicle (CAV 1, ramjet propulsion) at $M_\infty = 6.8$ and $H = 30.0\ km$ (SÄNGER lower stage), and (CAV 2, scramjet propulsion) at $M_\infty = 12$ and $H = 34.2\ km$. For the latter we have assumed a rather high dynamic pressure $q_\infty = 65.0\ kPa$. From this, we get with:

$$q_\infty = \frac{\rho_\infty v_\infty^2}{2} = \frac{\gamma}{2} p_\infty M_\infty^2 \ \Rightarrow\ p_\infty = \frac{2 q_\infty}{\gamma M_\infty^2} \ \Rightarrow\ H, \tag{6.1}$$

the pressure p_∞= 644.84 Pa, and the flight altitude[4] $H = 34.2\ km$, Table 12.1. The overall length of the vehicle is assumed to be $L_{ref} = 20.0\ m$. The product of dynamic pressure q_∞ times $c_{p_{e,onset}}$, Table 6.2, a measure of the lift force, is approximately the same for the two CAV-type vehicles.

In Table 6.1 the flight parameters of the four vehicle cases are collected.

Under onset flow data we understand the external (outer edge of the boundary layer) inviscid flow data immediately ahead of the control surface

[4] This is a very simple approach, the determination of a real trajectory point takes much more data into account, Section 2.2.

$(e, onset)$[5], and of course also those of the attached viscous flow, which in general can be considered as being of boundary layer type. Unfortunately, onset flow data, as we wish to consider them, are not available in the literature.

In order to get estimates of the onset flow properties, we approximate the configurations with the help of the RHPM flyer. This very simple approximation in the sense of shock-expansion theory yields constant one-dimensional inviscid flow data, which we also consider as external flow data of the onset flow boundary layer. However, we choose the flat-plate situation with the extended RHPM$^+$ approach, Section 10.1, taking into account the γ_{eff} influence, and also the total pressure loss due to the bow shock. The compressibility factor is chosen to be $Z = 1$. Therefore the affected data – temperatures and related entities – must be considered with extra care.

We understand, that the RHPM$^+$ approach is a rather crude way to estimate data. Therefore the resulting data in Table 6.2 are meant for illustrative purposes only.

The onset flow length, l_{onset}, which governs the onset flow boundary layer thickness, Table 6.5, was chosen to be equal to the reference length for the RV-W-type flight vehicle cases, where the body flap is considered. For the two CAV-type vehicles, where the elevator/aileron surfaces are considered, the length from the leading edge nose to mid-chord of the control surfaces was chosen. Although the flow situation on the lower (windward) side of a delta wing at angle of attack is complicated due to the primary attachment lines there, see, e.g., Sub-Section 3.3.2 of [10], this assumption nevertheless should serve sufficiently well[6].

Due to the surface radiation cooling, that is important in all four cases, the radiation-adiabatic surface temperature is not constant, [10]. It drops strongly from the forward stagnation point in the nose region towards the vehicle aft part. The same holds on the wing part of a vehicle, where it drops from the primary attachment line at the leading edge towards the wing's trailing edge. Despite this we choose for all cases a constant wall temperature and hence onset wall temperature $T_{w,onset}$, which is equal to the radiation-adiabatic temperature at the onset location. For the case RV-W 1 we choose $T_w = 1{,}000.0\ K$ according to data from [11] for the re-entry flight STS-2 of the Space Shuttle Orbiter. For case RV-W 2 we take $T_w = 900.0\ K$. For CAV 1 we choose $T_w = 1{,}000.0\ K$, too. Because of the shorter onset flow length of CAV 2 we take $T_w = 1{,}200.0\ K$.

---

[5] Actually, if the entropy layer, see below, has not been swallowed by the boundary layer, the inviscid flow above the boundary layer edge has a strong gradient in direction normal to the surface, even a small velocity minimum is possible, Sub-Section 6.2.2. This flow property has large influence on the flap effectiveness, too.

[6] The reader should note, that the onset flow lengths involved in the four cases are large compared to the lengths which are found on the vertical stabilizers, either single or dual surfaces on the upper side of the airframe, or on the fins at the wing tips of the same vehicles, Sub-Section 6.6.1.

Table 6.2. Data of the lower-side inviscid onset flow (RHPM$^+$ approximation, $Z = 1$) of the body flap for the cases RV-W 1, RV-W 2, and elevator/aileron surfaces for the cases CAV 1, CAV 2. The subscript 'e' stands for boundary layer edge condition.

Flight vehicle:	RV-W 1	RV-W 2	CAV 1	CAV 2
M_∞ [-]	20	6	6.8	12
H $[km]$	70.0	40.0	30.0	34.2
α [°]	40.0	25.0	5.0	5.0
l_{onset} $[m]$	32.0	32.0	25.0	6.0
γ_{eff} [-]	1.14	1.4	1.4	1.3
θ [°]	43.66	34.28	12.06	7.71
$M_{e,onset}$ [-]	3.93	2.83	5.95	11.4
$T_{e,onset}$ $[K]$	3,678.3	789.18	287.22	275,34
$v_{e,onset}$ $[m/s]$	4,308.3	1,593.4	2,021.7	3,652.9
$\rho_{e,onset}$ $[kg/m^3]$	$1.31 \cdot 10^{-3}$	$1.67 \cdot 10^{-2}$	$3.17 \cdot 10^{-2}$	$2.75 \cdot 10^{-2}$
$q_{e,onset}$ $[kPa]$	12.12	21.18	64.88	183.31
$p_{e,onset}$ $[kPa]$	1.38	3.78	2.62	2.17
$Re^u_{e,onset}$ $[m^{-1}]$	$5.83 \cdot 10^4$	$7.49 \cdot 10^5$	$3.49 \cdot 10^6$	$5.61 \cdot 10^6$
$c_{p_{e,onset}}$ [-]	0.89	0.48	0.037	0.023
$T_{w,onset} = T_{ra,onset}$ $[K]$	1,000.0	900.0	1,000.0	1,200.0

We discuss now only a few data in Table 6.2. We note that in case RV-W 1 the pressure coefficient with $c_{p_{e,onset}} = c_{p_w} = 0.89$ by chance compares well with $c_{p_w} \approx 0.89$ at the pressure plateau at the lower side of the Orbiter at $M_\infty = 24$ and $\alpha = 40°$, Sub-Section 3.5.2. Our simple approach however does not yield flight Mach number and high-temperature real-gas effects independence, Section 3.6.

The dynamic pressure of the (inviscid) onset flow, $q_{e,onset}$, represents the momentum flux and hence can be considered as a measure of the flap effectiveness. This is larger than q_∞, which is used for non-dimensionalization. Anyway, we see that $q_{e,onset}$ is small for the two RV-W-type vehicle cases, and distinctly larger for the CAV-type vehicle cases.

We will discuss the other data of the table in the next sub-section together with the total pressure loss corrected data.

In closing this part, we note some guiding principles regarding vehicle-configuration issues. The one-dimensional flow data given in Table 6.2, although being approximate data, represent the optimal onset flow situation. This follows from a simple flow-momentum consideration. The onset flow to

Table 6.3. Flight parameters of the numerical simulation of the flow past the HOPPER configuration, [12].

Case	M_∞	$Re_{L,\infty}$	L $[m]$	α [°]	$x/L\|_{trans}$	ε
A	14.4	$8.41 \cdot 10^6$	50.2	31.3	0.5	0.8
B	3.2	$2.31 \cdot 10^7$	50.2	15.0	fully turbulent	0.8

an aerodynamic trim or control surface definitely should be as two-dimensional as possible, Sub-Sections 3.2.2 and 4.2.2, and the hinge line of such a surface should be located orthogonally to the onset flow direction. We call this in the following the "optimal onset flow geometry". The trim or control surface itself should have a span, respectively aspect ratio, as large as possible in order to reduce side-edge losses[7]. For the same reason the deflection angles should be as small as possible. The side-edge losses become larger with increasing deflection angle.

We illustrate the matter of onset flow situation with data from a numerical simulation, [12], of the flow past the RV-W-type HOPPER/PHOENIX configuration, Sub-Section 3.3.6. The HOPPER configuration was devised and studied, [13], in the frame of the "Future European Space Transportation Investigations Programme (FESTIP)", and became later the reference concept of the German national technology programme ASTRA. PHOENIX is the 1:6 down-scaled experimental vehicle for low-speed tests in the ASTRA programme.

The flight parameters of two trajectory points (cases) are given in Table 6.3. Although in case A the flight Mach number $M_\infty = 14.4$ belongs to a flight altitude above 50.0 km, it was assumed that laminar-turbulent transition occurs. The transition location was chosen summarily as $x/L = 0.5$. The surface emissivity coefficient was taken to be $\varepsilon = 0.8$.

Fig. 6.3 gives the overall view on the lower side of the flight vehicle for case A. All skin-friction lines emanate from the forward stagnation point. The two primary attachment lines lie symmetrically below the forward part of the fuselage and at the wing at a small distance below the leading edge. The skin-friction lines depart (diverge) at these lines to the upper and the lower side of the configuration[8]. At the lower side of the fuselage, which is flat for $x/L > 0.45$, the body flap onset flow is to a good approximation two-dimensional. At the lower side of the wings this is not the case. The wall pressure coefficient varies slightly at the lower side of the configuration, with $c_{p_w} \approx 0.5$ just ahead of the hinge line.

[7] See the considerations in Section 6.6 regarding the influence of flap width and length on the vehicle pitching moment and the hinge moment.

[8] Compare the flow-field topology with that at the windward side of the Space Shuttle Orbiter, Sub-Section 3.2.2. Obviously the HOPPER configuration does not have a blunt-cone like front part.

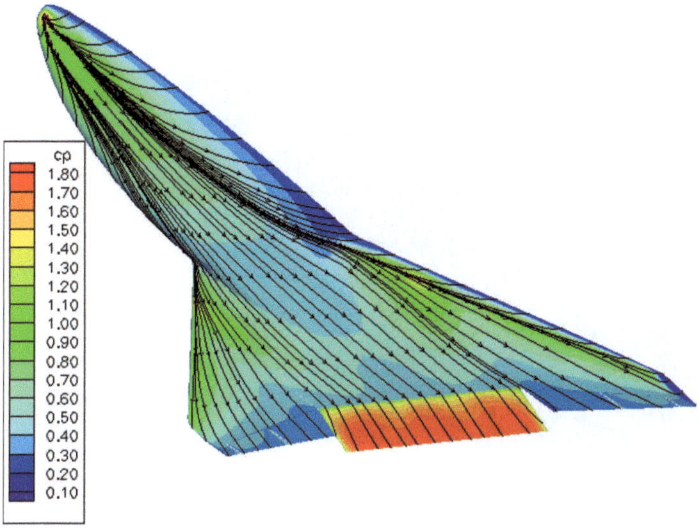

Fig. 6.3. Skin-friction lines and wall pressure coefficient ($c_p \equiv c_{p_w}$) distribution at the lower side of the HOPPER configuration, [12], case A, Table 6.3, downward body flap deflection $\eta_{bf} = +20°$. The body flap has the (chord) length $l_{bf} = 3.5$ m, and the width (span) $w_{bf} = 11.75$ m.

A closer look reveals that the body flap indeed has a nearly optimal onset flow geometry, Fig. 6.4, upper part. It is almost fully two-dimensional and orthogonal to the hinge line. The flow separates at the hinge line due to the extremely large deflection angle $\eta_{bf} = +40°$. Because the onset flow is turbulent, only a – relative to the flap length – small separation regime is present with small upstream influence. The actual distance from the separation line to the attachment line, however, is between ≈ 0.2 m at the symmetry line and ≈ 0.4 m further outboard. Towards the flap edge it becomes distinctly smaller.

Due to the optimal onset flow geometry, we have a nearly constant pressure distribution on a wide part of the flap surface. The pressure rises in downstream direction essentially monotonically towards a plateau value on the flap. The situation on a large part of the body flap resembles case b), Fig. 6.9 in Sub-Section 6.3.1. Near the side edge of the body flap the flow becomes three-dimensional and a pressure relaxation is discernible.

The situation is different at the two wing flaps, Fig. 6.4, lower part, which are multi-functional elevator/aileron surfaces, Sub-Section 6.6.1. The flow situation is clearly sub-optimal. While the flow is only weakly three-dimensional, the hinge lines are strongly swept. The inboard flap, deflected with $\eta_{iwf} = +20°$, anyway having only a small aspect ratio, shows accordingly a pressure-coefficient distribution which is not constant, except for a small part towards the outer edge and the trailing edge. A flow separation at the hinge line, however, is not detectable.

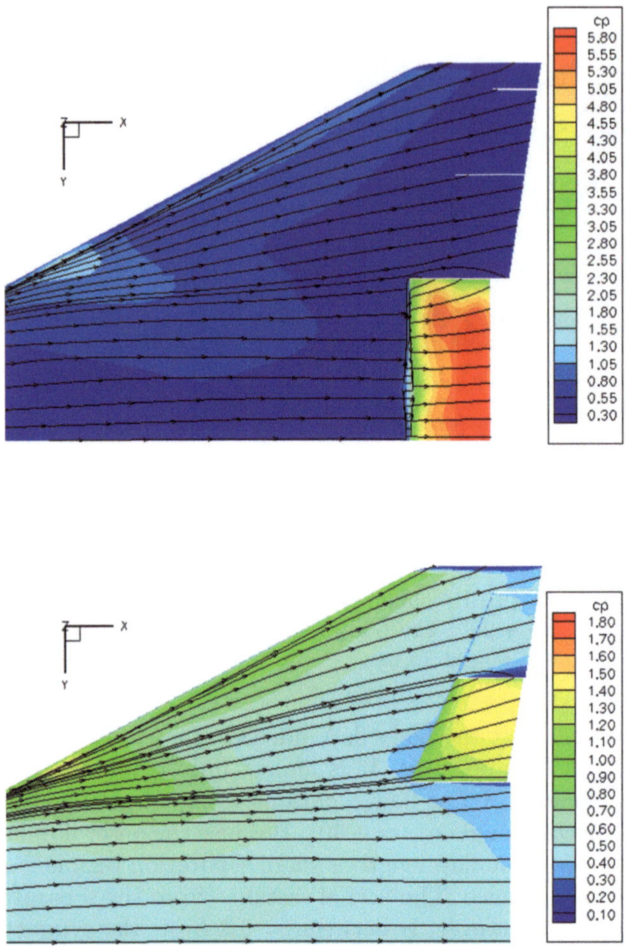

Fig. 6.4. Detailed view of skin-friction lines and wall pressure coefficient ($c_p \equiv c_{p_w}$) distribution at the left lower-side aft part of the HOPPER configuration, [12], case A, Table 6.3. Upper part: body flap at $\eta_{bf} = +40°$, lower part: inboard wing flap at $\eta_{iwf} = +20°$. Note that the color-coding is different for the two parts.

Overall systems demands usually will not permit sweeping configurational changes in order to improve a sub-optimal onset flow geometry. Local changes of the fuselage/wing shape, however, can improve the situation. Given the large flight speed and altitude span of either RV-W-type or CAV-type flight vehicles, an optimization in this sense generally will be very difficult.

6.2.2 The Onset Flow Entropy Layer

While looking at the literature on flap effectiveness, one can sometimes get the impression, that the entropy layer just leads to additional complications. That it is, regarding flap effectiveness, an issue besides others. Such an impression would be misleading. Hypersonic flight vehicles have always finite nose and leading edge radii because of the thermal loads they have to cope with. A finite radius leads to an entropy layer, whose thickness depends on the radius, on the free-stream Mach number, and on the high-temperature real-gas effects present.

If the configuration surface downstream of the nose or the leading edge has a large extension in main flow direction, the entropy layer may be "swallowed" by the developing attached viscous flow, the latter in general being of boundary layer type, i. e. hypersonic viscous interaction phenomena being weak. For a given blunt-body shape entropy-layer swallowing hence is a function of the Reynolds number in the flow regime behind the bow shock. A general discussion of the implications of the entropy layer can be found in, e.g., [10].

The entropy layer comes into being due to the curvature of a shock surface, here the bow shock surface. In our case the streamline, which crosses the normal-to-the flow portion of the bow shock, experiences the largest entropy rise or total pressure loss. Because the neighboring streamlines have a smaller total pressure loss, the inviscid flow has a velocity gradient in direction normal to the body surface. This is similar to the gradient in a boundary layer, but much weaker, because the inviscid flow does not have a no-slip wall boundary condition. The effect of the entropy layer on the effectiveness of an aerodynamic control surface hence is due to the total pressure loss, which manifests itself mainly in a reduced local dynamic pressure.

In the frame of our simple onset flow consideration we can approximately take into account the entropy-layer effect by impressing the maximum total pressure loss – that due to a normal shock – on the one-dimensional inviscid flow, Sub-Section 10.1.5. We use the observation, that with given free-stream conditions the surface pressure depends primarily only on the angle of attack of the considered surface portion (Newton flow property). We keep hence $p_{e,onset}$ from Table 6.2. Proceeding as sketched in Sub-Section 10.1.5, we obtain the onset flow data, corrected by the total pressure loss, given in Table 6.4.

The change of the inviscid onset flow parameters from Table 6.2 to Table 6.4 is dramatic, but different for the two flight vehicle classes, even if this is not yet the final picture, as will be discussed below:

- The reduced total pressure in all cases leads to a distinctly smaller velocity $v_{e,onset,tpl}$. This in turn – the total enthalpy at the boundary layer edge remains constant – results in larger temperatures $T_{e,onset,tpl}$ (remember that $Z = 1$ was chosen), and both together in reduced onset flow Mach numbers $M_{e,onset,tpl}$. In the case RV-W 1 the Mach number is now of the order of the actual Mach number at the pressure plateau of the Space Shuttle Orbiter, Sub-Section 3.5.2. While the onset flow Mach numbers of the CAV cases in Table 6.2 indicate that compressibility effects in the turbulent flow must be accounted for[9], this appears now no more to be the case.

- The reduced velocity $v_{e,onset,tpl}$, in concert with the reduced density $\rho_{e,onset,tpl}$, causes a much reduced dynamic pressure $q_{e,onset,tpl}$ in all cases. The pressure coefficients of course are not changed, they were chosen to be identical: $c_{p_{e,onset,tpl}} \equiv c_{p_{e,onset}}$. The unit Reynolds number $Re^u_{e,onset,tpl}$ is now so small even for the CAV-type vehicles, that it could be doubted, that the flow actually is turbulent.
- In the case RV-W 1 we still have a "cold wall" boundary layer situation ($T_{w,onset,tpl} < T_{e,onset,tpl}$), whereas the "hot wall" situation in the other cases has changed to a "cold wall" situation, except for the case CAV 1.

Table 6.4. Total pressure loss 'tpl' corrected data (RHPM$^+$ approximation, $Z = 1$) of the lower-side inviscid onset flow of the body flap for the cases RV-W 1, RV-W 2, and elevator/aileron surfaces for the cases CAV 1, CAV 2. The subscript 'e' stands for boundary layer edge condition.

Flight vehicle:	RV-W 1	RV-W 2	CAV 1	CAV 2
M_∞ [-]	20	6	6.8	12
H $[km]$	70.0	40.0	30.0	34.2
α [°]	40.0	25.0	5.0	5.0
γ_{eff} [-]	1.14	1.4	1.4	1.3
$M_{e,onset,tpl}$ [-]	1.19	1.48	2.81	3.2
$T_{e,onset,tpl}$ $[K]$	6,954.1	1,428.3	902.1	2,223.8
$v_{e,onset,tpl}$ $[m/s]$	1,802.0	1,120.2	1,689.2	2,914.8
$\rho_{e,onset,tpl}$ $[kg/m^3]$	$6.91 \cdot 10^{-4}$	$9.22 \cdot 10^{-3}$	$1.01 \cdot 10^{-2}$	$3.4 \cdot 10^{-3}$
$q_{e,onset,tpl}$ $[kPa]$	1.12	5.78	14.43	14.45
$p_{e,onset,tpl} \equiv p_{e,onset} [kPa]$	1.38	3.78	2.62	2.17
$Re^u_{e,onset,tpl}$ $[m^{-1}]$	$8.53 \cdot 10^3$	$1.98 \cdot 10^5$	$4.42 \cdot 10^5$	$1.42 \cdot 10^5$
$c_{p_{e,onset,tpl}} \equiv c_{p_{e,onset}}$ [-]	0.89	0.48	0.037	0.023
$T_{w,onset,tpl} \equiv T_{ra,onset}$ $[K]$	1,000.0	900.0	1,000.0	1,200.0

We have noted above that this is not yet the final picture. Why? We have actually considered only the worst picture, viz. the inviscid flow with total pressure loss at the body surface. Actually the inviscid flow away from the surface suffers a smaller total pressure loss than that at the surface,

[9] Compressibility effects in attached turbulent flow occur at boundary layer edge Mach numbers $M_e \gtrapprox 5$, Morkovin's hypothesis, see, e.g., [8], [14], and must be taken into account in turbulence models.

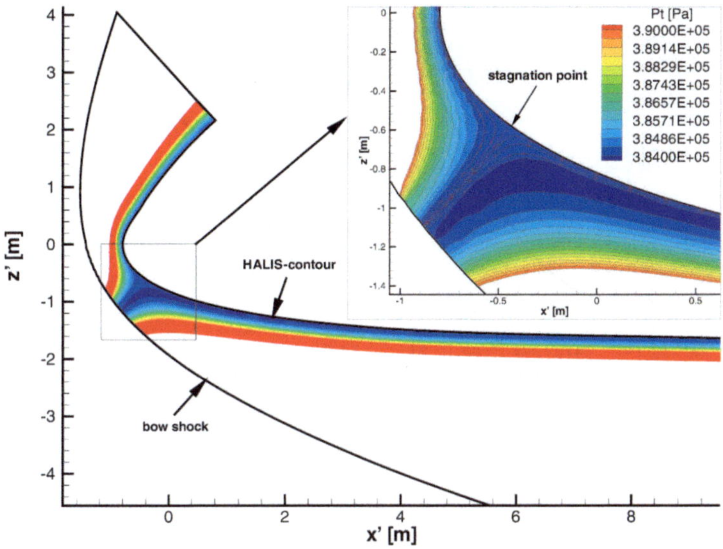

Fig. 6.5. Second form of the entropy layer: computed example, two-dimensional inviscid flow past the centerline of the HALIS configuration, $M_\infty = 10$, $H = 24.0$ km, $\alpha = 40°$, perfect gas, [15]. Total pressure contours and streamlines in the forward stagnation point region.

which decreases with increasing distance from the surface and results in the rotational nature of the entropy-layer flow, see, e.g., [10].

However, we must distinguish between two forms of the entropy layer. The first form, typical for symmetric flow, has a velocity profile normal to the body surface like a slip-flow boundary layer. The second form is typical for asymmetric flow, where the streamline hitting the forward stagnation point does not cross the normal-to-the flow portion of the bow shock surface, Fig. 6.5. The computation, [15], was made two-dimensionally for the lower centerline of the HALIS configuration, [16], see also Sub-Section 3.5.2. In this case the maximum entropy streamline lies off the body surface and is at the same time the minimum velocity streamline. The entropy layer hence has a wake-like structure. This form possibly is in general present at the windward side of a RV-W-type configuration at high angle of attack[10].

The profiles of total pressure p_t, pressure p, Mach number M, and velocity v in direction normal to the surface at $x' = 3.99$ m, see Fig. 6.5, are given in Fig. 6.6. The, however very shallow, wake-like structure of the entropy layer off the surface is well discernible.

Whether the second form of the entropy layer is really of influence in our case, cannot be decided. The swallowing of the entropy layer by the growing

[10] In [10] it was sketched, Fig. 6.20, the other way around. Whether we have the second form always on the windward side, cannot be answered.

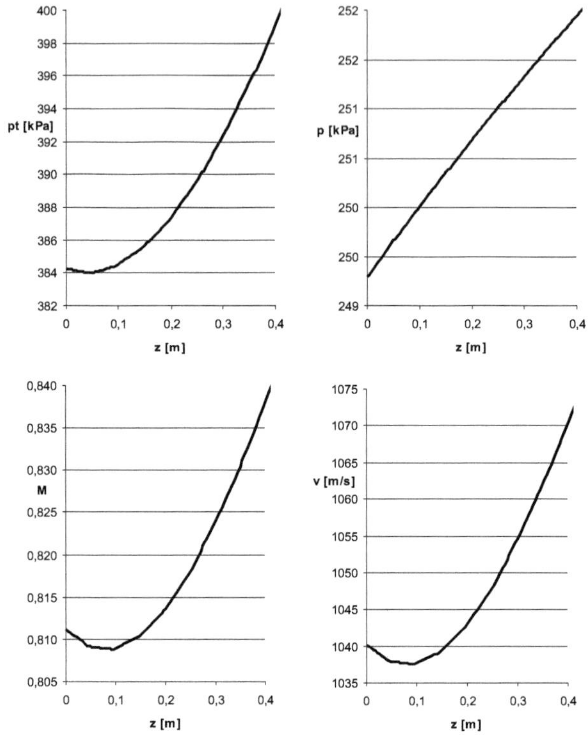

Fig. 6.6. Second form of the entropy layer: computed example, two-dimensional inviscid flow past the centerline of the HALIS configuration, $M_\infty = 10$, $H = 24.0$ km, $\alpha = 40°$, perfect gas, [15]. Profiles in direction normal to surface at $x' = 3.99$ m, Fig. 6.5: total pressure p_t, pressure p, Mach number M, and velocity v.

boundary layer[11], and the accordant change of its edge flow (v_e growing (first form) or first diminishing and than growing (second form)) possibly is of interest mainly for laminar-turbulent transition.

In reality, due to its growing thickness, the attached viscous flow – the boundary layer – swallows the entropy layer partly or even completely, depending on the thickness of the entropy layer and of the boundary layer. The unit Reynolds number and the boundary layer running length are important parameters in this regard, because they govern the boundary layer thickness.

Unfortunately, we cannot illustrate this with our estimated data, but if the entropy layer is swallowed partly or even completely, the external inviscid flow is no more that with the larger total pressure loss, but one with more favorable properties. The dynamic pressure of the onset flow is larger, the edge Mach

[11] This is the usual formulation. A better one would be "the growing of the boundary layer into the entropy layer", which describes the situation better.

number of the attached viscous flow, too, and the attached viscous flow itself is affected[12].

The consequence for the onset flow as a whole is, that in any case not the maximum total pressure loss governs the inviscid onset flow, but a smaller one, depending on how much the entropy layer has been swallowed. This is configuration dependent, and different for body flaps, elevators, ailerons and rudders. The boundary layer itself suffers a (strong) total pressure loss, too, which by far compensates the beneficial effects of entropy-layer swallowing. The thickness, and the velocity and density profile shapes (laminar or turbulent flow, thermal state of the surface, see next sub-section) of the attached viscous flow therefore are major factors regarding flap effectiveness, too.

We summarize, concerning the influence of the entropy layer, that the bluntness of a flight vehicle shape influences not only the wave drag and the effectiveness of radiation cooling, [10], but also the effectiveness of aerodynamic trim and control surfaces via the onset flow properties.

6.2.3 The Onset Flow Boundary Layer

The properties of the boundary layer – the attached viscous flow part – of the onset flow also influence the effectiveness of an aerodynamic control surface, and especially also the thermal state of the surface and the thermal loads on the flap structure, Sub-Section 6.3.3. To study the first point, we consider the optimal situation of two-dimensional onset flow and a non-swept hinge line.

The two most important properties of the boundary layer are the momentum, which it carries, and its thickness[13]. Both depend strongly on the state of the boundary layer (laminar, transitional or turbulent), and on the local properties of the inviscid flow, with high-temperature real-gas effects in the background. Of large importance too are the necessary and permissible surface properties, Chapter 9 and [10]. These regard especially transitional and turbulent boundary layers, which react in general strongly on the real-life irregularities of the surface (roughness, steps, gaps, waviness).

A turbulent boundary layer carries more momentum than a laminar one, because of the fuller velocity profile $v(y)$, Fig. 6.7, y being the coordinate

[12] If we take as phenomenological model of attached viscous flow the boundary layer, we can deduce from its characteristic properties, [10], that flow features depend strongly on the local outer (external inviscid flow) and inner (wall) boundary conditions, rather than on the (upstream) initial conditions (however, the local outer conditions are influenced by the swallowed entropy layer). This concerns especially the shapes of the profiles – in direction normal (y) to the surface – of the velocity $v(y)$, the temperature $T(y)$, and the density $\rho(y)$ (via the compatibility conditions). This means, that for the attached viscous onset flow, the locally present external inviscid flow indeed is predominantly of relevance.

[13] Depending on the kind of consideration or the employed phenomenological model, the determining boundary layer thickness can be the displacement thickness δ_1, the momentum thickness δ_2, or other entities.

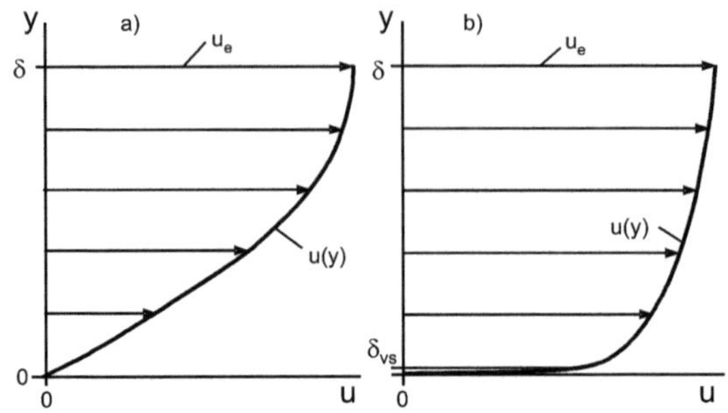

Fig. 6.7. Tangential velocity profiles of a two-dimensional boundary layer, a) laminar flow, b) turbulent flow, [10]. Edge of viscous sub-layer: δ_{vs}.

normal to the surface, provided the density profile $\rho(y)$ is similar in both cases. Therefore it can negotiate a stronger pressure rise in flow direction without separating. Consequently, the flap angles for incipient separation are far larger for turbulent than for laminar flow.

Another issue is the thickness of the onset flow boundary layer. If the onset flow boundary layer is thin compared to the frontal projection $h_{flap,proj}$ of the deflected control surface, Fig. 6.1, with given length l_{flap} and deflection angle η_{flap}, its influence will be small:

$$\delta_{bl,onset} \ll h_{flap,proj} = l_{flap} \sin \eta_{flap}. \tag{6.2}$$

The deflection angle η^*_{flap}, at which the flap would be fully "buried" in the onset flow boundary layer, is found from:

$$\sin \eta^*_{flap} = \frac{\delta_{bl,onset}}{l_{flap}}. \tag{6.3}$$

With other words – disregarding the state of the boundary layer, laminar or turbulent – the boundary layer thickness will not have much influence on flap effectiveness, if $\eta_{flap} \gg \eta^*_{flap}$.

In order to apply this to the four cases, which we introduced above in Table 6.1, we estimate the boundary layer thicknesses of the onset flow at the location of the hinge line for the flow with total pressure correction, Table 6.4. We assume fully laminar flow for the RV-W 1 case, and fully turbulent flow for the cases RV-W 2, CAV 1, and CAV 2. Although the involved temperatures are rather large, we use the reference-temperature approach, Sub-Section 10.4, rather than the reference enthalpy approach. The ensuing errors are acceptable for our illustrating purpose. For the viscosity the simple power-law approximation, eq. (10.61), is employed, which of course becomes unreliable above $T \gtrapprox 1{,}500.0\ K$. At such temperatures the viscosity is no more a variable of the temperature T only, but, due to high-temperature real-gas effects, a variable of, for instance, the internal energy and the density, see, e.g., [10].

Table 6.5. Onset flow boundary layer thicknesses $\delta_{bl,onset,tpl}$ at the lower vehicle sides of the cases RV-W 1, RV-W 2, CAV 1, and CAV 2, estimated with total pressure loss 'tpl' corrected data from Table 6.4. The second last line gives the (chord) lengths of the body flap, RV-W 1 and RV-W 2, respectively of the elevator/aileron combination (mean lengths), CAV 1 and CAV 2.

Flight vehicle:	RV-W 1	RV-W 2	CAV 1	CAV 2
M_∞ [-]	20	6	6.8	12
H [km]	70.0	40.0	30.0	34.2
α [°]	40.0	25.0	5.0	5.0
γ_{eff} [-]	1.14	1.4	1.4	1.3
l_{onset} [m]	32.0	32.0	25.0	6.0
$T_{w,onset,tpl} \equiv T_{ra,onset}$ [K]	1,000.0	900.0	1,000.0	1,200.0
$T^*_{onset,tpl}$ [K]	4,108.5	1,288.4	1,233.2	2,391.4
$Re^{u*}_{onset,tpl}$ [m^{-1}]	$2.03 \cdot 10^4$	$2.35 \cdot 10^5$	$2.63 \cdot 10^5$	$1.26 \cdot 10^5$
$\delta_{bl,onset,tpl}$ [m]	0.20 (lam)	0.50 (turb)	0.40 (turb)	0.15 (turb)
l_{flap} [m]	2.21	2.21	3.0	0.75
η^*_{flap} [°]	5.2	13.1	7.7	11.5

The resulting reference temperatures and unit Reynolds numbers are given in Table 6.5. Compared to the data without total pressure correction – not shown here – the reference temperature is in all cases larger, $T^*_{onset,tpl} > T^*_{onset}$, because $T_{e,onset,tpl} > T_{e,onset}$, and the reference unit Reynolds numbers are changed accordingly. The onset flow boundary layer thicknesses $\delta_{bl,onset,tpl}$ are all larger with total pressure correction, because the Reynolds numbers are smaller.

With the respective flap lengths l_{flap} in Table 6.5, which are estimated data from the reference vehicles, Sub-Section 6.2.1, and the estimated onset flow boundary layer thicknesses $\delta_{bl,onset,tpl}$, we arrive at the η^*_{flap} in the last line. Up to these deflection angles the flap is buried in the boundary layer. For turbulent flow these angles are less critical, because of the fuller velocity profile[14].

Consider case c) of Fig. 6.9. If $\eta_{flap} \lessapprox \eta^*_{flap}$, the interacting shock system, unlike shown in that figure, is buried in the boundary layer. This happens the more, if the flow is turbulent, because in that case the sonic line lies very close to the wall. In any case, of course, the ramp shock of the deflected inviscid flow, the "reattachment shock" in Fig. 6.9, is always there. No systematic studies are available regarding this phenomenon.

[14] We don't investigate here, whether the thickness of the viscous sub-layer is the characteristic one in such cases.

We note in passing, that the concave flow situation, which arises when the boundary layer flow enters the deflected flap surface, can lead to a centrifugal instability, with Görtler vortices appearing in the laminar boundary layer, see, e.g., [10]. Such pairwise counter-rotating vortices transport momentum and enthalpy towards the wall. They promote laminar-turbulent transition, enhance thermal loads in the form of streamwise striations, see Sub-Section 6.3.3, and can alter the separation and reattachment behavior of the boundary layer.

We have mentioned above the density profile $\rho(y)$. In this context the above cited "cold" and "hot" wall situation is important. Attached viscous flow is of boundary layer type, if hypersonic viscous interaction and/or low density effects as well as bluntness induced entropy-layer effects, are small or absent. If this is true, the pressure in it, in direction normal to the surface, is constant, $p\,(y) =$ const., at least approximately. This means $\rho_w T_w = \rho(y)T(y) = \rho_e T_e$. Hence for the same inviscid flow a hot wall will result in a small density at and near the wall, and a cold one in a large density. This holds for both laminar and turbulent flow. Since the momentum flux is proportional to the density, the cold-wall situation in general is favorable regarding flap effectiveness. More momentum is transported by the boundary layer flow and the possibility of flow separation is reduced.

The wall temperature, generally the thermal state of the surface, [10], influences not only the density at the wall. As we have seen above, a larger wall temperature also means a thicker boundary layer. It means further also a stronger displacement of the inviscid flow, an enhancement of flow three-dimensionality (the same transverse pressure gradient, but a smaller lateral momentum flux, hence a stronger flow deflection, [10]), but, especially for turbulent flow, a reduced wall shear stress, as we will discuss at the end of this sub-section.

Of course, here looms a simulation problem. In flight the vehicle surface is radiation cooled, and therefore we don't have a uniform hot-wall or cold-wall situation. But we definitely have not a cold-wall situation, as we have it in ground-facility simulation. Hence a proper experimental simulation of flap effectiveness can be difficult, if a large accuracy is demanded.

We illustrate the influence of the wall temperature of the onset flow with numerical simulation data again of the flow past the HOPPER configuration [12], Table 6.3, case B. The deflection angle of the body flap is $\eta_{bf} = +20.0°$. We compare two results, upper side in Fig. 6.8: computation with the radiation-adiabatic wall temperature belonging to case B, and lower side: computation with the radiation-adiabatic wall temperature belonging to case A. The wall temperature of the onset flow belonging to case B is about $T_{w,onset,B} = 500.0\ K$ (cold case), and that belonging to case A is about $T_{w,onset,A} = 1{,}500.0\ K$ (hot case). On the flap in both cases the temperatures are larger.

We don't contrast here a ground-simulation facility situation with a flight situation. We compare rather a case of possible thermal reversal, [10], where

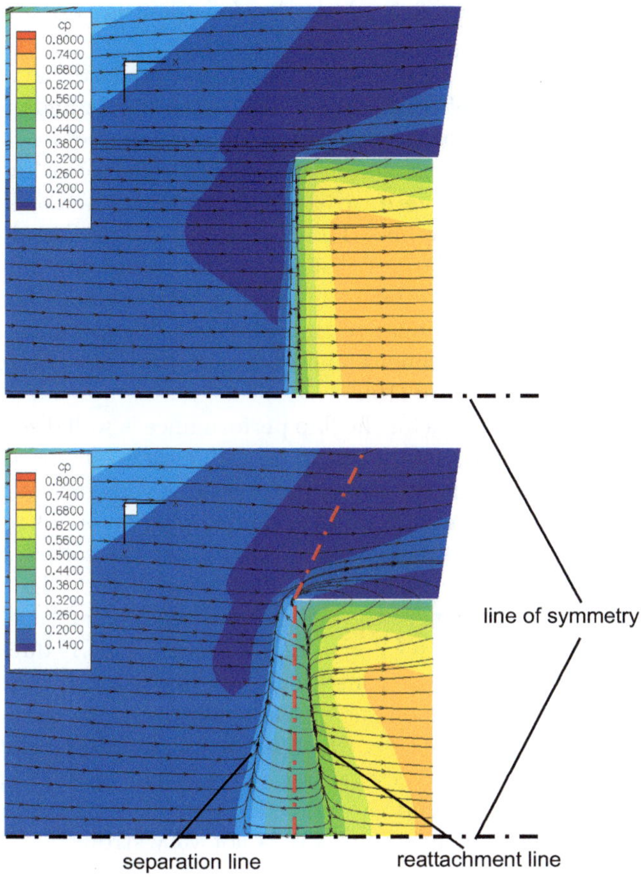

Fig. 6.8. View of skin-friction lines and wall pressure coefficient ($cp \equiv c_{p_w}$) distribution at the left lower-side aft part of the HOPPER configuration, case B, Table 6.3, [12]. Body flap deflection: $\eta_{bf} = +20°$. Upper part: computation results with the radiation-adiabatic wall temperature belonging to case B (cold case), lower part: computation results with the radiation-adiabatic wall temperature belonging to case A (hot case).

the temperature of the TPS surface due to the thermal inertia is still larger than the one which belongs to the momentary flight conditions. However, the situation considered is hardly representative for a real re-entry flight, because in general such a large thermal reversal does not occur.

For both cases we have a nearly optimal onset flow situation. The cold case exhibits a relative small separation region at the hinge line, with small upstream influence, see the discussion of Fig. 6.4. The pressure rises then monotonically towards the well extended pressure plateau on the flap, the situation corresponding on a large part of the flap to case b), Fig. 6.9 in Sub-Section 6.3.1. The three-dimensionality near the edge of the flap is like that which we have seen in Fig. 6.4, where the deflection is two times the present one.

The hot case shows some more, but still weak three-dimensionality in the onset flow[15]. Then we see the effect expected due to the low boundary layer momentum: strong separation around the hinge line compared to the cold case. The separated flow regime is characterized by a separation line lying well ahead of the hinge line (large upstream influence), with the distance largest at the vehicle's symmetry line. The reattachment line lies closer to the hinge line, but has also its largest distance at the symmetry line. The spreading of the distance of the separation and the reattachment line from the cold to the hot case is very similar to that observed in experimental and numerical investigations in two-dimensional (laminar) flat-plate/ramp flow, [17], [18], see also the discussion in [10].

The consequence of the hot case for the flap performance is well discernible. The flap pressure rise seems to be still monotonic, although we should have, at least approximately in the symmetry plane, the pressure overshoot associated with case c), Fig. 6.9, Sub-Section 6.3.1. An extended plateau is not reached, however, the maximum pressure coefficient is like that for the cold-wall case. The flow on the flap surface is highly three-dimensional, the edge effects are stronger. The resulting change in the vehicle pitching moment coefficient from the cold to the hot case is $\triangle C_m = +\ 22.9$ per cent. This is due to the smaller pitch-down flap moment as consequence of the reduced flap force, which counteracts the pitch-up moment of the vehicle, and to the smaller skin-friction on the vehicle's lower side, which has a similar effect. The hinge moment coefficient is changed by only $\triangle C_{m,h} = -\ 4.54$ per cent. It appears to be influenced by the flow around the side edge toward the flap's leeside, where at $M_\infty = 3.2$ the shadow effect anyway is not very strong.

6.3 Asymptotic Consideration of Ramp Flow

6.3.1 Basic Types of Ramp Flow

We consider again the aerodynamic control surface with optimal onset flow geometry. Such flow is classically studied in the form of flat-plate/ramp flow, Fig. 6.1. We assume, that the ranges of onset flow Mach numbers and ramp angles are such that the ramp shocks are only weak shock waves, i. e. in the inviscid case attached shock waves. We begin with looking at the three basic types of flat-plate/ramp flow, shown in Fig. 6.9.

Case a) is the inviscid flow case as it is sketched in Fig. 6.47. We regard its ramp wall-pressure as our asymptotic wall pressure. Case b), shock/boundary

[15] The three-dimensionality ahead of the separation line is caused by a weak transverse pressure gradient of the inviscid flow, and is enhanced by the large wall temperature, see the remark above. The onset flow Mach number ahead of the separation line is around $M_{e,onset} \approx 1.6$, such that the flow is still spatially hyperbolic, and no upstream influence of the deflected flap is present ahead of the primary separation line.

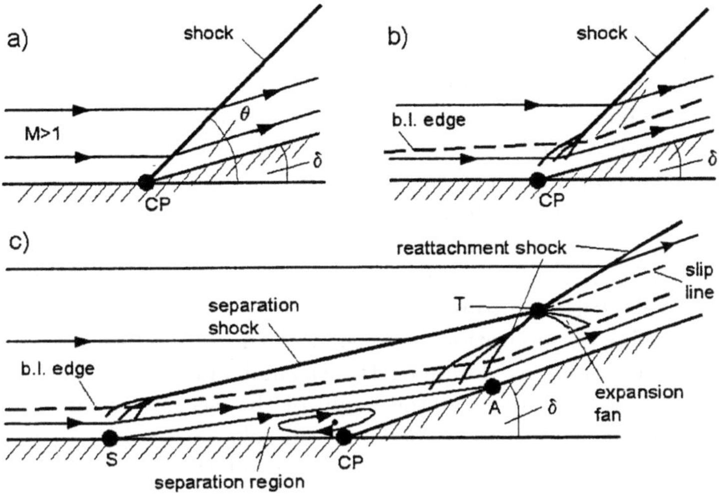

Fig. 6.9. Schematics of two-dimensional flows over a flat-plate/ramp configuration, [10]: **a)** inviscid flow, **b)** viscous flow with non-separating boundary layer, **c)** viscous flow with (local) separation. S denotes the separation point, A the reattachment point, T the triple shock point, and CP the corner point.

layer interaction without separation, is found at small ramp angles and high momentum onset boundary layer flow[16]. It is associated, with a monotonic wall-pressure rise on the ramp towards the plateau pressure (small ramp angles).

Finally case c) depicts the flow with significant separation, here sketched with a rather thin onset flow boundary layer. This case, see below, is characterized by a small pressure rise and a pressure plateau upstream of and over the hinge line and a further monotonic increase of the wall pressure to the ramp plateau. This is typical for small onset flow Mach numbers, or small ramp angles at large onset flow Mach numbers. At large ramp angles or at large onset flow Mach numbers this plateau is approached non-monotonically with a pressure overshoot.

This different behavior is due to the two-shock system (separation shock and reattachment shock in Fig. 6.9, case c), which coalesce in the outer flow to the single ramp shock of case a)). The separation and reattachment shocks have different shock angles and static pressure rise at different combinations of onset flow Mach number and ramp angle. The pressure overshoot just downstream of the reattachment shock becomes stronger with rising onset

[16] It appears that a small separation region is always present around the hinge line due to the shock/boundary layer interaction, even for very weak shocks. In the literature sometimes a distinction is made between *true* and *effective* incipient separation.

flow Mach number. With a simple consideration, [19], we find for instance for a 14.0° ramp, that the overshoot appears for $M_1 \gtrapprox 5$, becoming progressively larger with rising M_1.

The available experimental evidence points to the fact that downstream of the pressure overshoot the asymptotic pressure plateau with the pressure of case a) always is approached, provided the ramp is long enough. This happens through the expansion fan emanating at the triple shock point, apart from boundary layer displacement effects. In practice this is in general the case.

All three cases occur for laminar as well as turbulent boundary layer flow. Case c) occurs with a laminar onset flow boundary layer at a ramp angle smaller than with a turbulent onset flow.

Criteria for incipient separation, extension of the separation region, and peak heating correlations for laminar, transitional and turbulent flow can be found in the literature given at the begin of this section. Many of them stem from investigations in cold hypersonic tunnels with cold-wall models. The thickness δ_0 of the onset flow boundary layer at the hinge line (at zero ramp angle), often serving as characteristic length for the separation region upstream of the hinge line, was introduced in [20].

We note here only the criteria for incipient separation. For laminar cold-wall onset flow '1' we have for the incipient separation angle, η_{is}, [21], see also, e.g., [1]:

$$\eta_{is}[°] = \frac{80\,\sqrt{\overline{\chi}_1}}{M_1}, \tag{6.4}$$

with the hypersonic viscous interaction parameter for laminar flow

$$\overline{\chi}_1 = \frac{M_1^3\sqrt{C_1}}{\sqrt{Re_{1,l_{onset}}}}, \tag{6.5}$$

and the Chapman-Rubesin constant:

$$C_1 = \frac{\mu_w T_1}{\mu_1 T_w}. \tag{6.6}$$

In [22] experimental data of incipient separation of turbulent flat-plate/ ramp flow were correlated with the inviscid pressure jump belonging to Mach number and ramp-angle pairs. In terms of the pressure jump p_2/p_1 the correlations read:

$$M_1 \lessapprox 4.5: \quad \frac{p_{2,is}}{p_1} = 1 + 0.3\,M_1^2, \tag{6.7}$$

and:

$$M_1 \gtrapprox 4.5: \quad \frac{p_{2,is}}{p_1} = 0.17\,M_1^{2.5}. \tag{6.8}$$

From this the incipient-separation ramp angle η_{is}, which belongs to the pressure jump $p_{2,is}/p_1$, can be inferred.

These criteria are of limited value for real flight-vehicle configurations, because, besides the actual onset flow geometry, factors like the entropy layer, the wall temperature, the boundary layer thickness, and high-temperature real-gas effects effectively change the picture. Nevertheless, if the flow situation of the aerodynamic control surface considered meets at least approximately the flow situation, which is the basis of the above and other criteria, assertions can be made[17].

Here we are interested primarily in the effectiveness of aerodynamic control surfaces. Such surfaces have, in the range of the deflection angles employed, in general an extension of the separation region of only a couple of boundary layer thicknesses δ_0 upstream and downstream of the hinge line. Hence the length of the separation region is much smaller than that of the flap surface, l_{flap}, itself, Figs. 6.4 and 6.8.

In view of flap effectiveness we hence have in mind the larger part of the flap surface, which is not directly affected by the strong interaction phenomena around the hinge line. We look at the asymptotic behavior of wall pressure, the thermal state of the surface, and the wall shear stress. We assume optimal onset flow. We assume further, that on the ramp surface downstream of the strong interaction domain displacement effects of the boundary layer can be neglected (negligible hypersonic viscous interaction and low-density effects), and that thermo-chemical rate effects are absent. Of course, we develop the asymptotic considerations out of the experimentally observed behavior of the wall entities in the strong interaction domain around the hinge line. In doing this, we make use of the above criteria for incipient separation ramp angles.

6.3.2 Asymptotic Behavior of the Wall Pressure

In Figs. 6.4 and 6.8 we did observe on the body flap that apparently the pressure rises monotonically to a peak pressure plateau. We can imagine that this is the pressure of the ideal inviscid ramp flow, Fig. 6.9, case a), and sketched in Fig. 6.47, which we consider as the asymptotically achievable flap pressure. This pressure is a function of the Mach number of the onset flow, the thermo-chemical properties of the gas, and the ramp angle. These in turn are functions of flight speed, altitude and vehicle attitude, Sub-Section 6.1.1.

We turn now to other examples from the literature in order to study the wall-pressure behavior on flat-plate/ramp configurations. First we look at experimental data, Fig. 6.10, which were found with a sharp-edged flat-plate/ramp model, [24]. The plate's length from the leading edge to the hinge line was $L = 0.43\ m$, the length of the flap was $l_{flap} = 0.1\ m$, and its width $w_{flap} = 0.178\ m$. The onset flow parameters with a turbulent boundary layer approaching the ramp are given in Table 6.6.

We see in Fig. 6.10 that indeed, for all ramp angles, the ideal inviscid ramp pressure is asymptotically reached very fast at only 25.0 to 40.0 mm (four to six hinge-line boundary layer thicknesses δ_0) downstream of the hinge line.

---

[17] Detailed discussions of strong interaction flow situations present at real flight vehicle configurations can be found in, e.g., [23], [5], [7].

Table 6.6. Onset flow parameters of the turbulent flat-plate/ramp flow from [24]. The Reynolds number was found with the thickness $\delta_0 = 0.08\ m$ of the boundary layer at the hinge line. Test gas is nitrogen.

M_1	$T_t\ [K]$	$p_t\ [kPa]$	Re_{δ_0}	$T_w\ [K]$
9.22	1,070.0	$101,325.0 \cdot 10^3$	$4.0 \cdot 10^5$	295.0

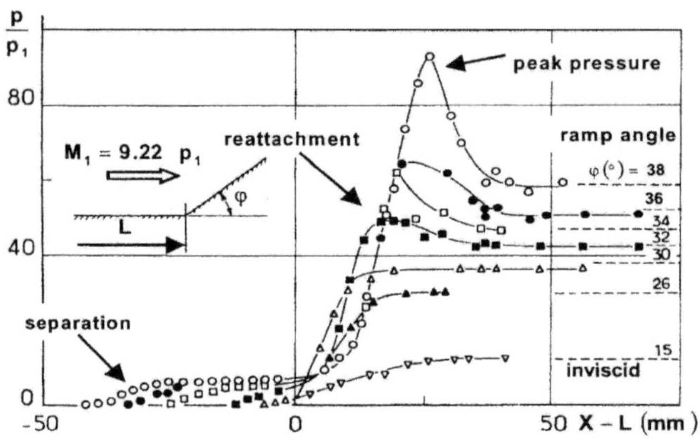

Fig. 6.10. Experimentally found wall-pressure ($p \equiv p_w$) distributions of a turbulent flat-plate/ramp flow as function of the ramp angle φ ($\equiv \eta$), [4], data source: [24]. The onset flow conditions are given in Table 6.6.

The ramp angle for incipient separation following eq. (6.8) is φ_{is} ($\equiv \eta_{is}$) $\approx 32°$. At the ramp angle $\varphi = 32°$ however, we have already case c), Fig. 6.9, with separation ahead of the hinge line. For $\varphi \leqq 30°$ we have clearly case b) with no or only small separation at the hinge line and a monotonic pressure rise just downstream of the hinge line towards the asymptotic value. The upstream influence for $\varphi \geqq 30°$ increases with increasing ramp angle, however, the plateau pressures reached in the separation region ahead of the hinge line are similar. On the ramp, once separation occurs, we see always a prominent overshoot with the peak pressure located in the vicinity (just downstream) of the reattachment point A, Fig. 6.9 c). This is typical for the Edney type IV interaction, [25]. The drop of the pressure behind the peak towards the (asymptotic) plateau pressure is due to the expansion starting at the triple point T indicated in Fig. 6.9 c).

In the literature one can find sometimes the summarizing statement, that flow separation reduces flap effectiveness. The pressure distributions shown in Fig. 6.10 for the different ramp angles demand a closer look. Compared to the ideal inviscid wall pressure distribution on the lower side of the flight vehicle and on the flap, sketched in Fig. 6.47, we find that the pressure plateaus indeed do not cover the whole flap surface. This we saw also in Figs. 6.4 and Fig. 6.8.

However in Fig. 6.10 we observe two interesting facts concerning the flow on the pressure side of the flap. The first is, that, if separation is present, the pressure is increased upstream of the hinge line which indirectly enhances the vehicle pitching moment M_{pitch}, Section 6.6. The second is, that, of course, a monotonic rise of the pressure towards the plateau pressure reduces the hinge moment M_{hinge}. This would happen already if no or only a weak separation is present. But, if an overshoot over the plateau pressure happens, it will even compensate partly for the reduced pressure – compared to the ideal inviscid pressure – just downstream of the hinge line.

The balance is delicate, but it would be more apt to say, that viscous effects in any case reduce flap effectiveness, but that separation does not so in a dramatic way, unless massive separation is present, as shown, for instance, in Fig. 6.8, lower part. This in general holds also for laminar flow, see the example below. Concerning the hinge moment M_{hinge}, we can say that the monotonic pressure rise reduces it, but that a non-monotonic rise with a pressure overshoot can compensate this to a degree, too. Of course also the length of the control surface compared to that of the affected region (separated flow) around the hinge line must be taken into account.

Before we look at such flow, we study how present-day numerical simulation methods can cope with turbulent flat-plate/ramp flow[18]. In [26] we find results which are representative for the state of the art regarding the capabilities of Reynolds-averaged Navier-Stokes methods. The onset flow parameters of the selected case, studied experimentally in [27] with the sharp-edged flat-plate/ramp model from [24], are given in Table 6.7.

Table 6.7. Onset flow parameters of the turbulent flat-plate/ramp flow study [26], experimental data from [27]. Test gas is nitrogen.

M_1	p_1 $[Pa]$	T_1 $[K]$	Re_1^u $[m^{-1}]$	T_w $[K]$	δ_0 $[m]$
9.22	$2.3 \cdot 10^3$	59.44	$4.7 \cdot 10^7$	295.0	0.008

Fig. 6.11 shows the comparison of the experimentally[19] and the numerically found wall pressure distributions for the ramp angle $\eta = 38°$. This would be a rather large flap angle for an aerodynamic control surface. The six applied statistical turbulence models therefore are severely strained.

The experimental data are nearly identical with those for the same flap angle shown in Fig. 6.10. No turbulence model reproduces well the pressure rise upstream of the hinge line regarding extent and plateau. On the ramp the asymptotic inviscid pressure plateau is lying at $p_w/p_1 \approx 58.8$. This is reached – with different accuracy – by all turbulence models, but not the magnitude and location of the pressure peak. All in all, given the extremely large ramp

---

[18] A short discussion of simulation issues is given in Sub-Section 6.7.2.

[19] The work reported in [27] is an extension of that in [24].

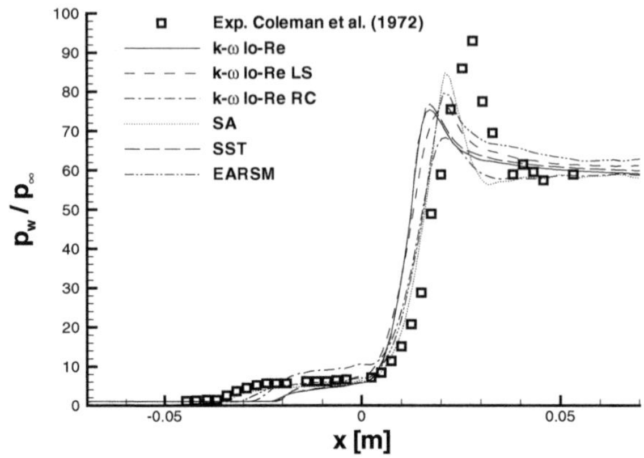

Fig. 6.11. Surface pressure $p_w(x)/p_\infty$ ($\equiv p_w/p_1$) of a turbulent $M_1 = 9.22$ flat-plate/ramp flow with ramp angle $\eta = 38°$, computed with several turbulence models, [26], and found experimentally, [27]. The onset flow conditions are given in Table 6.7.

angle, the computations are not so bad regarding the wall pressure, although they fail to capture correctly the separation and reattachment process.

Now to the laminar flat-plate/ramp flow. In principle we find the same features as for turbulent flow, but usually exaggerated. This is due to the smaller momentum transported by the onset flow in the laminar boundary layer. In Fig. 6.12 we show measured, [28], and computed, [17], wall-pressure (above) and Stanton numbers (below) distributions found for a sharp-edged flat-plate/ramp model. The plate's length from the leading edge to the hinge line was $l_{onset} = 0.4389\ m$, also the length of the flap was $l_{flap} = 0.4389\ m$. The onset flow parameters are given in Table 6.8.

Table 6.8. Onset flow parameters of the laminar flat-plate/ramp flow from [17]. Test gas is air.

M_1	T_t $[K]$	T_1 $[K]$	Re^u $[m^{-1}]$	T_w $[K]$
14.1	3,623.0	88.88	$2.362{\cdot}10^5$	297.22

The not constant wall pressure ahead of the hinge line (as well as the value $p_w/p_\infty = 1.96$ at $x = 0.395\ m$ for the lower ramp angle $\theta = 15°$) shows that hypersonic viscous interaction is present. If we take the reference temperature, eq. (10.78), with $T^* \approx 970.0\ K$ as a measure, we can speak about a cold-wall situation. With the respective criterion $\overline{\chi}_{crit} \approx 4$, [10], we get hypersonic viscous interaction for $x < x_{crit} \approx 1.9\ m$, which seems to confirm this. Nevertheless, we can use the data for the discussion of laminar flat-plate/ramp flow.

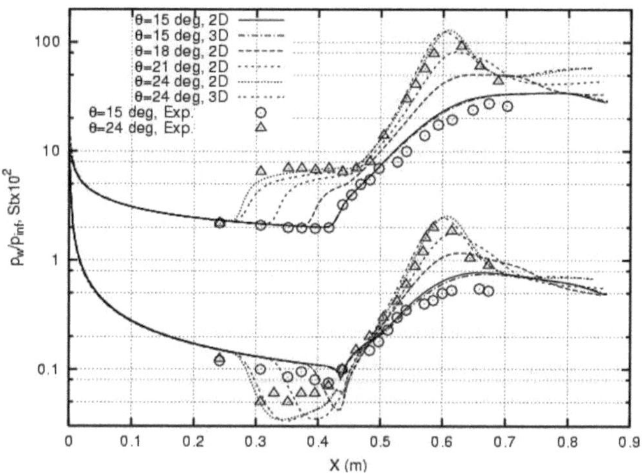

Fig. 6.12. Laminar $M_1 = 14.1$ flat-plate/ramp flow with angles from θ ($\equiv \eta$) $= 15.0°$ to $24.0°$, [17]. The onset flow parameters are given in Table 6.8. Surface pressure $p_w(x)/p_\infty$ ($\equiv p_w(x)/p_1$) (above) and Stanton number $St(x)$ ($\equiv St_{gw}(x)$) (below) were found computationally, [17], and experimentally, [28].

The computed wall-pressure data agree sufficiently well[20] with the measured data for both $\theta = 15°$ and $\theta = 24°$. The criterion eq. (6.4) yields the incipient separation angle $\theta_{is} = 16.3°$. For the $\theta = 15°$ case indeed separation is not indicated, the pressure rises monotonically towards the pressure plateau. For the ramp angles larger than 15.0° separation clearly is indicated with growing upstream influence at nearly constant plateau pressure, and the typical overshoot over the ramp plateau pressure behind the hinge line, Fig. 6.12. The asymptotic inviscid plateau pressure is $p_w/p_1 = 24.6$ for the $\theta = 15°$ case and $p_w/p_1 = 58.0$ for the $\theta = 24°$ case. Even with the present hypersonic viscous interaction it can be considered as a valid approximation of the ramp's plateau pressure.

Remains to look at the wall-pressure behavior in the presence of laminar-turbulent transition in the vicinity of the hinge line. In [29], for instance, experimental and computational investigations were performed of the flow past a flat-plate/ramp configuration with transition occurring near the reattachment point "A", Fig. 6.9, c). The results indicate, regarding the wall pressure distribution, that different assumptions about the state of the boundary layer result in changes around the hinge line, as was to be expected. However, the asymptotic plateau pressure is reached in all cases at a short distance behind the hinge line. Of more concern is the influence of transition on the thermal state of the surface, which is treated in the next sub-section.

Finally we have a glance at the influence of high-temperature real-gas effects. Two issues arise here. On the one hand, they may change the inviscid

[20] In [17] grid resolution problems are noted for $\theta > 18°$.

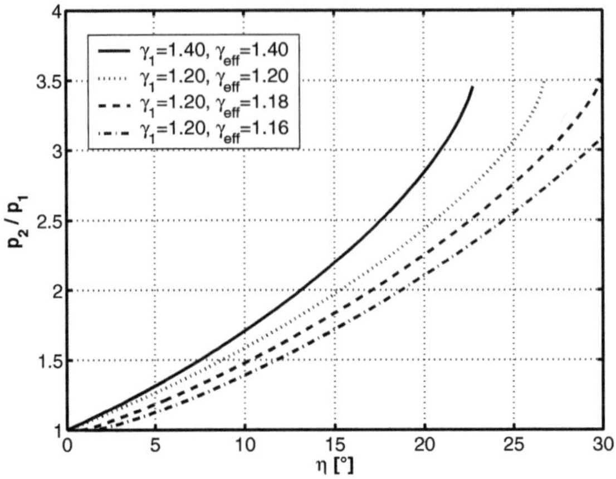

Fig. 6.13. Ratio of ramp pressure to onset flow pressure, p_2/p_1 ($\equiv p_{ramp}/p_{onset}$), as function of the ramp angle η and for $\gamma_1 = 1.2$ and different $\gamma_2 = \gamma_{eff}$. Flight vehicle RV-W 1, Table 6.1, $M_{e,onset} = 2$. If the curves end, this means that the respective ramp shocks become strong shocks and detach.

flow geometry (shock angle), interaction and separation patterns, and also the wall pressure around the hinge line. On the other hand they affect the asymptotic plateau pressure on the flap. Not very much is known about the first issue, see, e.g., [4]. Regarding the second issue, we study in a very simple parametric way, how the inviscid pressure on the flap, p_2 is affected.

For the re-entry type vehicle RV-W 1, Table 6.1, we choose somewhat arbitrarily the onset flow Mach number $M_{e,onset} = 2$, larger than that given in Table 6.4, smaller than that in Table 6.2, in order to take into account entropy-layer swallowing. Otherwise the range of possible ramp-deflection angles with weak ramp-shock would have been too much restricted. The maximum ramp angle was chosen to be $\eta_{ramp} = 30°$, Figs. 6.13 and 6.14.

High-temperature real-gas effects were simulated by choosing different ratios of effective specific heats. The influence of the ratio of specific heats in the onset flow ahead of the ramp was studied with $\gamma_1 = 1.2$ in Fig. 6.13 and $\gamma_1 = 1.16$ in Fig. 6.14. Accordingly smaller values of $\gamma_2 = \gamma_{eff}$ were chosen for the flow on the ramp.

What we see in Fig. 6.13 for a given ramp angle η, is a decrease of the asymptotic plateau pressure p_2 with decreasing γ_{eff}, i. e., increasing high-temperature real-gas effects. The pressure span for, say, $\eta = 20°$, is moderate with $p_{ramp}/p_{onset} \approx 2.8$ to 2 for the range of γ_{eff} considered. The percentage of the influence of γ_{eff} is rather large. With $\gamma_{eff} \approx 1.14$ reflecting reality, a simulation of the flow with, for instance, perfect gas, i. e. $\gamma = 1.4$, would lead to a too large flap effectiveness[21]. The effects are somewhat increased for the smaller γ_1, Fig. 6.13.

[21] The decreasing γ_{eff} has another effect. It shifts the ramp angle at which the oblique shock detaches to form a strong shock to significantly larger values.

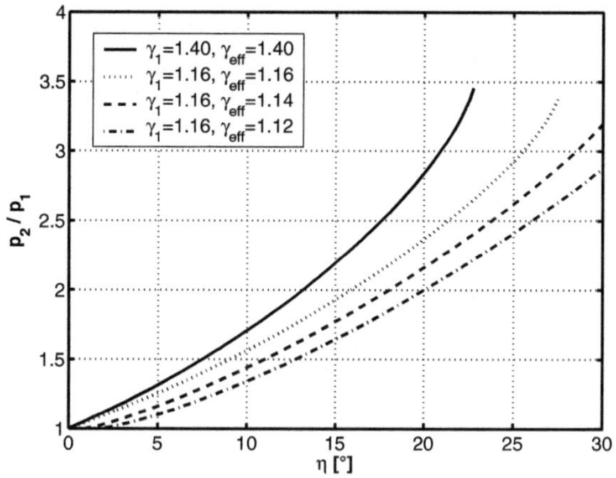

Fig. 6.14. Ratio of ramp pressure to onset flow pressure, p_2/p_1 ($\equiv p_{ramp}/p_{onset}$), as function of the ramp angle η and for $\gamma_1 = 1.16$ and different $\gamma_2 = \gamma_{eff}$. Flight vehicle RV-W 1, Table 6.1, $M_{e,onset} = 2$. If the curves end, this means that the respective ramp shocks become strong shocks and detach.

We note also in Figs. 6.13 and Fig. 6.14, that the pressure ratio increases faster with increasing ramp angle. We can show this explicitly for large Mach numbers and small ramp angles. With, [10]:

$$M_1 \rightarrow \infty,\ \eta\ small:\ c_{p_2} \rightarrow \frac{4 sin^2\theta}{\gamma+1},\ \theta \rightarrow \frac{\gamma+1}{2}\eta \tag{6.9}$$

we find, locally applied, assuming constant γ – with the Mach number retained due to the definition of c_p:

$$\begin{aligned} M_1 \rightarrow \infty,\ \eta\ small: \triangle\left(\frac{p_2}{p_1}\right) &= \frac{p_2 - p_1}{p_1} \rightarrow \frac{\gamma(\gamma+1)}{2} M_1^2 \eta^2, \\ \Rightarrow \triangle\left(\frac{p_2}{p_1}\right) &\sim \eta^2,\ \frac{d}{d\eta}\left[\triangle\left(\frac{p_2}{p_1}\right)\right] \sim \eta. \end{aligned} \tag{6.10}$$

The increase of a flap's pressure ratio and gradient with η for supersonic flow of course is reflected directly by any related moment and its gradient, see, e.g., [30] and also below.

The γ_{eff} effect found in Figs. 6.13 and Fig. 6.14, however, must be considered very carefully. The reality is much more complex. Assume a RV-W-type flight vehicle in the RHMP$^+$ approximation at given speed, altitude and angle of attack. If we would model the flow past it with different high-temperature real-gas approximations in terms of γ_{eff}, we would observe, that for any $\gamma_{eff} > \gamma_{eff,nominal}$ the temperature of the flow at the windward side of the vehicle is larger than that belonging to $\gamma_{eff,nominal}$. Because onset flow pressure and velocity are not much affected, the onset flow Mach number would be smaller in these cases than for $\gamma_{eff,nominal}$, which certainly changes the picture.

Table 6.9. Parameters of the computational study of the flow past the Orbiter for the flight (STS-1) and the tunnel situation, [31]. Laminar flow with no-slip boundary conditions was assumed, the flight situation was computed with a non-equilibrium real-gas model and a catalytic wall condition.

Computation case	M_∞ [−]	Altitude [km]	$L_{vehicle}$ [m]	v_∞ [km/s]	ρ_∞ [kg/m^3]	T_∞ [K]	T_{wall} [K]
Flight situation	23.68	73.1	32.77	6.81	$5.29 \cdot 10^{-5}$	205.0	rad.-adiab.
Tunnel situation	10	-	0.246	1.42	$788.0 \cdot 10^{-5}$	50.0	ambient

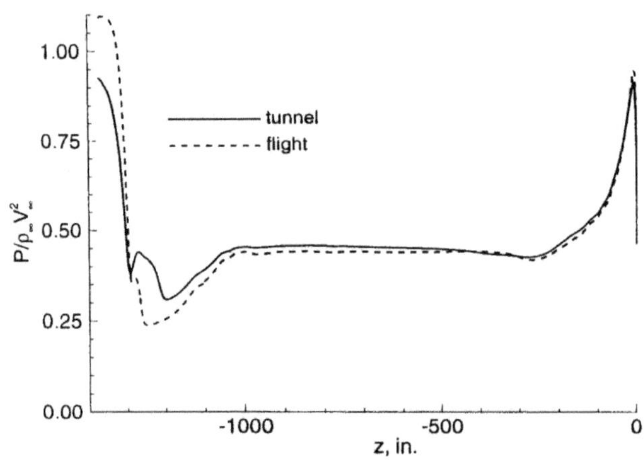

Fig. 6.15. Wall pressure distributions at the lower symmetry line of the Space Shuttle Orbiter at $\alpha = 40°$, body flap deflection $\eta_{bf} = 20°$, [31]. The broken line gives the pressure distribution in the flight situation, the full line that of a perfect-gas tunnel situation. The forward stagnation point is at the right, the body flap at the left.

We study this with an example available in the literature. In a computational study the flow past the Space Shuttle Orbiter was investigated with a Navier-Stokes code for the flight situation and the tunnel situation, [31]. Table 6.9 shows the flight parameters of the two computation cases.

We look first at the wall pressure ratio at the lower symmetry line of the Space Shuttle Orbiter. Fig. 6.15, [31], shows that in the flight situation the dimensionless pressure on the body flap (at the far left in the figure) is larger than in the tunnel situation (perfect gas). This is opposite to the result, which we see in Fig. 6.13. The cause is the change of the onset flow Mach number due to real-gas effects. This Mach number is larger in the flight situation. The real-gas effects hence dominate the pressure ratio indirectly more via the onset flow Mach number than directly, as shown in Fig. 6.13.

Table 6.10. Ramp-shock angles $\theta_{ramp\ shock}$ and pressure ratios p_{ramp}/p_{onset} of the laminar flat-plate/ramp flow for $\eta_{bf} = 20°$ obtained with onset flow parameters M_{onset}, γ_{eff} from [31].

Computation case	M_{onset}	γ_{eff}	$\theta_{ramp\ shock}$	p_{ramp}/p_{onset}
Tunnel situation	2.13	1.4	49.60°	2.90
Flight situation	3.45	1.2	32.75°	3.71

This is illustrated in Table 6.10, where ramp-shock angles $\theta_{ramp\ shock}$ and pressure ratios p_{ramp}/p_{onset} are given. They were obtained with the M_{onset} and γ_{eff} data determined in [31] from the computations[22]

The most important result from [31] regarding flap effectiveness and real-gas effects in view of the whole vehicle, the Space Shuttle Orbiter, is shown in Fig. 6.16. Plotted is the incremental pitching moment $\triangle C_m = C_{m,\eta_{bf}}$ - $C_{m,\eta_{bf}=0}$ as function of the body flap deflection angle η_{bf} and the angle of attack α. The results, again for the flight situation and for the perfect-gas tunnel situation, like in Fig. 6.15, show the influence of presumably the high-temperature real-gas effects on the pitching moment of the flight vehicle. Their influence implicitly increases for increasing angle of attack (body flap onset flow characteristics), and for increasing body flap angle. For $\alpha = 40°$ and $\eta_{bf} = 20°$ the body flap effectiveness is almost 50 per cent larger if real-gas effects are regarded. Also the (negative) gradient of the moment with regard to the deflection angle, $C_{m,\eta} = d C_m/d\eta$, increases with η_{bf}, see above, eq. (6.10), and that stronger for the flight case than for the tunnel case.

The authors of [31] discuss their results with all caution. The influence of the slight geometrical changes of the vehicle shape for the computations, of the aeroelastic deformation of the body flap, elevon gap flow, base pressure, leeside pressure on the body flap, and viscous effects is considered. The free-stream Reynolds numbers are $O(10^6)$ in both cases. The actual onset flow Reynolds numbers are not given. All in all, it appears that indeed the high-temperature real-gas effects, via the change of the onset flow Mach number, are the major cause of the discrepancies between flight and wind-tunnel situation.

---

[22] Keeping in mind the influence of entropy-layer swallowing on the onset flow Mach number, its exact determination from a Navier-Stokes computation is somewhat problematic, because there the boundary layer edge is not clearly indicated. We note further, that obviously the boattailing has a positive effect regarding the effectiveness of the body flap, because it increases the onset flow Mach number. This effect holds also for the elevators/ailerons at the wing's trailing edge, Fig. 6.44.. The present data are in fair agreement with the pressure rise on the body flap shown in Fig. 6.15 for both cases[23], and show in any case that indeed the onset flow Mach number is of large importance.

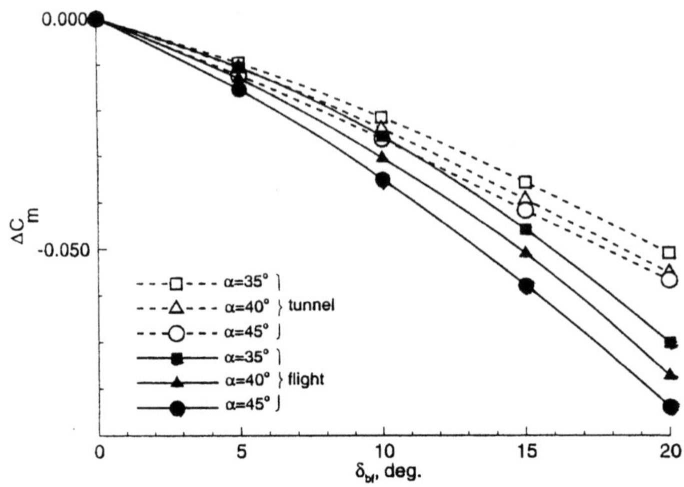

Fig. 6.16. Computed incremental pitching moment $\triangle C_m = C_{m,\eta_{bf}}$ - $C_{m,\eta_{bf}=0}$ of the Orbiter as function of the body flap deflection angle δ_{bf} $(\equiv \eta_{bf})$ and the angle of attack α, [31]. The full lines gives $\triangle C_m$ in the flight situation, the broken line that of the perfect-gas tunnel situation.

The basic conclusion is that the body flap of the Space Shuttle Orbiter is more effective in flight than was determined in perfect-gas ground-facility simulation. The hypersonic pitching moment anomaly, which was observed on STS-1, Section 3.5, thus appears not to be caused by reduced flap effectiveness in flight, as is sometimes asserted in the literature. In contrary, ... *had the body flap exhibited the same effectiveness in flight as in the tunnel, the vehicle may not have (been) trimmed at all*, [31].

Back to Fig. 6.13. We have seen there that for RV-W-type flight vehicles at large angle of attack – because then the onset flow Mach number is small – the (asymptotic inviscid) pressure rise on the flap as function of the deflection angle is not large, and that the influence of the parameter γ_{eff} is moderate. We have also learned, that onset flow Mach numbers themselves depend on high-temperature real-gas effects, and that this must be taken into account, because it can be, depending on the configuration under consideration, the deciding factor regarding flap effectiveness.

At CAV-type vehicles the onset flow Mach numbers are large, and the picture changes. We show this in Fig. 6.17, where for the hypersonic aircraft type vehicle CAV 2, Table 6.1, the pressure rise is given as function of η_{ramp} and three different γ_{eff}. Again we have chosen somewhat arbitrarily an onset flow Mach number, now $M_{e,onset} = 10$, larger than that given in Table 6.4, smaller than that in Table 6.2, in order to take into account entropy-layer swallowing.

We see a behavior qualitatively like that in Fig. 6.13. The pressure span for $\eta_{ramp} = 20°$ however is now much larger with $p_{ramp}/p_{onset} \approx 20$ to 22 for the range of γ_{eff} considered. The percentage of the γ_{eff} effect is a lot

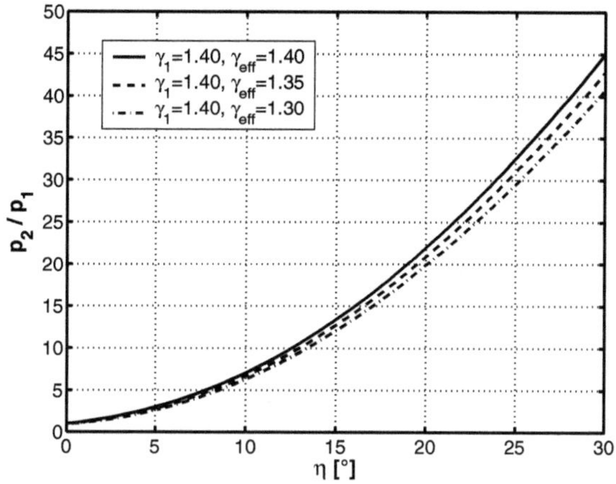

Fig. 6.17. Ratio of ramp pressure to onset flow pressure, p_2/p_1 ($\equiv p_{ramp}/p_{onset}$), as function of the ramp angle η and three different γ_{eff}. Flight vehicle CAV 2, Tables 6.1 and 6.2, $M_{e,onset} = 10$.

smaller. Also here an indirect influence of high-temperature real-gas effects via the onset flow Mach number exists. Even if it is smaller than in the case of RV-W-type flight vehicles, it must be regarded in vehicle design.

These exercises have shown in an approximate way the basic influence of high-temperature real-gas effects (indirectly also via the onset flow Mach number) on the effectiveness – in terms of the asymptotic plateau pressure due to a weak shock wave – of an aerodynamic trim or control surface with optimal onset flow geometry.

In reality a much more thorough investigation, like in [31], is necessary in order to get the influence of all relevant parameters right. Special problems arise if a sub-optimal onset flow geometry is given, for instance at elevators, ailerons, and rudders. Of course also the wider problem of stabilization surfaces, for instance vertical stabilizers, needs special attention.

6.3.3 Asymptotic Behavior of the Thermal State of the Surface

The asymptotic thermal state of the surface on an aerodynamic trim or control surface can be constructed at least approximately, if we consider the flat-plate/ramp configuration with optimal onset flow geometry.

In that case we have already seen, that the pressure on the ramp asymptotically reaches a plateau. Regarding, for instance, the heat flux in the gas at the wall, q_{gw}, we can imagine that its distribution on the ramp asymptotically is somehow proportional to the inverse of the boundary layer thickness for laminar flow, or to the inverse of the thickness of the viscous sub-layer for turbulent flow, and hence to the inverse – with some power – of the boundary layer running length x. This is sketched in Chapter 9, for more details see [10]. The radiation-adiabatic temperature will behave accordingly.

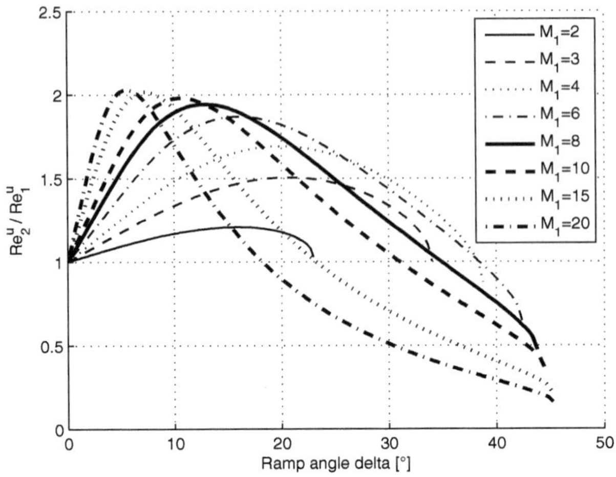

Fig. 6.18. Ratio of unit Reynolds numbers across the shock of a flat-plate('1')/ramp('2') configuration as function of the ramp angle η for different flat-plate Mach numbers M_1 and the ratio of the specific heats $\gamma = 1.4$, [32]. The viscosity μ in all cases was computed with the power-law relation, eq. (10.61), hence the figure holds for temperatures $T_1, T_2 \gtrapprox 200.0\ K$.

The large increase of both the heat flux in the gas at the wall and the radiation-adiabatic temperature on the ramp, compared to those on the flat plate, is mainly due to the decrease of the boundary layer thicknesses. These depend, in different ways, inversely on the local unit Reynolds number $Re_e^u = \rho_e v_e/\mu_e$, Sub-Section 10.4.2, hence $q_{gw} \sim (Re_e^u)^{1-n}$, and $T_{ra} \sim (Re_e^u)^{0.25(1-n)}$, $n = 0.5$ for laminar, $n = 0.2$ for turbulent flow, Sub-Section 10.4.3. Intuitively one would assume that Re_e^u rises always across the shock wave of a flat-plate/ramp configuration. In [10] it is shown and explained that this is not true. We illustrate this in Fig. 6.18.

Take for instance the Mach number $M_1 = 8$. For $0° < \eta < 35°$ we have $Re_{e,2}^u > Re_{e,1}^u$, with a maximum around $\eta \approx 12°$. This means, that in this interval indeed all boundary layer thicknesses on the ramp ('2') will be smaller than those on the flat plate ('1'). For $\eta > 35°$ the opposite is true. At smaller Mach numbers M_1 smaller maximum ramp angles with weak shock are reached, the ratios and their maxima become smaller, too. Interesting is that for larger Mach numbers the ramp angles decrease, for which the unit Reynolds number ratios become < 1. Then also the maxima are more pronounced, lying around two. High-temperature real-gas effects in general increase the maxima of the unit Reynolds number ratios and shift the ratios smaller than one to larger ramp angles.

Unfortunately, this does not influence the thermal state of the surface in the same way. Both the heat flux in the gas at the wall and the radiation-adiabatic temperature increase always with increasing ramp angle, because of the increase of the temperature, $T_{e,2} > T_{e,1}$, and hence of the heat-conduction

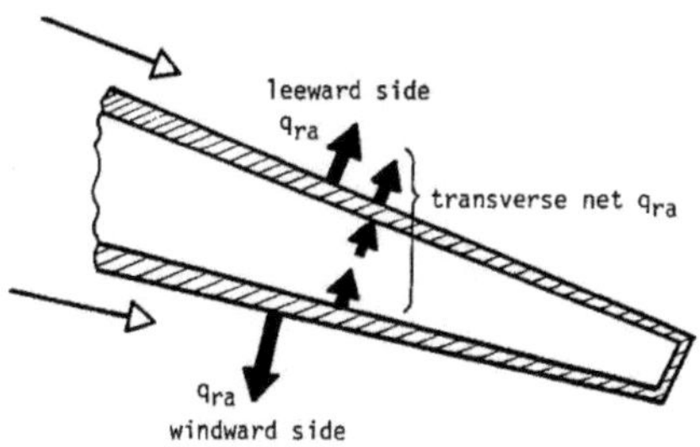

Fig. 6.19. Transverse heat transfer, by either thermal radiation q_{ra} or heat conduction q_w through a flap structure for cooling purposes, [33].

coefficient $k_{e,2}$ and of the reference temperature T^*, Sub-Section 10.4.3. Nevertheless, the particular behavior of the boundary layer thicknesses, as well as of the thermal state of the wall, and of the shear stress[24] with $\tau_w \sim (Re_e^u)^{1-n}$, can be of interest in view of several flow phenomena. Depending on M_1 and γ_{eff} one can choose the ramp angle either to make use or to avoid the peak value of Re_2^u/Re_1^u.

Prerequisite for the above postulated asymptotic behavior of the heat flux in the gas at the wall is a constant wall temperature, or a constant (small) heat flux into the wall (in the presence of surface radiation cooling). This means, for instance, negligible transverse heat conduction/radiation through the structure of the flap, Fig. 6.19, and negligible non-convex effects, i. e. no mutual radiative heat exchange between surface portions, Section 9.1.

We analyse now experimental flat-plate/ramp wall heat transfer data. For laminar flow these are given in Fig. 6.12 in terms of a Stanton number St, and for turbulent flow in Fig. 6.22 in terms of the normalized heat flux in the gas at the wall q_{gw}/q_{ref}.

Our analysis makes use of the reference-temperature relations given in Sub-Section 10.4.3 for the heat flux in the gas at the wall at flat surfaces (these relations indeed show the above mentioned asymptotic behavior). The employed pressure is the constant inviscid pressure on both the flat plate and on the ramp. On the latter this is the asymptotic plateau pressure downstream of the oblique shock wave. The relations for the determination of the virtual origin of the ramp boundary layer for the flat-plate/ramp configuration are provided in Sub-Section 10.4.4.

The application to the experimental 15° flat-plate/ramp data in Fig. 6.12, which is the case without separation, yields the result shown in Fig. 6.20. We observe, although we have only few experimental data points of the Stanton number, that they follow, ahead of the hinge line, as well as asymptotically on the flap, the $x^{-0.5}$ proportionality, Sub-Section 10.4.3. The increase around the hinge line is monotonic. The agreement between the experimental and the reference-temperature data appears to be as good as that between the

[24] We observe for the skin-friction coefficient $c_f \sim (Re_e^u)^{-n}$, Sub-Section 10.4.3.

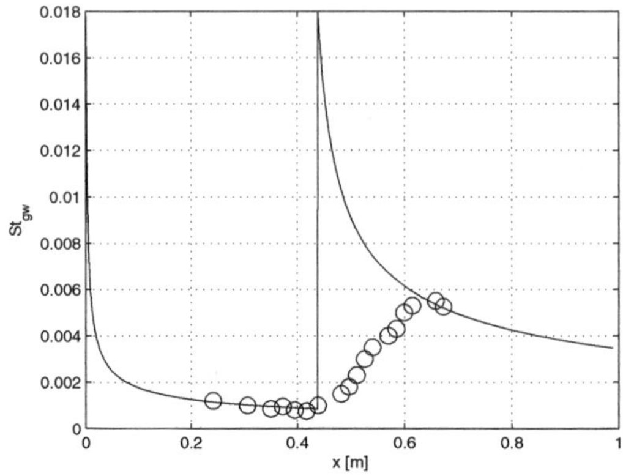

Fig. 6.20. Laminar $M_1 = 14.1$ flat-plate/ramp flow with $\theta\ (\equiv \eta) = 15°$, [17], see Fig. 6.12. Asymptotic behavior of the Stanton number $St_{gw} = q_{gw}/[\rho_\infty v_\infty (h_t - h_w)]$ as function of the location x, Fig. 6.12. The circles represent the experimental data, [28], the line the data found with the reference-temperature approach.

Navier-Stokes data and the experimental data in Fig. 6.12. However, we must keep in mind that hypersonic viscous interaction is present.

For the 24.0° flat-plate/ramp data in Fig. 6.21, the case with separation, the picture is not so good. Again we have not enough experimental data on the ramp. Ahead of the hinge line we see the expected $x^{-0.5}$ behavior of the Stanton number, followed by a steep rise towards a maximum, which lies in the vicinity of the reattachment point A, Fig. 6.9 c), where the characteristic boundary layer thickness has a minimum. Downstream of that the Stanton number tends towards towards the $x^{-0.5}$ behavior, however, without fully reaching it.

Data in terms of the normalized heat flux in the gas at the wall q_{gw}/q_{ref} for turbulent flow past a 38° flat-plate/ramp configuration are given in Fig. 6.22. They belong to the pressure data shown in Fig. 6.11. Other than for the pressure, the agreement between the computed, [26], and the measured, [27], data is not good. The large boundary layer edge Mach number leads to large compressibility effects, which, together with the strong compression at the ramp, show up stronger in the heat flux in the gas at the wall, than in the wall pressure.

All turbulence models predict the rise of the heat flux on the flat plate too close to the hinge line, i. e. underpredict the extent of separation. The experimentally observed rise lies at $x \approx$ - 0.055 m. The following increase of the heat flux downstream of the hinge line is predicted too early and too steep, with a peak too high, by all employed turbulence models, except for the $k - \omega$ low Reynolds number model with rapid compression (RC) correction ($k - \omega$ lo-Re RC). This model gives still a rise too steep, but a peak lower than the experimental one, but similar to that. Given the in [27] not specified experimental uncertainties, this result is acceptable.

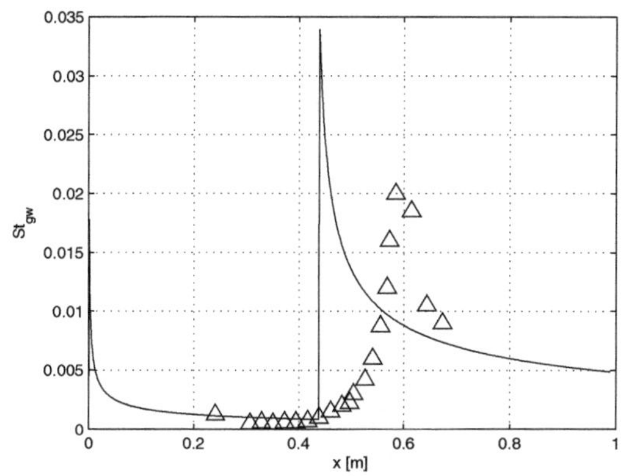

Fig. 6.21. Laminar $M_1 = 14.1$ flat-plate/ramp flow with θ ($\equiv \eta$) $= 24°$, [17], see Fig. 6.12. Asymptotic behavior of the Stanton number St_{gw} as function of the location x, Fig. 6.12. The triangles represent the experimental data, [28], the line the data found with the reference-temperature approach.

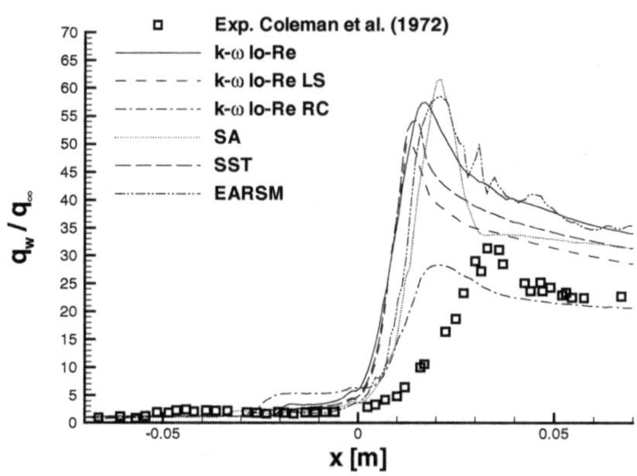

Fig. 6.22. With q_∞ non-dimensionalized heat transfer in the gas at the wall $q_w(x)$ ($\equiv q_{gw}(x)$) for turbulent $M_1 = 9.22$ flat-plate/ramp flow with $\eta = 38°$, [26], see also Fig. 6.11. Data computed with several turbulence models, and experimental data [27]. The reference data $q_{ref} = q_\infty \equiv q_{gw,ref}(x)$ are found in the whole x-interval for the ramp angle $\eta = 0°$.

The application of the reference-temperature relations – with the junction treated as in the laminar cases above – to the case in Fig. 6.22 gives, due to the chosen reference heat flux, a constant ratio equal to 1 ahead of the hinge line, Fig. 6.23. On the ramp, in the vicinity of the reattachment point A, Fig. 6.9 c), at $x \approx 0.04\ m$, the experimental heat-flux data begin to tend towards the asymptotic $x^{-0.2}$ behavior typical for turbulent flow. Results can be improved by using the approximate procedure proposed in [34], see, e.g., also [29], to put the virtual origin of the ramp boundary layer in the vicinity

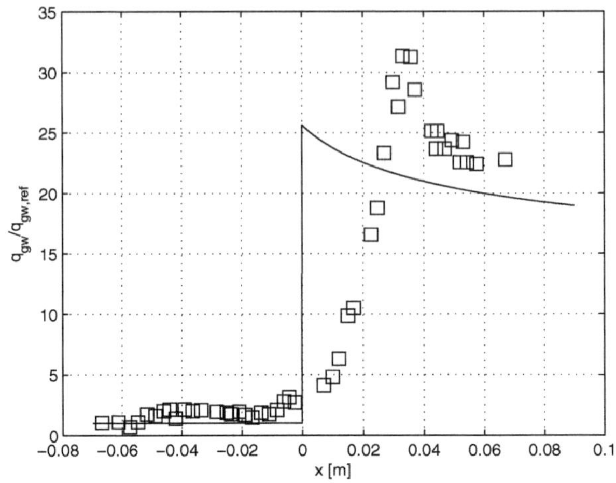

Fig. 6.23. Turbulent $M_1 = 9.22$ flat-plate/ramp flow with $\eta = 38°$, [26]. Asymptotic behavior of the non-dimensionalized heat transfer in the gas at the wall $q_{gw}(x)$ as function of the location x, Fig. 6.22. The reference data $q_{gw,ref}(x)$ again are found in the whole x-interval with $\eta = 0°$. The squares represent the experimental data, [27], the line the data found with the reference-temperature approach.

of the reattachment point on the ramp. That gives a quite good correlation of the experimental data, nevertheless, in contains a certain ambiguity, therefore we don't show such results.

The behavior of the data in Fig. 6.23, also that of the data with the procedure of [34], [29], is possibly a sign that very shortly downstream of the reattachment point the viscous sub-layer is fully reestablished, again an indication that boundary layer like flow depends stronger on the boundary conditions than on the initial conditions[25].

Although the turbulence model $k-\omega$ lo-Re RC in Fig. 6.22 meets the level of the experimental data, it appears not to show fully the proper asymptotic behavior which the other models seem to do, with the exception of the Spalart-Allmaras model (SA). Whether and how these establish then the level of the experimental data further downstream is not known. In any case, all but one turbulence model would predict thermal loads too large not only in the strong interaction region, but probably also on the remaining flap surface.

The test case is very demanding. Turbulence compressibility effects play a role but also the large pressure rise and the complex interacting flow field. The conclusion in [26] about the performance of the turbulence models – fair to good prediction of the pressure, see above, generally not so good one of the heat flux – therefore cannot be generalized. In general it holds that the

[25] This seems to contradict the observation of D. C. Wilcox, [14], that ... *the state of a turbulent flow at a given position depends upon upstream history and cannot be uniquely specified in terms of the local strain-rate tensor as in laminar flow.*

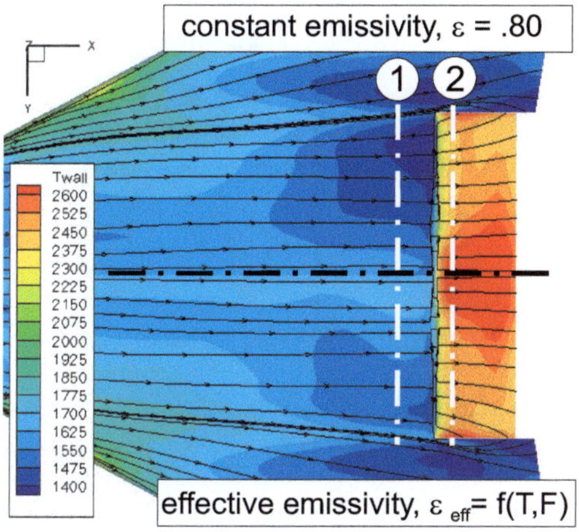

Fig. 6.24. Distribution of the radiation-adiabatic temperature T_{ra} ($\equiv T_{wall}$) and the skin-friction lines at the lower-side aft part of the HOPPER configuration, case A, Table 6.3, [12]. Body flap deflection: $\eta_{bf} = +40°$. Upper part: computed distribution with constant emissivity coefficient $\varepsilon = 0.8$. Lower part: computed distribution with effective emissivity ε_{eff}, taking into account non-convex effects. F is the geometrical factor which quantifies the non-convexity.

performance of turbulence models in strongly interacting flows, particularly with flow separation, in the high speed as well as in the low-speed domain, for the latter, see, e.g., [35], still need significant improvement.

Regarding the asymptotic behavior of the radiation-adiabatic temperature on the ramp for either laminar or turbulent flow no data are available to analyse. Ground-facility simulation of this temperature is not possible, [10], suitable flight data are not available. Following eq. (10.101) it is, however, permitted to assume, a $x^{-0.25n}$ behavior.

In this context non-convex phenomena, connected to surface-radiation cooling, need to be considered. At a control surface their influence will be the larger, the larger the angle of deflection is[26]. Figs. 6.24 and 6.25, [12], show results concerning these phenomena in the vicinity of the body flap of the HOPPER configuration, case A in Table 6.3, $\eta_{bf} = +40°$. In the upper part of Fig. 6.24 the computed radiation-adiabatic wall temperature distribution with constant emissivity coefficient $\varepsilon = 0.8$ is given, in the lower part that computed with the effective, or fictitious, emissivity coefficient ε_{eff}, taking into account – using the GETHRA code in the form derived in [36] – non-convex effects due to the deflected flap.

[26] We concentrate the following discussion on the primary problem which occurs with radiation cooling of the onset flow surface of the flight vehicle and of the flap surface. The problem of non-convex effects in the flap gaps or, like in the case of the X-38, on the back side of the split body flap/elevon in the fuselage cavity, needs special attention, see below.

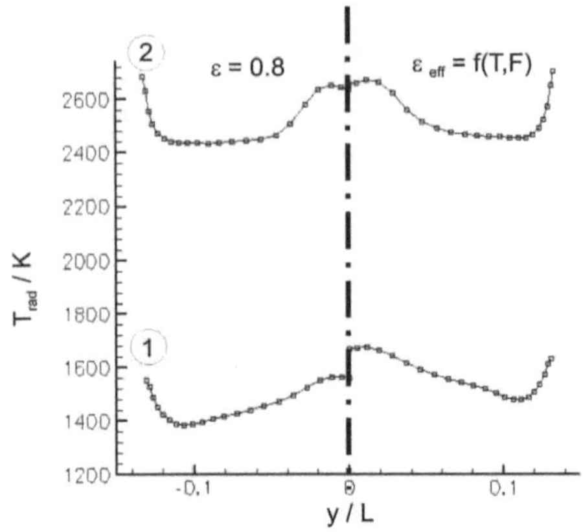

Fig. 6.25. Details of the solutions shown in Fig. 6.24, [12]: span-wise temperature distributions at position 1 in the onset flow regime, and at position 2 on the body flap, left side with constant emissivity, right side with non-convex effects, Figs. 6.24.

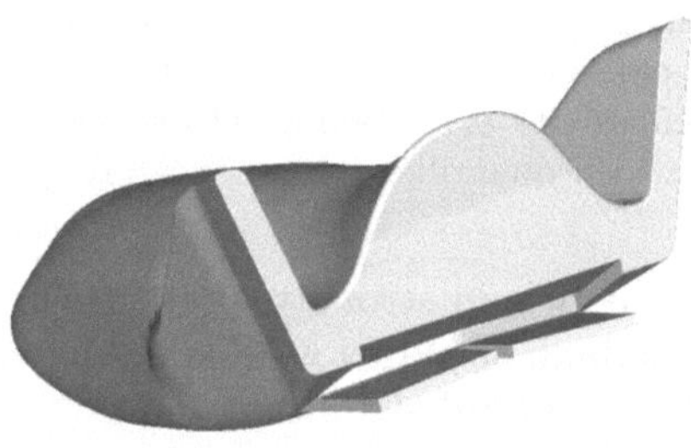

Fig. 6.26. Backside of the X-38 with the cavity above the split body flap/elevon, $\eta_{bf} = +20°$, [37].

On the lower side of the flight vehicle the radiation-adiabatic temperature decreases in both cases towards the hinge line of the body flap as expected. The very slight divergence of the skin-friction lines at the symmetry line results there in a somewhat smaller reduction of T_{ra}, Fig. 6.25. On the flap the temperature is about 1,000.0 K higher in both cases. The expected decrease of T_{ra} on the flap in downstream direction is indicated, however, with the given data we cannot verify the $x^{-0.05}$ behavior which we expect for turbulent flow.

The non-convex effects show up other than anticipated. On the flap the temperature is hardly affected at all. On the lower side of the vehicle ahead of the hinge line the effect is rather strong with $\triangle T_{ra} \approx 100.0$ to 150.0 K, with only a weak influence on the separation behavior. Obviously the high surface temperature on the flap causes a stronger reduction of the effective emissivity on the lower side of the vehicle than vice versa. For details in this regard see [10]. Because the separation behavior is not much affected by the increase of the wall temperature in the onset flow region, the flap effectiveness is not reduced. However, the layout of the TPS on the lower side of the vehicle would need to take into account such a large $\triangle T_{ra}$.

Fig. 6.25 with the span-wise temperature distributions at position 1 in the onset flow regime, and at position 2 on the body flap – left side with constant emissivity, right side with non-convex effects – shows also another effect. At the side edge of the flap, visible also in Fig. 6.25, a strong temperature rise occurs in both cases (upper curves). This is due to the acceleration of the flow into the gap which leads to a thinning of the boundary layer. The corresponding weaker rise at position 1 (lower curves) is caused by the three-dimensionality of the onset flow.

Even if we must have reservations regarding the performance of the employed basic $k-\omega$ model[27], and the mesh-independence of the solution, [12], the results show us what to expect from non-convex effects qualitatively and quantitatively. In any case they must be regarded, because thermal loads should be known as exact as possible for the TPS layout[28].

Strong non-convex effects occur at the upper (leeward) side of the split body flap/elevon of the X-38, Sub-Section 3.3.5, i. e., the cavity region at the rear, Fig. 6.26. We discuss some results of an exploratory study, [37], [38], keeping in mind the limitations due to turbulence modelling and grid sensitivity.

Table 6.11 shows the flight conditions of the from [37] selected flight cases 2 and 3. The basic emissivity coefficient of the surface material was chosen to be $\varepsilon_0 = 0.8$, the effective surface emissivity coefficient ε_{eff} again was determined with the help of the GETHRA code.

Table 6.11. Flight parameters of the numerical simulation of the flow past the X-38 configuration, [37].

Case	M_∞	H [km]	$Re_{L,\infty}$	L [m]	α [°]	η_{bf} [°]	$x/L\|_{trans}$	ε_0
2	15	54.1	$1.52 \cdot 10^6$	8.33	40.0	+20.0	fully turbulent	0.8
3	17.5	58.8	$1.02 \cdot 10^6$	8.33	40.0	+20.0	fully laminar/flap turbulent	0.8

The computed surface-pressure distribution and the skin-friction lines on the left part of the base area of the vehicle, on the left body flap, and in the cavity above the body flap are given in Fig. 6.27 for case 2. The skin-friction line topology in the cavity, although not well discernible, indicates the flow around the side edges of the flap and convergence to the attachment line (note the relative maximum of the surface pressure there) present at the rear

27 The deflection angle of the body flap with $\eta_{bf} = +40°$ is very large, however the onset flow Mach number is with $M_{e,onset} \lessapprox 3.5$ small enough such that compressibility effects should not play a role. Anyway, the main problem with turbulence models is to handle separated flow rather than compressibility.

28 Mind that at this unrealistically large deflection angle the surface temperatures at the flap, Fig. 6.25, are anyway at or beyond the limits of present-day materials.

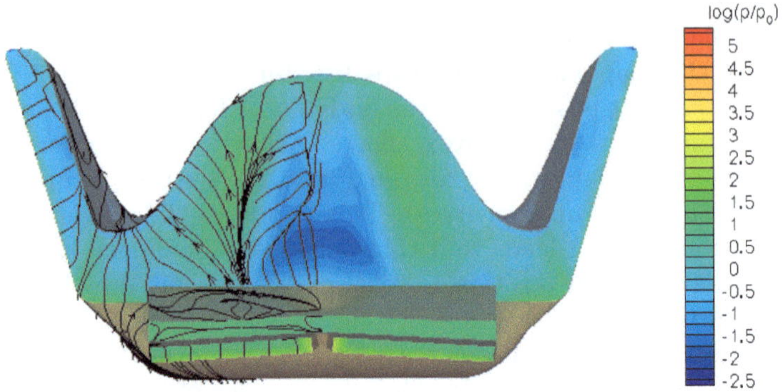

Fig. 6.27. Rear view of the X-38 with the surface-pressure distribution and skin-friction lines on the left part of the base area of the vehicle, on the body flap, and in the cavity above the split body flap/elevon, case 2, Table 6.11, [37]. Computation with ε_{eff}, $p_0 \equiv p_\infty$.

surface of the vehicle. The figure illustrates also well the hypersonic shadow effect at the base area of the flight vehicle with $p = p_w \approx 0.02$ to $0.3\ p_\infty$ at the middle axis of the area, and similarly at the blunt trailing edges of the winglets.

The resulting ε_{eff} distribution in the cavity is shown in Fig. 6.28. Note the low values along the cavity corners and especially in the vicinity of the middle gap, where even a small negative effective emissivity is indicated. At the middle of the forward wall of the cavity a region with larger ε_{eff} points to the possibility of a less impeded surface radiation.

However, looking at the ε_{eff} distribution one must remember, that the actual radiation-adiabatic wall temperature distribution affects the effective emissivity, [10]. This temperature distribution itself originates from the local boundary layer properties, which in turn are affected by the restricted radiation cooling. These – non-linear – couplings of several phenomena make a simple and comprehensive explanation of the result nearly impossible.

Finally, in order to emphasize the effect, we show in Fig. 6.29 the resulting $\triangle T_{wall} = T_{wall,\varepsilon_{eff}}$ - T_{wall,ε_0} in the cavity. For the cavity forward wall – with exception of a small peak at the symmetry plane – an average jump in wall temperature of about 500.0 K results, when we switch from $\varepsilon_0 = 0.8$ to ε_{eff}. Along the cavity ceiling the rise in temperature decreases from the forward outboard corners with $\triangle T_{wall} \approx 500.0\ K$ to values of about 100.0 K when approaching the rear end and the symmetry plane. The absolute values are $T_{wall,\varepsilon_{eff}} \approx 900.00\ K$ to 1,200.0 K.

The leeward side of the body-flap shows some variations of $\triangle T_{wall}$ and a maximum temperature increase of about 500.0 K located at the inboard half of the surface. At the outboard half the average temperature rise is about 250.0 K. Along the body flap inboard side-edge, like at the blunt trailing edge, $\triangle T_{wall}$ is in the range of 100.0 K to 250.0 K, with the maximum

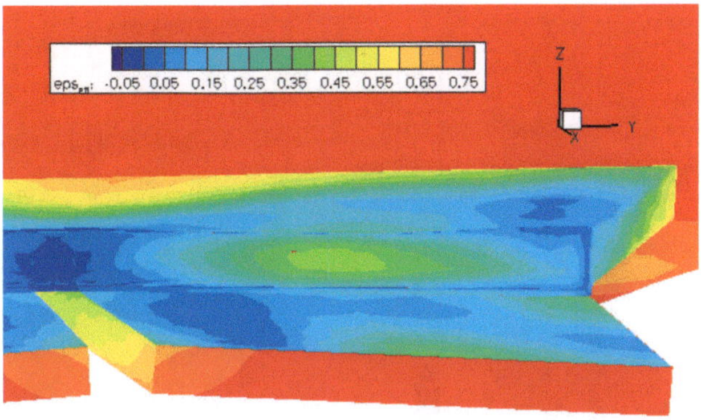

Fig. 6.28. Look from behind into the right-hand side part of the cavity of the X-38 with the distribution of the resulting effective surface emissivity ε_{eff}, case 2, Table 6.11, [37].

values reached when approaching the body flap hinge line. The largest $\triangle T_{wall}$ appears at a small portion of the cavity front wall just in front of the gap between the two split body flaps. In this place it reaches up to 1,000.0 K. This peak seems to be due to the particular surface and grid characteristics used for the computations, which were not problem-adequate there.

In [37] an attempt was made to simulate the influence of transverse heat transfer from the windward to the leeward sides of the split flaps for case 3 in Table 6.11. The windward side temperature, ranging from $T_{wall} \approx 1{,}600.0$ K to 1,950.0 K, was simply imposed on the leeside surface of the flap. This resulted, Fig. 6.30, in a partly large increase of the surface temperature at the cavity ceiling, with a number of severe hot-spot locations, reaching almost the radiation-adiabatic temperature at the forward stagnation point of the vehicle.

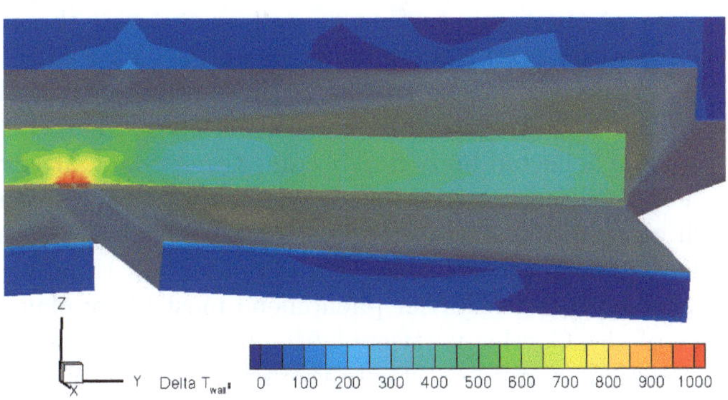

Fig. 6.29. Look from behind into the cavity on the right-hand side of the X-38 with the distribution of the resulting $\triangle T_{wall}$ due to non-convex effects, case 2, Table 6.11, [37]. The reference temperature distribution is that found with $\varepsilon_0 = 0.8$.

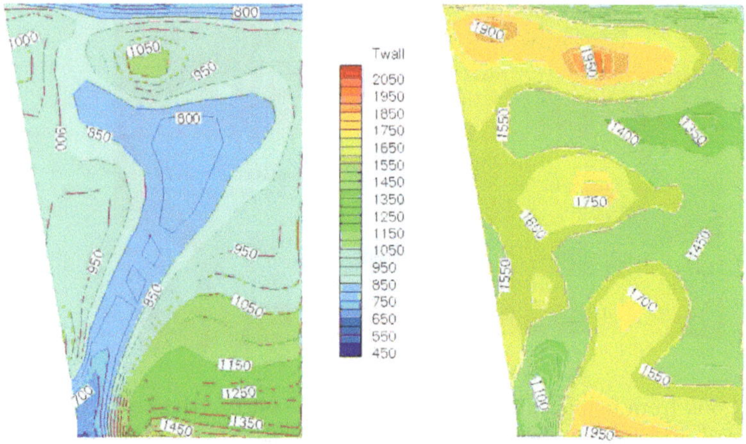

Fig. 6.30. Transverse heat transfer: distribution of the radiation-adiabatic wall temperature $T_{wall} \equiv T_{ra}$ at the ceiling of the cavity above the left body flap, case 3, Table 6.11, [37]. Left part of the figure: reference computation with ε_{eff}, right part: computation with ε_{eff} and temperature of the windward side of the body flap imposed on the lee side. View of the ceiling from below, the right-hand side edges of the figures corresponds to the flaps trailing edges, the lower edges to the (middle) gap sides.

So far we have considered either fully laminar or fully turbulent flow past control surfaces. If laminar-turbulent transition occurs in the interaction domain around the hinge line, the situation will become complicated. The experimental or theoretical/numerical prediction of the location and the extent of the transition region in attached viscous flow is still unreliable and inaccurate, see, e.g., [10]. This holds in particular when transition occurs in a flow domain with strong interaction and flow separation phenomena.

Transition in the interaction domain will affect especially the behavior of the thermal state of the ramp surface as well as the wall shear stress. If, for instance, separation occurs around the hinge line, case c), Fig. 6.9, transition may occur already in the free shear layer upstream or close to the reattachment point A[29]. This was observed at the Space Shuttle Orbiter, where the flow past the deflected body flap became transitional, at very large Mach numbers, long before transition occurred on the vehicle's windward side, [23]. In such cases the heat transfer may become even larger than for fully turbulent flow, [29]. Responsible for this is probably a thinning of the reattached boundary layer, also Görtler phenomena most likely play a role.

The occurrence of the Görtler vortices, [39], [40], [8], mentioned already on page 292, is a particular problem. These vortices are due to a centrifugal instability, see, e.g., [10]. They can appear in concave flow situations like at

---

[29] This is a phenomenon similar to bubble transition downstream of the suction peak of an airfoil.

a flat-plate/ramp (compression corner) configuration in laminar flow, transitional, and even turbulent flow, for the latter see, e.g., [3]. In hypersonic flow they can lead to the phenomenon of striation heating, i. e., transversal heat-flux variations. Results of a parametric study in [29] point to a bounding of these variations by the local turbulent heating level, if the flow would be turbulent. They decay to zero, when fully turbulent flow is attained over the entire ramp surface.

We close this part by showing the result of a numerical study, [41], [38], of the flow past the split body flap/elevon, $\eta_{bf} = 20°$, of the X-38 with Görtler vortices occurring, Fig. 6.31, [42]. It was possible to simulate the striation heating due to these vortices in laminar as well as transitional flow by employing in the flap region strongly refined computational grids in span-wise direction.

In the left part of Fig. 6.31 the striation heating on the flap surfaces, generated by the geometrically triggered vortices, is shown for a case with otherwise fully laminar flow. In the right part contour-surfaces of the streamwise perturbation velocity in the flow past the left flap are shown – with exaggerated features for a better view – which are the indicator of the presence of vortex-like disturbances. Note in the left part of the figure the large heat flux in the nose region as well as along the left and the right primary attachment lines, which is due to the thinning of the boundary layer, attributable to the flow divergence occurring there, [10].

The discussions in this part have shown that an asymptotic thermal state of the surface can be expected on a flap with optimal onset flow geometry, as long as the flow is two-dimensional, or at most only weakly three-dimensional. Unfortunately the basis of well suited experimental data is small and partly restricted to the immediate vicinity of the strong interaction domain around the hinge line.

The potentially important influence of non-convex effects of surface radiation cooling on the thermal state of the surface has been demonstrated with

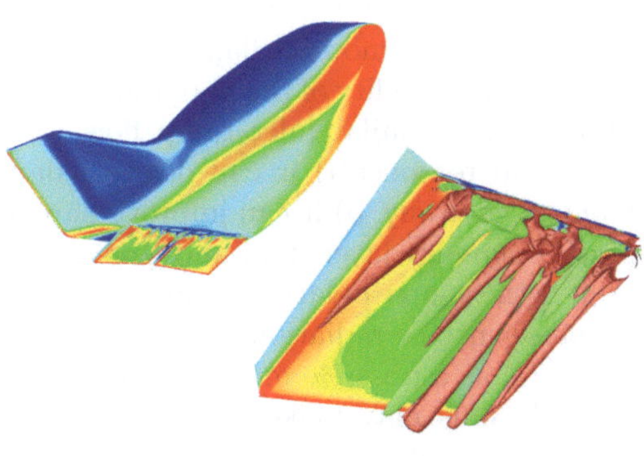

Fig. 6.31. Numerical simulation of striation heating on the split body flap, $\eta_{bf} = 20°$, of the X-38 (left part), and visualization (with exaggeration of the thickness of the flow features) of the streamwise Görtler vortices on the left flap (right part), [42].

computational data, especially for the complex cavity above the split body flap of the X-38.

In general it must be noted, that turbulent, and more so transitional flow states make the numerical, but to a large degree also the experimental simulation of control-surface flow problematic. Hence thermal loads cannot yet be predicted with the desired reliability and accuracy in all for flight-vehicle design important cases.

6.3.4 Asymptotic Behavior of the Wall-Shear Stress

The wall-shear stress exerted by the flow past the aerodynamic control surface is of large importance, if the flap material, and its coating, respectively, is thermally highly stressed. The coating of, for instance a body flap, has anti-oxidation, low catalycity, and high thermal emissivity properties, which must not be degraded by erosion processes due to the skin-friction.

If separation happens around the hinge line of the control surface, case c), Fig. 6.9, the wall-shear stress certainly is small in that region. However, downstream of the reattachment point A it will initially be large. The asymptotic behavior, found with the reference-temperature approximation, then is $\tau_w \sim x^{-n}$, Sub-Section 10.4.3. We note in addition, that the wall-shear stress depends on the wall temperature, via the reference temperature, too, with $\tau_w \sim (T^*/T_e)^m$, where $m = n(1+\omega)-1$, which is always smaller one. Increasing T_w, and hence T^*, reduces τ_w. The effect is much stronger for turbulent flow, the exponent being there m = - 0.670, than for laminar flow with m = - 0.175, ω in both cases being 0.65, Section 10.2. Experimental and numerical data regarding the asymptotic behavior are available for the domain just downstream of the hinge line, respectively the reattachment point, but we don't pursue this topic further.

6.4 Hinge-Line Gap Flow Issues

Between the vehicle structure and the movable flap is a gap of a certain width, Fig. 6.1. Its role is to insure the movability of the flap. Flow through the hinge-line gap, called bleed-flow in the similar inlet ramp flow problem, Section 4.5.5, is of interest for several reasons: a) it can degrade flap effectiveness because it takes flow momentum away, b) it can improve flap effectiveness, if it removes low momentum boundary layer flow, and thus reduces separation around the hinge line, c) it is the source of large thermal loads on the structure surrounding the gap, because surface radiation cooling is not possible there.

Point b) is discussed in [23] as a possible means *to destroy separation and increase flap effectiveness*[30]. This may make necessary a hinge gap width larger than needed for flap movability. Then also inviscid flow may enter the gap, and the thermal loads problem is enhanced. The reason is, that this

[30] The reader is asked to recall the discussion of Fig. 6.10 on page 299 regarding the effect of separation.

flow, other than the boundary layer flow, carries the original total enthalpy. Only the surface-near part of the boundary layer flow has lost total enthalpy due to surface radiation cooling.

A closer look at the problem shows, that one should – if this is possible at all due to the overall layout of the flight vehicle – bleed away low-momentum boundary layer flow material already ahead of the hinge gap, rather than through it, Section 4.5.5. This would improve the integral momentum of the onset flow boundary layer, like, as discussed in [23], a turbulence tripping of the boundary layer would, if it is laminar ahead of the flap. Turbulence tripping on the other hand is effective only at low boundary layer edge Mach numbers ($M_{e,onset} \lessapprox 5$). Therefore it could be a remedy for flap separation when employed at re-entry vehicles, but not at CAV- or ARV-type vehicles with large onset flow Mach numbers.

In the following we discuss modelling issues of gap flow heating. The flow past the reference configuration 'X-38 body flap with hinge-line gap' was studied experimentally and numerically in the German TETRA Programme, Sub-Section 8.4.3. Geometry and arrangement of the partially cooled generic flap model, [43], are shown in Fig. 8.19, the dimensions of the model and the (final) computational grid in Fig. 6.32, [44]. The width of the model was $w_{flap} = 0.144\ m$.

The blunt-nosed flat-plate part of the configuration was oriented at angle of attack, $\alpha = 15°$, with the flap deflected with $\eta_{bf} = +20°$. The flow parameters are given in Table 6.12. The selected computation results discussed in the following were obtained assuming two-dimensional laminar flow, no-slip boundary conditions, and a non-equilibrium real-gas model with non-catalytic wall. The radiation-adiabatic surface boundary condition was employed, and non-convex effects were taken into account.

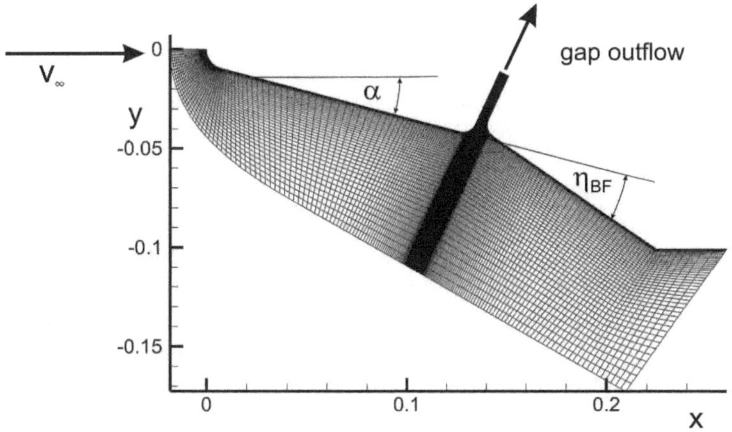

Fig. 6.32. Geometry and computational grid of the generic flap model, [44], $\alpha = 15°$, $\eta_{bf} = +20°$, gap width 5.0 mm. The axis dimensions are given in meters.

Table 6.12. Flow conditions of the tests with the generic flap model in the plasma tunnel L3K of DLR, [44]. Test gas is air.

M_∞	p_t $[kPa]$	T_t $[K]$	h_t $[MJ/kg]$	p_∞ $[Pa]$	ρ_∞ $[kg/m^3]$	T_∞ $[K]$	α $[°]$	η_{bf} $[°]$	ε_0
7.27	570.0	6,255.0	5.44	86	0.000387	620.0	15.0	+ 20.0	0.85

The distribution of the effective emissivity coefficient ε_{eff} on the whole surface of the model is shown in Fig. 6.33. This is the initial ε_{eff} distribution, which was found assuming $T_w = 1{,}000.0\ K$. It was updated during the computation with the actual computed radiation-adiabatic temperature distribution. The original material emissivity of the surface is $\varepsilon_0 = 0.85$. Radiation from the gap bottom was suppressed. Strong non-convex effects are observed only in the gap region, Fig. 6.34. The effective emissivity coefficient is seen to drop fast to almost zero deep in the gap, as is to be expected.

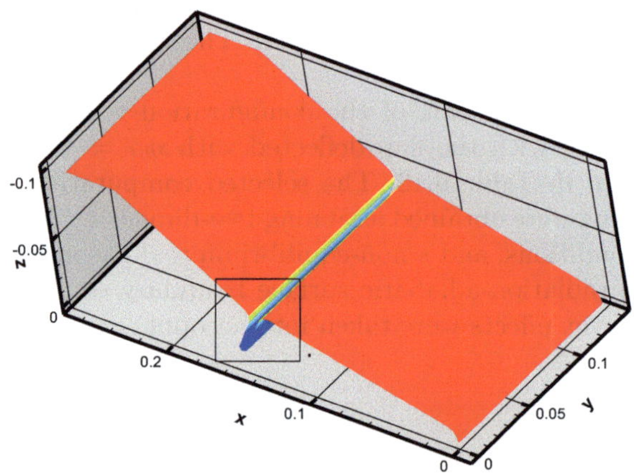

Fig. 6.33. Distribution of the initial effective emissivity coefficient ε_{eff} on the whole surface of the model, [44], [38]. The flow is coming from the right-hand side.

The open gap computations were made with an outlet pressure of 100.0 Pa. A change to 500.0 Pa had no noteworthy influence on the solution. We see in Fig. 6.35 at the left the high pressure at the blunt nose of the configuration. The small kink at $x \approx 0.025\ m$ indicates the recompression shock due to the jump of the curvature at the blunt-nose/flat-plate junction. The pressure decreases slightly in downstream direction, pointing to possible hypersonic viscous interaction and blunt-nose effects. With closed gap the wall pressure departs at $x \approx 0.075\ m$ – there the upstream separation occurs, Fig. 6.9 c) – to form the pressure plateau ahead of the hinge line. Downstream of the hinge line it rises then monotonically without reaching a ramp pressure plateau.

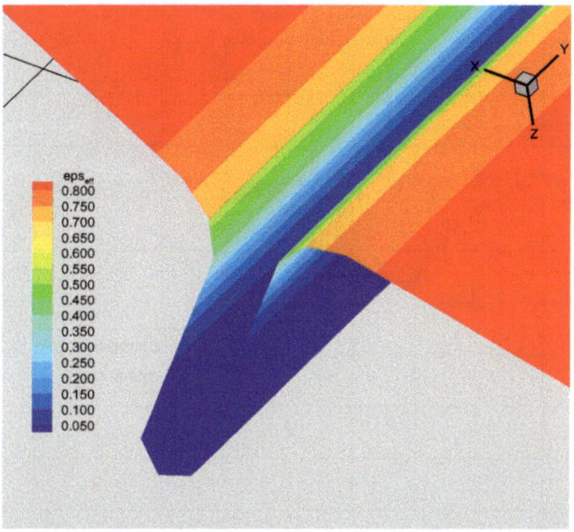

Fig. 6.34. Detail from Fig. 6.33: distribution of the initial effective emissivity coefficient ε_{eff} in the gap region, [44], [38].

With open gap the picture changes completely. Separation does not occur upstream of the hinge line, the wall pressure rises only weakly just ahead of the gap. In the gap a small separation bubble forms, see Fig. 8.20, Sub-Section 8.4.3. On the ramp then we have the pressure rise on a higher level than before. In general we find the situation mentioned on page 320 as point b), discussed in [23]: a wide enough gap around the hinge line takes away the upstream separation and, in addition, rises the flap pressure, thus increasing the flap effectiveness. A question not answered is, however, the asymptotic behavior, i. e. whether on a flap long enough with gap the same pressure plateau is reached as without gap.

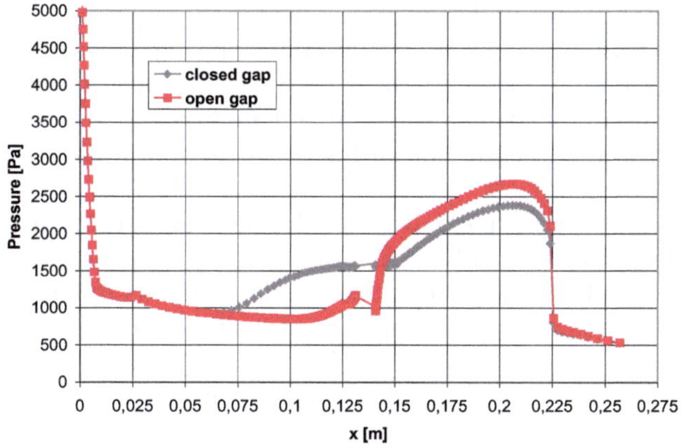

Fig. 6.35. Computed surface pressure distribution with closed and with open gap, gap region omitted, [44].

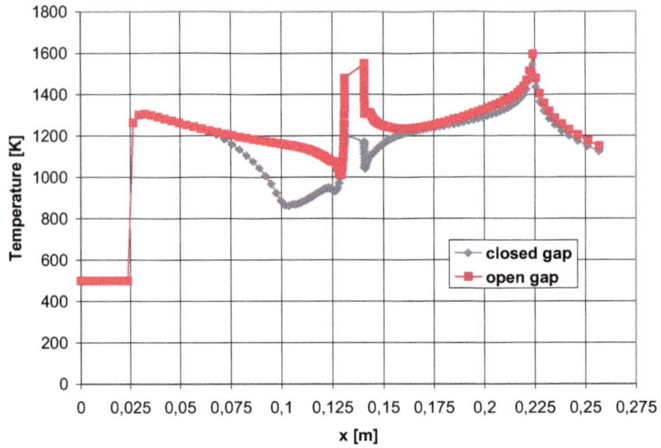

Fig. 6.36. Computed radiation-adiabatic surface temperature distribution with closed and with open gap, gap region omitted, [44].

This effect, however, is bought with the above mentioned increase of the thermal loads in the gap, point c), page 320. With closed gap, Fig. 6.36, we find the drop of the radiation-adiabatic wall temperature beginning at the location of separation, $x \approx 0.075\ m$ – compare with Fig. 6.12 – which rises downstream of $x \approx 0.1\ m$ again to about $T_{ra} \approx 1{,}000.0\ K$ just ahead of the gap. Downstream of the gap it rises to a peak at the end of the ramp part at $x = 0.225\ m$ of the configuration, without attaining on the ramp the asymptotic decrease as seen for instance in Figs. 6.20 and 6.21. At the right-hand side of the gap a small temperature peak, just below $1{,}200.0\ K$, is indicated.

With open gap the radiation-adiabatic temperature ahead of the hinge line follows longer than in the case without gap the $x^{-0.125}$ proportionality of laminar flow, Sub-Section 10.4.3, and then drops towards the small separation bubble at the forward wall of the gap, Fig. 8.20. Along the gap walls it reaches, as far it can be represented in Fig. 8.20, the expected high values, here $1{,}500.0$ to $1{,}600.0\ K$. Downstream of the gap the temperature first drops and then rises similar as in the closed gap case, but with slightly larger values, towards the peak.

The experiment with the partially cooled ramp configuration was performed in the plasma tunnel L3K of DLR, [43]. The steady state temperature distribution, reached after about $325.0\ s$, of course is different to that discussed above. To take the whole experimental situation into account, the multidisciplinary simulation sketched in Sub-Section 8.3.3 becomes necessary. It is shown there, see also, e.g., [45], that a three-dimensional flow-structure coupled simulation is necessary to capture the model-in-tunnel situation. Although the obtained accuracy was not yet satisfactory, see also Sub-Section 6.7.2, the modelling of the influence of heat conduction into the wind-tunnel model and a full description of flow three-dimensionality was deemed to be necessary. In a gap and its immediate vicinity, the radiation-adiabatic temperature, even with non-convex effects taken into account, is no more a sufficient and reliable approximation as it is in general for the thermal state of the wall of external radiation cooled vehicle surfaces, [10]. This holds also for gaps of, e.g., a TPS.

6.5 Aerothermodynamic Issues of Reaction Control Systems

Reaction control systems (RCS) of RV-W- and RV-NW-type flight vehicles have two purposes:

- manoeuvering of the vehicle in orbit, including the de-orbiting manoeuvre, where control forces cannot be generated aerodynamically,
- control of the vehicle during re-entry, Section 2.1, on at least a part of the trajectory (RV-W-type vehicles), or on the whole trajectory (RV-NW-type vehicles).

A RCS thruster produces either a force or a force and a moment on the vehicle. Side effects in orbital flight are, for instance:

- stray forces because of interaction with vehicle surfaces (jet impingement), Fig. 6.37 right,
- contamination of vehicle surfaces with plume constituents.

During re-entry flight these side effects can occur, but other, more important effects are:

- the amplification or even reduction of the thruster force by interaction with the flow field past the flight vehicle, Fig. 6.37 left,
- the amplification of thermal loads on the vehicle structure, especially the creation of hot-spot situations by the said interaction.

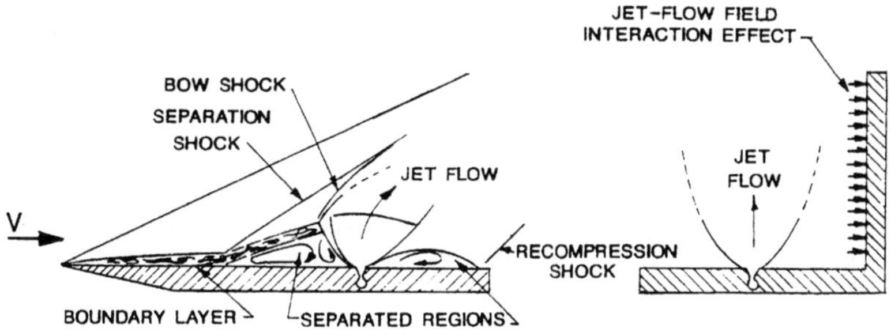

Fig. 6.37. Schematics of interactions of a RCS jet embedded in a supersonic flow field, [46]. Left: interaction of the jet with the surface flow field, [47], right: interaction of the jet with an adjacent surface.

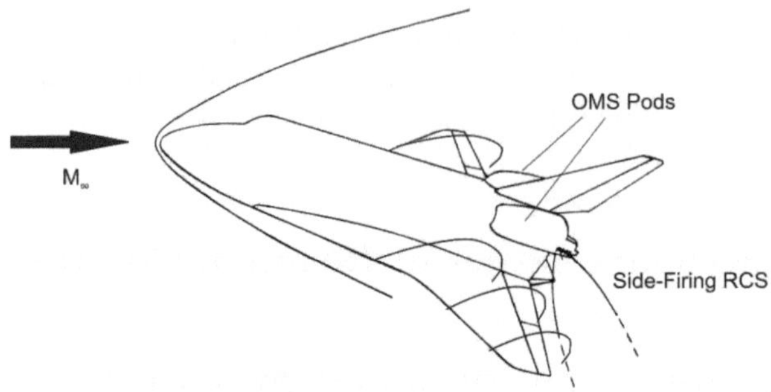

Fig. 6.38. Sketch of the side-firing jet situation at the Space Shuttle Orbiter at $\alpha \approx 40°$, after [46].

The RCS performance was shown for the Space Shuttle Orbiter in general to be predictable with the help of ground-facility simulation, [48]. Many studies of the jet/surface flow field interaction shown in Fig. 6.37 left, can be found in the literature, see, e.g., [47]. Much data is available for both laminar and turbulent surface flow fields.

The first Orbiter flight however has revealed a particular problem related to the flow situation sketched in Fig. 6.37 right. It arose during the initial bank maneuver with the side-firing RCS thruster housed in the Orbital Manoeuvering System (OMS) pod. The produced rolling moment was smaller than predicted. The sketch in Fig. 6.37 right, is in so far misleading, as in this case at large flight Mach number and $\alpha \approx 40°$, the surface in question was the upper side of the wing. The flow above and behind the wing at this flight attitude is massively separated, Fig. 6.38, the Reynolds number is very small, [46].

RCS performance for yaw and roll control of the Space Shuttle Orbiter, as we have noted in Chapter 2, is needed due to its central vertical stabilizer (and rudder) down to flight Mach numbers $M_\infty \approx 1$. Its ground simulation is complicated by the large Mach/Reynolds number span present together with changing free-stream dynamic pressure and large angles of attack.

We don't attempt to discuss here the flow phenomena and simulation issues of this particular problem. We restrict ourselves in the following to the discussion and illustration of the basic interaction phenomena, Fig. 6.37 left, of reaction control systems during re-entry flight of the vehicle. We do this in view of thruster force amplification/reduction and thermal loads amplification at two different locations of the X-38 configuration. The material and the data stem from computational studies performed in the frame of the German national technology programme ASTRA, [49] to [51].

The flight parameters are given in Table 6.13. The combinations of angle of attack, thruster and thruster locations, Tables 6.15 and 6.14, do not reflect flight situations. They were chosen – in view of the influence of the onset flow – such that different jet/flow field interactions occur, the strongest in case 1, the weakest in case 3.

Table 6.13. Flight parameters of the computational study of jet/flow field interaction at the X-38 configuration, [49]. The flow is assumed to be laminar and in thermo-chemical equilibrium. No-slip and radiation-adiabatic boundary conditions are applied at the vehicle's surface.

Case	M_∞	H $[km]$	p_∞ $[Pa]$	T_∞ $[K]$	ρ_∞ $[kg/m^3]$	L $[m]$	ε	Thruster	α [°]
1	17.5	58.8	26.19	258.08	$3.546 \cdot 10^{-4}$	8.33	0.8	1	0.0
2	17.5	58.8	26.19	258.08	$3.546 \cdot 10^{-4}$	8.33	0.8	2	0.0
3	17.5	58.8	26.19	258.08	$3.546 \cdot 10^{-4}$	8.33	0.8	2	40.0

Two thrusters at different locations at the front part of the vehicle were considered, Fig. 6.39. Thruster 1 has a nominal design thrust of 110.0 N, and thruster 2 of 400.0 N, Table 6.15. The location of thruster 1 is the original one. The location of thruster 2, of course is an unlikely one, but it was chosen in comparison to that of thruster 1 such that:

- the vehicle surface at that position is less curved longitudinally,
- a different jet/flow field interaction occurs,
- the vehicle reaction moments are similar.

The geometrical data of the thrusters are shown in Table 6.14. The coordinates of the reference point, the center-of-gravity of the flight vehicle, are: $x_{ref} = 4.7951$ m, $y_{ref} = 0.0$ m, $z_{ref} = 0.8042$ m, with the origin of the coordinate system lying 15.24 cm ahead of the nose point in the lower-side plane of the vehicle.

The exit planes of the thrusters are flush with the vehicle surface, the nozzle axes are oriented parallel to the z-axis of the vehicle. The thruster jets are assumed to leave their nozzles without profile and boundary layer. The

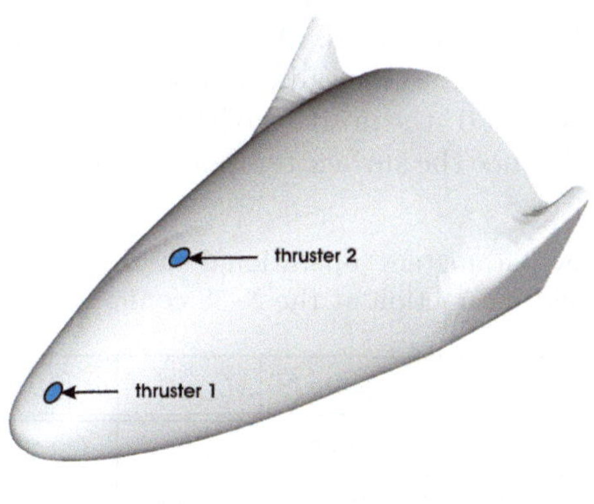

Fig. 6.39. Locations of the thrusters at the X-38 configuration, [49].

Table 6.14. The geometrical data of the thrusters of the computational study of jet/flow field interaction at the X-38 configuration, [49]. The distance to the reference point is denoted with $\triangle x$.

Thruster	Exit diameter	Exit area A_{exit}	x-position	$\triangle x$
1	17.13 mm	$2.30 \cdot 10^{-4}\ m^2$	0.6351 m	4.16 m
2	244.0 mm	$4.68 \cdot 10^{-2}\ m^2$	2.5237 m	2.2714 m

thruster flow parameters are given in Table 6.15. The thrust $F_{z,jet} = |\underline{F}_{z,jet}|$ was computed with:

$$F_{z,jet} = \dot{m}\, u_{exit} + (p_{exit} - p_{env})\, A_{exit}, \tag{6.11}$$

where the environment (ambient) pressure was chosen to be $p_{env} = 0.0\ Pa$.

In the following we discuss selected results from [49], [50], details can be found there. We have first a look at results from a test case presented in [50]. It was computed with the parameters from case 2, Table 6.13, with the thruster located on a flat surface. The jet's exit pressure was ten times enlarged in order to enhance its influence on the surface flow field.

In Fig. 6.40 the surface pressure, skin-friction lines and, in the symmetry plane, the jet and streamlines are shown. The strong jet leaves the surface perpendicular, the surface flow comes from the lower right side. We see – compare to the cross-section view in Fig. 6.37 left – the primary separation line ahead of the jet, the separation shock, the (primary) flat horse-shoe vortex, the primary attachment line, and then, directly ahead of the jet, a second small horse-shoe vortex. This vortex is directly driven by the jet. The skin-friction lines go around the jet, and finally a separation line indicates the separation region behind the jet. Well discernible is the pressure minimum induced by the primary horse-shoe vortex, as well as the relative pressure maximum along and around the attachment line, which is typical for any attachment line.

The flow situation of the test case is found approximately also in case 1 of Table 6.13, Fig. 6.41. Here we look from above on the surface. On the right-hand side the skin-friction lines and the radiation-adiabatic temperature without thruster are shown. At $\alpha = 0°$ the position of thruster 1, Fig. 6.39, is directly impacted by the flow and hence the surface temperatures are around 1,600.0 K.

Table 6.15. Thruster flow parameters (operational environment) of the computational study of jet/flow field interaction at the X-38 configuration, [49].

Thruster	p_{exit} $[Pa]$	T_{exit} $[K]$	u_{exit} $[m/s]$	M_{exit} $[-]$	$\dot{m}$ $[kg/s]$	$F_{z,jet}$ $[N]$
1	16,210.0	56.5	640.5	4.25	0.148	98.5
2	192.1	728.5	3,243.4	5.99	0.139	459.8

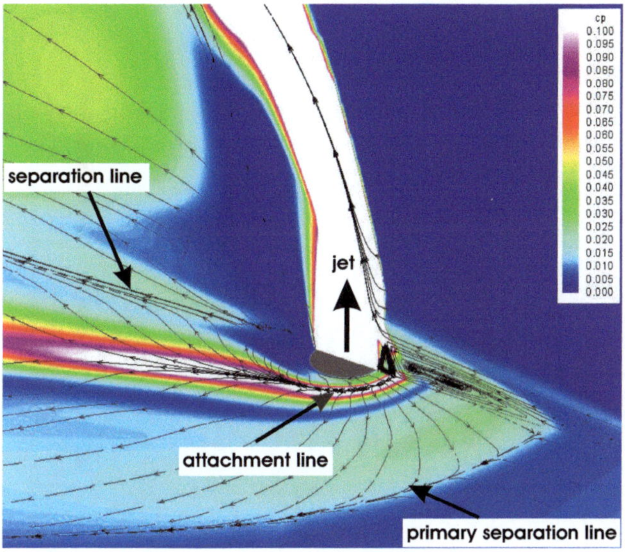

Fig. 6.40. Case 2: thruster 2 with p_{exit} = 1,921.0 Pa, perfect gas assumption and flat surface (test case): pressure iso-surfaces, streamlines and skin-friction lines, [49], [50]. The main flow direction is from the lower right to the upper left side.

The left-hand side of the figure now shows a remarkable amplification – and variation – of the surface temperature due to the jet/flow field interaction. The divergent flow at the (primary) attachment line causes a strong thinning of the viscous flow, which leads to a local hot-spot situation with radiation-adiabatic temperatures around 2,500.0 K, whereas on a large surface portion downstream of it temperatures around 2,000.0 K are found. At the primary separation line ahead of the jet, below the primary horse-shoe vortex, and also directly behind the jet, cold-spot situations with temperatures around 1,300.0 K are indicated.

For case 2 of Table 6.13, Fig. 6.42, we observe a pattern different to that of case 1, Fig. 6.41. The temperature field on the right-hand side indicates that thruster 2, with a jet stronger than that of case 1, Table 6.15, lies already somewhat in the hypersonic shadow of the fuselage. Consequently no upstream jet/flow field interaction occurs (left side of Fig. 6.42). The streamlines directly ahead of the jet are straight. This points to a low-momentum onset flow and an ejector-like effect of the jet.

Of particular interest are here the hot-spot situations. We find such a situation directly at the jet, and then behind the jet along part of the symmetry line, where the attachment line shows a strong divergence of the skin-friction lines. The ensuing temperatures are about 500.0 K larger than without jet.

We observe low-momentum onset flow and an ejector-like effect of the jet also for case 3 of Table 6.13 in Fig. 6.43. The onset flow of the jet is even less

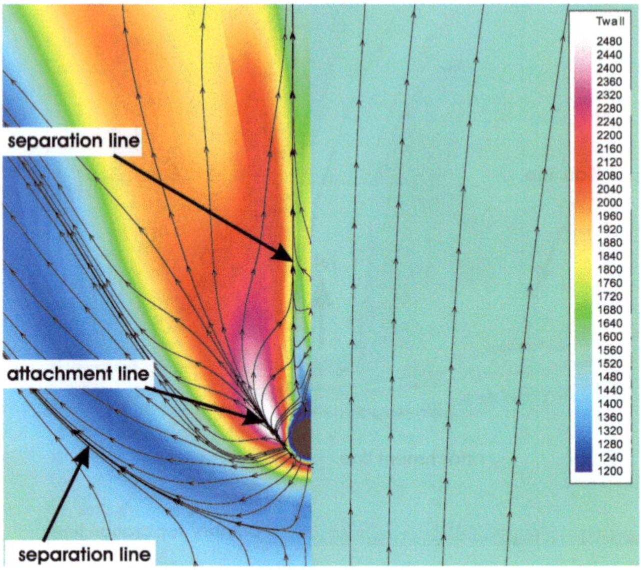

Fig. 6.41. Case 1: X-38 surface with temperature iso-surfaces and skin-friction lines, $\alpha = 0°$. Left side: flow field with thruster 1, right side: flow field without thruster, [49] to [51].

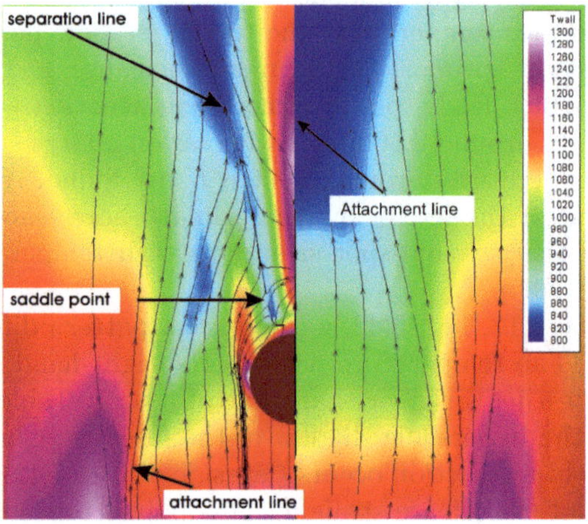

Fig. 6.42. Case 2: X-38 surface with temperature iso-surfaces and skin-friction lines, $\alpha = 0°$. Left side: flow field with thruster 2, right side: flow field without thruster, [49], [50]. The color code of the surface temperature is different from that in the other figures.

affected by the jet. The reason is, that now, at $\alpha = 40°$, the thruster location is fully in the hypersonic shadow of the fuselage, and, as indicated at the right-hand side of the figure, at the beginning of a lee-side separation domain.

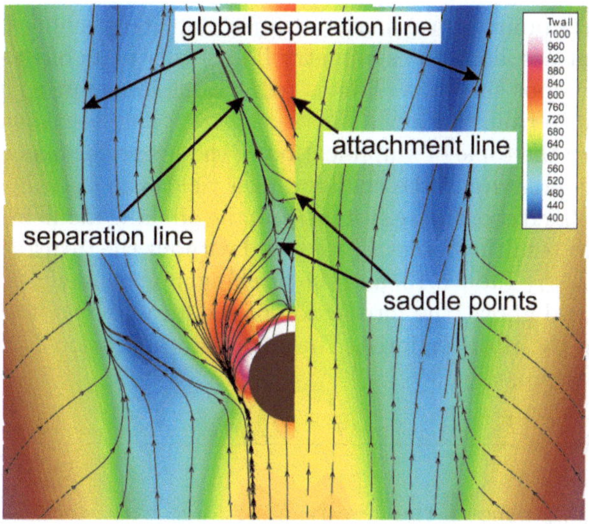

Fig. 6.43. Case 3: X-38 surface with temperature iso-surfaces and skin-friction lines, $\alpha = 40°$. Left side: flow field with thruster 2, right side: flow field without thruster, [49], [50]. The color code of the surface temperature is different from that in the other figures.

Table 6.16. Normal forces F_z ($\equiv N$) and pitching moment coefficients C_m of the X-38 configuration with and without thruster jet, [49].

Case	α [°]	Thruster	$F_{z,jet}$ $[N]$	F_z $[N]$	C_m $[-]$
-	0.0	no	0.0	-3,920.88	0.00213
1	0.0	1	- 98.5	-4,155.80	0.00196
2	0.0	2	- 459.8	-4,312.10	0.00129
-	40.0	no	0.0	65,955,20	0.0653
3	40.0	2	-459.8	65,475.62	0.0643

The global separation line, being the primary separation line at the upper part of the fuselage, Fig. 6.43, right side, leads to a well pronounced cold-spot situation. This is asymmetric, because of the different properties of the two boundary layers involved in this primary separation, which is of "open separation" type, [52], see also the discussion of the Blunt Delta Wing (BDW) example in [10]. The left-hand side of Fig. 6.43 still shows the open-separation type, although distorted by the jet action.

At the side of the jet we see an attachment line like pattern of the skin-friction lines, accompanied by a hot-spot situation, which increases towards the rear part of the jet. Due to the particular skin-friction line topology, with an attachment point approximately 1.5 jet diameters behind the jet, an attachment line ensues along the symmetry line downstream of this point. It leads also to a hot-spot situation. This is remarkable in so far, as it lies at a rather large distance from the jet.

We come now to the topic of amplification of the jet performance. Integration of the surface pressure of the configuration yields the force F_z ($\equiv N$) in z-direction (body-fixed coordinate system) and the pitching moment M. In Table 6.16 we give the resulting forces F_z with and without the thruster jet forces $F_{z,jet} = - F_{thruster\,jet}$, Table 6.15, and the pitching moment coefficient C_m.

As was mentioned above, the interaction of the jet with the flow field can amplify or reduce the thruster force. The thrust amplification factor A_{F_z} is defined in the following way:

$$A_{F_z} = \frac{F_{z,config,with\,jet} - F_{z,config,without\,jet}}{F_{z,jet}}, \tag{6.12}$$

here $F_{z,config,with\,jet}$ denotes the aerodynamic force with working thruster, $F_{z,config,without\,jet}$ that without working thruster.

With the forces from Table 6.16 we find the amplification factors in Table 6.17. In case 1, with strong jet/flow field interaction, we find an amplification larger than two. The smaller interaction in case 2, where the jet is deflected such that it impinges on the surface, [49], leads to a de-amplification. In case 3, where the jet is fully in the hypersonic shadow, the interaction is weaker, but the jet is not deflected down to the surface. Here we get a small amplification. The pitching moment coefficient C_m is reduced accordingly with approximately seven per cent in case 1, forty per cent in case 2, and two per cent in case 3.

Table 6.17. Thruster amplification factors A_{F_z} from the computational study of jet/flow field interaction at the X-38 configuration, [49].

Case	α [°]	Thruster	A_{F_z} [−]
1	0.0	1	2.38
2	0.0	2	0.85
3	40.0	2	1.04

6.6 Configurational Considerations

Hypersonic vehicles of either type basically have the same stabilization, trim, and control surfaces which ordinary aircraft have. Elevators, ailerons, and rudders are control surfaces for pitch, roll, and yaw motion, respectively. Functions may be combined in multi-functional control surfaces, while in special arrangements control surfaces may serve as airbrakes for energy modulation, too. Longitudinal trim can be achieved with elevators and ailerons together. For RV-W-type vehicles, however, the Space Shuttle Orbiter did set a precedent by using the body flap as primary longitudinal trim device. Originally it was to act as a heat shield for the main engines nozzles only.

Both the blunt RV-W-type and the slender CAV-type flight vehicles usually have delta or double-delta shaped wings[31] which are fully blended with the fuselage at least on the lower (windward) side. The limiting case of RV-W-type vehicles is the lifting body with, because of a small lift-to-drag ratio, restricted cross-range capabilities, Section 2.1, and limited low-speed properties. A good example is the X-38 vehicle, Sub-Section 3.3.5.

In neither flight vehicle class is a horizontal tail unit found as we have it at subsonic and transonic aircraft. The reasons are the large thermal loads, structural weight considerations and, with airbreathing CAV-type vehicles, propulsion integration issues, and last but not least the low and critical stability contributions of tail units in combination with low aspect-ratio wings.

6.6.1 Examples of Stabilization, Trim and Control Devices

Aerodynamic pitch trim of RV-W-type vehicles usually is made with the body flap, Fig. 6.44 a) and b). Longitudinal control is achieved by means of the so called elevons. These are multi-functional devices which act as elevators (symmetric deflection for pitch control) or ailerons (asymmetric deflection for roll control). On the Orbiter and the HOPPER configuration they are located at the wing's trailing edges, and split into outboard and inboard surfaces, Fig. 6.44 a) and b). A special case is the X-38, Fig. 6.44 c). Here two surfaces at the lower end of the vehicle act as split body flap and elevons.

Aerodynamic yaw (directional) stability and control of RV-W-type vehicles is achieved with the help of single or dual vertical stabilizer surfaces and rudders, located at the back of the upper side of the vehicle, Fig. 6.44. The Orbiter and the HOPPER configuration have a single vertical stabilizer, the X-38 has a dual vertical stabilizer in the form of wing-tip surfaces. Such wing-tip solutions are also found on a number of older vehicle projects, see, e.g., [23], and also on the European HERMES vehicle.

Roll control is made, with the elevons in aileron function, in addition with the rudder(s) at high angles of attack (avoidance of adverse yaw-roll coupling), roll damping is mainly achieved with the wing. The rudder of the Orbiter is a split control surface. Both panels deflected together in the same direction act as rudder for yaw control, and deflected symmetrically to the left and the right as speed brake (air brake) for drag modulation. The HOPPER configuration has speed brakes at the sides of the aft fuselage.

As was noted and discussed in Sub-Section 2.1.2 and in Section 6.5, RV-W-type flight vehicles need a reaction control system for different stabilization

[31] Such shapes provide large leading edge sweep for high speed (supersonic and hypersonic) flight, but lead to the necessary low-speed properties of the low-aspect ratio wings, and in the case of CAV-type vehicles permit also the control of the neutral-point shift with flight Mach number.

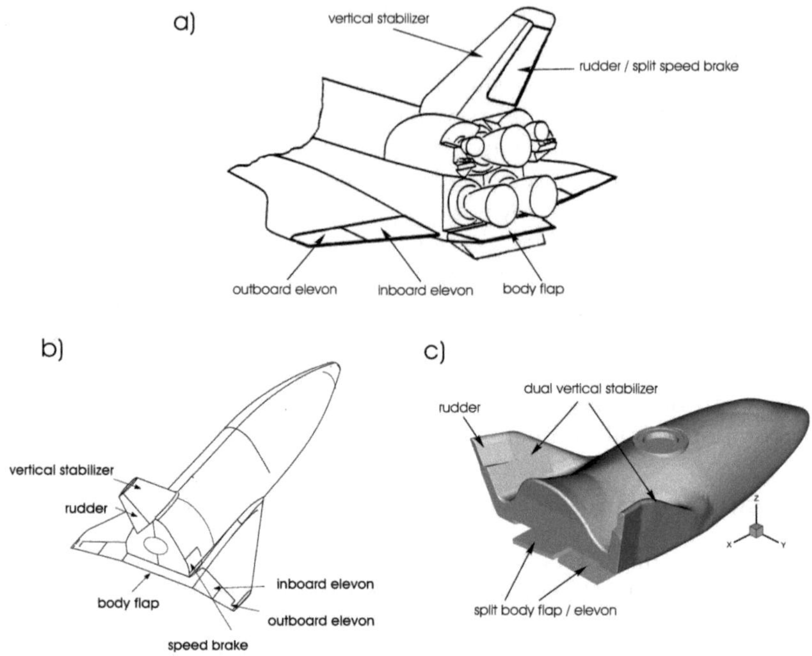

Fig. 6.44. Trim, stabilization, and control surfaces of a) the Space Shuttle Orbiter, [53] (body flap shown in two positions), b) HOPPER, and c) X-38. Deflection and hinge moments characteristics of the Orbiter surfaces are given in Table 6.18.

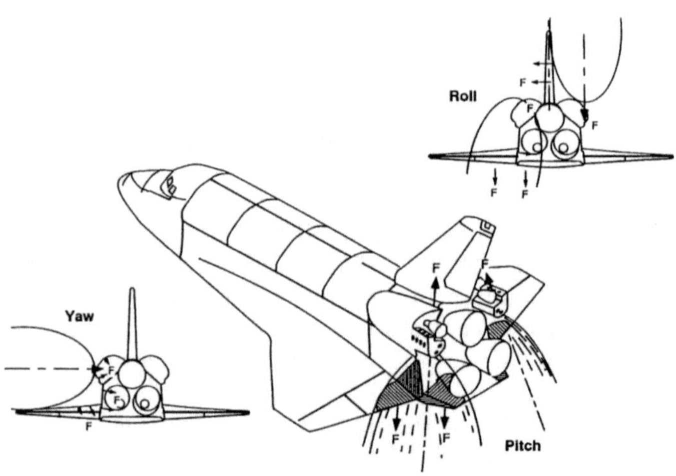

Fig. 6.45. The aft mounted reaction control system of the Space Shuttle Orbiter, [54].

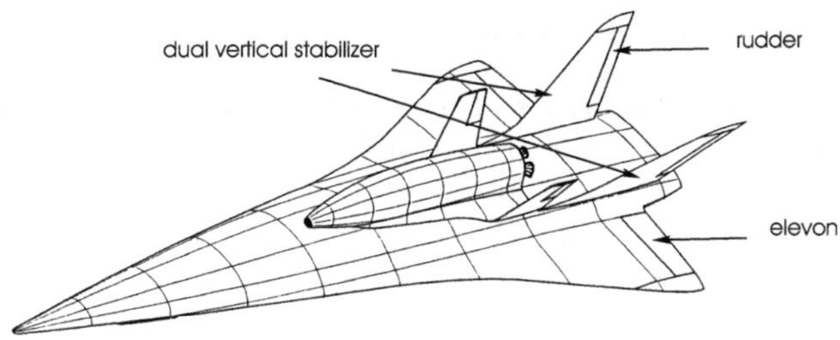

Fig. 6.46. Stabilization and trim/control surfaces of the lower stage of the TSTO system SÄNGER, [55]. The upper side of the vehicle is shown with the upper stage HORUS.

and control objectives on parts of the re-entry trajectory. We show as example in Fig. 6.45 the configuration of the Space Shuttle Orbiter's RCS for pitch, yaw, and roll control.

Aerodynamic trim, stabilization and control of CAV-type and ARV-type flight vehicles is made basically with the same surfaces as we have seen on RV-W-type vehicles. However, the trim surface (body flap) function usually is taken over by the elevon surfaces, with a strong coupling with the thrust vector, Sub-Section 2.2.3. A dual vertical stabilizer was foreseen, for instance, for the lower stage of SÄNGER, Fig. 6.46.

Table 6.18. Deflection and hinge moments characteristics of the Space Shuttle Orbiter's aerodynamic trim and control surfaces, [57]. Upward moments like downward deflection angles are counted positive, see also Fig. 6.47.

Device	Deflection angle [°]	Deflection rate [°/s]	Hinge Moment [mN]
Body flap	+ 22.5 to - 11.7	1 to 3	> - 158,172.0
Inboard elevon			
as elevator	+ 20 to - 35	20	- 103,150.7 to + 89,932.1
as aileron	+ 10 to - 10	20	- 103,150.7 to + 89,932.1
Outboard elevon			
as elevator	+ 20 to - 35	20	- 49,372.0 to + 43,610.0
as aileron	+ 10 to - 10	20	- 49,372.0 to + 43,610.0
Rudder	+ 27.1 to - 27.1	14	< + 90,384.0
Speed brake	0 to + 98.6	6 to 11	> + 282,450.0

6.6.2 Geometrical Considerations

The overall geometrical peculiarities of hypersonic vehicles tell us, that the so called flap volume, the product of flap surface (plan area) times the distance (moment arm) from its center-of-pressure to the respective axis, which is oriented at the center-of-gravity of the flight vehicle, as a rule is rather small. This in general holds for all three axes and for stabilization surfaces, too. For either vehicle class the center-of-gravity is located rather far back. Characteristic numbers for RV-W-type vehicles are 57 to 70 per cent aft position, Table 3.3. HORUS, the upper stage of the TSTO space transportation system SÄNGER, carries both hydrogen and oxygen for its ascent after separation from the lower airbreathing stage, and thus was projected to have a center-of-gravity location of approximately 54 per cent aft position at launch and approximately 60 per cent at re-entry, [55].

The situation is different with airbreathing CAV-type flight vehicles. Here the center-of-gravity position is more aft due to the location of the propulsion package. For the SÄNGER lower stage alone but with fuel, for instance, we find approximately 70 per cent, and for the whole TSTO-system at upper stage separation (nominal at $M_\infty = 6.7$) approximately 73 to 74 per cent, [55].

For faster hypersonic aircraft this may change, Fig. 4.25, because the surface of the asymmetric external nozzle (single expansion ramp) of airbreathing CAV-type vehicles increases with increasing flight Mach number, [56]. Therefore with higher design flight Mach number the propulsion package, and hence the center-of-gravity will be located more forward.

The small moment arms, which we generally find on aerospace vehicles, compared to those of ordinary aircraft, demand stabilization or control surfaces which generate large forces. This can be achieved with a large size and/or, regarding only control and trim surfaces, with large deflection angles[32]. Concerning their size, stabilization and control surfaces must fit into the general layout of the flight vehicle. To get sufficient flap effectiveness at large deflection angles demands appropriate flap sizes and aspect ratios, too. Structural weight is another concern, and finally weight, volume, performance (hinge moments, flap deflection rates), and the energy demand of the actuators of the control surfaces require special attention.

As example for the demands on the actuators of a RV-W-type flight vehicle, deflection and hinge moment characteristics of the Space Shuttle Orbiter[33] are given in Table 6.18. Deflection rates are generally small. In contrast to them, those of modern, unstable flying fighter aircraft are up to $80°/s$.

---

[32] In general flap effectiveness encompasses the need for trim drag as small as possible in order not to compromise the effective lift-to-drag ratio of the flight vehicle, Chapter 2.

[33] Some of the characteristics, for instance the deflection angles, are cited somewhat differently in [53].

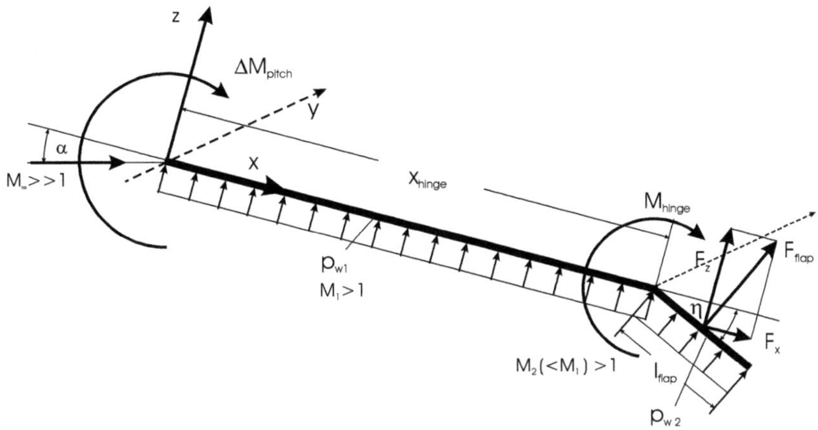

Fig. 6.47. Idealized hypersonic flight vehicle with deflected control surface, the generated flap force F_{flap}, the increment of the vehicle pitching moment $\triangle M_{pitch}$, and the hinge moment M_{hinge}. It is assumed that a constant pressure p_{w_1} acts on the lower side of the flight vehicle, whereas a constant $p_{w_2} = p_{flap}$ acts on the lower side only of the deflected (η) control surface which has the dimensions length l_{flap} and width (span) w_{flap}. The lateral vehicle axis (y) is normal to the x-z plane.

6.6.3 Flap Width Versus Flap Length

Regarding the geometry of an aerodynamic control surface, it is a question how to shape its planform best. In the following we study in particular the relation between its width and length (chord). A simple consideration reveals how moment arms, flap forces and moments are mutually connected. In Fig. 6.47 an idealized flight vehicle with a deflected control surface is shown. The latter has the length l_{flap} and the width (span) w_{flap}, the flap surface (plan area) is $l_{flap} w_{flap}$. The deflection angle is η.

We note that on a real vehicle x_{hinge} is the distance to the center-of-gravity, not to the vehicle nose, and that the center-of-pressure of the flap is not necessarily located at 0.5 l_{flap}.

We assume constant pressure p_{w_1} on the lower side of the flight vehicle. The pressure $p_{w_2} = p_{flap}$, acting only on the lower side of the flap, is also assumed to be constant[34]. The flap force F_{flap} then is:

$$F_{flap} = l_{flap} w_{flap} p_{flap}. \tag{6.13}$$

It has the two components:

$$F_z = \cos\eta \, F_{flap}, \quad F_x = \sin\eta \, F_{flap}. \tag{6.14}$$

[34] This is the ideal inviscid case. In reality the pressure on the flap will not be constant, and, if separation occurs around the hinge line, also the pressure on the lower side of the vehicle, just upstream of the hinge line, will not be constant, Sub-Section 6.3.2.

The flap force times its lever arm with respect to the vehicle's y-axis is balanced by the increment[35] of the vehicle moment $\triangle M_{pitch}$:

$$\triangle M_{pitch} = -(x_{hinge} + \cos\eta \frac{l_{flap}}{2}) F_z - \sin\eta \frac{l_{flap}}{2} F_x. \quad (6.15)$$

If $l_{flap}/2 \ll x_{hinge}$ and for small to moderate deflection angles η, this reduces, with eq. (6.14) to:

$$\triangle M_{pitch} = -x_{hinge}\, l_{flap}\, w_{flap}\, p_{flap}. \quad (6.16)$$

The flap force times its lever arm with respect to the flap axis now must be balanced by the hinge moment M_{hinge}:

$$M_{hinge} = -\frac{l_{flap}}{2} F_{flap} = -\frac{l_{flap}^2}{2}\, w_{flap}\, p_{flap}. \quad (6.17)$$

We see from this simple consideration that $\triangle M_{pitch} \sim x_{hinge}\, l_{flap}\, w_{flap}$, whereas the hinge moment is $M_{hinge} \sim l_{flap}^2\, w_{flap}$. With regard to the demands on the actuator, this means that the flap length l_{flap}, because it appears quadratically, is more critical than the flap width w_{flap}. For the vehicle pitching moment the moment arm x_{hinge} is the governing length.

We conclude that a large x_{hinge} is beneficial for the vehicle pitching moment. Simultaneously, regarding the hinge moment, and hence the needed actuator performance, a large flap width w_{flap} is to be preferred rather than a large flap length l_{flap}. However, in any case a trade-off is necessary[36], because with increasing flap length in general smaller deflection angles are needed (see above: influence of the onset flow boundary layer, asymptotic behavior of the ramp pressure), with the benefit of smaller thermal loads. We note in this context, that a large aspect ratio of the flap reduces flow three-dimensionality and associated losses in effectiveness.

6.6.4 Volumes of Stabilization and Control Surfaces

The so called volume of a stabilizer or control surface – plan area (size) times moment arm (e.g. distance from mid chord of control surface to center-of-gravity) – is a measure for its overall performance, i. e. the moment it

---

[35] It is customary to include, in the aerodynamic model of a flight vehicle, the increment due to a deflected flap in the overall moment around the respective axis. We deviate here from that custom in order to show more clearly the influence of the geometrical properties of a flap on the vehicle moment and the hinge moment.

[36] We point in this regard to the general flight mechanical layout of the vehicle. Reducing the static stability margin, Chapter 7, would reduce the trim moment. With the small lever arms of hypersonic vehicles, in contrast to classical aircraft, this still means a rather large trim force and hence a considerable trim drag. A reduced stability margin, on the other hand, makes a flight vehicle more susceptible to atmospheric and other perturbations.

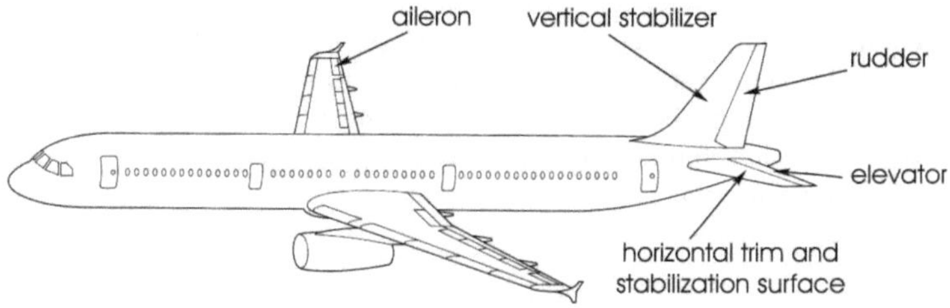

Fig. 6.48. Trim, stabilization, and control surfaces of a typical passenger aircraft.

generates. In order to get a feeling for the orders of magnitude of volumes of flight vehicles in this regard, we consider in Table 6.19 selected geometrical data[37], and in Table 6.20 the plan areas, moment arms, and volumes of aerodynamic trim, stabilization and control surfaces, of a RV-W-type vehicle (Space Shuttle Orbiter), a CAV-type vehicle (SÄNGER lower stage), and a typical small/medium-range passenger aircraft. For the latter we show a typical configuration in Fig. 6.48, which exhibits nicely the large moment arms, hence volumes, which the different aerodynamic trim, stabilization and control surfaces have. The center-of-gravity lies at $x/L \approx 0.43$.

Regarding the Space Shuttle Orbiter, Figs. 3.2 and 3.40, we give the data in relation to the plan area $A_{ref} = 361.3\ m^2$ (usually the wing area alone, $249.91\ m^2$, is used as the reference surface for force and moment coefficients), the vehicle length up to the body flap, and the half span, see, e.g., [54]. For the SÄNGER lower stage, Fig. 3.2, we give the data in relation to the plan area, the vehicle length, and the half span, see, e.g., [55]. For the passenger aircraft we take the wing area as reference area, the fuselage length as reference length, and also the wing half span.

The selected data in Table 6.20 for the RV-W-type and the CAV-type vehicle are partly estimated data. Since we are here interested in the relations of the volumes only, we don't look at the whole problem of flyability and controllability of the vehicles, also not at pitch, roll, and yaw damping. We also do not regard that, for instance, in the case of the Space Shuttle Orbiter, Sub-Section 2.1.2, the RCS is active in pitch for altitudes down to $H \approx 70.0$ km, the rudder becomes active only at flight Mach numbers $M_\infty < 5$, and that the RCS in yaw is active down to $M_\infty \approx 1$. Concerning the SÄNGER lower stage, we disregard the influence of the thrust vector on the longitudinal motion, Sub-Section 2.2.3. We also note, that the configuration at the end of the German Hypersonic Technology Programme was still a preliminary one. The demands from the side of flyability and controllability were not yet taken fully into account.

Anyway, the numbers alone in Table 6.20 do not say all. They must be seen in view of the issues of stable and controlled flight, e.g., (longitudinal) trim, longitudinal and lateral stability, moments of inertia around the three

[37] We choose in each case the volume $V_{ref} = A_{ref} \cdot L_{ref}$ as reference flap volume.

Table 6.19. Geometrical reference data of a RV-W-type vehicle (Space Shuttle Orbiter, [54]), a CAV-type vehicle (SÄNGER lower stage, [55]), and a typical small/medium-range passenger aircraft: plan area, length, half span, reference flap volume, and the chosen x-reference position of the center-of-gravity. The center-of-gravity position given for the SÄNGER lower stage is that of the TSTO-system just before upper stage launch.

Reference data	RV-W-type vehicle	CAV-type vehicle	Small/medium-range passenger aircraft
Area A_{ref}	361.3 m^2 total plan area	1,435.0 m^2 total plan area	122.0 m^2 wing plan area
Length L_{ref}	32.77 m	82.5 m	38.0 m
Half span s_{ref}	11.92 m	22.55 m	17.0 m
Reference flap volume $V_{ref} = A_{ref}\,L_{ref}$	11,841.10 m^3	118,387.50 m^3	4,636.0 m^3
center-of-gravity reference position x_{cog}/L_{ref}	0.66	0.74	0.43

axes, angular turn rates, and damping properties. Therefore we contrast them only with the (estimated) data of the typical small/medium-range passenger aircraft.

We observe summarily, that the volume of the longitudinal trim and stabilization device of the passenger aircraft is roughly one order of magnitude larger than that of the RV-W-type and the CAV-type vehicle. The other volumes concerning the longitudinal motion are also significantly larger for the passenger aircraft, mostly simply due to the more forward location of the center-of-gravity. The volumes for roll control are of similar size. All volumes of the CAV-type vehicle are significantly smaller than that of the other ones. This probably is due to the preliminary state of design and because the influence of the propulsion system was not yet taken into account.

Table 6.20 underlines for RV-W-type and CAV-type flight vehicles - in contrast to the passenger aircraft – the need of high effectiveness of all aerodynamic trim, stabilization and control surfaces which must be ensured despite the harsh aerothermodynamic environment at hypersonic flight.

In general the volume of a stabilization, trim or control surface cannot be enlarged in a simple manner. Two basic possibilities are given: one is to enlarge the plan area, the other to enlarge the moment arm.

If the moment arm is fixed, an increase of the plan area would be a way to get a desired performance. This must be considered in connection with the overall layout of the vehicle. If the plan area of a given stabilization or control surface cannot be enlarged, utilization of two surfaces is a possibility.

Table 6.20. Aerodynamic trim, stabilization and control surfaces of a RV-W-type vehicle (Space Shuttle Orbiter), a CAV-type vehicle (TSTO-system SÄNGER lower stage), and a typical small/medium-range passenger aircraft, and selected relative estimated plan areas, moment arms, and volumes of trim, stabilization and control surfaces. For the reference values see Table 6.19.

Relative data	RV-W-type vehicle	CAV-type vehicle	Typical small/medium-range passenger aircraft
Longitudinal trim and stabilization devices:	body flap	elevators	horizontal tail plane
Plane area A/A_{ref}	0.035	0.048	0.25
x-moment arm L/L_{ref}	0.38	0.17	0.5
Volume V/V_{ref}	0.013	0.0082	0.125
Pitch control devices:	elevators	elevators	elevators
Area A/A_{ref}	0.098	0.048	0.074
x-moment arm L/L_{ref}	0.27	0.17	0.52
Volume V/V_{ref}	0.026	0.0082	0.038
Lateral stabilization devices:	single vertical tail surface	dual vertical tail surfaces	single vertical tail surface
Area A/A_{ref}	0.11	0.12	0.18
x-moment arm L/L_{ref}	0.32	0.20	0.48
Volume V/V_{ref}	0.035	0.024	0.086
Lateral control devices:	rudder	dual rudders	rudder
Area A/A_{ref}	0.021	0.021	0.055
x-moment arm L/L_{ref}	0.41	0.25	0.51
Volume V/V_{ref}	0.0086	0.0053	0.028
Roll control devices:	ailerons	ailerons	ailerons
Area A/A_{ref}	0.037	0.022	0.035
y-moment arm L/s_{ref}	0.83	0.76	0.86
Volume V/V_{ref}	0.031	0.017	0.03

For instance, for vertical stabilizers, and hence also rudders, dual arrangements, Sub-Section 6.6.1, are a typical means to increase their volume in view of moment-arm restrictions. Such means can also be wing-tip fins. At RV-W-type vehicles, like at the HERMES and the X-38 configurations, wing-tip fins have the other benefit that they are not completely shadowed by the fuselage at the large angles of attack during re-entry flight.

However, such arrangements pose other problems. Dual vertical stabilizers may be prone to structural fatigue problems due to impingement of lee-side vortices of the delta wing. The anyway highly thermally loaded wing-tip fins and rudders, [58], must avoid during re-entry the impingement of the bow shock and hence must be wrapped inside the bow shock surface, [59]. Otherwise the interaction would reduce effectiveness, and the – by strong interaction phenomena dramatically enlarged – thermal loads would increase structural weight due to insulation needs.

To enlarge moment arms in general is not a valid option[38]. A proposal in this regard was the configuration of the British SSTO space transportation vehicle HOTOL (*Horizontal Take-Off and Landing*) project from 1986. The about 63.0 m long ARV-type vehicle was to have a fuselage diameter of 7.0 m and a span of the delta wing of 28.0 m. Because the center-of-gravity of the airbreathing propulsion system – with oxygen collection for the rocket cycle above $M_\infty \approx 7$ at $H \gtrapprox 30.0\ km$ altitude – was located very far back, Fig. 6.49, an unconventional solution was considered for the placement of the aerodynamic stabilization and control devices, see, e.g., [60].

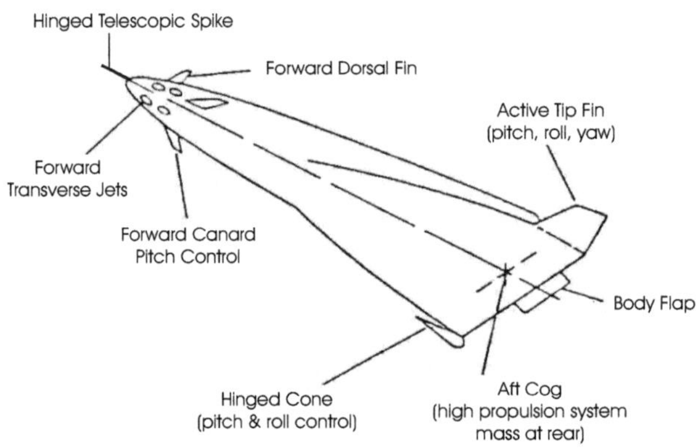

Fig. 6.49. Configuration of the British SSTO project HOTOL from 1986 with unconventional aerodynamic stabilization and control devices at the forward portion of the airframe, [60].

At the aft body a conventional body flap was foreseen and further at the wing-tips “active” fins for pitch, roll, and yaw control as well as hinged cones for pitch and roll control. At the forebody of the vehicle a canard for pitch control, a dorsal fin for yaw stabilization, and an aerospike was proposed, in addition to RCS thrusters. Of course the arrangement near the vehicle nose yields the desired large moment arms. However, the classical weathercock stability for both the longitudinal and the lateral motion is given up. Instead the vehicle is to fly unstable because of the very aft location of the center-of-gravity.

[38] In contrast to aerodynamic trim and control surfaces, a RCS has fewer restrictions regarding the placement of the individual thrusters.

This is fundamentally different to the only longitudinal statically unstable hypersonic re-entry flight of the Space Shuttle Orbiter, Sub-Section 3.4.2, because HOTOL is an ascent and return vehicle (ARV) which must cover the whole flight Mach number span from take-off to orbit and back. Unstable flight in principle is possible, see modern fighter aircraft, but the connected problems of flight control of such a large airbreathing vehicle make the proposal questionable.

From the side of aerothermodynamics it must be noted, that many of the issues of forward located stabilization and control surfaces are not well understood, see, e.g., [23]. On the one hand the surfaces are located in regions with small boundary layer thickness (favorable property of the onset flow), which in principle enhances their effectiveness. On the other hand the interaction of the flow past these surfaces with the flow field behind them over the vehicle may trip premature laminar-turbulent transition, and induce shock interaction phenomena. Both will cause adverse thermal load increments. A solution as proposed for HOTOL eventually will lead to design and development problems and risks, which should be avoided by all means if possible.

6.7 Concluding Remarks

Effectiveness of trim, stabilization and control devices depends on a number of factors. A major one is their location on the flight vehicle. For aerodynamic control surfaces the optimal onset flow geometry is obvious: two-dimensional onset flow orthogonal to the hinge line. Flight parameters, the vehicle shape and flight attitude govern the resulting entropy layer, the thickness of the onset flow boundary layer, and the inviscid onset flow Mach number.

Experience and the available data base regarding hypersonic flight-vehicle shapes are still small. Detailed ground-facility test and computational data, correlated with flight data, is only available for the Space Shuttle Orbiter. It can be dangerous to generalize flight and also ground-simulation results found with only one vehicle shape. In any case, different flight vehicle types put different demands on devices, and the involved aerothermodynamic phenomena can be vastly different.

In the following sub-sections we summarize important results of this chapter, with a short discussion of ground-simulation issues.

6.7.1 Summary of Results

Important results of the above discussions are listed with selected references, a general review is not intended.

- Effectiveness of aerodynamic trim and control surfaces is governed to a large degree by the onset flow properties, Section 6.2. Important is the optimal onset flow geometry.

- Viscous and high-temperature real-gas effects have their influence primarily via the onset flow. The vehicle's windward side surface temperature, governed by radiation cooling, has a strong influence on the boundary layer thickness and its separation behavior, Sub-Section 6.2.3.
- The governing similarity parameters are not those of the free-stream but those locally at the edge of the boundary layer. This is important especially for RV-W-type flight vehicles, where despite large flight Mach numbers at large angles of attack the onset flow Mach numbers are small, and Reynolds numbers are relatively high, Sub-Section 6.2.1.
- At RV-W-type vehicles a boattailing seems to be beneficial for the body flap effectiveness, because it increases the onset flow Mach number, Sub-Section 6.3.2. No systematic studies of this effect are known. On the other hand, boattailing introduces a sensitivity against high-temperature real-gas effects, Sub-Section 6.2.1.
- Space Shuttle Orbiter data indicate that real-gas effects, via the onset flow properties, are the major factor regarding flap effectiveness, Sub-Section 6.3.2. This may hold in general for RV-W and RV-NW-type vehicles, but not for CAV and ARV-type vehicles.
- The increase of the flap deflection angle η in supersonic flow in general increases both the vehicle moment and the hinge moment, and also the moment gradient $C_{m,\eta} = d\,C_m/d\,\eta$, Sub-Section 6.3.2.
- The asymptotic pressure plateau on the flap surface is that of the inviscid pressure jump depending on deflection (ramp) angle, onset flow Mach number, and real-gas effects, Sub-Section 6.3.1. This holds at flight altitudes, where the local Reynolds number is not too small and hypersonic viscous interaction is negligible. Space Shuttle Orbiter data suggest that this may hold already at the large speeds and altitudes of the initial re-entry trajectory, even if there, due to the small dynamic pressure, flap effectiveness is very small, and the RCS therefore is active.
- The flap (chord) length should be large enough to make use of the asymptotic pressure plateau, Sub-Section 6.3.1. In general this is no problem, because even at large deflection angles the extent of the separation zone is small. Space Shuttle Orbiter experience shows that in a ground-simulation facility this may be predicted wrongly, [31].
- Separation ahead of the hinge line, case c) in Fig. 6.9, and the resulting separation shock, are not necessarily leading to a dramatic decrease of the flap effectiveness, Sub-Section 6.3.2. On the other hand, these effects, which appear – depending on the onset flow Mach number – at large deflection angles, don't enhance flap effectiveness, if the flap length is large enough, such that the asymptotic pressure plateau governs the flap force. The hinge moment in any case may be more strongly influenced than the vehicle pitching moment.
- The hinge moment depends stronger on the flap length than on its width, Section 6.6. It rises quadratically with rising flap length. A large aspect

ratio of the control surface reduces losses due to three-dimensional flow effects.

- A small flap length makes larger deflection angles necessary, with the danger that case c) in Fig. 6.9 occurs. In any case, not only the flap effectiveness must be considered, but in conjunction with it also the hinge moment with the resulting actuator performance, weight, volume, and energy demand, Section 6.6.
- Any flap deflection leads to an increase of the thermal loads on the flap, with a considerable enhancement of them around the flap's hinge line, if case c) in Fig. 6.9 is present. Regarding the thermal state of the surface – surface temperature in the presence of radiation cooling and heat flux in the gas at the wall – an asymptotic behavior is observed too, if the flap length is large enough, Sub-Section 6.3.3.
- Non-convex surface radiation effects due to flap deflection seem to play a role more for the onset flow via the surface temperature, [12]. Transverse heat transfer through the flap structure can reduce overall thermal loads, but if a cavity situation like on the X-38 is given, thermal loads there will increase, Sub-Section 6.3.3.
- The wall shear stress away from the hinge line shows an asymptotic behavior, too, Sub-Section 6.3.4. Its magnitude is important in view of possible surface material erosion effects.
- Experimental and numerical investigations so far concentrate always only on the phenomena present around the flap's hinge line. Because the asymptotic behavior of mechanical and thermal loads on the flap is important for vehicle layout, data are needed to check and validate numerical prediction methods. This concerns especially the prediction of thermal loads in presence of turbulent flow.
- The gap at the hinge line is of concern, because gap flow may influence flap effectiveness. Large gap flow at large deflection angles can reduce or even delete separation around the hinge line, however, with an increase of thermal loads in the gap, Section 6.4. Gaps between neighboring control surfaces decrease effectiveness and locally rise thermal loads.
- On real configurations several 3-D effects can be present. Besides the already mentioned effect of finite flap aspect ratio we can have a sub-optimal onset flow situation, side-wall effects, and interactions with, for instance, the vehicle's bow shock in off-design flight conditions. In all such cases the basic aerothermodynamic phenomena sketched so far will be present, however, modulated accordingly. In general control surface effectiveness is reduced, especially thermal loads can be increased.
- Active control of adverse effects on a flight configuration in general is hardly possible. Layout and optimization of the aerodynamic shape, and the structure and materials concept of the vehicle, as well as the flight trajectory, must take care of such effects. Control means like surface bleed flow applied for inlet ramp flow, Section 4.5.5, are in general not an option for aerodynamic control surfaces.

- Interaction of the jet of a reaction control system with the surface flow, Section 6.5, poses special problems. These are the amplification or de-amplification of the jet effectiveness, the amplification of thermal loads (hot-spots), even at some distance from the jet, and induction of stray forces due to jet impingement. Again it is a matter of overall vehicle layout to fashion and optimize RCS effectiveness and to cope with adverse effects.

6.7.2 Simulation Issues

For general issues of ground-facility and computational simulation see, e.g., Sub-Section 9.3.1, also [10]. Regarding trim and control devices of hypersonic vehicles the major problem is the characterization of the onset flow properties. With the given shape of the vehicle in a design cycle the accurate and reliable estimation of the effectiveness of a device to a large degree depends on the correct quantification of these properties. This also holds for the quantification of secondary effects, like trim drag, thermal loads increments, et cetera.

In view of RV-W and RV-NW-type flight vehicles simulation has the particular problem of high-temperature real-gas effects. Vehicle shape dependent, these might be of secondary importance for overall lift and drag (longitudinal motion), if Oswatitsch's independence principle holds, but not for the pitching moment, and the lateral motion. Reynolds number, thermal state of the surface (thermal surface effects, [61]) as well as (real-life) surface properties affect the viscous part of the onset flow of a device. The state of the boundary layer – laminar, transitional, turbulent – can be of very large importance, too.

For the viscous effects dominated CAV and ARV-type (ascent mode) flight vehicles Reynolds number, surface temperature, and surface properties play the major role in view of the fact, that the flow predominantly is turbulent, with the location of the transition zone additionally being of large importance. Real-gas effects in general may play a smaller role.

Ground-facility simulation now has the problem of simultaneously keeping the relevant scaling laws right, with adequate model sizes and surface properties, as well as with adequate instrumentation. Computational simulation suffers more from deficits of transition and turbulence modelling than from deficits of transport property and high-temperature real-gas models.

When looking at the basic effects and properties of control-device flow, obviously much has been studied, especially regarding the involved strong interaction phenomena. However, building-block experiments in the sense of those discussed in [62] are rare. They should show the asymptotic behavior of the flow parameters on, for instance, a control surface, and at the same time take into account the different domains of similarity parameters, which we have at re-entry vehicles in contrast to airbreathing cruise and acceleration vehicles. Suitable and reliable validation data, including especially minutely reconstructed skin-friction line and velocity field topologies, are needed to improve flow-physics and thermodynamic models for computational methods and their validation.

These are the general problems. We have seen, when going into details of design problems as well as building-block experiments, that multidisciplinary simulation issues of large importance arise, too. The radiation-adiabatic temperature is a good approximation of the actual temperature during flight on regular radiation cooled surfaces. In hinge and side gaps of control surfaces, gaps between TPS tiles, between nose caps and tiles, between TPS panels, also at sharp leading edges, this is no more the case, even if non-convex effects are regarded. The reason is that in the wall material temperature gradients normal to the surface, as well as tangentially, occur which can no more be neglected. This concerns generally heat transport by conduction and/or radiation in the structure and the material.

If, for the sake of lightweight design and reduction of design margins, a very accurate prediction of thermal and mechanical loads, besides that of functionality, becomes necessary, this can only be done with the help of multidisciplinary computational simulation. Flow phenomena, and structure and material phenomena must be treated in a coupled manner, Sub-Section 8.4.3, in order to describe thermo-mechanical fluid-structure interactions.

Building-block experiments have to overcome a host of challenges. Hot experimental techniques, [61], hot model-surface problems (in [63] for instance, thermal wave phenomena on C/C-SiC surfaces in hot air flow are reported), sensor falsification (to be overcome with non-intrusive opto-electronic measurement methods), et cetera must be mastered. Today this is possible in laboratory experiments, but the modelling and verification of multidisciplinary simulation and optimization methods makes much more efforts necessary. Finally in-flight testing with dedicated experimental vehicles is required in order to calibrate and verify physical models and data.

6.8 Problems

Problem 6.1. Compute with the relations given in Sub-Section 6.3.1 the incipient separation angle $\eta_{is} \equiv \theta_{is}$ for the laminar flat-plate/ramp flow case from Table 6.8. Use the viscosity relation eq. (10.61).

Problem 6.2. Compute with the relations given in Sub-Section 6.3.1 the incipient separation angle $\eta_{is} \equiv \varphi_{is}$ for the turbulent flat-plate/ramp flow case from Table 6.7.

Problem 6.3. Derive eq. (6.10) from eq. (6.9).

Problem 6.4. Check up to what angles η the proportionality

$$\frac{p_2 - p_1}{p_1} \sim \eta^2$$

gives reasonable results in a) Fig. 6.13 and b) Fig. 6.17. Choose the cases $\gamma_1 = \gamma_{eff} = 1.4$ and take $\eta = 5°$ in each figure as reference point.

Problem 6.5. In Fig. 6.27 we find at the base above the body flap a minimum surface pressure $log(p/p_0) \approx -1.8$, with $p_0 \equiv p_\infty$. How large is the appendant pressure coefficient c_p? How large is the vacuum pressure coefficient $c_{p_{vac}}$ which was defined in Sub-Section 3.2.3? Is the base pressure close to vacuum?

References

1. Delery, J.M., Marvin, J.G.: Shock Wave/Boundary Layer Interactions. AGARDograph 280 (1986)
2. Dolling, D.S.: Fifty Years of Shock-Wave/boundary layer Interaction Research: What Next? AIAA J. 39(8), 1517–1531 (2001)
3. Knight, D., Yan, H., Panaras, A.G., Zheltovodov, A.: Advances of CFD Prediction of Shock Wave Turbulent Boundary Layer Interactions. Progress in Aerospace Sciences 39, 121–184 (2003)
4. Arnal, D., Delery, J.M.: Laminar-Turbulent Transition and Shock Wave/Boundary Layer Interaction. RTO-EN-AVT-116, pp. 4-1–4-46 (2005)
5. Simeonides, G.: Hypersonic Shock Wave Boundary Layer Interactions over Simplified Deflected Control Surface Configurations. AGARD-FDP/VKI Special Course, AGARD R-792, pp. 7-1–7-47 (1993)
6. Settles, G.S., Dolling, D.S.: Swept Shock Wave Boundary Layer Interactions. In: Hemsch, M.J., Nielsen, J.N. (eds.) Tactical Missiles Aerodynamics. AIAA Progress in Aeronautics, vol. 104, pp. 297–379 (1986)
7. Panaras, A.G.: Review of the Physics of Swept-Shock/Boundary Layer Interactions. Progress in Aerospace Sciences 32, 173–244 (1996)
8. Smits, A.J., Dussauge, J.-P.: Turbulent Shear Layers in Supersonic Flow, 2nd edn. AIP/Springer, New York (2004)
9. Delery, J.M.: Shock Wave/Turbulent Boundary Layer Interaction and its Control. Progress in Aerospace Sciences 22, 209–280 (1985)
10. Hirschel, E.H.: Basics of Aerothermodynamics. Progress in Astronautics and Aeronautics, AIAA, Reston, Va, vol. 204. Springer, Heidelberg (2004)
11. Wüthrich, S., Sawley, M.L., Perruchoud, G.: The Coupled Euler/Boundary Layer Method as a Design Tool for Hypersonic Re-Entry Vehicles. Zeit-schrift für Flugwissenschaften und Weltraumforschung (ZFW) 20(3), 137–144 (1996)
12. Häberle, J.: Einfluss heisser Oberflächen auf aerothermodynamische Flugeigenschaften von HOPPER/PHOENIX (Influence of Hot Surfaces on Aerothermodynamic Flight Properties of HOPPER/PHOENIX). Diploma Thesis, Institut für Aerodynamik und Gasdynamik, Universität Stuttgart, Germany (2004)
13. Daimler-Benz Aerospace Space Infrastructure, FESTIP System Study Proceedings. FFSC-15 Suborbital HTO-HL System Concept Family. Dasa München/Ottobrunn, Germany (1999)
14. Wilcox, D.C.: Turbulence Modelling for CFD. DCW Industries, La Cañada, CAL, USA (1998)
15. Kliche, D.: Personal communication. München, Germany (2008)
16. Griffith, B.F., Maus, J.R., Best, J.T.: Explanation of the Hypersonic Longitudinal Stability Problem – Lessons Learned. In: Arrington, J.P., Jones, J.J. (eds.) Shuttle Performance: Lessons Learned. NASA CP-2283, Part 1, pp. 347–380 (1983)

17. Marini, M.: Analysis of Hypersonic Compression Ramp Laminar Flows under Sharp Leading Edge Conditions. Aerospace Science and Technology 5, 257–271 (2001)
18. Henze, A., Schröder, W., Bleilebens, M., Olivier, H.: Numerical and Experimental Investigations on the Influence of Thermal Boundary Conditions on Shock Boundary Layer Interaction. Computational Fluid Dynamics J. 12(2), 401–407 (2003)
19. Simeonides, G.: Personal communication (2006)
20. Drougge, G.: An Experimental Investigation of the Influence of Strong Adverse Pressure Gradients on Turbulent Boundary Layers at Supersonic Speeds. FFA Report 47, Stockholm, Sweden (1953)
21. Needham, D.A.: Laminar Separation in Hypersonic Flow. Doctoral thesis, University London, U.K (1965)
22. Korkegi, R.H.: Comparison of Shock-Induced Two- and Three-Dimensional Incipient Turbulent Separation. AIAA J. 13(4), 534–535 (1975)
23. Neumann, R.D.: Defining the Aerothermodynamic Methodology. In: Bertin, J.J., Glowinski, R., Periaux, J. (eds.) Hypersonics. Defining the Hypersonic Environment, vol. 1, pp. 125–204. Birkhäuser, Boston (1989)
24. Elfstrom, G.M.: Turbulent Hypersonic Flow at a Wedge-Compression Corner. J. of Fluid Mechanics 53, Part 1, 113–127 (1972)
25. Edney, B.: Anomalous Heat Transfer and Pressure Distributions on Blunt Bodies at Hypersonic Speeds in the Presence of an Impinging Shock. FFA Rep. 115 (1968)
26. Coratekin, T., van Keuk, J., Ballmann, J.: On the Performance of Upwind Schemes and Turbulence Models in Hypersonic Flows. AIAA J. 42(5), 945–957 (2004)
27. Coleman, G.T., Stollery, J.L.: Heat Transfer from Hypersonic Turbulent Flow at a Wedge Compression Corner. J. Fluid Mechanics 56, 741–752 (1972)
28. Holden, M.S., Moselle, J.R.: Theoretical and Experimental Studies for the Shock Wave-Boundary Layer Interaction on Compression Surfaces in Hypersonic Flow. ARL 70-0002, Aerospace Research Laboratories, Wright-Patterson AFB, Ohio (1970)
29. Simeonides, G., Haase, W.: Experimental and Computational Investigations of Hypersonic Flow About Compression Ramps. J. Fluid Mechanics 283, 17–42 (1995)
30. Reisinger, D., Müller, J., Olejak, D., Staudacher, W.: Effect of Mach Number and Reynolds Number on Flap Efficiencies in the Supersonic Regime. In: Proc. 85th Semi-Annual STA Meeting, Atlanta, GA (1996)
31. Weilmuenster, K.J., Gnoffo, P.A., Greene, F.A.: Navier-Stokes Simulations of Orbiter Aerodynamic Characteristics. In: Throckmorton, D.A. (ed.) Orbiter Experiments (OEX) Aerothermodynamics Symposium. NASA CP-3248, Part 1, pp. 447–501 (1995)
32. Thorwald, B.: Personal communication (2006)
33. Hirschel, E.H.: Heat Loads as Key Problem of Hypersonic Flight. Zeitschrift für Flugwissenschaften und Weltraumforschung (ZFW) 16(6), 349–356 (1992)
34. Bushnell, D.M., Weinstein, L.M.: Correlation of Peak Heating for Reattachment of Separated Flow. J. Spacecraft and Rockets 5, 1111–1112 (196 Radiation Cooled8)
35. Celic, A., Hirschel, E.H.: Comparison of Eddy-Viscosity Turbulence Models in Flows with Adverse Pressure Gradient. AIAA J. 44(10), 2156–2169 (2006)

36. Zeiss, W.: Modelling of Hot, Radiation Cooled Surfaces. TET-DASA-21-TN-2402, Dasa, München/Ottobrunn, Germany (1999)
37. Behr, R.: Hot, Radiation Cooled Surfaces. TET-DASA-21-TN-2410, Dasa, München/Ottobrunn, Germany (2002)
38. Weiland, C., Longo, J.M.A., Gülhan, A., Decker, K.: Aerothermodynamics for Reusable Launch Systems. Aerospace Science and Technology 8, 101–110 (2004)
39. Görtler, H.: Über eine dreidimensionale Instabilität laminarer Grenzschichten an konkaven Wänden. Nachrichten der Gesellschaft der Wissenschaften zu Göttingen, Mathematisch-Physikalische Klasse, Neue Folge, Fachgruppe 1, Bd. 2, pp. 1–26 (1940)
40. Ginoux, J.J.: Streamwise Vortices in Reattaching High-Speed Flows: A Suggested Approach. AIAA J. 9(4), 759–760 (1971)
41. Lüdecke, H., Radespiel, R., Schülein, E.: Simulation of Streamwise Vortices at the Flaps of Re-Entry Vehicles. Aerospace Science and Technology 8, 703–714 (2004)
42. Lüdecke, H.: Computation of Görtler Vortices in Separated Hypersonic Flows. In: Satofuka, N. (ed.) Computational Fluid Dynamics 2000, pp. 171–176. Springer, Heidelberg (2001)
43. Gülhan, A.: Investigation of Gap Heating on a Flap Model in the Arc Heated Facility L3K. TET-DLR-21-TN-3101, DLR/Köln-Porz, Germany (1999)
44. Behr, R., Görgen, J.: CFD Analysis of Gap Flow Phenomena. TET-DASA-21-TN-2403, Dasa München/Ottobrunn, Germany (2000)
45. Mack, A., Schäfer, R., Gülhan, A., Esser, B.: Flowfield Topology Changes Due to Fluid-Structure Interaction in Hypersonic Flow Using ANSYS and TAU. In: Breitsamter, C., Laschka, B., Heinemann, H.-J., Hilbig, R. (eds.) New Results in Numerical and Experimental Fluid Mechanics IV. Notes on Numerical Fluid Mechanics and Multidisciplinary Design, NNFM 87, pp. 196–203. Springer, Heidelberg (2004)
46. Scallion, W.I.: Space Shuttle Reaction Control System – Flow Field Interactions During Entry. In: Throckmorton, D.A. (ed.) Orbiter Experiments (OEX) Aerothermodynamics Symposium. NASA CP-3248, Part 1, pp. 345–370 (1995)
47. Spais, F.W., Cassel, L.A.: Aerodynamic Interference Induced by Reaction Controls. AGARD-AG-173 (1973)
48. Monta, W.J., Rausch, J.R.: Effects of Reaction Control System Jet-Flow Field Interactions on a 0.015 Scale Model Space Shuttle Orbiter Aerodynamic Characteristics. NASA-CR-134074 (1973)
49. Zeiss, W.: Jet/Airflow Interaction. TET-DASA-21-TN-2409, Dasa, München/Ottobrunn, Germany (2002)
50. Zeiss, W., Behr, R.: Vortical Type Flow with Respect to Jet-External Airflow Interaction. In: Breitsamter, C., Laschka, B., Heinemann, H.-J., Hilbig, R. (eds.) New Results in Numerical and Experimental Fluid Mechanics IV. Notes on Numerical Fluid Mechanics and Multidisciplinary Design, NNFM 87, pp. 188–195. Springer, Heidelberg (2004)
51. Weiland, C.: Typical Vortex Phenomena in Flow Fields Past Space Vehicles. In: Blackmore, D., Krause, E., Tung, C. (eds.) Vortex Dominated Flows, pp. 251–269. World Scientific Publishing, Singapore (2005)
52. Wang, K.C.: Separating Patterns of Boundary Layer Over an Inclined Body of Revolution. AIAA Journal 10, 1044–1050 (1972)
53. Allen, M.F., Romere, P.O.: Space Shuttle Aerodynamics Flight Test Program. In: Throckmorton, D.A. (ed.) Orbiter Experiments (OEX) Aerothermodynamics Symposium. NASA CP-3248, Part 1, pp. 281–298 (1995)

54. Romere, P.O.: Orbiter (Pre STS-1) Aerodynamic Design Data Book Development and Methodology. In: Throckmorton, D.A. (ed.) Orbiter Experiments (OEX) Aerothermodynamics Symposium. NASA CP-3248, Part 1, pp. 249–280 (1995)
55. Hauck, H.: Leitkonzept SÄNGER – Referenz-Daten-Buch. Issue 1, Revision 2, Dasa, München/Ottobrunn, Germany (1993)
56. Edwards, C.L.W., Small, W.J., Weidner, J.P.: Studies of Scramjet/Air-frame Integration Techniques for Hypersonic Aircraft. AIAA-Paper 75-58 (1975)
57. Kafer, G.C.: Space Shuttle Entry/Landing Flight Control Design Description. AIAA-Paper 82-1601 (1982)
58. Herrmann, U., Radespiel, R., Longo, J.M.A.: Critical Flow Phenomena on the Winglet of Winged Reentry Vehicles. Zeitschrift für Flugwissenschaften und Weltraumforschung (ZFW) 19, 309–319 (1995)
59. Perrier, P.C.: Concepts of Hypersonic Aircraft. In: Bertin, J.J., Glowinski, R., Periaux, J. (eds.) Hypersonics. Defining the Hypersonic Environment, vol. 1, pp. 40–71. Birkhäuser, Boston (1989)
60. East, R.A., Hutt, G.A.: Vehicle Geometry Effects on Hypersonic Stability and Control. In: Proc. Symposium on Fluid Dynamics and Space, VKI, Rhode Saint Genese, Belgium (1986)
61. Hirschel, E.H.: Thermal Surface Effects in Aerothermodynamics. In: Proc. 3rd European Symposium on Aerothermodynamics for Space Vehicles, Noordwijk, The Netherlands, ESA SP-426, pp. 17–31 (1999)
62. Marvin, J.G.: A CFD Validation Roadmap for Hypersonic Flows. AGARD CP-514, pp. 17-1–17-16 (1992)
63. Esser, B., Gülhan, A.: Thermal Fluid-Structure Interaction in Different Atmospheres. In: Proc. 5th European Workshop on Thermal Protection Systems and Hot Structures, Noordwijk, The Netherlands, ESA SP-631 (2006)

7 Forces, Moments, Center-of-Pressure, Trim, and Stability in General Formulation

In this chapter we provide the general formulations of forces, moments, the center-of-pressure, trim, and static stability for the three vehicle classes considered in this book. We think that this is necessary, because usually in the literature these entities are given only for slender aircraft configurations at small angle of attack.

We are beginning with the moment equation and give then the general formulation of the center-of-pressure. Special considerations follow of the static stability of bluff configurations, i. e. RV-NW-type vehicles, and of slender configurations, i. e. CAV-type (and ARV-type) vehicles. The chapter closes with some further contemplations of static stability and with the provision of coordinate transformations of forces.

7.1 Moment Equation

The coordinate systems used in aerodynamics and in flight mechanics as well as the directions in which the aerodynamic forces are counted positive, differ in the literature, but also from country to country and sometimes from aerospace company to aerospace company. Therefore the aerodynamicist and/or the aerospace engineer has in each case precisely to check, in which coordinate system and how the forces and moments are defined, see, e.g., [1] to [4]. This holds also, and more so, for symbols, units and the nomenclature, see also Sub-Section 2.1.3.

We present here in Fig. 7.1 coordinate systems which are widely used for winged aerospace vehicles. In order not to overburden this figure we have drawn a second one, Fig. 7.2, with the definitions of the forces and moments in the various coordinate systems.

Finally, Fig. 7.3 shows the coordinate systems and the definitions of force and moment coefficients actually applied in this book – with some exceptions – for aerodynamics and flight mechanics of the three vehicle classes.

Hence we refer in the following derivations to Fig. 7.3. This is the most general case, since it includes the z-offset of the center-of-gravity, which plays a

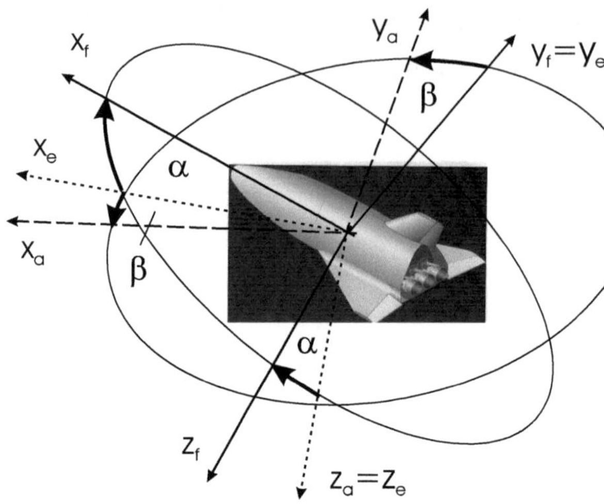

Fig. 7.1. Coordinate systems generally used for winged aerospace vehicles. Body axis system (x_f, y_f, z_f), experimental axis system (x_e, y_e, z_e), air-path axis system (x_a, y_a, z_a).

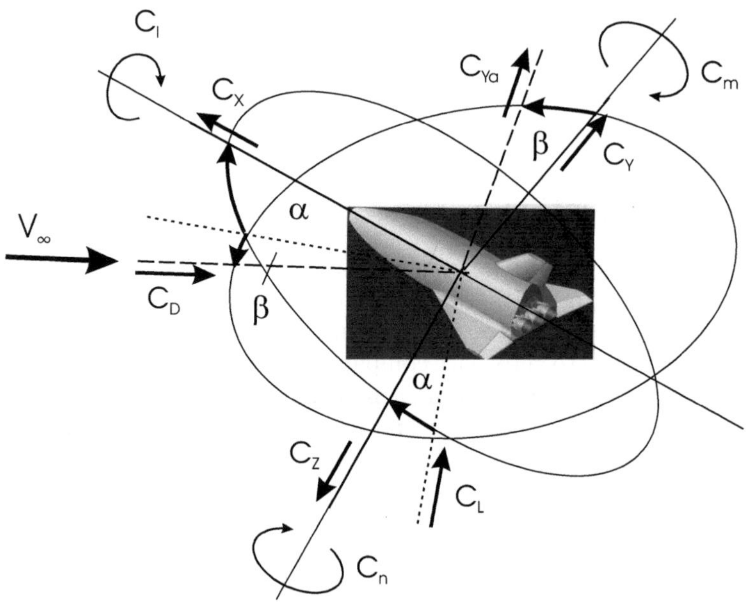

Fig. 7.2. Definitions of aerodynamic force and moment coefficients for winged aerospace vehicles with respect to the coordinate systems defined in Fig. 7.1.

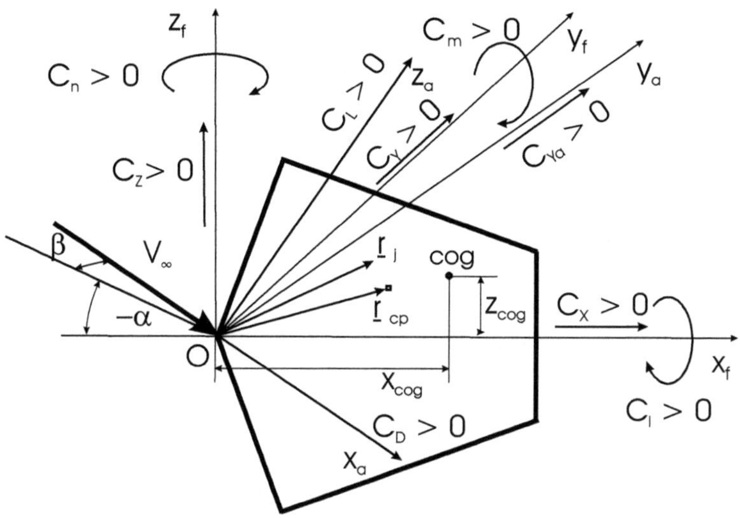

Fig. 7.3. Coordinate systems and definitions of aerodynamic force and moment coefficients applied for aerodynamics and flight mechanics in this book. Body axis system (x_f, y_f, z_f), air-path axis system (x_a, y_a, z_a).

major role for RV-NW-type vehicles. This holds also for RV-W-type vehicles, but not necessarily for CAV-type and ARV-type vehicles. All the formulas derived below, including the signs, are valid also for the definitions given in Figs. 7.1, 7.2.

The equation for the torque vector reads in terms of the body-oriented axis system:

$$\underline{M}_{cp} = \underline{M}_j + \Delta\underline{r} \times \underline{F}_{aero} = 0, \tag{7.1}$$

with

$$\Delta\underline{r} = \underline{r}_j - \underline{r}_{cp}\ , \ \underline{M}_j = q_\infty A_{ref} L_{ref} \begin{Bmatrix} C_l \\ C_m \\ C_n \end{Bmatrix}_j , \ \underline{F}_{aero} = q_\infty A_{ref} \begin{Bmatrix} C_X \\ C_Y \\ C_Z \end{Bmatrix},$$

and C_X, C_Y, C_Z[1] the coefficients of the axial, the lateral and the normal aerodynamic force, C_l, C_m, C_n the coefficients of the roll, pitch and yaw moment, Figs. 7.2, 7.3, $q_\infty = 0.5\,\rho_\infty v_\infty^2$ the dynamic pressure, A_{ref} the reference area, L_{ref} the reference length, and $\underline{r}$ the local position vector. The subscript *cp* denotes the center-of-pressure, *cog* the center-of-gravity and j an arbitrary reference point.

---

[1] For the sake of completeness we mention, that the axial and normal force coefficients C_X and C_Z are often denoted by C_A and C_N.

We obtain from eq. (7.1) after rearrangement[2]:

$$L_{ref}\left\{\begin{array}{c} C_l \\ C_m \\ C_n \end{array}\right\}_j + \left\{\begin{array}{r} C_Z\left(y_j - y_{cp}\right) - C_Y\left(z_j - z_{cp}\right) \\ -C_Z\left(x_j - x_{cp}\right) + C_X\left(z_j - z_{cp}\right) \\ C_Y\left(x_j - x_{cp}\right) - C_X\left(y_j - y_{cp}\right) \end{array}\right\} = 0. \tag{7.2}$$

The trim conditions are found, if the reference point j is chosen to be the center-of-gravity *cog*. For trim all moments about the *cog* are zero:

$$C_{l|cog} = C_{m|cog} = C_{n|cog} = 0.$$

From eq. (7.2) we obtain then:

$$\begin{aligned} z_{cog} - z_{cp} &= \frac{C_Z}{C_Y}\left(y_{cog} - y_{cp}\right), \\ z_{cog} - z_{cp} &= \frac{C_Z}{C_X}\left(x_{cog} - x_{cp}\right), \\ x_{cog} - x_{cp} &= \frac{C_X}{C_Y}\left(y_{cog} - y_{cp}\right). \end{aligned} \tag{7.3}$$

The flight vehicle can be trimmed, if these center-of-gravity positions are fulfilled[3]. The general solution of eq. (7.3) is indicated – see also eq. (7.1) – by $\Delta\underline{r} \parallel \underline{F}_{aero}$, a specific one by $\Delta\underline{r} = 0$ and a trivial one by $\underline{F}_{aero} = 0$. Below, these correlations are explained in more detail.

7.2 General Formulation of the Center-of-Pressure

The coordinates of the center-of-pressure *cp* are determined in the following way. The reference point j is positioned in the origin of the coordinate system O, which means $\underline{r_j} = 0$, Fig. 7.3. Then in eq. (7.2) the three moments are separated each into two parts based on the contributing coordinates[4]:

$$L_{ref}\left\{\begin{array}{c} C_l|_o^y + C_l|_o^z \\ C_m|_o^x + C_m|_o^z \\ C_n|_o^x + C_n|_o^y \end{array}\right\} = \left\{\begin{array}{c} +C_Z y_{cp} - C_Y z_{cp} \\ -C_Z x_{cp} + C_X z_{cp} \\ +C_Y x_{cp} - C_X y_{cp} \end{array}\right\}. \tag{7.4}$$

By equating coefficients, the coordinates of the center-of-pressure are now:

$$x_{cp} = -L_{ref}\frac{C_m|_o^x}{C_Z} = +L_{ref}\frac{C_n|_o^x}{C_Y}, \tag{7.5}$$

[2] It is customary to employ (dimensionless) coefficients of forces and moments, but to keep dimensional coordinates, L_{ref} appears in this relation.

[3] As usual for cross products the rank of the related matrix is 2, which means that eq. (7.3) has only two linear independent components.

[4] Note that the subscript O indicates that the moment is related to the origin of the coordinate system.

$$y_{cp} = +L_{ref}\frac{C_l|_o^y}{C_Z} = -L_{ref}\frac{C_n|_o^y}{C_X}, \tag{7.6}$$

$$z_{cp} = +L_{ref}\frac{C_m|_o^z}{C_X} = -L_{ref}\frac{C_l|_o^z}{C_Y}. \tag{7.7}$$

Each of the two components of the moments around the origin of the coordinate system, Figs. 7.3 and 7.4, are found by integration of the components of the respective unit-surface forces $f_{i,aero}$ ($i = x,\ y,\ z$) acting on the surface element dS times the respective lever arm, measured from the origin O. ΔF_{aero} is composed of the surface pressure force ΔF_{pw} and the skin-friction force $\Delta F_{\tau w}$ and further we have $f_{i,aero}\,\Delta S = \Delta F_{i,aero}$, Fig. 7.4:

$$\begin{aligned} C_l|_o^y &= \frac{1}{q_\infty A_{ref} L_{ref}} \iint_S f_{z,aero}\, y\, dS, \\ C_l|_o^z &= -\frac{1}{q_\infty A_{ref} L_{ref}} \iint_S f_{y,aero}\, z\, dS, \end{aligned} \tag{7.8}$$

$$\begin{aligned} C_m|_o^x &= -\frac{1}{q_\infty A_{ref} L_{ref}} \iint_S f_{z,aero}\, x\, dS, \\ C_m|_o^z &= \frac{1}{q_\infty A_{ref} L_{ref}} \iint_S f_{x,aero}\, z\, dS, \end{aligned} \tag{7.9}$$

$$\begin{aligned} C_n|_o^x &= \frac{1}{q_\infty A_{ref} L_{ref}} \iint_S f_{y,aero}\, x\, dS, \\ C_n|_o^y &= -\frac{1}{q_\infty A_{ref} L_{ref}} \iint_S f_{x,aero}\, y\, dS. \end{aligned} \tag{7.10}$$

For the interpretation of the center-of-pressure cp we consider the flow in the symmetry plane of an axisymmetric body, where $y_{cp} = 0$, Fig. 7.4.

The center-of-pressure lies on the action line of the resultant aerodynamic force. The two parts $C_m|_o^x$ and $C_m|_o^z$ of the pitching moment M are equal to the moment of the normal force Z times the x-location of the center-of-pressure, x_{cp}, and that of the axial force X times the z-location of the center-of-pressure, z_{cp}. The components of the aerodynamic force (X, Y, Z) are obtained by integration of pressure and shear stress along the entire surface of the considered configuration. The center-of-pressure thus is defined such that it is the location where the moments of the axial and the normal force are equal to the moments due to the integrated differential moments in x and z direction.

We consider now an example. For a wedge-like (2-D) body with no lateral forces ($\beta = 0°$, $C_Y = 0$), eq. (7.3) yields $y_{cog} = y_{cp}$, and the center-of-gravity is restricted to the plane of symmetry ($y_{cog} = 0$). So, only the second equation of eq. (7.3) remains. This means that for a given aerodynamic state

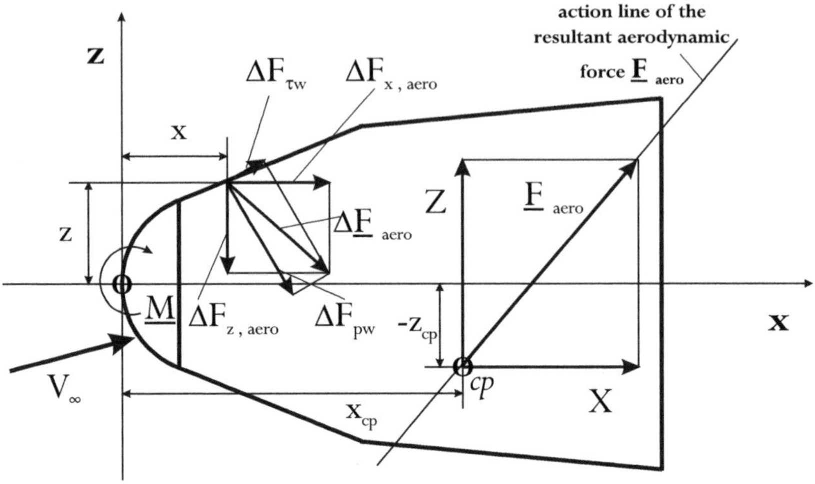

Fig. 7.4. Schematic of the center-of-pressure, body axis system, symmetric case with $y_{cp} = 0$ (axial force $X \equiv A$, normal force $Z \equiv N$). Note: $f_{i,aero}\, \Delta S = \Delta F_{i,aero}$.

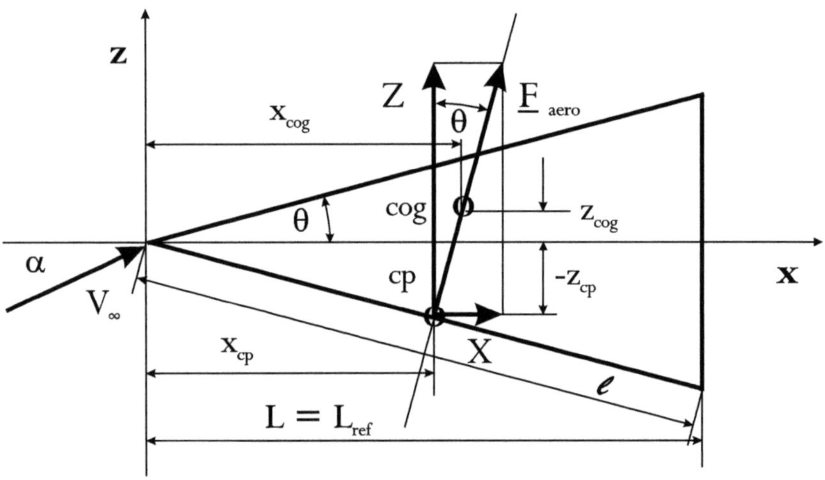

Fig. 7.5. Wedge-like shape with the center-of-gravity *cog* away from the center-of-pressure *cp*.

(C_X, C_Z, x_{cp}, z_{cp} are known) the vehicle can be trimmed by all the center-of-gravity positions satisfying:

$$z_{cog} = z_{cp} + \frac{C_Z}{C_X} \left(x_{cog} - x_{cp} \right) . \tag{7.11}$$

We show this by considering the problem sketched in Fig. 7.5 (see also Problem 5.1). Since in the Newtonian limit for $\alpha > \Theta$ the surface force is only given by the wall pressure p_w of the windward side, we have $X/Z = tan\,\Theta$ (mind that in this example $z_{cp} < 0$).

With eqs. (7.5), (7.7), (7.9) we can calculate the fixed position of the center-of-pressure to[5]:

$$z_{cp} = -\frac{1}{2}\, l \sin\Theta,$$
$$x_{cp} = \frac{1}{2}\, l \cos\Theta. \qquad (7.12)$$

The vector of the resultant aerodynamic force $\underline{F}_{aero}$ can be thought to have its origin at the fixed center-of-pressure position and is perpendicular to the wedge contour for $\alpha > \Theta$. Significant values for z_{cog} must meet the condition $z_{cog} \gtreqqless z_{cp}$ and with eq. (7.11) it yields $x_{cog} \gtreqqless x_{cp}$. Note: Trim with $z_{cog} = 0$ is only possible in this simple example for $\alpha > \Theta$. Generally for axisymmetric shapes trim with $z_{cog} = 0$ cannot be achieved, see Sub-Section 5.3.3.

7.3 Static Stability Considerations for Bluff Configurations

It may be mentioned here that for classical RV-NW-type vehicles like Apollo and Viking or similar, $|C_Z/C_X| \ll 1$ (see Figs. 5.10, 5.11, 5.12), and that therefore the influence of x_{cog} on the trim condition is low just in opposite to the influence of z_{cog}. Fig. 7.6 highlights the situation.

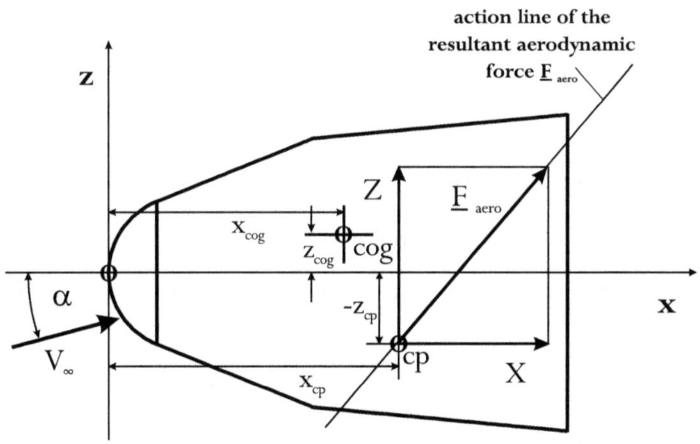

Fig. 7.6. Center-of-pressure *cp* and center-of-gravity *cog* definitions for bluff bodies on the basis of Fig. 7.4.

[5] In the example Fig. 7.5 the following holds: $S \Rightarrow l$, $f_{i,aero} = const.$, $x(l) = l\,\cos\Theta$, $z(l) = -\,l\,\sin\Theta$.

Let us consider now the static stability for the longitudinal movement. The second component of eq. (7.2) with $j \equiv cog$ reads:

$$L_{ref}C_m - C_Z(x_{cog} - x_{cp}) + C_X(z_{cog} - z_{cp}) = 0, \tag{7.13}$$

and we then find

$$L_{ref}\frac{\partial C_m}{\partial \alpha} = -C_Z\frac{\partial x_{cp}}{\partial \alpha} + C_X\frac{\partial z_{cp}}{\partial \alpha} + \frac{\partial C_Z}{\partial \alpha}(x_{cog} - x_{cp}) - \frac{\partial C_X}{\partial \alpha}(z_{og} - z_{cp})\,. \tag{7.14}$$

The vehicle flies stable, if $\partial C_m/\partial\alpha < 0$, Fig. 7.9. The right-hand side of eq. (7.14) combined with eq. (7.11) can then be written:

$$C_Z\frac{\partial x_{cp}}{\partial \alpha} - C_X\frac{\partial z_{cp}}{\partial \alpha} + \frac{x_{cog} - x_{cp}}{C_X}\left(C_Z\frac{\partial C_X}{\partial \alpha} - C_X\frac{\partial C_Z}{\partial \alpha}\right) > 0. \tag{7.15}$$

For the particular case that the center-of-pressure and the center-of-gravity are identical ($\underline{r}_{\,cp} = \underline{r}_{\,cog}$), we obtain for the stable state:

$$\frac{\partial x_{cp}}{\partial \alpha} > \frac{C_X}{C_Z}\frac{\partial z_{cp}}{\partial \alpha}. \tag{7.16}$$

The vehicle system is unconditionally stable, if $\partial z_{cp}/\partial\alpha < 0$ and $\partial x_{cp}/\partial\alpha > 0$ (for $C_X/C_Z > 0$ and α positive). This means, that an increase of α must shift the line of action of the resultant aerodynamic force behind the center-of-gravity, hence causing an incremental restoring (nose-down) moment. A decrease of α must shift the line of action ahead of the center-of-gravity, hence causing an incremental restoring (nose-up) moment.

7.4 Static Stability Considerations for Slender Configurations

An important quantity for the assessment of longitudinal static stability for slender vehicles of CAV-type is the neutral point.

The general definition of the neutral point reads, [3]:

<u>The location of the neutral point x_N is the location of the center-of-gravity x_{cog} for which $dC_m/d\alpha = 0$.</u>

Since slender vehicles fly usually with small angles of attack, we assume that $C_Z \approx C_L$. Fig. 7.7 illustrates the relations of the center-of-gravity and the center-of-pressure for this kind of vehicles.

One definition of the neutral point concept assumes that the pitching moment is the sum of the moment at zero lift and the moment of the lift force L times the distance between the neutral point x_N and the center-of-gravity x_{cog} :

$$C_m = C_m|_{C_L=0} - \frac{x_N - x_{cog}}{L_{ref}}\,C_L. \tag{7.17}$$

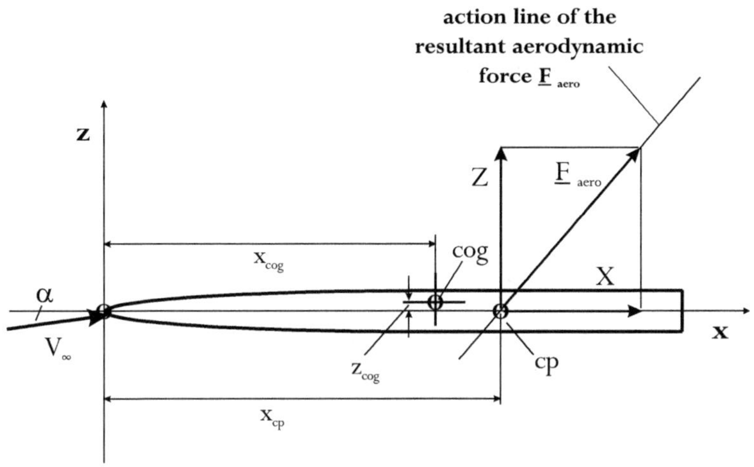

Fig. 7.7. Center-of-pressure *cp* and center-of-gravity *cog* definitions for slender flight vehicles.

We compare now the components of eqs. (7.13) and (7.17) for slender CAV-type vehicles:

$$L_{ref}\,C_m = C_X(z_{cp} - z_{cog}) - C_Z(x_{cp} - x_{cog}),$$
$$L_{ref}\,C_m = L_{ref}C_m|_{C_L=0} - C_L(x_N - x_{cog}), \tag{7.18}$$

and obtain with $z_{cp} = 0$, what we have extracted from Fig. 7.7:

$$L_{ref}\,C_m|_{C_L=0} = -\,C_X z_{cog} \quad \text{for} \quad x_N = x_{cp}\,.$$

Another definition of the neutral point concept expresses the observation, that the pitching moment coefficient can be split in a part at zero lift $C_m|_{C_L=0}$ and a part described by the pitching moment slope with respect to lift times the lift coefficient, [4]:

$$C_m = C_m|_{C_L=0} + \frac{dC_m}{dC_L}\,C_L. \tag{7.19}$$

By equating coefficients in eqs. (7.17) and (7.19) we directly find:

$$\frac{dC_m}{dC_L} = -\frac{x_N - x_{cog}}{L_{ref}}. \tag{7.20}$$

From eq. (7.20) it follows directly the general definition of the neutral point (see above). The vehicle is trimmed, if $C_m = 0$, and flies longitudinally stable, if $dC_m/d\alpha < 0$, respectively $dC_m/dC_L < 0$:

$$\frac{dC_m}{dC_L} = -\frac{1}{L_{ref}}(x_N - x_{cog}) < 0 \quad \Longrightarrow \quad x_N > x_{cog}. \tag{7.21}$$

With other words, the neutral point of the untrimmed vehicle has to be located downstream of the center-of-gravity. The sketch in Fig. 7.8 serves for clarification. The magnitude of $x_N - x_{cog}$ is called the static margin.

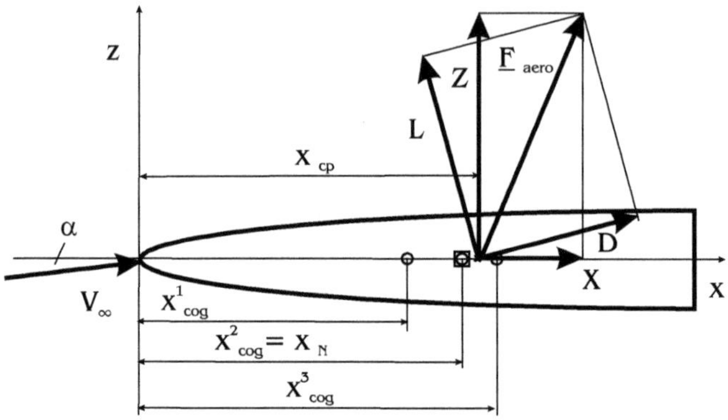

$$x^1_{cog}: L_{ref}\, dC_m/dC_L = -(x_N - x^1_{cog}) < 0 \Longrightarrow \text{stable}$$

$$x^2_{cog}: L_{ref}\, dC_m/dC_L = -(x_N - x^2_{cog}) = 0 \Longrightarrow \text{indifferent}$$

$$x^3_{cog}: L_{ref}\, dC_m/dC_L = -(x_N - x^3_{cog}) > 0 \Longrightarrow \text{unstable}$$

Fig. 7.8. Sketch illustrating the neutral point definition, x_{cp}: center-of-pressure. x_{cog}: center-of-gravity, x_N: neutral point.

If the neutral point lies ahead of the center-of-gravity, $x_N < x_{cog}$, the static margin is negative, and the vehicle is unstable, regardless of whether it is trimmed or untrimmed.

7.5 Further Contemplations of Static Stability

The condition $\partial C_m/\partial\alpha < 0$ at α_{trim} is not sufficient to fly the vehicle stable. It must further be demanded, that this condition holds for a sufficiently large angle of attack interval around the trim point T: $\Delta\alpha^-_{trim} \leqq \alpha_{trim} \leqq \Delta\alpha^+_{trim}$, Fig. 7.9, and that C_m becomes zero at reasonable values of α_{trim}.

The size of the interval is not defined exactly, it is governed at least by the dynamic stability properties of the flight vehicle[6].

[6] A RV-NW-type vehicle enters the reentry trajectory with an angle of attack α and a flight path angle γ. This attitude is brought about with the help of rocket thrusters. During the descent, at lower altitudes, the vehicle usually will begin to oscillate around the pitch axis. Even if the vehicle is dynamically stable, the angle of attack interval $\Delta\alpha^+_{trim} \leqq \alpha_{trim} \leqq \Delta\alpha^-_{trim}$ must be larger than the possible α-amplitudes.

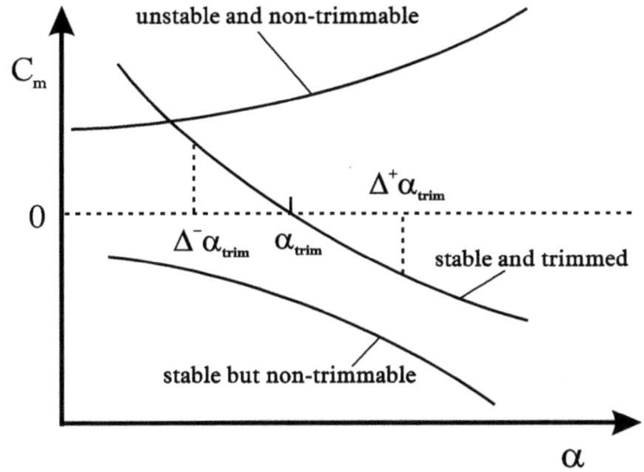

Fig. 7.9. Schematic of stability demands. The condition $\partial C_m/\partial\alpha < 0$ must hold not only in the trim point $C_m(\alpha_{trim}) = 0$, but also in a sufficiently large interval around α_{trim}: $\Delta\alpha^-_{trim} \leqq \alpha_{trim} \leqq \Delta\alpha^+_{trim}$.

The magnitude of $|\Delta\underline{r}|$, eq. (7.1), is a measure for the static margin (see also Chapters 3 and 6). It is given for the in y-direction symmetric case in eq. (7.14) by the third and the fourth term on the right-hand side: $x_{cog} - x_{cp}$ and $z_{cog} - z_{cp}$. It means that for the reference angle of attack condition the center-of-gravity must lie "ahead" of the center-of-pressure, such that the line of action of the resultant aerodynamics is to be moved forward to meet the trim condition α_t.

7.6 Coordinate Transformations of Force Coefficients

In closing the considerations we give the transformation of the force coefficients defined in the body-fixed coordinate system, with C_X the axial, C_Z the normal, C_Y the lateral force coefficient, into the coefficients of the aerodynamic (air-path) system, with C_D the drag, C_L the lift, C_{ya} the side force:

$$\begin{aligned} C_X &= C_D \cos\alpha\cos\beta - C_{ya}\cos\alpha\sin\beta - C_L\sin\alpha, \\ C_Z &= C_D \sin\alpha\cos\beta - C_{ya}\sin\alpha\sin\beta + C_L\cos\alpha, \\ C_Y &= C_D\sin\beta + C_{ya}\cos\beta. \end{aligned} \tag{7.22}$$

The angle β is the sideslip or yaw angle, (Figs. 7.1, 7.3). For a vehicle with a symmetry in the $x - z$ plane, we obtain for $\beta = 0$ from eq. (7.22):

$$\begin{aligned} C_X &= C_D\cos\alpha - C_L\sin\alpha, \\ C_Z &= C_D\sin\alpha + C_L\cos\alpha. \end{aligned} \tag{7.23}$$

The inverse of this relation reads:

$$
\begin{aligned}
C_D &= C_X \cos\alpha + C_Z \sin\alpha, \\
C_L &= C_Z \cos\alpha - C_X \sin\alpha.
\end{aligned} \tag{7.24}
$$

Everybody who has worked with lifting capsules has made the experience, that the direction of lift L is negative for positive angles of attack in opposite to other winged and non-winged vehicles (e.g. bicones). The reason for that is, as already mentioned earlier, that bluff bodies have a large axial-force coefficient C_X and a small normal-force coefficient C_Z, which is different for slender vehicles. Hence the lift coefficient C_L, eq. (7.24), changes sign when $C_X \sin\alpha$ becomes larger than $C_Z \cos\alpha$. We show this with an example.

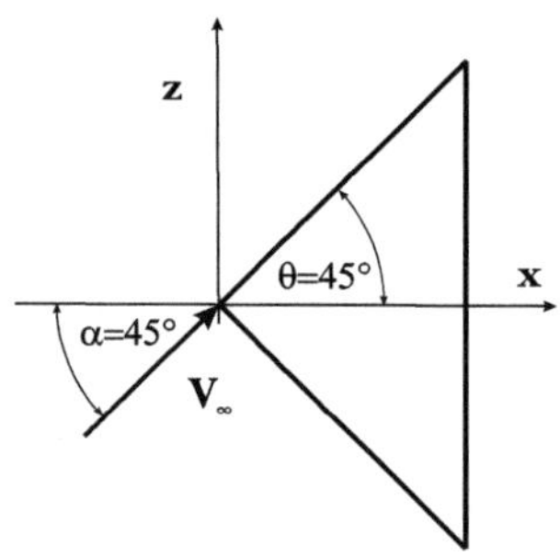

Fig. 7.10. Double wedge with no lift for an angle of attack $\alpha = 45°$ and a wedge angle $\Theta = 45°$ in the Newtonian limit.

A double wedge with a wedge angle of $\Theta = 45°$ has in the Newtonian limit (hypersonic limit) no lift for an angle of attack $\alpha = 45°$, Fig. 7.10, Problem 5.1. At this angle of attack[7] the lift is positive for $\Theta < 45°$ and negative for $\Theta > 45°$. For lower angles of attack the transmission from positive to negative lift takes place at higher wedge angles Θ. We demonstrate that with the double wedge of Fig. 7.5.

In Fig. 7.11 the evolutions of lift as function of α for the wedge angles $\Theta = 42°$, $45°$, $52°$, $54.736°$ are plotted. For positive angles of attack α the lift is completely positive for $\Theta = 42°$ and reaches, as noted before, the zero point for $\Theta = 45°$ at $\alpha = 45°$. Increasing the wedge angle to $\Theta = 52°$ leads to negative lift for $\alpha \gtrapprox 31°$. The limiting value for the wedge angle, where negative lift for all $\alpha > 0°$ exists, is given by $\Theta = 54.736°$. A good example is the Viking type shape, despite the fact that it is axisymmetric, with a forebody consisting of a blunted cone with $\Theta = \theta_1 > 70°$, Fig. 5.6, which produces negative lift for all reasonable angles of attack $\alpha > 0°$.

[7] For cones this is only approximately true, since the leeward side is placed in the Newtonian shadow for $\alpha = \Theta = 45°$ and the windward side produces a small lift contribution.

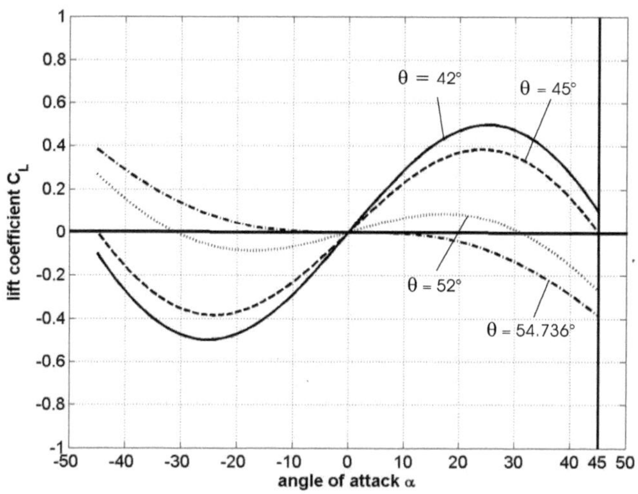

Fig. 7.11. Lift behavior of the double wedge, Fig. 7.5, as function of the angle of attack α and wedge angle Θ.

7.7 Problems

Problem 7.1. From Fig. 5.12 for the SOYUZ capsule, we find for $M_\infty = 0.95$ and $\alpha =$ - 20° with the reference point coordinates $x_{ref} = 0.370\ D_1$, $z_{ref} = 0.039\ D_1$ approximately the force coefficients:

$$\begin{aligned} C_Z &= -0.01, \\ C_X &= \quad 1.21, \\ C_m &= -0.07. \end{aligned}$$

1. Calculate the lift and drag coefficients C_L and C_D.
2. Shift the moment reference point to $x_{ref} = 0.40 D_1$ and $z_{ref} = 0.045 D_1$ and compute the corresponding pitching moment coefficient C_m.
3. What are the effects of shifts of x_{ref} and z_{ref} on the pitching moment coefficient?

References

1. American National Standards Institute. Recommended Practice for Atmosphere and Space Flight Vehicle Coordinate Systems. American National Standard ANSI/AIAA R-0004-1992 (1992)
2. Brockhaus, R.: Flugregelung. Springer, Heidelberg (2001)
3. McCormick, B.W.: Aerodynamics, Aeronautics, and Flight Mechanics. John Wiley& Sons, New York (1979)
4. Schlichting, H., Truckenbrodt, E.: Aerodynamics of the Aeroplane, 2nd edition (revised). McGraw Hill Higher Education, New York (1979)

8 Multidisciplinary Design Aspects

The design and development approaches of aerospace vehicles of either kind are basically the same as those of conventional aircraft. Their background is given by Cayley's design paradigm. Sir George Cayley (1773 - 1857) was an early British aviation pioneer, who conceived the essentials of the aircraft as we know it today, [1].

Cayley's design paradigm, [2], [3], reads (not in his own words):

- Assign functions plainly to corresponding sub-systems, e.g.
  - wing ⇒ provision of lift,
  - propulsion system ⇒ overcoming of drag,
  - horizontal stabilizer and elevator ⇒ longitudinal trim, stabilization and control,
  - vertical stabilizer and rudder ⇒ lateral (directional) stabilization and control,
  - fuselage ⇒ payload accommodation,
  - et cetera.
- Have the different functions and the corresponding sub-systems only weakly and linearly coupled, then you can treat and optimize each function and sub-system more or less independently of the others, and nevertheless treat and optimize in this way the whole aircraft, which integrates all functions and sub-systems.

This paradigm has been proven to be very effective (ideally it should hold for every technical apparatus). However, the quest for more performance and efficiency, the opening of new flight-speed domains et cetera, has led over the years to higher and higher integrated functions and sub-systems, i. e., a weakening of Cayley's design paradigm. Of course, this is different for different kinds of flight vehicles, and does in each case not necessarily encompass all major functions and sub-systems.

In the context of our considerations it is interesting and important to note, that a differentiation similar to that of sub-systems and functions has taken place also of the engineering disciplines, which are involved in the design and development of flight vehicles[1]. This is natural, and was and is indeed also a strong technological driving factor.

The differentiation of the engineering disciplines, however, had also adverse effects. It led, for instance, to the presently strongly established sequential and iterative design cycles with a weak interaction of the disciplines. It further led in some cases to autonomy drives of disciplines by duplicating – under the euphemism "adaptation" – skills and tools of other disciplines, which then often did not participate in the sub-sequential developments of the mother disciplines. In view of these developments, the meaning of Cayley's design paradigm is expanded to cover also the disciplines, i. e. both the differentiation of functions/sub-systems (first aspect), and of disciplines (second aspect). The significant observation is now, that both aspects of Cayley's design paradigm are, as already indicated, persistently weakening in modern aircraft design, which holds even more for aerospace flight vehicles.

An important issue for aircraft, as well as aerospace vehicles, for the latter especially for CAV/ARV-type flight vehicles, is that with the present design and development approach the actual quantification of the aero-thermo-servo-elastic properties of the vehicle's airframe is made only very late in the development process, after the first – already completely defined and developed – airframe has been assembled[2]. Partly this quantification process extends deep into the flight envelope opening process. Changes – actually "repair solutions" – which must be made of the airframe, if the said properties don't meet the requirements, can be very costly and in any case will increase the structural weight of the vehicle.

Similar problems exist with regard to the structures and materials layout of hypersonic vehicles in view of the thermal loads associated with high speed flight. Thermal protection systems of either kind cost much weight, not to mention maintenance and repair efforts during vehicle lifetime. The real performance of a TPS will become apparent only after the first flight(s). Again "repair solutions" to either improve a TPS or to shed unnecessary mass will be costly and time consuming.

All this must be seen in view of the payload fraction (pay load/ dry mass empty), which, for instance, a TSTO space transportation system has, [4]. For the SÄNGER system this was estimated to be approximately 4.4 per cent (rocket launcher: approximately 2 per cent). Compared to payload fractions of modern transport aircraft of 30 to 45 per cent, these are very small numbers. This means that the design is much more critical for hypersonic space transportation systems than for transonic aircraft. Hence the development risk is very much larger, which must be seen in the context of the also very much larger development costs.

---

[1] This differentiation also holds for university education and for research at universities and research establishments.

[2] The reader is asked to recall in this regard the particular design problems of airbreathing hypersonic vehicles, Section 4.5.4.

A good chance to overcome these problems, basically the weakening of Cayley's design paradigm (first aspect), lies in the fantastic developments of computational simulation, which we have seen since about two decades, [2]. In general it appears that a strong and non-linear coupling of functions/sub-systems asks also for a strong discipline coupling. But there are also cases, where a strong disciplines coupling is mandatory, even if functions/sub-systems obey Cayley's design paradigm. The treatment of these design problems in the classical way is partly possible, but often it leads to increasingly large time and cost increments. Design risks can become large, and, especially with very strong functions/sub-systems couplings, they can become untenable. The discussion shows, that the weakening of Cayley's design paradigm, in both aspects, asks for new approaches. This holds strongly for aerospace vehicles, especially airbreathing aerospace planes, for both space transportation and military purposes.

The still enormous growth of computer power and the capabilities of information technologies make a post-Cayley design paradigm (integrated design) a viable option. This usually is understood when speaking of multidisciplinary design and optimization. Of course present day design and development is multidisciplinary, however, usually only in the sense, that the parent discipline brings in the other disciplines in terms of simple and partly very approximate methods.

True multidisciplinary design and optimization must overcome the second aspect of Cayley's design paradigm. It must bring together the best suited tools in a strong coupling of the disciplines. The methods of numerical aerodynamics are a key element of such approaches. Large advancements, however, are necessary in flow-physics and thermodynamics modelling, [5].

Large advancements – and a new thinking – are necessary, too, in structural mechanics. New structure-physics models are necessary in order to permit to quantify the influence of joints of all kind (non-linearities, damping), of non-linear deformations (buckling), et cetera. This is true in particular when static and dynamic aeroelastic properties of the airframe are to be described and optimized with high accuracy and reliability, [2].

Actually a shift from the *perfect-elastic* to the high-fidelity *real-elastic* airframe consideration and modelling is necessary already in the early design phases, and not late in the development process, [3]. This holds analogously also for thermal protection, propulsion integration, guidance and control, et cetera.

8.1 Introduction and Short Overview of the Objectives of Multidisciplinary Design Work

In the last decade the work on multidisciplinary design methods was motivated by the recognition that the development of such complex systems as aerospace vehicles can no longer be conducted by an isolated treatment of the various components (first aspect of Cayley's design paradigm). It is a very complex task to conduct a multidisciplinary design process for such vehicles with the goal to optimize their performances and missions. This strictly implies the strong interference between various disciplines (second aspect of Cayley's design paradigm). The global process encompasses a technological and an economical[3] part. The technological part consists mainly of the

- aerodynamic shape,
- aerothermodynamic performance including stability, controllability and maneuverability,
- ascent, descent, and contingency trajectories,
- guidance and control concept,
- masses: vehicle, propellants, payload,
- internal lay-out with the definition of the center-of-gravity and the moments of inertia,
- structures and materials layout including thermal protection systems,
- main propulsion system,
- reaction control system.

There exists for every technical system the challenge to deal with a large number of design variables and constraints which come from the disciplines contributing to the definition and synthesis of the system. These contributing disciplines are in general not independent from each other. Some of the interactions between the disciplines are so intense that only a closely coupled consideration of the physical behavior leads to really reliable and valuable results, [6], [7].

We don't have the intention here to give a comprehensive survey about all the various methods for dealing with the multidisciplinary design and optimization problem. We shortly address some of the analysis tools, [8].

First, however we state that the design variables forming the design space can be classified as entities being

- either continuous, such as the thrust of a rocket as function of the physics of the flow and the material under the condition of minimized side loads,
- or non-continuous (integer, discrete), such as the cost of the design and development of a highly sophisticated space-transportation system.

If the design variables are continuous, gradient-based and search methods are employed. Response-surface methods are used, when the whole vehicle is treated as a system with non-continuous (integer, discrete) design variables.

Gradient-based methods should be applied whenever possible, since they are much more effective than search methods. Gradient-based methods use the information of the function derivatives to locate local extremes. Search methods use only the function information to determine a global extreme.

In case that some elements of the design space are non-continuous (mixed-variable design space), gradient-based optimization algorithms are not the

[3] Customarily the main items considered are costs, operations, and rate of return of investment.

methods of choice. Typical disciplines, which do not provide continuous design variables, are material definition, manufacturing, maintenance, and cost determination.

Methods are available which are well suited for dealing with mixed-variable design spaces. These are the response-surface methods[4] with approximations based on

- neural networks, see, e.g., [9], [10],
- fuzzy logics, see, e.g., [11], [12],
- genetic algorithms, see, e.g., [13], [14].

A response surface is generated by the experts of the disciplines who treat the single parts (sub-systems) of the whole system. Each discipline contributes design solutions for statistically selected combinations of design parameters within the design space. Therefore, the response-surface method generally performs optimization on system level rather than on discipline level.

Exemplarily, Fig. 8.1 shows a simplified flow chart of the optimization process for the design of an aerospace vehicle with an inner loop generating a single design point.

We distinguish two interactions[5] between the various disciplines. The first type of interaction is the weak one, in the sense that the disciplines involved each have a separate physical description. Examples are:

- Steady contour changes of elements of vehicle shape $\Longrightarrow$ change of flow field,
- Change of aerodynamic performance $\Longrightarrow$ change of flight trajectory,
- Change of characteristics of propulsion system $\Longrightarrow$ change of vehicle's flight mechanics.

The second type of interaction is the strong one, where the disciplines involved require a coupled physical description, if advanced technology aspects with new capabilities like enhanced performance, flight safety, reliability and cost effectiveness are taken into account. Examples for that case are:

- Dynamic contour change of elements of vehicle shape due to aerodynamic (mechanical) loads $\Longleftrightarrow$ unsteady change of aerodynamic flow field (aeroelasticity),
- Heat transfer in structures $\Longleftrightarrow$ change of aerothermodynamic flow field (thermal surface effects),
- Contour change of elements of vehicle shape by thermal stresses $\Longleftrightarrow$ deformation of structure and change of aerothermodynamic flow field.

[4] These methods are also called search methods.

[5] With interaction we mean the mutual influence of physical behavior, i.e. the exchange of physical information. With coupling we mean the mathematical formulation of the exchange of information.

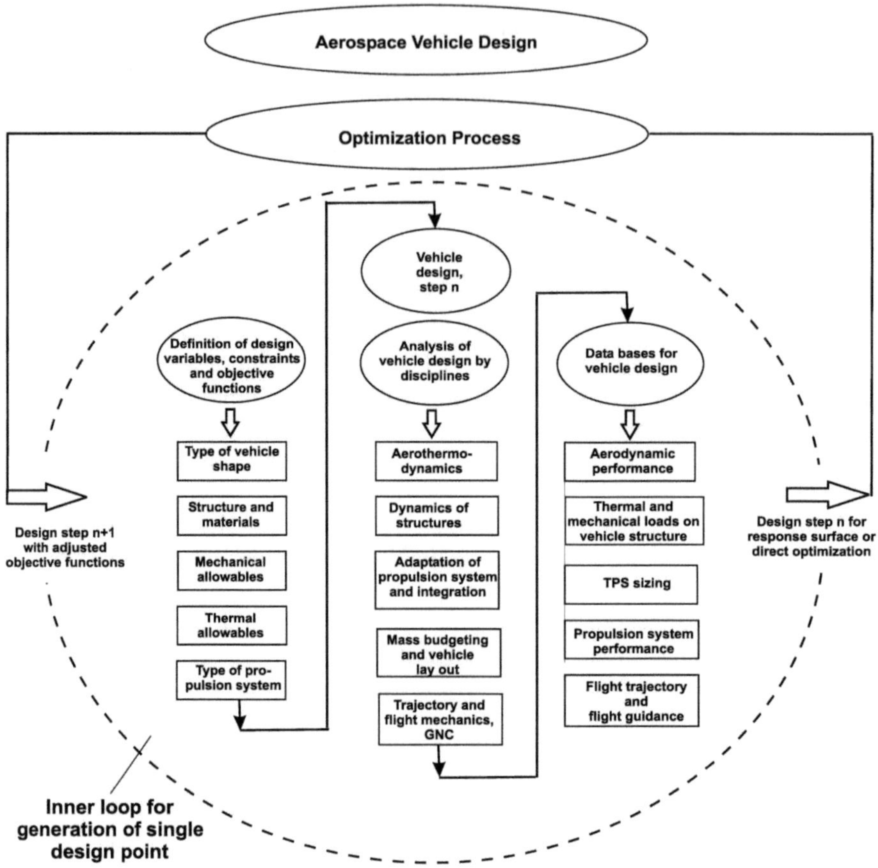

Fig. 8.1. Simplified flow chart illustrating the optimization process for aerospace vehicle design[5].

In agreement with the scope of this book we focus now our attention on approaches to describe the second type of interaction, viz. strong interactions.

8.2 Equations for Fluid-Structure Interaction Domains

8.2.1 Fluid Dynamics Equations

In Chapter 11 we present the fluid dynamic equations for unsteady viscous flows in Cartesian coordinates x, y, z, eq. (11.1). For flows past elastically

[5] "Type" denotes, for example the class of an aerospace vehicle, either a capsule or a bicone or a winged re-entry vehicle, and similar for the propulsion system.

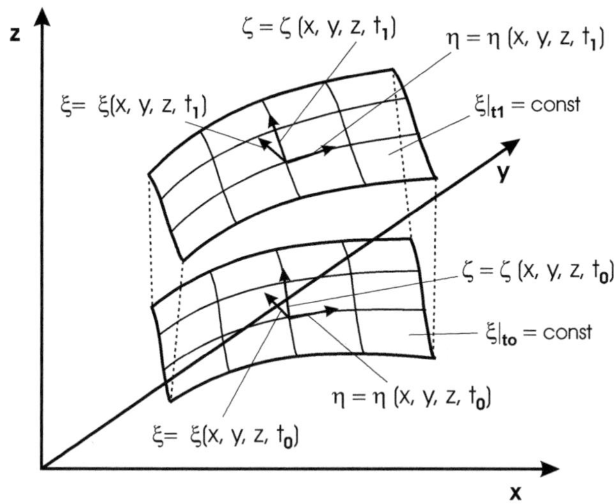

Fig. 8.2. Arbitrary non-orthogonal time-dependent coordinates defining iso-surfaces $\xi = const$ at times t_0 and t_1.

deforming configurations (e.g. due to unsteady aerothermodynamic loads), the body surface as "inner" boundary of the flow-computation domain deforms in a time-dependent manner. Due to that, any grid, either structured on unstructured, of the numerical flow solution procedure is also time-dependent. Therefore a coordinate transformation to arbitrary non-orthogonal time-dependent coordinates ξ, η, ζ, Fig. 8.2, is the method of choice for dealing with this kind of problems:

$$\begin{aligned} \xi &= \xi(x, y, z, t), \\ \eta &= \eta(x, y, z, t), \\ \zeta &= \zeta(x, y, z, t), \\ \tau &= t. \end{aligned} \tag{8.1}$$

Eq. (11.1) then reads:

$$\frac{\partial \underline{\hat{Q}}}{\partial \tau} + \frac{\partial(\underline{\hat{E}} - \underline{\hat{E}}_{visc})}{\partial \xi} + \frac{\partial(\underline{\hat{F}} - \underline{\hat{F}}_{visc})}{\partial \eta} + \frac{\partial(\underline{\hat{G}} - \underline{\hat{G}}_{visc})}{\partial \zeta} = 0, \tag{8.2}$$

with:

$$\underline{\hat{Q}} = J^{-1}\underline{Q}, \tag{8.3}$$

$$\underline{\hat{E}} = J^{-1}\begin{pmatrix} \rho U \\ \rho u U + \xi_x p \\ \rho v U + \xi_y p \\ \rho w U + \xi_z p \\ (\rho e_t + p) U - \xi_t p \end{pmatrix}, \quad \underline{\hat{E}}_{visc} = J^{-1}\begin{pmatrix} 0 \\ \xi_x \tau_{xx} + \xi_y \tau_{xy} + \xi_z \tau_{xz} \\ \xi_x \tau_{yx} + \xi_y \tau_{yy} + \xi_z \tau_{yz} \\ \xi_x \tau_{zx} + \xi_y \tau_{zy} + \xi_z \tau_{zz} \\ \xi_x \beta_x + \xi_y \beta_y + \xi_z \beta_z \end{pmatrix},$$

$$\underline{\hat{F}} = J^{-1}\begin{pmatrix} \rho V \\ \rho u V + \eta_x p \\ \rho v V + \eta_y p \\ \rho w V + \eta_z p \\ (\rho e_t + p) V - \eta_t p \end{pmatrix}, \quad \underline{\hat{F}}_{visc} = J^{-1}\begin{pmatrix} 0 \\ \eta_x \tau_{xx} + \eta_y \tau_{xy} + \eta_z \tau_{xz} \\ \eta_x \tau_{yx} + \eta_y \tau_{yy} + \eta_z \tau_{yz} \\ \eta_x \tau_{zx} + \eta_y \tau_{zy} + \eta_z \tau_{zz} \\ \eta_x \beta_x + \eta_y \beta_y + \eta_z \beta_z \end{pmatrix}, \tag{8.4}$$

$$\underline{\hat{G}} = J^{-1}\begin{pmatrix} \rho W \\ \rho u W + \zeta_x p \\ \rho v W + \zeta_y p \\ \rho w W + \zeta_z p \\ (\rho e_t + p) W - \zeta_t p \end{pmatrix}, \quad \underline{\hat{G}}_{visc} = J^{-1}\begin{pmatrix} 0 \\ \zeta_x \tau_{xx} + \zeta_y \tau_{xy} + \zeta_z \tau_{xz} \\ \zeta_x \tau_{yx} + \zeta_y \tau_{yy} + \zeta_z \tau_{yz} \\ \zeta_x \tau_{zx} + \zeta_y \tau_{zy} + \zeta_z \tau_{zz} \\ \zeta_x \beta_x + \zeta_y \beta_y + \zeta_z \beta_z \end{pmatrix},$$

$$\begin{aligned} \beta_x &= u\tau_{xx} + v\tau_{xy} + w\tau_{xz} - q_x, \\ \beta_y &= u\tau_{yx} + v\tau_{yy} + w\tau_{yz} - q_y, \\ \beta_z &= u\tau_{zx} + v\tau_{zy} + w\tau_{zz} - q_z. \end{aligned} \tag{8.5}$$

Here $J^{-1} = \left|\dfrac{\partial(x,y,z,t)}{\partial(\xi,\eta,\zeta,\tau)}\right|$ is the determinant of the Jacobi matrix of the geometry.

Note that the Cartesian derivatives of the velocity components in the definition of τ_{ij} are transformed using the chain rule. The contravariant velocities are defined by, see, e.g., [15]:

$$\begin{aligned} U &= \xi_t + \xi_x u + \xi_y v + \xi_z w, \\ V &= \eta_t + \eta_x u + \eta_y v + \eta_z w, \\ W &= \zeta_t + \zeta_x u + \zeta_y v + \zeta_z w. \end{aligned} \tag{8.6}$$

Another form of eq. (8.6) is given by

$$U = \xi_x(u - x_\tau) + \xi_y(v - y_\tau) + \xi_z(w - z_\tau), \tag{8.7}$$

and similar for V and W, where x_τ, y_τ, z_τ are the Cartesian components of the grid velocity.

In the literature, [16] to [19], for the Navier-Stokes equations often the following form is used:

$$\frac{\partial(J^{-1}\underline{Q})}{\partial t} + J^{-1}\left(\frac{\partial(\underline{E} - x_\tau \underline{Q})}{\partial x} + \frac{\partial(\underline{F} - y_\tau \underline{Q})}{\partial y} + \frac{\partial(\underline{G} - z_\tau \underline{Q})}{\partial z}\right) = \\ = J^{-1}\left(\frac{\partial \underline{E}_{visc}}{\partial x} + \frac{\partial \underline{F}_{visc}}{\partial y} + \frac{\partial \underline{G}_{visc}}{\partial z}\right), \tag{8.8}$$

which is identical to eq. (8.2).

Eq. (8.2) combined with eq. (8.7) as well as eq. (8.8) is named Arbitrary Lagrange-Euler formulation (ALE), which means that grid points move with velocities other than the velocity of the local fluid element[7].

8.2.2 Structure Dynamics Equations

The main task of an analysis of a structural system consists in the calculation of the deformations and stresses due to externally applied mechanical and thermal loads. In the theory of elasticity, considering a linear orthotropic material, the relationship between stresses and strains is given by, [20]:

$$\underline{\sigma} = D(\underline{\varepsilon} - \underline{\varepsilon}_T - \underline{\varepsilon}_I), \tag{8.9}$$

with $\underline{\sigma}$ denoting the stresses, D the elasticity matrix containing the appropriate material properties, $\underline{\varepsilon}$ the resulting strains, $\underline{\varepsilon}_T$ the thermal strains, and $\underline{\varepsilon}_I$ the initial strains.

Further we have:

$$\underline{\sigma} = \begin{pmatrix} \sigma_{xx} \\ \sigma_{yy} \\ \sigma_{xx} \\ \sigma_{xy} \\ \sigma_{yz} \\ \sigma_{zx} \end{pmatrix}, \ \underline{\varepsilon}_I = \begin{pmatrix} \varepsilon_{Ixx} \\ \varepsilon_{Iyy} \\ \varepsilon_{Izz} \\ \varepsilon_{Ixy} \\ \varepsilon_{Iyz} \\ \varepsilon_{Izx} \end{pmatrix}, \ \underline{\varepsilon}_T = \begin{pmatrix} \alpha_{xx} \\ \alpha_{yy} \\ \alpha_{zz} \\ 0 \\ 0 \\ 0 \end{pmatrix} \Delta T, \tag{8.10}$$

$$\underline{\varepsilon} = \begin{pmatrix} \varepsilon_{xx} \\ \varepsilon_{yy} \\ \varepsilon_{xx} \\ \varepsilon_{xy} \\ \varepsilon_{yz} \\ \varepsilon_{zx} \end{pmatrix} = \begin{pmatrix} \frac{\partial}{\partial x} & 0 & 0 \\ 0 & \frac{\partial}{\partial y} & 0 \\ 0 & 0 & \frac{\partial}{\partial z} \\ \frac{\partial}{\partial y} & \frac{\partial}{\partial x} & 0 \\ 0 & \frac{\partial}{\partial z} & \frac{\partial}{\partial y} \\ \frac{\partial}{\partial z} & 0 & \frac{\partial}{\partial x} \end{pmatrix} \begin{pmatrix} u \\ v \\ w \end{pmatrix} = S\underline{u}, \tag{8.11}$$

[7] If the grid point moves with the fluid element, this is called a Lagrange formulation. If the grid point does not move, while the fluid element travels through the grid, we speak of an Euler formulation.

where $\underline{u} = (u,\ v,\ w)^T$ are the displacements, $\underline{\alpha} = (\alpha_{xx},\ \alpha_{yy},\ \alpha_{zz},\ 0,\ 0,\ 0)^T$ the thermal expansion coefficients, and $\Delta T = T - T_{ref}$ is the temperature difference[8].

The equilibrium condition $S^T \underline{\sigma} + \underline{b} = 0$, saying that the internal forces have to be equal to the external forces (body forces) in connection with the principle of virtual work, leads to the relation, [21]:

$$\int_V \delta\underline{\varepsilon}^T \underline{\sigma}\, dV - \int_V \delta\underline{u}^T \underline{b}\, dV - \int_S \delta\underline{u}^T \underline{t}\, dS - \sum_a \delta\underline{u}\,{}_a^T \underline{f}^a = 0, \tag{8.12}$$

with $\delta\underline{\varepsilon}$ being the virtual strains, $\delta\underline{u}$ the virtual displacements, $\underline{b}$ the body forces, $\underline{t}$ the tractions (e.g. flow shear stress) on the surface, and $\underline{f}^a$ the single forces at nodal points a.

Transferring eq. (8.12) to a finite element solution, the integrals are divided to yield sums over individual elements 'e':

$$\sum_e \int_{V^e} \delta\underline{\varepsilon}^T D(\underline{\varepsilon} - \underline{\varepsilon}_T - \underline{\varepsilon}_I)\, dV - \sum_e \int_{V^e} \delta\underline{u}^T \underline{b}\, dV -$$
$$- \sum_e \int_{S^e} \delta\underline{u}^T \underline{t}\, dS - \sum_a \delta\underline{u}\,{}_a^T \underline{f}^a = 0. \tag{8.13}$$

We apply now the usual assumptions of finite element theory, [21], [22], namely that the displacements in an element are defined by the displacements $\underline{\tilde{u}}$ at the nodes of this element:

$$\delta\underline{u} = \sum_a \underline{N}_a\, \delta\underline{\tilde{u}}_a \equiv \underline{N}\, \delta\underline{\tilde{u}}, \tag{8.14}$$

and that the strains are formulated by:

$$\delta\underline{\varepsilon} = \sum_a S\underline{N}_a\, \delta\underline{\tilde{u}}_a = \sum_a \underline{B}_a\, \delta\underline{\tilde{u}}_a \equiv \underline{B}\, \delta\underline{\tilde{u}}, \tag{8.15}$$

where $\underline{N}_a$ are the shape functions at the nodal point a. Then with eq. (8.13) we obtain:

$$\sum_e \delta\underline{\tilde{u}}\,{}_a^T \left(\int_{V^e} \underline{B}\,{}_a^T D\, (\underline{B}_b \underline{\tilde{u}}_b - \underline{\alpha}\Delta T - \underline{\varepsilon}_I)\, dV - \right.$$
$$\left. - \int_{V^e} \underline{N}\,{}_a^T \underline{b}\, dV - \int_{S^e} \underline{N}\,{}_a^T \underline{t}\, dS - \underline{f}^a \right) = 0. \tag{8.16}$$

The displacements $\delta\underline{\tilde{u}}_a$ and $\underline{\tilde{u}}_b$ at the nodal points are independent from the integration variable and can be written outside of the integrals. With

$$K\underline{\tilde{u}} = \underline{F}, \tag{8.17}$$

[8] In order to avoid confusion, we note that T, used as superscript in this and the next sub-section, denotes the transpose of the corresponding vector or matrix.

we find:

$$K = \sum_e \int_{V^e} \underline{B}^T D \underline{B}\, dV,$$

and:

$$\underline{F} = \sum_e \int_{V^e} \underline{B}^T D \underline{\alpha}\, \Delta T dV + \sum_e \int_{V^e} \underline{B}^T \underline{\varepsilon}_I\, dV + \\ + \sum_e \int_{V^e} \underline{N}^T \underline{b}\, dV + \sum_e \int_{S^e} \underline{N}^T \underline{t}\, dS + \underline{f}. \qquad (8.18)$$

K is called the stiffness matrix and $\underline{F}$ the load vector. Eq. (8.17) describes the static equilibrium of a structural system, [21].

In cases where the loads change rapidly with time, inertial forces as well as damping forces due to energy dissipation inside the material (e.g. microstructure movement) need to be considered. Replacing in eq. (8.17) the body force $\underline{b}$ by $\underline{\bar{b}} - \rho \underline{\ddot{u}} - \mu \underline{\dot{u}}$ yields:

$$M \underline{\ddot{\tilde{u}}} + C \underline{\dot{\tilde{u}}} + K \underline{\tilde{u}} = \underline{F}, \qquad (8.19)$$

with the mass matrix:

$$M = \sum_e \int_{V^e} \rho \underline{N}^T \underline{N} dV,$$

and the damping matrix[9]:

$$C = \sum_e \int_{V^e} \mu \underline{N}^T \underline{N} dV.$$

Eq. (8.19) represents the mathematical formulation for the dynamic behavior of elastic structures, however with homogenous damping properties. This points to the difficulty to describe *real-elastic* structures with point- and line-wise distributed joints (rivets, screws, gluing and welding zones), which introduce non-linearities and damping, and of non-linear deformations, as discussed shortly in the opening remarks of this chapter. If these could be modelled to the needed degree, the *real-elastic* properties of a structure could be determined in the design and development process of a flight vehicle much earlier than is currently possible, [2].

However, this topic presents enormous challenges. Scale discrepancies as large as in flow with turbulent boundary layers past whole flight-vehicle configurations would have to be mastered. It is not clear yet how to proceed. Possible approaches may use statistical models based on parameter identification as in statistical turbulence theory, combined with methods similar to direct numerical or large-eddy simulations. It appears therefore, that the classical approach of structural mechanics based on the computational determination only of the *perfect-elastic* properties, combined with structural tests late in the development process to find the *real-elastic* properties, has to be kept for quite some time.

[9] This derivation assumes that the resistance is linear to the velocity $\underline{\dot{\tilde{u}}}$.

8.2.3 Heat Transport Equation

The thermal energy equilibrium in a solid or fluid can be expressed by the diffusive equation[10]:

$$\rho c \frac{\partial T}{\partial t} + \nabla^T \underline{q} - \Omega = 0, \tag{8.20}$$

with ρ the density, c the specific heat, $\underline{q}$ = - $k\nabla T$ Fourier's law of heat conduction[11], k the thermal conductivity matrix and Ω the internal heat generation.

Using the principle of virtual temperatures (which acts in the same way as the virtual displacements), the Gauss' integral theorem and the proper considerations of boundary conditions, the finite element equilibrium equation for heat transport on the basis of eq. (8.20) has the form, [20], [23]:

$$\underbrace{\int_V \rho c \underline{N}^T \underline{N} dV}_{C} \; \dot{\mathbf{T}} + \underbrace{\int_V \underline{B}^T \lambda \underline{B} \, dV}_{K_c} \; \mathbf{T} + \underbrace{\int_{Sh} h \underline{N}^T \underline{N} dS}_{K_h} \; \mathbf{T} = \tag{8.21}$$

$$\underbrace{\int_V \Omega \underline{N}^T dV}_{P_\Omega} + \underbrace{\int_{Ss} q_s \underline{N}^T dS}_{P_q} + \underbrace{\int_{Sh} h T_e \underline{N}^T dS}_{P_h} - \underbrace{\int_{Sr} \hat{\sigma} \epsilon \, (T_s^4 - T_a^4) \underline{N}^T dS}_{P_r},$$

where the following boundary conditions[12] are applied:

$\underline{q} \cdot \underline{n} = -q_s$	specified surface heat flow on S_s,
$\underline{q} \cdot \underline{n} = h \, (T_s - T_e)$	convective heat exchange on S_h,
$\underline{q} \cdot \underline{n} = \hat{\sigma} \epsilon \, (T_s^4 - T_a^4)$	radiation heat exchange on S_r.

In eq. (8.21) $\mathbf{T}$ is the element nodal temperature vector and $\dot{\mathbf{T}}$ the time derivative of $\mathbf{T}$, T_s the surface temperature, T_e the temperature at the boundary layer edge, T_a the ambient temperature, h the convective heat transfer coefficient (see Section 10.3), $\hat{\sigma}$ the Stefan-Boltzmann constant[13] and ϵ the emission coefficient.

Eq. (8.21) can be written in matrix form:

$$C \, \dot{\mathbf{T}} + (K_c + K_h) \, \mathbf{T} = P_\Omega + P_q + P_h - P_r = \mathbf{P} \, , \tag{8.22}$$

---

[10] Unsteady heat conduction equation.

[11] For the generalized aerothermodynamic heat transfer formulation, see, e.g., [5].

[12] Generally the surfaces S_s, S_h and S_r are different, but in some specific application cases we may have $S_s \equiv S_h \equiv S_r$.

[13] The symbol for the Stefan-Boltzmann constant with the circumflex $\hat{\sigma}$ is used here in order to distinguish it from the symbol of the stresses σ.

with C being the heat capacity matrix, K_c the conduction matrix and K_h the convection matrix. The vector $\mathbf{P}$ contains the heat inputs arising from several sources defined above and the heat radiation from the surface[14]

8.3 Coupling Procedures

The main problem for disciplines analyzing physical states, which interact strongly with each other, is that they have to provide for a fast and precise exchange (or transfer) of data at the corresponding boundaries, in the way that an equilibrium state, either static or quasi-static (dynamic), is achieved[15].

The coupling of the solutions of various disciplines is easiest and provides best results, if in the different domains the same kind of numerical approximation methods (e.g. finite element methods (FEM)) as well as conformal meshes (e.g. unstructured grids) along the common boundaries are used. This has the consequence that at most an interpolation in the frame of the, e.g., finite element solution during the data transfer at the boundaries has to be conducted. But in real applications this is often not practical, since in general the fluid domain needs a finer grid resolution than the structural domain and the solution procedure itself (e.g. FEM) is not always appropriate for integrating the fluid dynamical equations, in particular in the hypersonic regime.

Therefore in the last years it could be observed that methods were developed which are based in each case on the best approximation in the corresponding application domain. This has some advantages. Firstly, methods for the various disciplines can be used, which are well tested and verified, which also includes commercial products in particular for the structural domain. Secondly, the grid can be more precisely adapted where it is required by the physics. It might be in one domain unstructured and in another domain structured as best suited to the solution method employed. However, the disadvantage consists in the more complex data transfer across the common interface boundary, which requires a special treatment, since in general a sophisticated interpolation method is necessary.

The coupling process depends strongly on the physical situation to be solved. For example, the requirements on the solution process for an aeroelastic problem are quite different from those of a heat transfer problem. Our main interest here consists in describing the coupling procedure for fluid - heat transfer problems (thermal fluid-structure interaction). Nevertheless we will address also some other applications like aeroelasticity (mechanical fluid-structure interaction), ablation, et cetera. We begin with the description of the *mechanical-fluid structure interaction* (classical aeroelastic

[14] In case of concave contours, the heat radiation depends on the local view factor (non-convex effects, Section 9.1) which describes for a given surface element the ratio of heat emission and absorption and leads to an adjustment of P_r, [22], [23].

[15] We consider here only solutions based on numerical methods for fluid dynamics (CFD) and for structural dynamics (CSD).

approaches), look then at the *thermal-fluid structure interaction* (without structural response) and finally at the *thermal-mechanical fluid-structure interaction* (with structural response).

8.3.1 Mechanical Fluid-Structure Interaction Aeroelastic Approach I

For the treatment of the aeroelastic problem, over the years, the authors of [24] to [29] have introduced a specific system of methods. Therein, both the equations for the dynamic structural motion and for fluid dynamics are approximated and solved by a finite element approach. Used are the fluid dynamic equations in the Euler formulation, eq. (11.1), where no grid movement is incorporated. As already mentioned, the data transfer between the disciplines along the common boundary is simplified by employing a monolithic approach. Since the unsteady CFD Euler method needs very significant computer time, a system-identification procedure was developed, based on some unsteady master CFD solutions for provision of the unsteady aerodynamic loads, which are required in the dynamic structural solver, eq. (8.19), where $F = f_a(t)$ represents the unsteady aerodynamic loads.

8.3.2 Mechanical Fluid-Structure Interaction Aeroelastic Approach II

Another strategy for the treatment of aeroelastic problems was pursued by the authors of [16] to [19], [30] to [32]. They consider besides the fluid and structure fields the moving mesh as a third field. The corresponding equations for the three fields have to be solved simultaneously.

The set of equations is compiled by the semi-discrete form of eq. (8.8) or eq. (8.2) for fluid dynamics, which could, for example, be approximated by a finite-volume approach, by eq. (8.19) for structural dynamics, and by

$$\tilde{M}\ddot{\underline{x}} + \tilde{C}\dot{\underline{x}} + \tilde{K}\underline{x} = K_t\tilde{\underline{u}} \tag{8.23}$$

for the moving mesh, where $\underline{x}$ describes the position of a moving fluid grid point. $\tilde{M}, \tilde{C}, \tilde{K}$ are fictitious mass, damping and stiffness matrices, and K_t is designed as transfer matrix describing the continuity between structural displacement and moving fluid mesh along the common interface boundary.

The coupling of the fluid and structure equations at the interface boundary Γ is conducted by imposing:

$$\begin{aligned} \sigma_S \cdot \underline{n} &= -p\,\underline{n} + \sigma_F \cdot \underline{n}, && \\ \frac{\partial \underline{u}_S}{\partial t} &= \frac{\partial \underline{u}_F}{\partial t} && \text{no-slip condition,} \\ \frac{\partial \underline{u}_S}{\partial t} \cdot \underline{n} &= \frac{\partial \underline{u}_F}{\partial t} \cdot \underline{n} && \text{slip condition,} \end{aligned} \tag{8.24}$$

with σ_S, σ_F the tensors of structural and fluid viscous stresses, $\underline{u}_S$, $\underline{u}_F$ the displacements of the structural and fluid fields, p the fluid pressure field and $\underline{n}$ the normal on a given point on Γ.

Further the coupling between the structure and the moving grid on Γ takes place via the relations:

$$\underline{x} = \underline{u}_S, \\ \frac{\partial \underline{x}}{\partial t} = \frac{\partial \underline{u}_S}{\partial t}. \tag{8.25}$$

For accuracy, stability and convergence reasons the solution of eq. (8.23) must meet the so called geometric conservation law, which can be achieved by an appropriate time discretization of the time-dependent grid position $\underline{x}$ representing the mesh velocity $\dot{\underline{x}}$, [16].

The three-fields approach allows for an employment of the partitioned solution procedure, where best suited numerical simulation methods can be applied for the various disciplines, like finite-volume or finite-difference methods in the fluid domain and finite element methods in the structural and moving mesh domains.

A solution strategy, regarding the time coupling for the set of the coupled equations (8.8), (8.19), (8.23), with a powerful capacity concerning the numerical stability and the accuracy of the results, is given by the staggered scheme, Fig. 8.3, [33].

This scheme can be described by :

1. Update the fluid mesh coordinates with structural displacements u^n and the velocities $\dot{u}^n$ to conform to the structural boundary at t^n.
2. Advance the flow from t^n to $t^{n+1/2}$.
3. Advance the structure using the pressure and the stress field at t^n from t^n to t^{n+1}.
4. Transfer the pressure and the stress field at $t^{n+1/2}$ to the structural code and transfer the structural displacements u^{n+1} and the velocities $\dot{u}^{n+1}$ to the fluid code.
5. Advance the flow from $t^{n+1/2}$ to t^{n+1}.
6. Re-compute the structure using the pressure and the stress field at $t^{n+1/2}$ from t^n to t^{n+1}.

The above procedure is a variation of the scheme reported in [18], which was successfully applied also to space applications, [34].

In general the mesh applied to the structural system is coarser than the mesh required for a proper resolution of the fluid domain. This means that typically the common boundaries Γ of the fluid and structural domains have non-matching discrete interfaces. Therefore, the transfer of the displacements from the structural domain to the fluid mesh as well as the transfer of the pressure and stress field to the structural mesh can be critical with respect to consistency and load conservation[16].

[16] Load conservation is considered here in the sense that the forces and energies (for example displacement work) on the fluid/structure interface Γ, evaluated by a suitable interpolation procedure, are consistent.

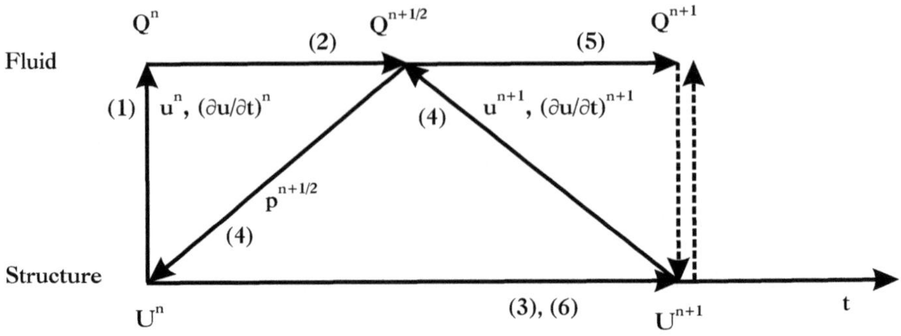

Fig. 8.3. Staggered scheme with inter-field parallelism for the coupling of the solutions of a three-fields fluid-structure interaction problem, $U = (u, \dot{u})^T$, [33].

A promising approach to solve this problem, sometimes indicated as the space-coupling strategy, is given in [19]. The main idea behind it is the definition of Gauss points on the finite elements, which are in contact with the structural boundary interface Γ_S, and the connection of these with appropriate cells or elements of the fluid boundary interface Γ_F, in the sense to provide these points with the pressure and the shear stress field values of the fluid. This procedure is known as "pairing" and several algorithms are available for solving this task, [35], [36].

A similar approach is reported in [37], where a neutral interface, defined by two Gauss parameters, is constructed, which acts as data-transfer medium between the boundary interfaces[17] Γ_S and Γ_F. These procedures are often used for practical applications conducted by the aerospace industry, [39] to [41].

8.3.3 Thermal-Mechanical Fluid-Structure Interaction

One of the earliest if not the first paper dealing with the simulation of coupled thermal-mechanical fluid-structure interaction is [42] from the year 1988. There a monolithic finite element environment was developed for solving the Navier-Stokes equations for the flow field, the structural equations for the mechanical response and the heat conduction equation for the thermal field in the solid. Further the finite element grids, created for the flow and the structure field, had at the common interface the same nodes which render any interpolation for transferring the boundary conditions unnecessary.

Generally, supersonic and hypersonic flows past RV-W- and RV-NW-type, as well as CAV/ARV-type flight vehicles, heat up the – in general radiation-cooled – surface material of the vehicles. This has two major consequences. Firstly, the structure responds to the severe surface-material temperature

[17] Another interesting solution regarding the data transfer between non-matching boundaries can be found in [38]. There, the construction of a virtual grid is proposed, which has a similar function as the neutral interface mentioned above.

field with thermal stresses and/or deformations. Secondly, responding to the severe thermal state of the surface – surface-temperature and temperature gradients in the gas at the wall, Chapter 9 – and the deformation, the general properties of the flow field (thermal surface effects, shock waves, expansion zones, local separation with vortex phenomena, et cetera) and the properties of the thermal boundary layer may change dramatically.

Therefore for an advanced design of aerospace vehicles coupled solutions of the corresponding disciplines are indispensable.

In the more general case, where thermal-mechanical fluid-structure interactions with moving grids are considered, the set of governing equations consists of

- the Navier-Stokes equations in ALE-formulation for the flow field, eq. (8.2) or (8.8),
- the finite element representation of structure dynamics, eq. (8.19),
- the finite element representation of the heat conduction in solids, eq. (8.22),
- the finite element representation for the moving mesh, eq. (8.23).

This is a four-fields approach in the sense described also in [31]. Another approach, where the determination of the moving mesh by eq. (8.23) is replaced by various mesh tracking procedures, is reported in [43]. For aerospace vehicles of RV-W- and RV-NW-type often the deformations due to thermal and mechanical (pressure and stress field) loads are small. This is the reason why the investigations reported in [23], [44] to [46] do not care for a moving mesh capability in the fluid domain.

To include the thermal conditions, the interface boundary conditions on Γ as formulated for the aeroelastic case, eqs. (8.24) and (8.25), have to be extended. They consist now of:

- the structural compatibility conditions, eq. (8.24):

$$\begin{aligned} \sigma_S \cdot \underline{n} &= -p\,\underline{n} + \sigma_F \cdot \underline{n}, && \\ \frac{\partial \underline{u}_S}{\partial t} &= \frac{\partial \underline{u}_F}{\partial t} && \text{no-slip condition,} \\ \frac{\partial \underline{u}_S}{\partial t} \cdot \underline{n} &= \frac{\partial \underline{u}_F}{\partial t} \cdot \underline{n} && \text{slip condition,} \end{aligned}$$

- the temperature continuity and heat flux equilibrium conditions:

$$\begin{aligned} \lambda_S \nabla T_S \cdot \underline{n} &= \lambda_F \nabla T_F \cdot \underline{n}, \\ T_S &= T_F, \end{aligned} \tag{8.26}$$

- the mesh motion continuity conditions, eq. (8.25):

$$\begin{aligned} \underline{x} &= \underline{u}_S, \\ \frac{\partial \underline{x}}{\partial t} &= \frac{\partial \underline{u}_S}{\partial t}. \end{aligned}$$

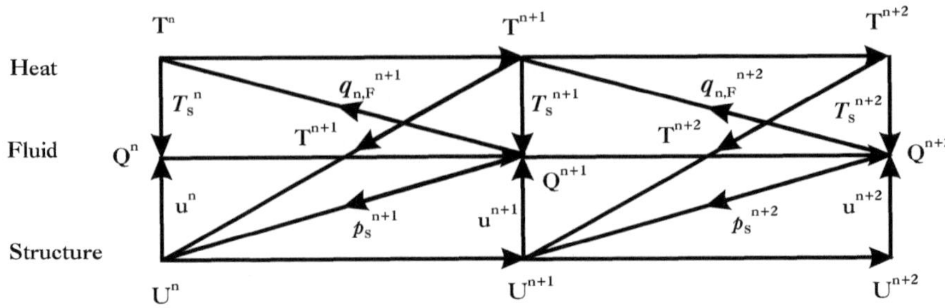

Fig. 8.4. Conventional serial staggered scheme (CSS) for the coupling of the solutions of thermal-mechanical fluid-structure interaction problems, [31].

As already indicated, for aeroelastic problems the evolution with respect to the development of simulation systems for thermal-mechanical fluid-structure interaction issues exhibits, that loosely coupled strategies are favored. This allows the combination of independent numerical methods for the single disciplines as well as the application of different grid structures.

One promising method for solving the system of coupled equations (8.8), (8.19), (8.22), (8.23), is to use a staggered scheme[18], similar to that shown Fig. 8.3. We present here the one proposed in [31] which is called the conventional serial staggered procedure (CSS), Fig. 8.4.

For pure fluid-thermal coupling (without mechanical response due to thermal loads) a similar staggered scheme can be found in [45], [47].

Another practical scheme for solving the system of coupled equations is the classical Dirichlet-Neumann iteration which can be applied also for dynamic thermal-mechanical fluid-structure interactions, [47], [48]. This scheme consists of the sequence of flux calculations on the interface boundary Γ_F by the fluid equations and the state calculations on the interface boundary Γ_S by the thermal-structural equations. Dirichlet boundary conditions are used in the fluid domain ($\underline{u}$, $\dot{\underline{u}}$, T), and Neumann conditions in the structural domain (q_n, $p \cdot \underline{n}$, $\sigma_F \cdot \underline{n}$).

Let the state in the structural field be denoted by Ψ, which contains deformations and temperatures, and the state at the structural interface boundary Γ_S by Ψ_S, then the iteration for $k+1$ is performed by, [44], [47]:

$$\Psi_S^{k+1} = \omega \, \Psi_S^k + (1-\omega) \, \Psi_S^k, \tag{8.27}$$

with the relaxation coefficient $\omega \leq 1.0$. The iteration has succeeded if a prescribed convergence limit is satisfied.

The other question is how to transfer the data on the interface boundary Γ from one domain to the other, when the grid nodes on the fluid and structure domain do not coincide (non-matching meshes). This is supported by the fact

[18] This is applicable only for transient problems, where the thermal behavior is transient and the mechanical and fluid behaviors are stationary.

that normally the grid spacing on the fluid side is finer compared to that of the structural side. So the need for an interpolation procedure arises, with the ability to conserve, for example, the sum of loads[19] (pressure and shear stress field) as well as the energy in terms of the heat fluxes along the common interface boundary.

A first possibility of performing the load transfer is given by introducing the Lagrange multipliers δ_i and an additional state variable z, whereby the homogeneity of the virtual work (or virtual power) is preserved, which leads to the relation, [49]:

$$\int_{\Gamma} \delta_i \left(\Psi_i - z \right) d\Gamma = 0, \qquad i = F, S. \tag{8.28}$$

In order to simplify the procedure, one can choose $z = \Psi_S|_\Gamma$, reducing the number of Lagrange multipliers δ_i to one, which makes the numerics easier-going.

A second method is given by the definition of a virtual surface grid, where the interface boundaries Γ_S and Γ_F are projected to this surface, which is described by two Gauss parameters, [37], [43]. Once the virtual surface is constructed, conventional interpolation routines are used for transferring the data from one interface boundary to the other, but this procedure is not in a fully conservative fashion.

A third method is based on the conservation of energy[20] for both the conversion of the fluid pressure and the shear stress fields into a mechanical load and the transfer of the heat fluxes to the structural interface boundary Γ_S, [31].

Recent investigations apply the commercial interpolation software MpCCI (Mesh-based parallel Code Coupling Interface), [50]. The standard technique of the commercial MpCCI software can be perceived as a particular formulation of the Lagrange multiplier method and is therefore conservative, [23], [45], [48]. In addition this software package includes non-conservative interpolation routines.

8.4 Examples of Coupled Solutions

8.4.1 Mechanical Fluid-Structure Interaction Aeroelastic Approach I

The method described in Sub-Section 8.3.1 was tested and applied to some aerospace vehicle shapes like the X-33 re-entry vehicle, Fig. 8.5, and NASA's generic hypersonic vehicle (GHV), which looks similar to Germany's lower stage SÄNGER wing-body configuration, Fig. 8.6, [51]. A typical result is

---

19 Conservation is satisfied if the sum of loads on the interface boundary of the structure Γ_S is equal to the sum of loads on the interface boundary of the fluid Γ_F.

20 For the pressure and shear stress field the virtual displacement work is considered, whereas for the thermal loads the integral of the heat fluxes over the interface boundaries is used, which has the dimension of a power, $[ML^2/t^3]$.

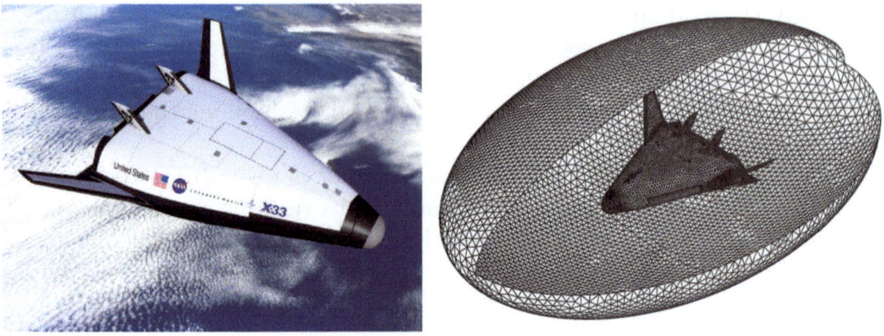

Fig. 8.5. X-33 re-entry vehicle shape (left), and unstructured mesh for a finite element fluid dynamics solver (right), [20].

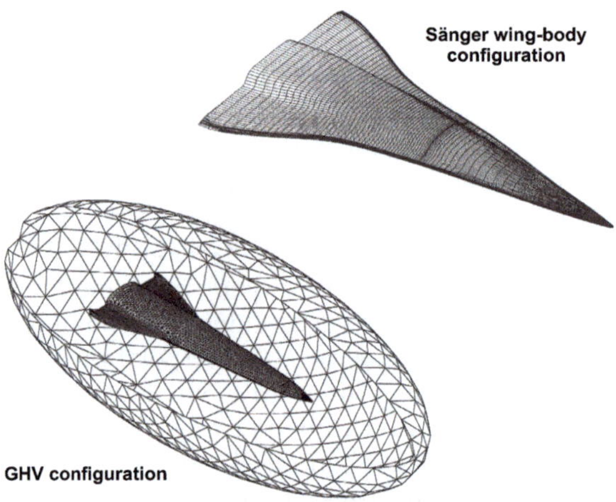

Fig. 8.6. Generic hypersonic vehicle (GHV) configuration with the unstructured finite element mesh for a fluid dynamics solver (lower left), [20]. Wing-body combination with the structured finite-volume surface mesh of Germany's SÄNGER lower stage configuration (upper right), [51].

plotted in Fig. 8.7, where the shape of the first symmetric bending mode is shown.

8.4.2 Mechanical Fluid-Structure Interaction Aeroelastic Approach II

The results obtained in the UNSI project of the European Union, performed in the years 1998 to 2000, are presented in [39]. In UNSI several contributors employed methods like the ones described in Sub-Section 8.3.2. Some of the results are presented below.

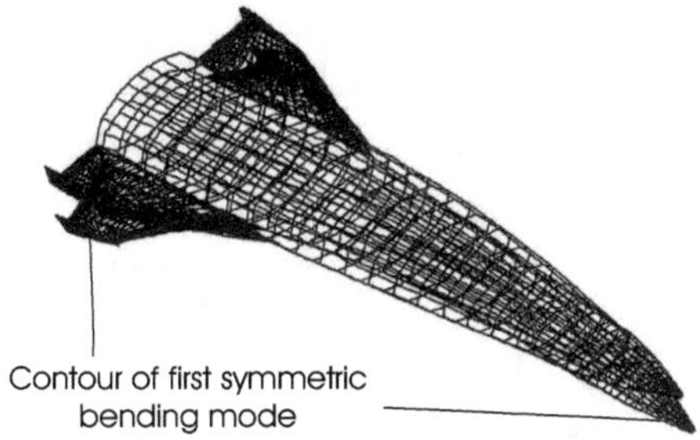

Fig. 8.7. Generic hypersonic vehicle (GHV). Typical structural mode shape: first symmetric bending mode, calculated with the method reported in [24] to [28] and [52].

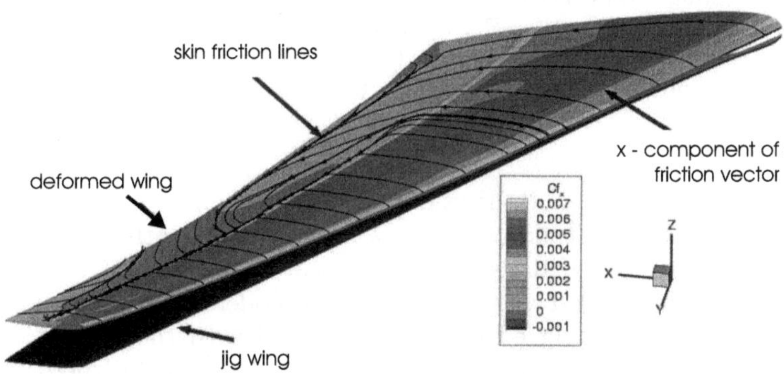

Fig. 8.8. AMP wing: static aeroelastic solution, [39]. Wing deformations and skin-friction lines for $M_\infty = 0.819$, $\alpha = 3.98°$, $Re = 1.99 \cdot 10^6$, $C_L = 0.58$.

We start with results of a study of the AMP[21] wing, which was designed jointly by AEROSPATIALE, DASA, DLR and ONERA in 1990 with the goal to undertake flutter studies for a modern aircraft in the transonic flight regime. There exists a broad experimental data base from investigations in ONERA's S2 wind tunnel in Modane. Every aeroelastic simulation (static or dynamic) of a wing configuration needs as initial geometry the so called jig shape ($\Longrightarrow$ wind-off shape, $n_z = 0$), which differs from the shape designed for cruise conditions[22]. Fig. 8.8 shows the static aeroelastic solution of the AMP wing performed with ONERA's inviscid-viscous interaction code VIS25 for the flow and the NASTRAN code for the structure.

---

[21] AMP $\Longrightarrow$ Aeroelastic Model Program.

[22] The jig shape has the following properties: a) the twist distribution of the design wing is reproduced as a result of a static mechanical fluid-structure simulation at cruise conditions, and b) the vertical locations of the jig wing sections (heights) coincide with the ones of the design wing in the no load case.

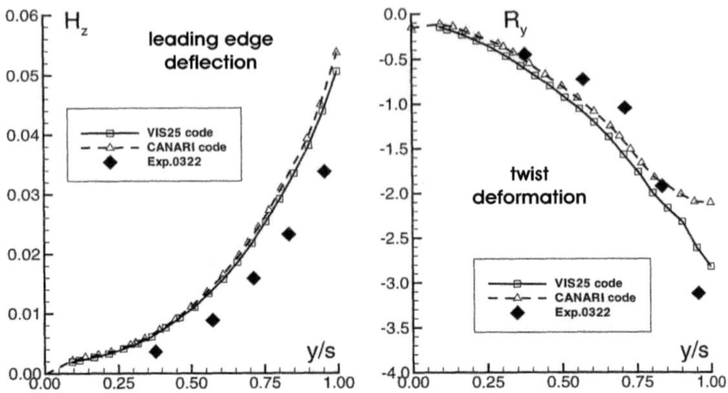

Fig. 8.9. AMP wing: static aeroelastic solution, [39]. Span-wise leading edge deflection H_z and twist deformation R_y. Flow field found with Euler (CANARI) and viscous-inviscid interaction code (VIS25). Comparison with experimental data of the ONERA S2 wind tunnel, Modane. $M_\infty = 0.862$, $\alpha = 1.6°$, $Re = 3.64 \cdot 10^6$, $C_L = 0.30$.

The span-wise deformation and the leading edge deflection for another case is shown in Fig. 8.9, where experimental data are compared with data from the VIS25 solver and from ONERA's more sophisticated fluid solver CANARI (unsteady Euler) again coupled with NASTRAN. A moderate overprediction can be observed of the leading edge deflection found with the simulation methods in contrast to the experimental data. This is also true for the wing twist, except at the wing tip, where larger deviations are observed (note the negative coordinate scale in the twist plot.).

The aeroelastic simulation procedure referred to above with the CANARI code for the flow field is also applied to the coupled dynamic fluid-structure problem. In that case the structural analysis is performed by a modal approach, where the six first mode shapes of the wing are taken into account. In Fig. 8.10 the time evolution of the modal coordinates is displayed for a low transonic test case with $M_\infty = 0.78$, $\alpha = 1.79°$, $Re = 3.49 \cdot 10^6$, $C_L = 0.30$. The stagnation pressure amounts to $p_{stag} = 90.0\ kPa$ for which obviously the unsteady responses are damped. This is also true for the first mode, where the damping occurs after a certain time delay, whereas the third mode looks indifferently. Beyond the stagnation pressure of $p_{stag} = 90.0\ kPa$, the modal coordinates are amplified indicating flutter onset.

A second test case, again in the frame of the UNSI project, considers the so called MDO[23] wing-fuselage configuration. This shape was designed in a Brite-Euram project of the EU performed in the years 1996 and 1997. Static aeroelastic computations were conducted by three contributors (two industrial companies, one research institute) using in total eight different

[23] MDO $\Longrightarrow$ Multidisciplinary Design Optimization.

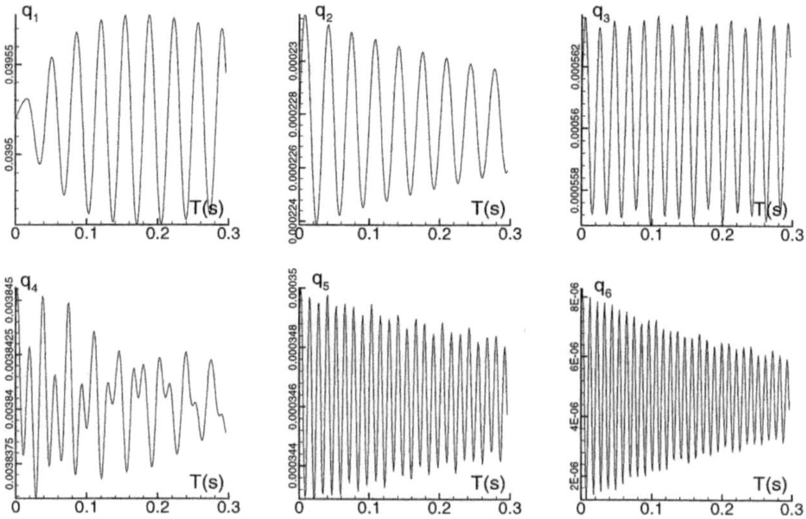

Fig. 8.10. AMP wing: dynamic aeroelastic solution, [39]. Flow field found with Euler code CANARI. Time evolution of the six first modal coordinates q_i. $M_\infty = 0.78$, $\alpha = 1.79°$, $p_{stag} = 90.0\ kPa$, $Re = 3.49 \cdot 10^6$, $C_L = 0.30$.

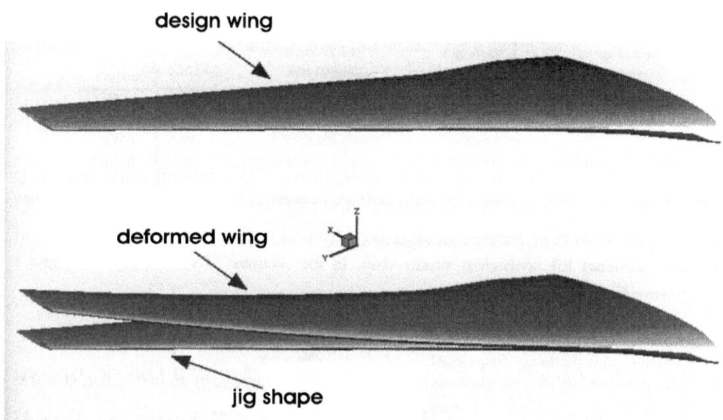

Fig. 8.11. MDO wing: static aeroelastic solution, [39]. Jig shape definition with ONERA's inviscid-viscous interaction code VIS25. Design flight conditions: $M_\infty = 0.85$, flight altitude $H = 11,280.0\, m$, total aircraft lift $C_L = 0.458$.

methods, including five Euler approaches for the flow field. Fig. 8.11 presents a three-dimensional view on the design, the deformed, and the jig shape of the MDO wing as a result of a static aeroelastic solution.

The quality of today's simulation capacity for aeroelastic problems is revealed by the two diagrams in Fig. 8.12. In the upper diagram the vertical bending deformations of the MDO wing, as predicted by the eight

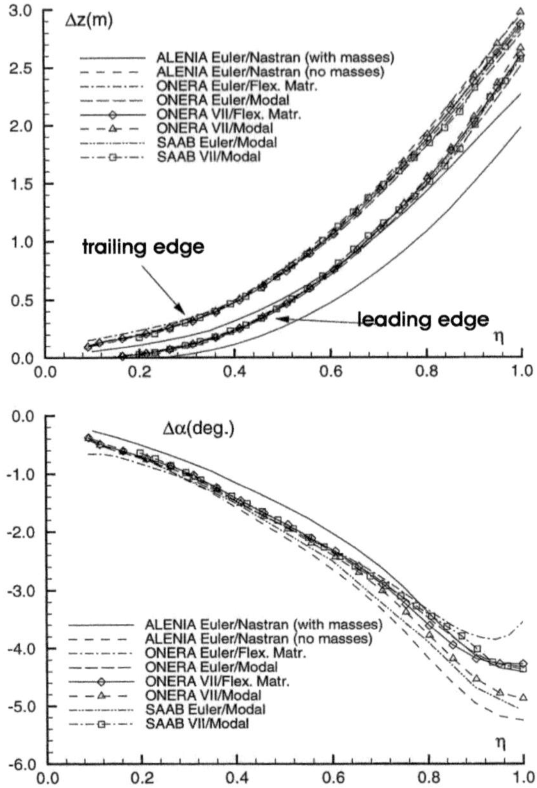

Fig. 8.12. MDO wing: static aeroelastic solution, [39]. Span-wise leading and trailing edge deflection $\triangle z$ (above) as well as twist deformation $\triangle\alpha$ (below). Design flight conditions: $M_\infty = 0.85$, flight altitude $H = 11,280.0\,m$, total aircraft lift $C_L = 0.458$.

methods mentioned before, are plotted. They show, despite the complexity of the problem, surprisingly good agreement for the trailing edge as well as the leading edge deflections. Note, that the deflection amounts to approximately $3.0\,m$ at the wing tip for a wing half-span of $37.5\,m$. The predictions of the twist deformation versus span (lower part of Fig. 8.12) agree well in the inner part of the wing, whereas at the tip larger differences appear, which we have similarly identified also in Fig. 8.9 (right).

Turning to simulations of the dynamic aeroelasticity of the MDO wing, two test conditions, apart from the optimized cruise conditions, were defined. For these the Mach number is $M_\infty = 0.88$, and in the first case the altitude amounts to $H = 7.0\,km$, corresponding to a dynamic pressure $q_\infty = 22.25\,kPa$, and in the second case the altitude is $H = 2.0\,km$, corresponding to $q_\infty = 43.10\,kPa$.

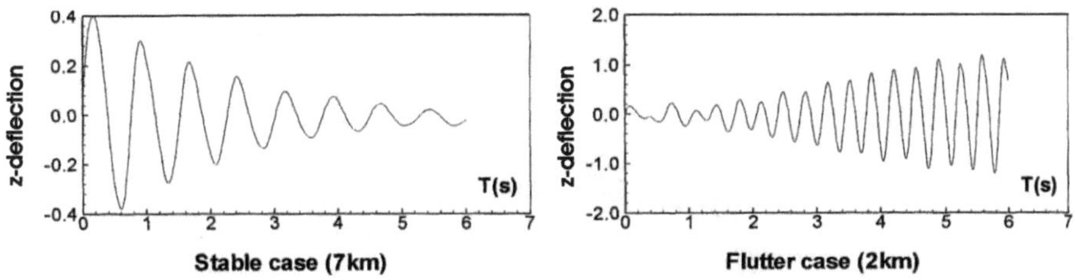

Fig. 8.13. MDO wing: dynamic aeroelastic solution (Euler solution by SAAB), [39]. z-deflection time series for a stable case at the flight altitude $H = 7.0\ km$ (left), and for a flutter case at the flight altitude $H = 2.0\ km$ (right). $M_\infty = 0.88$.

Four institutions (three industrial companies, one research institute) delivered results of six different methods for these test cases. The structural analysis is represented by a modal approach taking into account the twenty lowest wing-normal vibration modes. All the simulations predict for the $H = 7.0\,km$ case a damping of the excitations (stable flight) and for the $H = 2.0\,km$ case an increase of the oscillations (flutter response). As an example, Fig. 8.13 shows the time evolution of the deflection of the leading edge at 99 per cent half span. For the reader interested in more details, we refer to [39], [40].

8.4.3 Thermal Fluid-Structure Interaction

In Sub-Section 8.3.3 we have discussed the description of thermal-mechanical fluid-structure interactions characterized by a mechanical response in the form of mechanical and thermal stresses. In this Sub-Section we look now at interaction cases without mechanical response. Two cases with mechanical response are presented in Sub-Section 8.4.4.

Let us consider a winged aerospace vehicle. We are concerned with the question at what locations and under what conditions thermal fluid-structure interactions play a role. As we know from Chapters 3 to 6, the thermal loads are particularly high at forward stagnation points, along leading edges of wings, fins and winglets, at deflected aerodynamic control surfaces, and in regimes where a strong expansion of the flow leads to a drastic reduction of the boundary layer thickness (e.g., shoulder of capsules, Chapter 5).

What we also know is that at a surface part with small contour radius/radii the heat flux in the gas at the wall, q_{gw}, is larger than for the case with large radius/radii[24], [5]. If the vehicle surface is radiation cooled, which is the rule, this holds also for the wall temperature. However, this temperature then drops surface-tangentially very fast with increasing running length s of the boundary layer, and so does the heat flux in the gas at the wall.

High temperatures at strongly curved surfaces and strong temperature gradients in the surface-tangential direction, lead to a special phenomenon,

[24] On a sphere we have $q_{gw} \sim 1/\sqrt{R}$, Section 10.3.

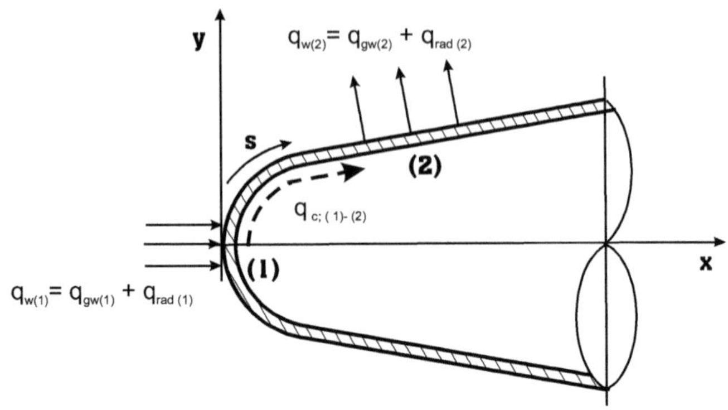

Fig. 8.14. Sketch of tangential heat transport $q_{c;(1)-(2)}$ by conduction inside a structural shell (hot primary structure of a nose cone). It diminishes the temperature at the stagnation point (1) and increases the temperature more downstream (2), leading there to a higher surface heat radiation $q_{rad(2)}$.

viz. heat transport by conduction tangentially through the structure. This is indicated at the shape shown in Fig. 8.14, representing the hot primary structure of a nose cone. Heat is conducted from location (1) to location (2), consequently the wall temperature at location (1) is reduced, whereas it is increased at location (2). As a result less heat is radiated away at (1) and more at (2) than would be the case if the surface material did not allow any tangential heat conduction.

Hence we can say that in regions with strong tangential heat conduction within the structure the surface-temperature distribution along the surface coordinate is smoothed out to some extent. With increasing curvature the effect is getting stronger. Accordingly we must note, that in such regimes on a flight vehicle, Fig. 8.15, the radiation-adiabatic wall temperature gradually is no more an adequate approximation of the actual surface temperature.

Similar effects can be observed in gaps, existing for example around the hinge area of flaps, elevons, et cetera. Turning to the heating of deflected aerodynamic control surfaces, in our case the body flaps of the X-38 vehicle, Fig. 8.16, we are confronted with the situation that the wall thermal flux $q_w = q_{gw} + q_{rad}$ (the value is negative for heat transferred into the structure!) conducts heat from the wind side to the lee side of the flap – transverse heat transport, Sub-Section 6.3.3 – and radiates into the cavity, heating up strongly the lower side of the fuselage.

For all these problems a coupled thermal fluid-structure simulation is indispensable, otherwise the thermal loads and the thermal state of the surface could be predicted with dramatic errors.

We study now quantitatively the situation of tangential heat transport at the two-dimensional representation of the leading edge of a hypersonic flight vehicle [53], like the SÄNGER lower stage, Chapter 4. In Fig. 8.17 (upper part) we find details of the geometry and the structure of the leading edge.

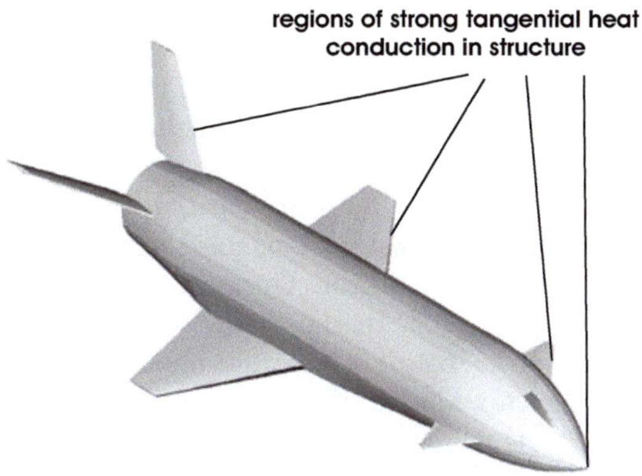

Fig. 8.15. Regimes on an aerospace vehicle with potentially strong thermal fluid-structure interactions.

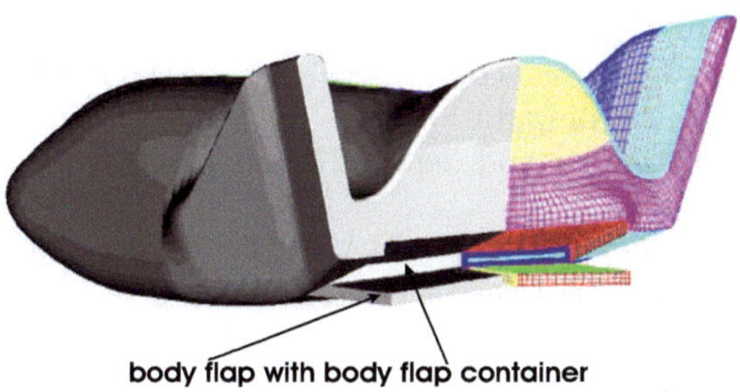

Fig. 8.16. X-38, view from behind showing the deflected body flaps with the body flap container as part of the fuselage. The right part of the figure shows the surface grid for flow computation.

For the nose radius $R_2 = 4.0\ mm$ iso-lines of the temperature are given in the lower part of the figure[25], found with a coupled Navier-Stokes/FEM solution for the flight conditions $M_\infty = 6.5$ and $H = 30.0\ km$.

The coupling of the discipline codes was conducted by the Dirichlet-Neumann method, eq. (8.27), and the data at the common boundary interface were transferred by means of a virtual surface.

For different nose radii the wall temperatures in the vicinity of the stagnation point are plotted in Fig. 8.18 (upper part). As expected, the peak temperatures increase with decreasing nose radius. In addition the wall heat flux $q_w = q_{gw} + q_{rad}$ is displayed in Fig. 8.18 (lower part), showing two interesting trends.

First, for large radii the wall heat flux q_w in the stagnation regime tends to zero, indicating that the wall behaves radiation-adiabatic. Second, for small radii the wall heat flux q_w is large at the stagnation point, while heat is

[25] The slight asymmetry of the temperature cannot be explained.

transferred tangentially in the structure along the coordinate s (not shown here). This leads to the effect, that some distance away from the stagnation point more heat (q_{rad}) is radiated from the surface than the flow is able to transport (q_{gw}) to the surface. This effect, although not very large in this case, is well discernible in Fig. 8.18 (below).

We focus now our attention on another case, the gap model, which was designed for experimental investigations of thermal loads in DLR's plasma tunnel L3K in the frame of the German technology programme TETRA[26], [54], see also Section 6.4. In the follow-on programme ASTRA[27] intensive work was devoted to coupled thermal-mechanical fluid-structure interaction problems. The gap model, Fig. 8.19, was fitted to obtain validation data for the newly developed coupled simulation environment. This environment, used for the test case, which we will discuss below, was based, [48], [49], on:

- the finite-volume τ code (DLR) for the solution of the Navier-Stokes equations on unstructured grids,
- the finite element commercial code ANSYS for the structural and thermal equilibrium mechanics computations,
- the Dirichlet-Neumann method, eq. (8.27), for the coupling of the discipline codes,
- the commercial interpolation software MpCCI, [50], for the data transfer between the non-matching boundary interfaces.

The two-dimensional Navier-Stokes solution – for the test conditions of the plasma tunnel L3K with $M_\infty = 7.27$, $\alpha = 15°$, $T_\infty = 620.0\ K$, $Re = 12.1 \cdot 10^4$ - with open gap[28] – gives an impression of the topology of the streamlines, Fig. 8.20, see also [55]. The figure shows, that despite the fact, that in reality the flow behaves three-dimensional, Fig. 8.21, a small vortex is created due to the flow separation at the upstream part of the gap inflow. This prevents a direct flow into the gap. The fluid material, which reaches the gap, streams around the vortex. Some of it has travelled up to an attachment point on the ramp before turning back to the gap.

The influence of the thermal fluid-structure coupling, with non-convex effects taken into account, see also Section 6.4 and Sub-Section 6.7.2, is shown in Fig. 8.21. The comparison of the wall temperatures exhibits clearly the coupling effect. Generally, the temperatures of the coupled solution are lower than the uncoupled (radiation-adiabatic surface boundary condition) ones

---

[26] **TE**chnologien für zukünftige Raum**TRA**nsportsysteme (Technologies for Future Space Transportation Systems), 1998-2001.

[27] **A**usgewählte **S**ysteme und **T**echnologien für zukünftige **R**TS-**A**nwendungen (Selected Systems and Technologies for Future Space Transportation Systems Applications), 2001-2003.

[28] Open and closed gaps represent different states of flow, for example near the hinge line of aerodynamic control surfaces (flaps, elevons, et cetera), Section 6.4.

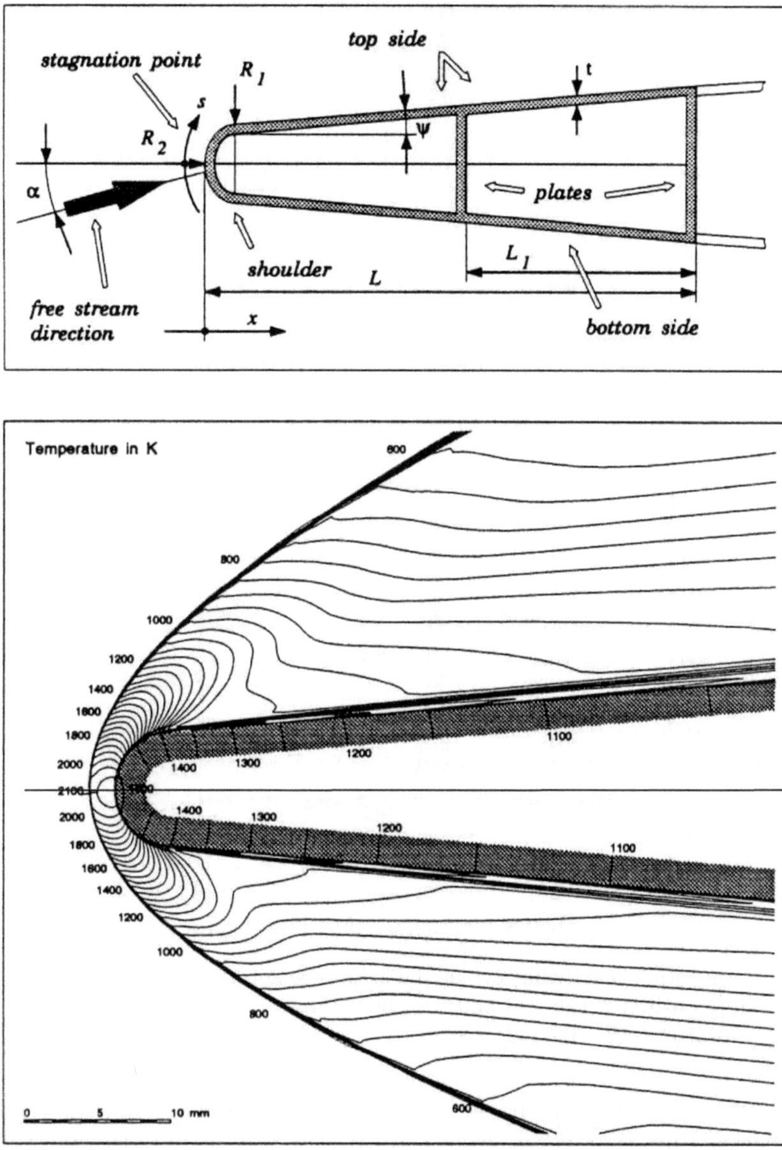

Fig. 8.17. Tangential heat transport: geometry of the 2-D leading edge test case (above), and lines of constant temperatures found with a coupled Navier-Stokes solution for $M_\infty = 6.5$ and $H = 30.0\ km$ (below), nose radius $R_2 = 4.0\ mm$, [53].

and this is more pronounced at the side edges and in regions with high surface curvature (e.g. gap inlet), Fig. 8.21 (left).

Wall temperatures in the symmetry plane, calculated with the coupled and the uncoupled approach, are compared with L3K plasma tunnel data in

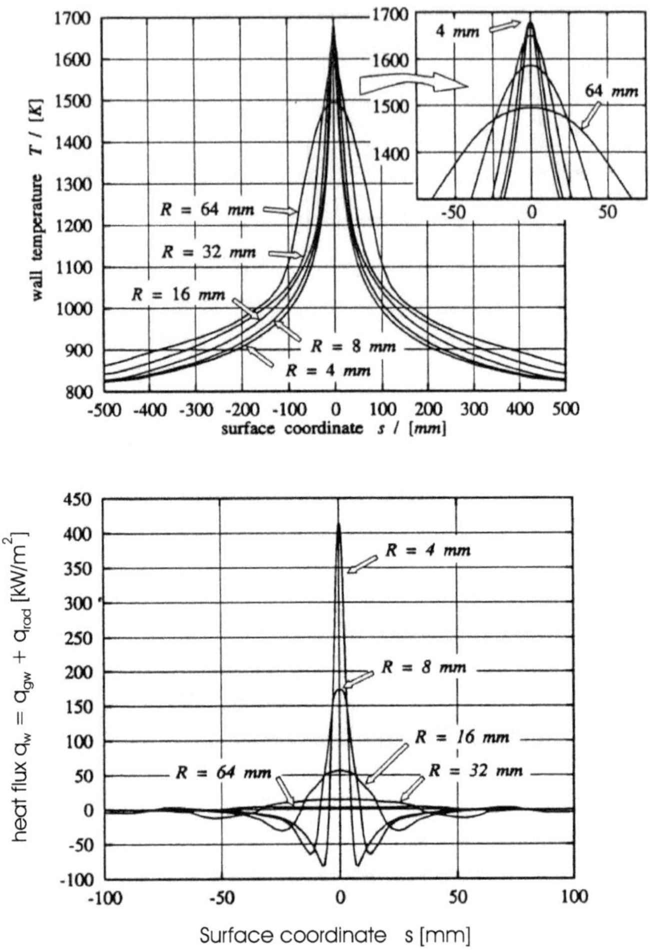

Fig. 8.18. Tangential heat transport in the structure shown in Fig. 8.17 (upper part): wall temperatures T (above) and wall heat fluxes into the structure q_w for different nose radii $R \equiv R_2$ (below), [53].

Fig. 8.21 (right). The level of the experimental data is higher than the level of the numerical ones. Further it seems that at least upstream of the gap the uncoupled data agree better with the experimental ones compared with the results of the coupled solution.

The next considered test case in this context is a generic body flap model, which was to model the situation at the rear of the X-38, see also Sub-Section 6.3.3. With it essentially the fluid and thermal behavior in the cavity between the lee side of the flap and the ground of the flap box (lower side of the fuselage) was investigated, Fig. 8.22. All plates of the model were manufactured from the ceramic material $C/C - SiC$, except for the blunt nose part, which was a water-cooled steel part.

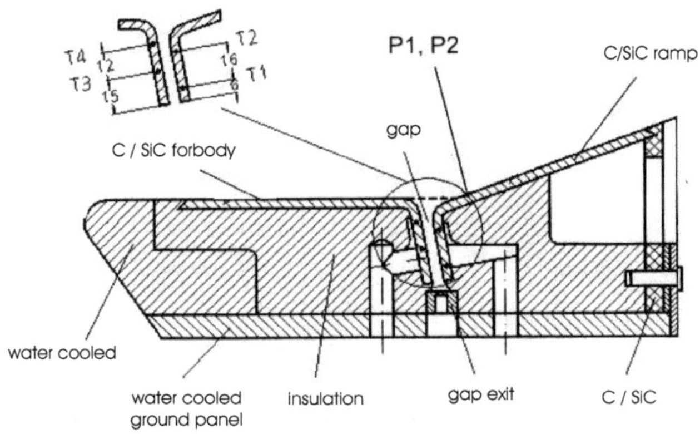

Fig. 8.19. Sketch of the configuration of the gap model, [54].

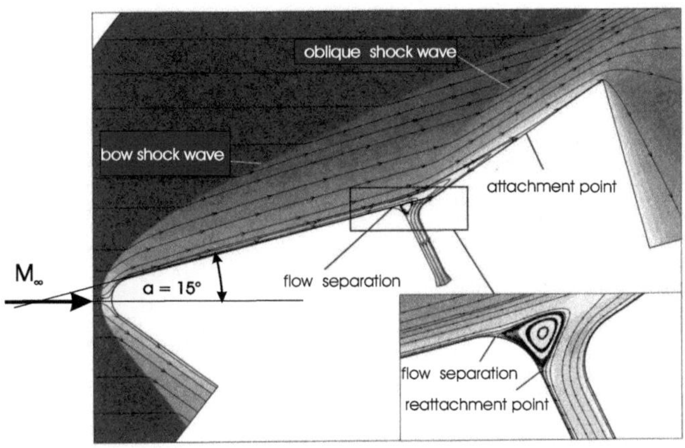

Fig. 8.20. Gap model: topology of the 2-D velocity field (streamlines) for open gap (lower right), [48], [56]. Plasma tunnel test conditions: $M_\infty = 7.27$, $\alpha = 15°$, $T_\infty = 620.0K$, $Re = 12.1 \cdot 10^4$.

We highlight in particular the heat conduction in the structure from the wind side to the lee side of the flap (transverse heat transfer) and the subsequent radiation exchange with the ground plate. In Fig. 8.23 the surface temperatures and the streamlines are displayed in an explosion view for the uncoupled (radiation-adiabatic surface boundary condition) and the coupled numerical simulation.

There are three main differences between the two solutions:

1. The temperatures along the boundaries of the flap and on the ground plate ('flap box ground' in Fig. 8.23) as well as at the junction between the forebody and the nose part are relatively high in the uncoupled solution. In the coupled solution these strong temperature gradients are

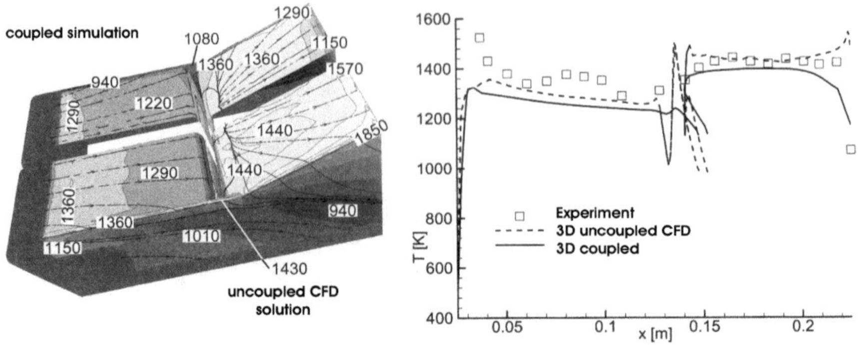

Fig. 8.21. Gap model: comparison of coupled and uncoupled (radiation-adiabatic) 3-D solution for model with open gap, [48], [56]. Wall temperature distribution (left), wall temperature distribution in symmetry plane (right).

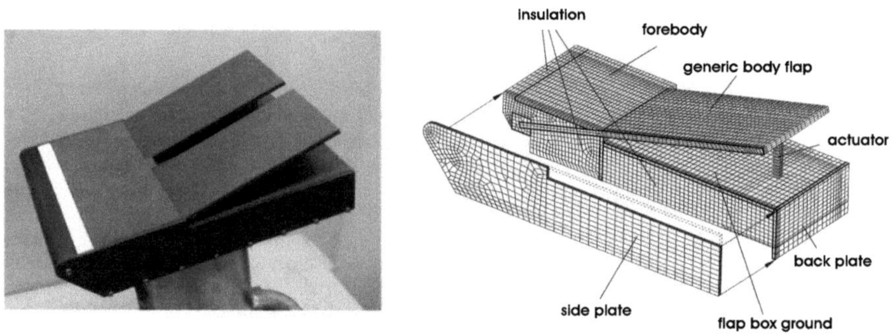

Fig. 8.22. Body flap model. Plasma tunnel model (left), sketch of the construction (right), [23], [48]. Material of the plates is $C/C-SiC$, and material of the insulation Al_2O_3.

damped out by the heat conduction in the structure. This leads to more homogeneous wall temperature fields.

2. The lee side of the flap has higher wall temperatures in the coupled solution compared to the uncoupled one due to the intensive transverse heat conduction from the wind side to the lee side.
3. The wall temperature level at the ground plate in the coupled solution (except at the side edges) is increased by the heat radiation from the lee side of the flap. Note in this context also the changes of the topology of the skin-friction lines.

Finally we consider the surface temperature along the middle section of the flap and the ground plate ($y = 50.0\,mm$), Fig. 8.24. Compared are the uncoupled and the coupled solutions with the experimental data. The experimental data on the wind side were obtained with an infrared imaging system whereas on the ground plate thermocouples were used. As before

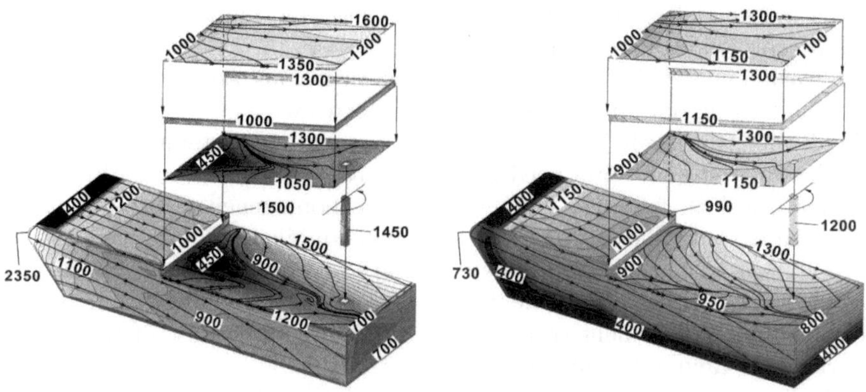

Fig. 8.23. Surface temperatures and streamlines on the body flap model. Radiation-adiabatic (uncoupled) solution (left), coupled solution (right), [23], [48], [57]. $M_\infty = 7.36$, $\alpha = 10°$, $T_\infty = 552.0\ K$, $Re = 11.9 \cdot 10^4$.

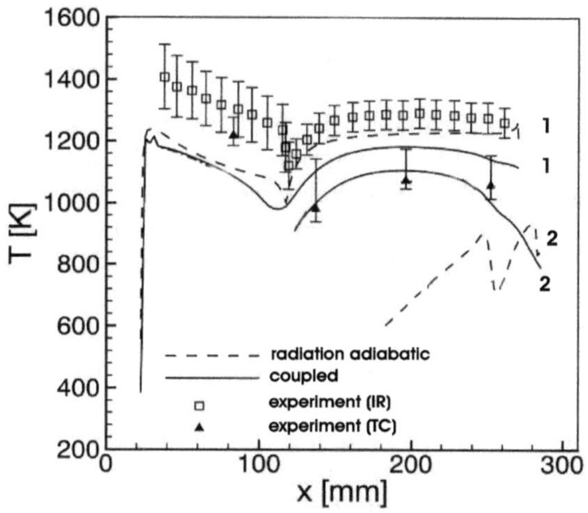

Fig. 8.24. Surface temperatures along the middle section of the flap ($y = 50.0\ mm$), comparison of results of radiation-adiabatic and coupled simulations with experimental data, [23]. $M_\infty = 7.36$, $\alpha = 10°$, $T_\infty = 552.0K$, $Re = 11.9 \cdot 10^4$, 1 denotes the windward side of the flap, 2 the ground plate of the flap box.

on the gap model the results of the uncoupled simulation for the wind side are closer to the experimental data than the coupled ones. The temperature level on the ground plate of the uncoupled solution is relatively low, what we already mentioned, but increases remarkably in the coupled simulation. The agreement of the data from this solution with the thermocouple data is more convincing than that with the infrared imaging data discussed before.

In order to assess the quality of the experimental and numerical data one should have the following in mind:

- First, such hypersonic, high-enthalpy flow situations are probably the most challenging ones from a physical point of view.
- Second, there are likely uncertainties in the experiment, stemming from, e.g., the "free-stream" produced by the conical nozzle, the heat absorption and emission from the tunnel walls, the homogeneity of the onset flow, the degree of chemical and thermal non-equilibrium of the gas, the gas pollution by the heating system of the tunnel, the accuracy of the measurement methods, in particular of the infrared imaging system, and others. Laminar/turbulent transition is very unlikely because of the low experimental Reynolds number.
- Third, the uncertainties in the numerical simulation of the flow field, e.g., in the modelling of the catalytic effects at the wall, of the chemical and thermal non-equilibrium effects, of the wall radiation effects including non-convex effects, the accuracy of the interpolation along the interface boundary, and others.

8.4.4 Thermal-Mechanical Fluid-Structure Interaction

We present here a result of probably the first work, [42], which dealt with a coupled thermal-mechanical fluid-structure interaction problem, as formulated in Sub-Section 8.3.3. A thin panel is considered, which is fixed in a panel holder, which is declined by an angle $\alpha = 15°$ to the free-stream with $M_\infty = 6.57$, and $Re^u = 1.214 \cdot 10^6 m^{-1}$, Fig. 8.25 a). The oblique shock produces a boundary layer flow with a Mach number $M_\infty = 4.24$, and an unit Reynolds number $Re^u = 2.16 \cdot 10^6 m^{-1}$, Fig. 8.25 b). All the sets of equations, given in Sub-Section 8.3.3 for the flow, the structure and the heat conduction in the solid are solved by using the finite element approach.

Due to the grid-point coincidence along the common boundary interface no interpolation was necessary for the data transfer. The coupled simulation was carried out for a time period of 30.0 s, where the system was yet in a transient phase. Equilibrium was estimated for times larger than 600.0 s.

We show in Fig. 8.26 for the first 30.0 s the time evolution of the deformation of the panel and the density distribution (left), as well as the wall temperature distribution (right)[29]. The temperature profiles reflect also the influence of the heat transfer into the cooled panel holder.

The same problem was treated with the more advanced coupling environment reported in [43]. Again the sets of equations were solved by finite element approaches, but in an iterative way (loosely coupled algorithm). Since different mesh sizes were used in the fluid and the structural domain, the data transfer at the interface boundary was conducted by interpolation via a virtual surface.

[29] Note: $1.0\, kg/m^3 = 1.94032 \cdot 10^{-3}\, lb_f s^2/ft^4$, $1.0\, K = 1.8°\, R$, $1.0\, in = 2.54\, cm$, Chapter 13.

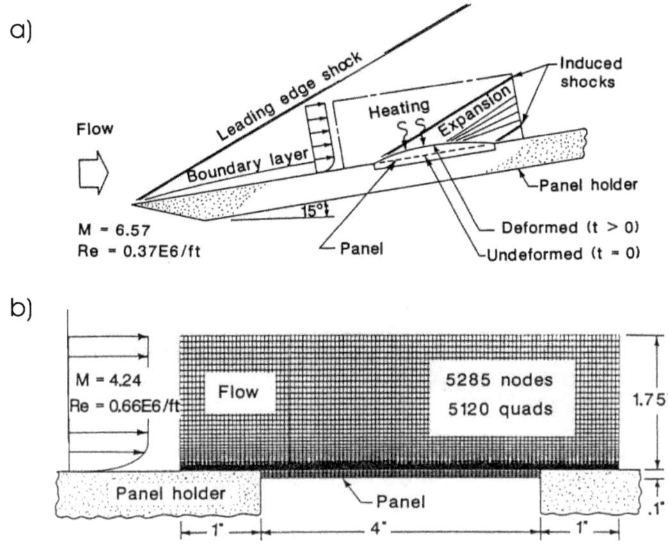

Fig. 8.25. Flow over a panel fixed in a panel holder, [42]. Coupled thermal-mechanical fluid-structure finite element model, panel holder inclination $\alpha = 15°$, free-stream conditions $M_\infty = 6.57$, $Re^u = 1.214 \cdot 10^6 m^{-1}$, upper figure a), and panel onset-flow conditions $M_\infty = 4.24$, $Re^u = 2.16 \cdot 10^6 m^{-1}$, lower figure b).

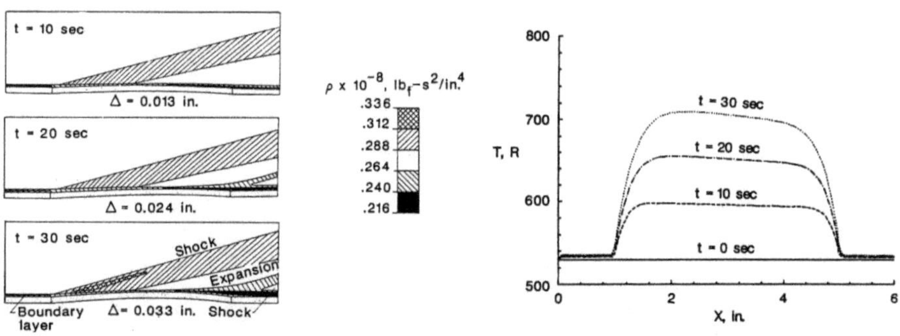

Fig. 8.26. Flow over a panel fixed in a panel holder, [42]. Panel deformation and density distribution (left), wall temperature distribution (right), $M_\infty = 4.24$, $\alpha = 15°$, $Re = 2.16 \cdot 10^6$.

As a last example we consider the case of a thermal-mechanical fluid-structure interaction with ablation, [58], [59]. Ablating materials are often employed on expendable RV-NW-type vehicles (re-)entering the atmosphere, e.g., capsules.

The theoretical treatment of ablation requires a complex multi-phase physical and chemical modelling. In the fluid domain a thermal-chemical non-equilibrium Navier-Stokes code on finite-volume basis was used, which is able

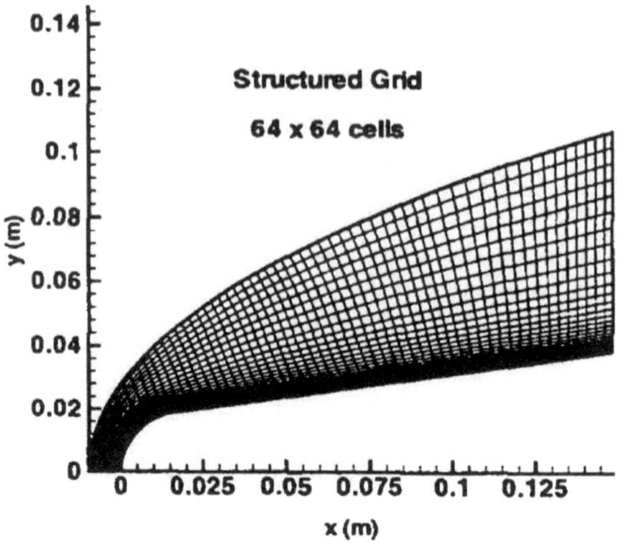

Fig. 8.27. Sphere-cone nose of a re-entry test vehicle, geometry and grid in the fluid domain, [58].

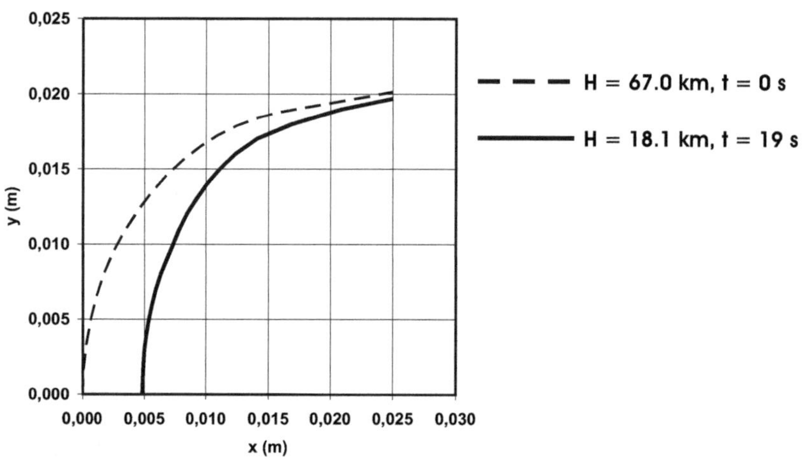

Fig. 8.28. Sphere-cone nose of a re-entry test vehicle, [59]. Ablated nose shape after a flight along a re-entry trajectory of 19.0 s.

to consider in the chemical model also the species from the ablated material injected into the air.

On the structural side a generalized finite element code for solving diffusion problems was applied to determine the heat conduction in the solid. This code includes the capability to update the structural grid which is changed due to the mass loss rate of ablation. To avoid complex interpolations, the meshes of the different domains have coincident nodes on the interface boundary. With the displacements provided by the structural code an updated mesh in the fluid domain was calculated.

A sphere-cone configuration as part of a simple space vehicle was investigated axisymmetrically, Fig. 8.27. The ablating material was carbon-carbon.

Ballistic flight was simulated from the altitude $H = 67.0\,km$ down to $H = 18.1\,km$ in a time period of $19.0\,s$. Fig. 8.28 shows the computed change of shape in the nose regime owing to ablation.

8.5 Conclusion

Multidisciplinary simulation and optimization is a relatively young branch in particular for the flow, the structure, and the heat conduction disciplines in combination with advanced numerical simulation methods. Of course, theoretical investigations of the aeroelasticity of airplanes or parts of airplanes have a long tradition, but in the earlier days the theoretical approach was usually based on analytical solutions. With the advent of high-performance computers and competitive computer codes for the numerical simulation of various disciplines, this situation has changed. Now we observe the trend that simulation environments for a coupled treatment of:

- mechanical fluid-structure (aeroelastic, static and dynamic (flutter), aeroacustics),
- thermal fluid-structure (aerothermal, quasi-steady),
- thermal-mechanical fluid-structure (aerothermoelastic, quasi-steady)

interactions are developed.

Any work on the creation of simulation codes needs data for the verification of the methods. These data can come from experiments, free flights or from other theoretical simulation methods. Often such data are barely available. Helpful for our discussions in Section 8.4 were data from the project IMENS[30], where Germany's DLR together with the industry (EADS-ST) combined experimental and theoretical work in that field, [23], [45], [46], [47], [48], [49].

The quality of the experimental and theoretical data are not in all aspects satisfactory, but the knowledge about the possible drawbacks and influences has a good basis and was further advanced after the end of the project. Looking at the worldwide ongoing efforts, it can be expected that in the near future the methods of multidisciplinary design and optimization will become accurate and reliable tools for the design and development of aerospace flight vehicles.

8.6 Problems

Problem 8.1 Consider expansion of air as perfect gas with a total temperature $T_t = 1{,}500.0\,K$. How large is the maximum possible speed V_m?

[30] "Integrated Multidisciplinary Design of Hot Structures for Space Vehicles".

Problem 8.2. Prove that the formulations eq. (8.2) and eq. (8.8) are identical, by employing the unsteady, two-dimensional inviscid part of these equations. It is sufficient to consider one component, for example the continuity equation.

Problem 8.3. The radiation-adiabatic temperature at a spherical nose with radius R is proportional to $R^{-0.125}$, [5]. We can determine this also from Section 10.3, when putting $q_{gw} \sim T_{ra}^4$ into the relations given there. Apply this to the results shown in the upper part of Fig. 8.18. Take as reference case the case with $R = 64.0\ mm$, where the tangential heat transfer is small and the wall temperature can be considered approximately as radiation-adiabatic, obtain results for the smaller radii, and compare with the data in the figure.

Problem 8.4 We have learned in this chapter that a thermal-mechanical coupling exists between aerothermodynamics and the structure of aerospace vehicles flying with high Mach numbers. We found that this coupling is strong in forward stagnation point regions, at leading edges of wings, winglets and tails and at body flaps which are located totally or partly underneath of a fuselage. At other parts of the vehicle structure only a weak coupling is observed, and the wall temperature there is close to the radiation-adiabatic temperature.

At CAV- or ARV-type vehicles with their airbreathing propulsion systems (see Figs. 4.5 and 4.14), consisting of the three main parts inlet, combustion chamber (burner) and expansion nozzle, the outer flow field is very complex, particularly with respect to the aerodynamic forces and moments as well as to the thermal loads on some surface elements.

It is obvious that at some of the main parts of the propulsion system strong thermal-mechanical fluid-structure interactions occur. Indicate two of them.

References

1. Anderson Jr., J.D.: The Airplane – A History of Its Technology, AIAA, Reston (2002)
2. Hirschel, E.H.: Towards the Virtual Product in Aircraft Design? In: Periaux, J., Champion, M., Gagnepain, J.-J., Pironneau, O., Stouflet, B., Thomas, P. (eds.) Fluid Dynamics and Aeronautics New Challenges. CIMNE Handbooks on Theory and Engineering Applications of Computational Methods, Barcelona, Spain, pp. 453–464 (2003)
3. Vos, J.B., Rizzi, A., Darracq, D., Hirschel, E.H.: Navier-Stokes Solvers in European Aircraft Design. Progress in Aerospace Sciences 38, 601–697 (2002)
4. Hirschel, E.H.: Historical Perspective on Programs, Vehicles and Technology Issues. In: Proc. RTO/AVT/VKI Lecture Series AVT-116 Critical Technologies for Hypersonic Vehicle Development, Rhode-Saint-Gènèse, Belgium, May 10-14. RTO-EN-AVT-116, pp. 1-1–1-22 (2005)
5. Hirschel, E.H.: Basics of Aerothermodynamics. Progress in Astronautics and Aeronautics, AIAA, Reston, Va, vol. 204. Springer, Heidelberg (2004)

6. Hammond, W.E.: Space Transportation: A System Approach to Analysis and Design. AIAA Educations Series, Reston, Va (1999)
7. Hajela, P.: Soft Computing Multidisciplinary Aerospace Design – New Directions for Research. RTO MP-35, paper 17 (1999)
8. Alexandrov, N., Hussaini, Y. (eds.): Multidisciplinary Design Optimization: State of the Art. SIAM Publications, Philadelphia (1992)
9. Haykin, S.: Neural Networks – A Comprehensive Foundation. Macmillan Publishing Company, Englewood (1994)
10. Hajela, P.: Neural Networks – Applications in Modeling and Design of Structural Systems. CISM Lecture Notes, Udine, Italy (1998)
11. Zadeh, L.: Fuzzy Sets. Information and Control 8, 338–353 (1965)
12. Yuan, W.G., Quan, W.W.: Fuzzy Optimum Design of Structures. Engineering Optimization 8, 291–300 (1985)
13. Haftka, R.T., Gurdal, Z.: Elements of Structural Optimization. Kluwer Academic Publishers, Dordrecht (1993)
14. Hajela, P., Yoo, J.: Constraint Handling in Genetic Search Using Expression Strategies. AIAA Journal 34(11), 2414–2420 (1996)
15. Aris, R.: Vectors, Tensors and the Basic Equations of Fluid Mechanics. Prentice Hall, Englewood Cliffs (1962)
16. Farhat, C., Lesoinne, M., Chen, P.S.: Parallel Heterogeneous Algorithms for the Solution of Three-Dimensional Transient Coupled Aeroelastic Problems. AIAA-Paper 95-1290 (1995)
17. Lesoinne, M., Farhat, C.: Geometric Conservation Laws for Aeroelastic Computations Using Unstructured Dynamic Meshes. AIAA-Paper 95-1709-CP (1995)
18. Farhat, C., Lesoinne, M.: Higher-Order Staggered and Subiteration Free Algorithms for Coupled Dynamic Aeroelasticity Problems. AIAA-Paper 98-0516 (1998)
19. Farhat, C., Lesoinne, M., LeTallec, P.: Load and Motion Transfer Algorithms for Fluid/Structure Interaction Problems with Non-Matching Discrete Interfaces: Momentum and Energy Conservation, Optimal Discretisation and Application to Aeroelasticity. Comput. Methods Appl. Mech. Eng. 157, 95–114 (1998)
20. Gupta, K.K., Meek, J.L.: Finite Element Multidisciplinary Analysis. AIAA Education Series, Reston, Va (2000)
21. Zienkiewics, O.C., Taylor, R.L., Zhu, J.Z.: Finite Element Method: Its Basis and Fundamentals. Elsevier Butterworth-Heinemann, New York (2005)
22. Bathe, K.J.: Finite Element-Methoden. Springer, Heidelberg (2002)
23. Schäfer, R.: Thermisch-mechanisches Verhalten heisser Strukturen in der Wechselwirkung mit einem umströmenden Fluid (Thermo-Mechanical Behavior of Hot Structures in Interaction with a Fluid). Doctoral Thesis, Technische Universität Stuttgart, Germany, DLR Forschungsbericht 2005 - 02 (2005)
24. Gupta, K.K., Petersen, K.L., Lawson, C.L.: On Some Recent Advances in Multidisciplinary Analysis of Hypersonic Vehicles. AIAA-Paper 92-5026 (1992)
25. Gupta, K.K.: Development of a Finite Element Aeroelastic Analysis Capability. Journal of Aircraft 33(5), 995–1002 (1996)
26. Cowan, T.J., Arena Jr., A.S., Gupta, K.K.: Accelerating CFD-Based Aeroelastic Predictions Using System Identification. AIAA-Paper 98-4152 (1998)
27. Cowan, T.J., Arena Jr., A.S., Gupta, K.K.: Development of a Discrete-Time Aerodynamic Model for CFD-Based Aeroelastic Analysis. AIAA-Paper 99-0765 (1999)

28. Gupta, K.K., Voelker, L.S., Bach, C., Doyle, T., Hahn, E.: CFD-Based Aeroelastic Analysis of the X-43 Hypersonic Flight Vehicle. AIAA-Paper 2001-0712 (2001)
29. Gupta, K.K., Bach, C., Doyle, T., Hahn, E.: CFD-Based Aeroservoelastic Analysis with Hyper-X Applications. AIAA-Paper 2004-0884 (2004)
30. Maute, K., Nikbay, M., Farhat, C.: Analytically Based Sensitivity Analysis and Optimization of Nonlinear Aeroelastic Systems. AIAA-Paper 2000-4825 (2000)
31. Tran, H., Farhat, C.: An Integrated Platform for the Simulation of Fluid-Structure-Thermal Interaction Problems. AIAA-Paper 2002-1307 (2002)
32. Selmin, V.: Coupled Fluid-Structure System. In: Haase, W., Selmin, V., Wingzell, B. (eds.) Progress in Computational Flow-Structure Interactions. Notes on Numerical Fluid Mechanics and Multidisciplinary Design, vol. 81, pp. 13–20. Springer, Heidelberg (2003)
33. Patel, A., Hirsch, C.: All-hexahedra Unstructured Flow Solver for External Aerodynamics with Application to Aeroelasticity. In: Haase, W., Selmin, V., Wingzell, B. (eds.) Progress in Computational Flow-Structure Interactions. Notes on Numerical Fluid Mechanics and Multidisciplinary Design, vol. 81, pp. 105–116. Springer, Heidelberg (2003)
34. Dervieux, A., Koobus, B., Schall, E., Lardat, R., Farhat, C.: Application of Unsteady Fluid-Structure Methods to Problems in Aeronautics and Space. In: Barton, N.G., Periaux, J. (eds.) Coupling of Fluids, Structures and Waves in Aeronautics. Notes on Numerical Fluid Mechanics and Multidisciplinary Design, vol. 85, pp. 57–70. Springer, Heidelberg (2003)
35. Maman, N., Farhat, C.: Matching Fluid and Structure Meshes for Aeroelastic Computations: A Parallel Approach. Computer & Structures 54, 779–785 (1995)
36. Löhner, R.: Robust, Vectorized Search Algorithms for Interpolation of Unstructured Grids. J. of Comp. Phys. 118, 380–387 (1995)
37. Grashof, J., Haase, W., Schneider, M., Schweiger, J., Stettner, M.: Static and Dynamic Aeroelastic Simulations in Transonic and Supersonic Flow. In: Haase, W., Selmin, V., Wingzell, B. (eds.) Progress in Computational Flow-Structure Interactions. Notes on Numerical Fluid Mechanics and Multidisciplinary Design, vol. 81, pp. 39–46. Springer, Heidelberg (2003)
38. Lepage, C.H., Habashi, W.G.: Conservative Interpolation of Aerodynamic Loads for Aeroelastic Computations. AIAA-Paper 2000-1449 (2000)
39. Haase, W., Selmin, V., Wingzell, B. (eds.): Progress in Computational Flow-Structure Interactions. Notes on Numerical Fluid Mechanics and Multidisciplinary Design, vol. 81. Springer, Heidelberg (2003)
40. Barton, N.G., Periaux, J. (eds.): Coupling of Fluids, Structures and Waves in Aeronautics. Notes on Numerical Fluid Mechanics and Multidisciplinary Design, vol. 85. Springer, Heidelberg (2003)
41. Ballmann, J.: Flow Modulation and Fluid-Structure Interaction at Airplane Wings. Notes on Numerical Fluid Mechanics and Multidisciplinary Design, vol. 84. Springer, Heidelberg (2003)
42. Thornton, E.A., Dechaumphai, P.: Coupled Flow, Thermal, and Structural Analysis of Aerodynamically Heated Panels. J. of Aircraft 25(11), 1052–1059 (1988)
43. Löhner, R., Yang, C., Cebral, J., Baum, J.D., Luo, H., Pelessone, D., Charman, C.: Fluid-Structure-Thermal Interaction Using a Loose Coupling Algorithm and Adaptive Unstructered Grids. AIAA-Paper 98-2419 (1998)

44. Haupt, M., Horst, P.: Coupling of Fluid and Structure Analysis Codes for Air- and Spacecraft Applications. In: Bathe, K.J. (ed.) Proceedings First MIT Conference on Computational Fluid and Solid Mechanics. Computational Fluid and Solid Mechanics, pp. 1226–1231. Elsevier, Amsterdam (2001)
45. Haupt, M., Niesner, R., Horst, P.: Flexible Software Environment for the Coupled Aerothermodynamic-Thermal-Mechanical Analysis of Structures. ESA SP-563 (2005)
46. Schäfer, R., Mack, A., Esser, B., Gülhan, A.: Fluid-Structure Interaction on a Generic Model of a Reentry Vehicle Nosecap. In: Librescu, L., Marzocca, P. (eds.) Proceedings 5th International Congress on Thermal Stresses and Related Topics. Blacksburg, Virginia (2003)
47. Haupt, M., Niesner, R., Unger, R., Horst, P.: Computational Aero-Structural Coupling for Hypersonic Applications. AIAA Paper 2006-3252 (2006)
48. Mack, A.: Analyse von heissen Hyperschallströmungen um Steuerklappen mit Fluid-Struktur Wechselwirkung (Analysis of Hot Hypersonic Flow Past Control Surfaces with Fluid-Structure Coupling). Doctoral Thesis, Technische Universität Braunschweig, Germany, DLR Forschungsbericht 2005 - 23 (2005)
49. Haupt, M., Niesner, R., Horst, P., Hannemann, V., Mack, A., Brandl, A.: Numerical and Software Concepts for the Coupling of Structural Thermal-Mechanical and Fluid-Dynamic Codes. In: Librescu, L., Marzocca, P. (eds.) Proceedings 5th International Congress on Thermal Stresses and Related Topics. Blacksburg, Virginia (2003)
50. N.N.: MpCCI, Mesh-based parallel Code Coupling Interface, Specification of MpCCI Version 1.3. Fraunhofer Institute for Algorithm and Scientific Computing SCAI (2002)
51. Weiland, C.: Stage Separation Aerothermodynamics. AGARD-R-813, pp. 11-1–11-28 (1996)
52. Gupta, K.K.: Development and Application of an Integrated Multidisciplinary Analysis Capability. International Journal for Numerical Methods in Engineering 40, 533–550 (1997)
53. Haupt, M., Kossira, H.: Analysis of Structures in Hypersonic Fluid Flow Including the Fluid-Structure Interaction. ESA SP-428 (1998)
54. Gülhan, A.: Investigation of Gap Heating on a Flap Model in the Arc Heated Facility L3K. Deutsches Zentrum für Luft- und Raumfahrt, TETRA Programme, TET-DLR-21-TN-3101 (1999)
55. Behr, R., Görgen, J.: CFD Analysis of Gap Flow Phenomena. Deutsches Zentrum für Luft- und Raumfahrt, TETRA Programme, TET-DASA-21-TN-2403 (2000)
56. Mack, A., Schäfer, R.: Fluid Structure Interaction on a Generic Body-Flap Model in Hypersonic Flow. J. of Spacecraft and Rockets 42(5), 769–779 (2005)
57. Esser, B., Gülhan, A., Schäfer, R.: Experimental Investigation of Thermal Fluid – Structure Interaction in High Enthalpy Flow. ESA SP-563 (2005)
58. Kuntz, D.W., Hassan, B., Potter, D.L.: An Iterative Approach for Coupling Fluid/Thermal Predictions of Ablating Hypersonic Vehicles. AIAA-Paper 99-3460 (1999)
59. Hassan, B., Kuntz, D.W., Salguero, D.E., Potter, D.L.: A Coupled Fluid/Thermal/Flight Dynamics Approach for Predicting Hypersonic Vehicle Performance. AIAA-Paper 2001-2903 (2001)

9 The Thermal State of a Hypersonic Vehicle Surface

The thermal state of the hypersonic vehicle surface governs on the one hand thermal surface effects (these are the influence of wall-temperature and temperature gradient in the gas at the wall on viscous and thermo-chemical phenomena at and near the vehicle surface) and on the other hand the thermal loads on the vehicle surface (regarding the structure and material layout of a TPS or a hot primary structure), Fig. 9.1, [1]. The thermal state of the surface indirectly governs also mechanical (pressure and shear stress) loads via its influence on primarily the wall-shear stress, for instance, in erosion processes at a TPS. This is, of course, a typical viscous thermal surface effect.

The thermal state of the surface and hence both thermal surface effects and thermal loads are strongly influenced by the surface properties, too, [1].

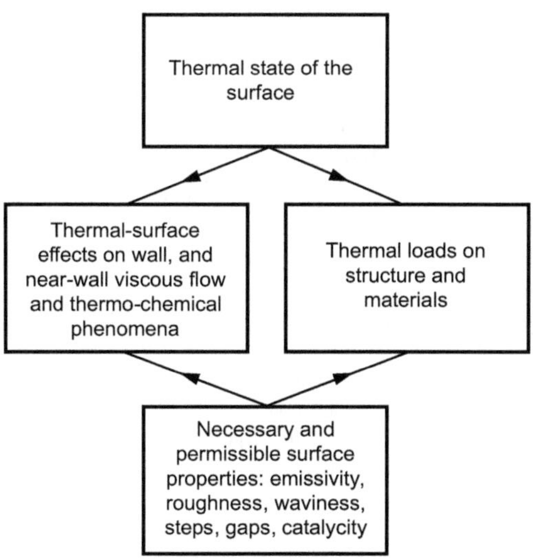

Fig. 9.1. The thermal state of the hypersonic vehicle surface and its different aero-thermal design implications, [1].

The most important necessary property is a high surface emissivity in order to achieve sufficient (thermal) radiation cooling. Permissible properties are surface catalycity, surface roughness et cetera. These should be sub-critical, i. e. should have no detrimental influence on the surface-near flow.

A particular problem is the determination of permissible surface roughness, waviness, step and gap height. The permissible geometrical scales should be as large as possible, because, if necessary, a high surface quality would drive up manufacturing costs. On the other hand, it can be very problematic to model them in experimental and/or computational investigations.

In this chapter we look first at the thermal situation at the surface of a hypersonic flight vehicle, define then the thermal state of the surface, and present finally an aerothermodynamic simulation compendium with special regard to the thermal state of the surface.

9.1 Heat Transport at a Vehicle Surface

Hypersonic flight vehicles either have a cold primary structure with a thermal protection system (typically RV-W and RV-NW flight vehicles), or a hot primary structure with internal insulation (typically some CAV-type flight vehicles). In reality mixtures of such structure and materials concepts are present. A RV-W-type vehicle, for instance, may have aerodynamic control surfaces with a hot primary structure without internal insulation.

A hypersonic vehicle surface typically is cooled by thermal surface radiation, [1]. On surface portions and/or in flight speed/altitude domains, where this is not sufficient, additionally active cooling, for instance by ablation (often employed on non-reusable flight systems) or transpiration cooling, is employed. For the following considerations we assume surfaces, which do not change shape, and also no transpiration cooling. Active cooling can be allowed for by prescribing a heat flux q_w into the wall. Assuming a smooth surface[1] and reducing the situation to a one-dimensional one[2], we find in the continuum regime the general thermal situation shown in Fig. 9.2.

Considering the situation shown in Fig. 9.2 in detail, we note, see also [1]:

- The sum of all heat fluxes is zero:

$$q_{gw} + q_{rad,w,nc} + q_{rad,g} - q_{rad,w} - q_w = 0, \tag{9.1}$$

[1] In general the surface must be sufficiently smooth only for airbreathing CAV-type flight vehicles, because they are drag critical, "permissible surface properties", see above and [1]. The classical TPS tile surface of the Space Shuttle Orbiter, for instance, has roughnesses which may be caused by tile misalignment, gaps, gap-filler protrusions et cetera, see, e.g., [2].

[2] The temperature gradients, and hence the heat fluxes tangential to the surface, which in downstream direction appear especially with radiation cooled surfaces, are neglected in this consideration. However, there might be situations, where this is not permitted, see Sub-Section 8.4.3.

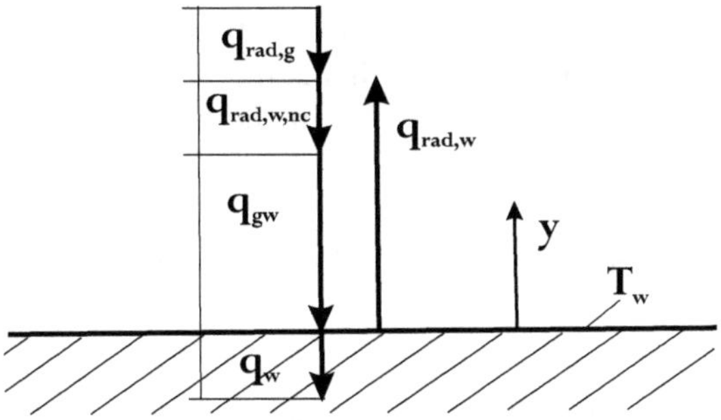

Fig. 9.2. Schematic description of the thermal state of the surface in the continuum regime, hence $T_{gw} \equiv T_w$. Tangential heat fluxes are neglected, y is the surface-normal coordinate. q_{gw}: heat flux in the gas at the wall, q_w: heat flux into the wall, $q_{rad,w}$: surface radiation (cooling) heat flux, $q_{rad,w,nc}$: surface radiation heat flux from non-convex surface portions, $q_{rad,g}$: shock-layer gas radiation heat flux.

and the wall temperature T_w is resulting from this balance. We assume $T_{gw} \equiv T_w$, neglecting a possible temperature jump in the slip-flow regime.

- During hypersonic flight heat is transported to the vehicle surface by three physical processes:
  - Diffusive heat transport in the gas towards the wall, more precisely, the heat flux in the gas at the wall, q_{gw}, is the first one[3]. In general it consists of classical heat conduction, heat transfer due to mass diffusion in thermo-chemical non-equilibrium, but also of heat transfer due to slip-flow, [1].
  - Transport of heat by thermal radiation from other surface portions of the flight vehicle, $q_{rad,w,nc}$, due to non-convex effects is the second one. This transport happens, for instance, between the rear of a body flap and the fuselage (see Sections 6.3 and 6.4 for PHOENIX and X-38 examples), et cetera. It can be described with the help of a fictitious or effective emissivity coefficient ε_{eff}, see, e.g. [1]. It takes into account the mutual visibility of the involved surface portions and especially the characteristic boundary layer thicknesses, which govern locally the wall radiation heat fluxes $q_{rad,w}$. Radiation heat coming from the sun can be neglected in our considerations.
  - Thermal radiation, $q_{rad,g}$, of the vibrational excited, dissociated and ionized gas in the shock layer is the third one. For classical Low Earth Orbit (LEO) re-entry (RV-W- and RV-NW-type flight vehicles) and also for hypersonic flight (CAV-type flight vehicles) with $v_\infty \lessapprox 8.0$ km/s, at $H \lessapprox 100.0$ km, $q_{rad,g}$ usually can be neglected, because the diffusive heat transfer q_{gw} is dominating. But, for example, for Lunar return of a RV-NW vehicle with an entry speed of approximately 10.6 km/s, the heat transfer due to thermal gas radiation, $q_{rad,g}$, becomes

[3] In the literature this is usually called convective heating.

very important and approaches values of 50 per cent of the total heat flux towards the vehicle surface. This is due to the strong increase of ionized particles which are generated in the bow shock layer.

- Heat is transported away from the vehicle surface by two processes[4]:
  - Thermal radiation, $q_{rad,w}$ – which we usually denote in this book with q_{rad} – away from the surface for cooling purposes:

$$q_{rad}(= q_{rad,w}) = \varepsilon \, \sigma \, T_w^4, \tag{9.2}$$

    with ε being the emissivity of the vehicle surface, and σ the Stefan-Boltzmann constant, Section 13.1.
    This is the major cooling mode for outer surfaces of classical Low Earth Orbit re-entry, and also for hypersonic flight, with speeds below approximately 8.0 km/s, at altitudes below approximately 100.0 km.
  - Heat transfer, q_w, into the surface material. This is the natural heat conduction process, which governs the thickness, for instance, of a TPS. In general $q_w \ll q_{rad,w}$. However, there are situations, where this assumption is not valid. In early vehicle design processes (vehicle definition) it usually can be neglected. The resulting wall temperature T_w is then the radiation-adiabatic or radiation equilibrium temperature: $T_w \equiv T_{ra}$. In the vehicle development process the complete situation shown in Fig. 9.2 should be taken into account in order to arrive at precise thermal loads predictions. Looking at the potential of the modern numerical simulation methods, this becomes increasingly possible, Chapter 8. If active cooling by a heat exchanger below the surface is employed, q_w will anyway be large according to the cooling needs at that surface portion.

9.2 The Thermal State of a Vehicle Surface

Under the "thermal state of the surface" we understand the temperature of the gas at the surface, *and* the temperature gradient in it normal to the surface. As we have seen above, these are not necessarily those of and in the surface material.

The thermal state of the surface thus is defined by, [1]:

1. The actual temperature of the gas at the wall surface, T_{gw}, and the temperature of the wall, T_w. If low-density effects (temperature jump) are not present, we have $T_{gw} \equiv T_w$, which we assume, for convenience, in the following considerations.
2. The temperature gradient in the gas at the wall, $\partial T/\partial y|_{gw}$, in direction normal to the surface, respectively the heat flux in the gas at the wall, q_{gw}, if the gas is a perfect gas or in thermo-chemical equilibrium.

[4] We don't consider here heat transport away from the surface due to ablation, transpiration, et cetera.

3. The temperature gradient in the material at the wall surface, $\partial T/\partial y|_w$, in (negative) direction normal to the surface, respectively the heat flux q_w (tangential gradients are also neglected). The heat flux q_w is not equal to q_{gw}, if radiation cooling is employed, Fig. 9.2.

We do not discuss here the implications of the thermal state of the surface for hypersonic flight vehicles of all kind. The reader instead is referred to [1]. We note, however, that in general $\partial T/\partial y|_{gw}$ or q_{gw} and T_{ra} decrease strongly for laminar flow from the front part of a hypersonic vehicle towards its aft part. The reason is, that the characteristic thickness of a laminar boundary layer, δ_{lam}, increases strongly with the boundary layer running length x, Sub-Section 10.4.2: $\delta_{lam} \sim x^{0.5} \rightarrow q_{gw,lam} \sim x^{-0.5}$, $T_{w,lam} \approx T_{ra,lam} \sim x^{-0.125}$.

If the boundary layer is turbulent, the effect is not so pronounced, because then the characteristic thickness is the thickness of the viscous sub-layer δ_{vs}, which increases much slower with the boundary layer[5] running length x: $\delta_{vs} \sim x^{0.2} \rightarrow q_{gw,turb} \sim x^{-0.2}$, $T_{w,turb} \approx T_{ra,turb} \sim x^{-0.05}$, [1]. These and the above relations are of qualitative nature, but approximate reality to a good degree, if they are applied in main-flow direction, with flow three-dimensionality and high-temperature real-gas effects being small.

The different characteristic thicknesses of laminar and turbulent boundary layers, Fig. 9.3, must be taken into account in grid generation for numerical methods. A special problem poses the turbulent boundary layer. The characteristic thickness of turbulent flow is that of the viscous sub-layer δ_{vs}, [1], which is much smaller than the boundary layer thickness δ_{turb}.

It is expressed in terms of the non-dimensional wall distance y^+, see, e.g., [3], [4]:

$$y^+ = \frac{y \rho u_\tau}{\mu}, \tag{9.3}$$

with u_τ being the friction velocity:

$$u_\tau = \sqrt{\frac{\tau_w}{\rho_w}}. \tag{9.4}$$

With $y^+ \approx 5$ defining the thickness of the viscous sub-layer[6], the first grid line away from the surface usually is located at $y^+ \lessapprox 1$. Because the velocity profile is linear in that regime, it is sufficient to put two to three grid lines below $y^+ \approx 5$.

This holds, if the transport equations of turbulence are integrated down to the body surface, which is the case with the so-called low-Reynolds number formulations. If a law-of-the-wall formulation is used, the first grid line can be located at $y^+ \approx 50$ to 100, and the distance down to the wall is bridged with the law-of-the-wall.

---

[5] Actually the proportionality $x^{0.2}$ is that of the scaling thickness δ_{sc} discussed in [1]. More correctly holds $\delta_{vs} \sim x^{0.1}$.

[6] See, e.g., [1], for the explicit relation for δ_{vs}.

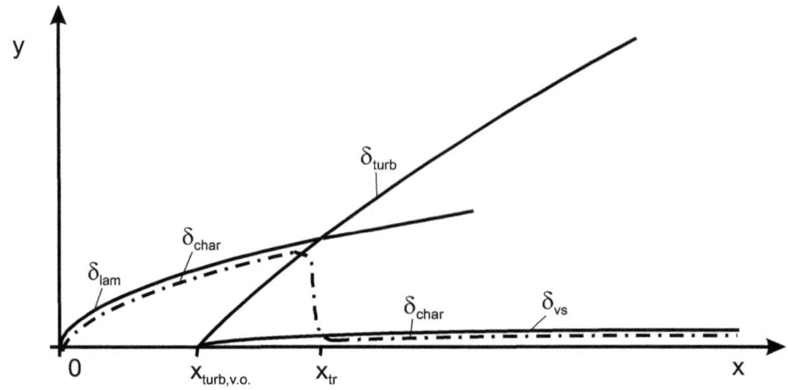

Fig. 9.3. Characteristic thickness δ_{char} of attached viscous flow of boundary layer type. In the laminar flow domain the boundary layer thickness δ_{lam} – usually defined by the approach of 99 per cent of u_e – is the characteristic thickness, in the turbulent domain it is the thickness of the viscous sub-layer, δ_{vs}, not the thickness δ_{turb}. The location of the virtual origin of the turbulent boundary layer, Sub-Section 10.4.4, is denoted with $x_{turb,v.o.}$, the transition location with x_{tr}.

However, for both formulations one must keep in mind that with radiation cooled surfaces large surface-normal gradients and wall-near extrema of the temperature T or the internal energy e occur, [1]. If the attached viscous flow is of boundary layer type, then the density ρ accordingly has a large surface-normal gradient and a wall-near extremum, too. This must be taken into account, when defining the computation grid.

We note further that the thermal state of the surface in general plays an important role for (airbreathing) CAV-type flight vehicles, which like any aircraft are drag-sensitive. There all viscous phenomena are affected, partly even very strongly, by the thermal state of the surface (viscous thermal surface effects).

If one considers RV-W- and RV-NW-type flight vehicles, the thermal state of the surface concerns predominantly the structure and materials layout of the vehicle (thermo-chemical thermal surface effects with regard to thermal loads), and not so much its aerodynamic performance. This is because the RV-type vehicle flies a "braking" mission, where the drag is on purpose large (blunt configuration, high angle of attack). Of course, if a flight mission demands large down-range or cross-range capabilities in the atmosphere, this may change.

9.3 Aerothermodynamic Simulation Compendium

The following compendium lists aerothermodynamic simulation means. About principal simulation topics see, e.g., [1]. We list here the general capabilities of simulation means, however with special regard to the simulation of the thermal state of a vehicle surface in view of thermal loads and thermal surface effects. Especially regarding the computational simulation, it is recommended to look at overviews, reviews et cetera, e.g., [5] to [10]. However, the discrete numerical methods of aerodynamics and aerothermodynamics have made large progress since these papers and books were written. This holds also for experimental simulation, there especially for image-based opto-electronic measurement methods.

9.3.1 Ground-Facility Simulation

- Facilities
  - "Cold" facilities: Blow-down and continuous facilities, total temperatures from ambient up to approximately 1,500.0 K. Limitations: Restricted similarity-parameter reproduction (usually only true Mach number), cold model surfaces.
  - "Hot" (high-enthalpy) facilities: Pulse facilities (shock tunnels of all kind), plasma tunnels (usually for materials and structure tests). Limitations: In general restricted similarity-parameter reproduction, high-enthalpy similarity usually means a flight-velocity but not a flight Mach number similarity (Mach number independence necessary), thermo-chemical freezing effects during nozzle expansion, cold model surfaces.

- Use for hypersonic flight vehicle definition and development
  - Applications: Aerodynamic shape definition, design verification, data set generation.
  - Results: Forces and moments. Discrete surface points measurements: Pressure p_w, shear stress τ_w, temperature T_w, heat flux q_w. Because radiation cooling cannot be simulated in ground-simulation facilities, this means $q_w \equiv q_{gw}$. Distributed surface data (image-based measurement techniques): Pressure p_w (pressure-sensitive paint techniques), shear-stress τ_w (skin-friction) lines (oil-film pattern), temperature T_w and heat flux q_w (temperature-sensitive liquid-crystal techniques). Off-surface flow visualization (schlieren, shadowgraph, interferometry): Shock locations, separation structures, wakes. A hot experimental technique in general is not available for ground-facility simulation, although it is in use for in-flight testing.

    Complete vehicle models due to their rather small size usually don't allow the reproduction of real-life surface properties (irregularities, waviness, steps, gaps, et cetera), which may influence the near-surface flow[7]. For the Space Shuttle Orbiter, however, tile gap patterns have been reproduced on the model surface for transition measurements, [2].
  - Thermal state of the surface: Temperature T_w and heat flux into the wall q_w (usually indirectly found) can be measured, and for perfect

[7] The influence of such properties in general is simulated and studied with the help of dedicated generic models.

gas or gas in thermo-chemical equilibrium the gradient $\partial T/\partial n|_{gw}$ can be determined, all, however, as a rule only for cold models. Hot-spot and cold-spot situations can only be detected with surface-mapping techniques, if resolution is adequately high. As a rule the true flight situation cannot be simulated, neither in cold nor in hot facilities: Cold model surfaces, no simulation of surface radiation cooling, restricted similarity-parameter reproduction (usually only true Mach number). If turbulent flow is present on parts of the vehicle surface in reality: True laminar-turbulent transition simulation is not possible, turbulence tripping is problematic.

Summary: With the present capabilities thermal loads can be measured with acceptable accuracy and reliability, to meet future higher demands will be problematic. The true in-flight thermal state of the surface and hence thermal surface effects cannot be simulated. "Hot model" techniques have been tried, but in general are not established.

9.3.2 Computational Simulation

- Numerical methods
  - □ Computer codes[8] (steady state and time-accurate solutions)
    - Equations: Viscous shock-layer equations (laminar), Euler equations, coupled Euler/higher-order boundary layer equations (laminar and turbulent), full Navier-Stokes equations (laminar and turbulent (Reynolds/Favre-averaged Navier-Stokes equations (RANS)), for simulation problems in non-continuum flow regimes the Boltzmann equation (Monte-Carlo methods).
    - Flow-physics models: Laminar-flow transport properties (with some exceptions good quality), statistical turbulence models are acceptable, except for strong interaction phenomena (turbulent shock/boundary layer interaction, separation, corner flow, et cetera), criteria for laminar-turbulent transition are neither accurate nor reliable.
    - High-temperature real-gas models: Thermal and chemical equilibrium, thermal non-equilibrium, chemical non-equilibrium. Gas models with different numbers of species and reactions, including ionization, multi-temperature models with different numbers of vibration temperatures are available. All models have, with some exceptions, good quality (problem of accumulated inaccuracies?).
    - Type of surface boundary conditions (viscous codes): No-slip boundary conditions in the continuum regime, wall (and shock) slip effects in the rarefied gas regime $H \approx 80.0$ to 100.0 km; prescribed wall temperature or heat flux including adiabatic wall,

[8] Regarding mesh generation, new approaches find only slowly their place. In general the problem can be considered as almost solved.

radiation-adiabatic wall (in multidisciplinary methods coupling with heat flux into the wall); non-catalytic, finite-rate catalytic, and fully catalytic wall; ablation pyrolysis, gas-mass flux, surface degradation.

The codes, as a rule, are applied to rigid flight-vehicle configurations, new multidisciplinary methods permit to take into account deforming and movable surfaces, i. e. aerodynamic trim and control surfaces, et cetera.

☐ Use for hypersonic flight vehicle definition and development
  - Applications: Aerodynamic shape definition, design verification, data set generation, problem diagnosis.
  - Results: Forces and moments, surface and off-surface distributions of pressure, shear-stress, temperature(s), heat fluxes, gas species, velocity components, shock locations, entropy-layer phenomena, et cetera. Only with Navier-Stokes methods: strong interaction flows including hypersonic viscous interaction, low-density (slip flow) effects, flow separation, wakes. Flow topologies of all kind, as well as hot-spot and cold-spot situations can be identified.

    Not possible in general is the simulation of real-life surface properties (irregularities, waviness, steps, gaps, et cetera) in view of laminar-turbulent transition and turbulent flow. If a sensitivity exists with regard to laminar-turbulent transition, the situation becomes critical. Results of computations of separated turbulent flow in general are not reliable.
  - Thermal state of the surface: The complete thermal state of the surface, i. e. the true flight situation, can be computed, including non-convex effects, $q_{rad,w,nc}$, and shock-layer gas radiation, $q_{rad,g}$. In multidisciplinary simulations heat transport into the wall, q_w, can be added. Hot-spot and cold-spot situations can be detected. Limitations are due to the insufficient modelling of laminar-turbulent transition and turbulent separation.

Summary: With the present capabilities viscous and thermo-chemical thermal surface effects as well as thermal loads can be computed with good accuracy and reliability, as long as no large sensitivity is given regarding laminar-turbulent transition and turbulent separation.

- Approximate (engineering) methods

☐ Computation Methods
  - Type: a) Inviscid methods: Newton-type, shock-expansion type, et cetera, b) viscous methods (laminar and turbulent): stagnation point relations (laminar), relations for flat surface portions (hypersonic viscous interaction can be taken into account in some methods), swept cylinders (attachment lines) et cetera, see, e.g., [11],

simple (2-D, axisymmetric analogue) boundary layer methods (finite difference and integral methods).
- Flow-physics models: Capabilities and restrictions like for numerical methods.
- High-temperature real-gas models: Capabilities and restrictions like for numerical methods. In general rather simple models are employed.
- Type of surface boundary conditions (viscous methods): In general only no-slip boundary conditions; prescribed wall temperature or heat flux including adiabatic and radiation-adiabatic walls.

☐ Use for hypersonic flight vehicle definition and development
- Applications: Aerodynamic shape definition, data set generation, trajectory definition and optimization.
- Results: Forces and moments, 1-D and quasi 2-D surface distributions of pressure, shear-stress, temperature, heat fluxes. Flow topologies of all kind, as well as hot-spot and cold-spot situations cannot be identified. If a sensitivity exists with regard to laminar-turbulent transition, the situation again becomes critical. The boundary layer methods permit to indicate flow separation.
- Thermal state of the surface: In principle the complete thermal state of the surface, although for simple surface configurations, can be computed, including non-convex effects, $q_{rad,w,nc}$, and shock-layer gas radiation, $q_{rad,g}$. In general, however, engineering methods do not have these capabilities. In simple multidisciplinary simulations heat transport into the wall, q_w, can be added. Hot-spot and cold-spot situations cannot be detected. Limitations again are due to the insufficient modelling of laminar-turbulent transition and turbulent separation.

Summary: Thermal loads can be computed for simple surface configurations, as long as no large sensitivity is given regarding laminar-turbulent transition. The description of viscous and thermo-chemical thermal surface effects in general is restricted, although in principle possible to a certain degree.

9.3.3 In-Flight Simulation

- Capabilities: In discrete surface points measurement of wall pressure, wall temperature, heat flux into the wall, q_w, et cetera are feasible with acceptable results. Avoided must be a local falsification of the actual vehicle surface properties (in terms of radiation emissivity and catalycity) by the material and surface properties of the inserted gauges and, in addition, their possible pollution.

  In-flight distributed wall-temperature and other measurements are not feasible in general. Initial wall-temperature mappings have been performed on the Space Shuttle Orbiter's windward side surface by means of infrared imagery from the Kuiper Astronomical Observatory aircraft, [12]. The project was rated a limited success. Space Shuttle Orbiter lee-side infrared temperature imagery was performed with the help of a pod mounted atop the vertical fin of the vehicle, [13]. This obviously more successful experiment was performed during five missions.

A special topic of in-flight simulation is parameter or systems identification, i. e. the determination of the true aerothermodynamic performance and properties of the real, elastic, controlled flying vehicle. The most completely studied vehicle in this regard is the Space Shuttle Orbiter, with a host of data available in [14].

An important prerequisite for all kind of in-flight measurements is an accurate air-data system, because the measured data must be correlated with the flight speed and the attitude of the vehicle, as well as with the thermodynamic data of the atmosphere ahead of the vehicle. Good accuracy is attained for the velocity v_∞, the pressure p_∞, and the attitude. A special problem in this regard is the atmospheric density ρ_∞. Opto-electronic/spectroscopic measurement systems should be able in the future to provide highly accurate measurements of ρ_∞ and T_∞.

- Applications: Acquisition of aerothermodynamic data for code modelling and validation, laminar-turbulent transition data, et cetera. Tests of heat-protection systems including flexible external insulations (FEI), hot structures, ablators, surface degradation, et cetera.
- Thermal state of the surface: The thermal state of the surface can be determined, if the emission and catalytic properties of the surface (possible falsification problem due to sensor material) are known. If that is the case, thermal surface effects can be identified by data analysis with the help of numerical aerothermodynamic computation methods.

References

1. Hirschel, E.H.: Basics of Aerothermodynamics. Progress in Astronautics and Aeronautics, AIAA, Reston, Va, vol. 204. Springer, Heidelberg (2004)
2. Goodrich, W.D., Derry, S.M., Bertin, J.J.: Shuttle Orbiter boundary layer Transition: A Comparison of Flight and Wind-Tunnel Data. AIAA-Paper 83-0485 (1983)
3. Wilcox, D.C.: Turbulence Modelling for CFD. DCW Industries, La Cañada, CAL, USA (1998)
4. Smits, A.J., Dussauge, J.-P.: Turbulent Shear Layers in Supersonic Flow, 2nd edn. AIP/Springer, New York (2004)
5. Neumann, R.D.: Defining the Aerothermodynamic Methodology. In: Bertin, J.J., Glowinski, R., Periaux, J. (eds.) Hypersonics. Defining the Hypersonic Environment, vol. 1, pp. 125–204. Birkhäuser, Boston (1989)
6. Hayes, J.R., Neumann, R.D.: Introduction to the Aerodynamic Heating Analysis of Supersonic Missiles. In: Mendenhall, M.R. (ed.) Tactical Missile Aerodynamics: Prediction Methodology. Progress in Astronautics and Aeronautics, AIAA, Reston, Va, pp. 63–110 (1992)

7. Bertin, J.J.: Hypersonic Aerothermodynamics. AIAA Education Series, Washington, D.C (1994)
8. Streit, T., Martin, S., Eggers, T.: Approximate Heat Transfer Methods for Hypersonic Flow in Comparison with Results Provided by Numerical Navier-Stokes Solutions. DLR FB 94-36 (1994)
9. Simeonides, G.: Generalized Reference-Enthalpy Formulation and Simulation of Viscous Effects in Hypersonic Flow. Shock Waves 8(3), 161–172 (1998)
10. Fujii, K.: Progress and Future Prospects of CFD in Aerospace – Wind Tunnel and Beyond. Progress in Aerospace Sciences 41(6), 455–470 (2005)
11. Haney, J.W.: Orbiter (Pre STS-1) Aeroheating Design Data Base Development Methodology: Comparison of Wind Tunnel and Flight Test Data. In: Throckmorton, D.A. (ed.) Orbiter Experiments (OEX) Aerothermodynamics Symposium. NASA CP-3248, Part 1, pp. 607–675 (1995)
12. Davy, W.C., Green, M.J.: A Review of the Infrared Imagery of Shuttle (IRIS) Experiment. In: Throckmorton, D. A. (ed.), Orbiter Experiments (OEX) Aerothermodynamics Symposium. NASA CP-3248, Part 1, pp. 215–232 (1995)
13. Throckmorton, D.A., Zoby, E.V.: Shuttle Infrared Leeside Temperature Sensing (SILTS) Experiment. In: Throckmorton, D. A. (ed.), Orbiter Experiments (OEX) Aerothermodynamics Symposium. NASA CP-3248, Part 1, pp. 233–248 (1995)
14. Throckmorton, D.A. (ed.): Orbiter Experiments (OEX) Aerothermodynamics Symposium. NASA CP-3248, Part 1 and 2 (1995)

10 Approximate Relations for the Determination of Aerothermodynamic Parameters Referred to and/or Employed in This Book

In this book we employ in some chapters approximate relations for the determination of aerothermodynamic data. These are used in order to give quantitative information and to illustrate phenomena. In order to have the book self-consistent to a good degree, we give in this chapter these relations as well as others, which we only refer to. First elements of the RHPM$^+$ flyer and the γ_{eff} approach together with the bow shock total pressure loss are presented, and some γ_{eff} results in the large Mach number limit are given. Then the used or referred-to relations for the estimation of transport properties are provided, and finally formulas for stagnation point heating and flat surface boundary layer parameters. References are provided in all instances. For more general and detailed information the reader is referred to [1].

10.1 Elements of the RHPM$^+$ Flyer: γ_{eff} Approach and Bow Shock Total Pressure Loss

10.1.1 Introduction and Delineation

In [1] the RHPM[1] flyer was introduced, which can be used to illustrate, and even to quantify with care and only to a certain degree, characteristic aerothermodynamic properties of the different vehicle classes. It is essentially a two-dimensional approximation of the sufficiently flat windward side of the vehicle[2] by a flat plate – naturally neglecting the nose bluntness –, or a succession of flat plates, if aerodynamic trim and/or control surfaces, or inlet ramps are considered. To this geometrical approximation the shock/expansion theory is applied, therefore the name RHPM flyer.

The RHPM flyer can be a fair or a rather crude approximation of real flight vehicles, depending on the geometry under consideration. The more slender a configuration is and the smaller the angle of attack, the better is the

[1] RHPM stands for Rankine-Hugoniot-Prandtl-Meyer.

[2] This means, that the angle of attack of the RHPM flyer is not necessarily the nominal angle of attack α of the approximated vehicle, but the inclination angle $\overline{\alpha}$, eq. (4.4), or a suitable mean angle.

approximation regarding for example the wall pressure coefficient c_{p_w}. Flow properties influenced by the fact that a blunt nose is present in reality and therefore a total pressure loss, for example flow velocity and Mach number at the body surface, may show larger errors. This holds also if high-temperature real-gas effects are present in reality. All should be checked beforehand for the case considered.

If more exact data are needed, simplified configurations can be employed, which approximate the lower symmetry line of, for instance, a RV-W-type vehicle. Examples are the asymptotic half-angle hyperboloid approximating the windward centerline of the Space Shuttle Orbiter, [2], and the hyperboloid-flare approximation of the windward centerline with bodyflap deflection of the HERMES vehicle, [3]. Of course, simple estimates are no more possible for these axisymmetric configurations, since solutions of the Euler or Navier-Stokes equations must be performed.

In this book, in Chapter 3, we use a sphere-cone (blunt cone) approximation for the windward symmetry line of the Space Shuttle Orbiter. We must accept the curvature jump at the sphere/cone junction, but obtain better asymptotic data for the flat part of the vehicle than with the asymptotic half-angle hyperboloid approximation. The nose radius is a function of the angle of attack, [2], [4].

We give now elements of an extension of the original RHPM flyer to the RHPM$^+$ flyer, taking into account a) a γ_{eff} for the flow past the vehicle including the bow shock[3], different from γ_∞, Sub-Sections 10.1.2 to 10.1.4, and b) the influence of the total pressure loss due to the vehicle's bow shock, approximated via the flow-normal portion of the bow shock, Sub-Section 10.1.5.

While using the RHPM$^+$ flyer approximation we must be aware of the limitations discussed above. Especially we must note that the error due to the basic RHPM approximation can be larger than that we get without taking into account high-temperature real-gas effects. Nevertheless, in suitable cases we are able to quantify approximately the latter, at least parametrically.

10.1.2 γ_{eff} Approach: General Considerations

The use of an effective ratio of the specific heats γ_{eff} permits to determine approximately high-temperature real-gas effects. At the Space Shuttle Orbiter for instance, for $M_\infty = 24$ and $\alpha = 40°$, ratios of specific heats computed with the help of numerical methods are $\gamma \approx 1.3$ just behind the bow shock in the nose region, $\gamma \approx 1.12$ close to the body in the nose region, and $\gamma \approx 1.14$ along the lower body surface, [5]. Although aerothermodynamic parameters found with approximate methods (shock-expansion theory, Newton-derived methods) employing the γ_{eff} approach must be considered with care, such methods can be used to quantify, parameterize, and illustrate real-gas effects in a simple way.

[3] γ_{eff} of course must be either be known beforehand from other sources, or is applied parametrically.

Approximate methods need first of all the relations for flow-parameter changes across a shock wave. The shock relations usually found in the literature are derived for a constant ratio of specific heats ($\gamma =$ const.) in the whole flow domain considered (nominal case). If we wish to consider high-temperature real-gas effects in approximate flow field relations for a flight vehicle, of course we have ahead of the vehicle the original atmosphere with $\gamma = \gamma_\infty$ ($= 1.4$ for air). If we chose an effective ratio of specific heats $\gamma_{eff} < \gamma_\infty$, which then is constant (!) behind the shock wave and on the body surface, we have two different ratios of specific heats in the whole flow domain.

In this case the shock relations are different from those for the nominal case found in the literature, e.g., [6]. Of course the conservation of mass flux, momentum flux and enthalpy flux across a shock wave must be fulfilled also here. We show first the detailed derivation of the relations for the normal shock wave and give then the relations for the oblique shock wave. Unfortunately for the γ_{eff} approach compact-form relations as those given in [6] are no more achievable.

We use the classical approach with '1' designating the parameters ahead of the shock wave, and '2' those behind it (in applications '1' would denote the parameters ahead of the shock with γ_∞, and '2' the parameters behind the shock with γ_{eff}).

The familiar pressure coefficient c_{p_2} reads generalized:

$$c_{p_2} = \frac{p_2 - p_1}{q_1} = \frac{2}{\gamma_1 M_1^2}\left(\frac{p_2}{p_1}|_{\gamma_2} - 1\right) = \frac{2}{\gamma_\infty M_\infty^2}\left(\frac{p_2}{p_\infty}|_{\gamma_{eff}} - 1\right). \tag{10.1}$$

10.1.3 Normal Shock Wave

For a normal shock wave the equations for the conservation of the three entities read, with h_t being the total enthalpy:

$$\rho_1 v_1 = \rho_2 v_2, \tag{10.2}$$

$$\rho_1 v_1^2 + p_1 = \rho_2 v_2^2 + p_2, \tag{10.3}$$

$$(h_{t_1} =)\ h_1 + \frac{1}{2}v_1^2 = h_2 + \frac{1}{2}v_2^2\ (= h_{t_2}). \tag{10.4}$$

In the γ_{eff} approach, the latter relation is changed. With (note that c_{p_1} and c_{p_2} denote here the specific heats at constant pressure):

$$h_1 = c_{p_1} T_1 = \frac{\gamma_1}{\gamma_1 - 1} R_1 T_1 = \frac{\gamma_1}{\gamma_1 - 1}\frac{p_1}{\rho_1}, \tag{10.5}$$

and

$$h_2 = c_{p_2} T_2 = \frac{\gamma_2}{\gamma_2 - 1} R_2 T_2 = \frac{\gamma_2}{\gamma_2 - 1}\frac{p_2}{\rho_2}, \tag{10.6}$$

we get instead of eq. (10.4):

$$(h_{t_1} =)\ \frac{\gamma_1}{\gamma_1 - 1}\frac{p_1}{\rho_1} + \frac{1}{2}v_1^2 = \frac{\gamma_2}{\gamma_2 - 1}\frac{p_2}{\rho_2} + \frac{1}{2}v_2^2\ (= h_{t_2}). \tag{10.7}$$

Note that $p_1 = \rho_1 T_1 R_1$ and $p_2 = \rho_2 T_2 R_2$. Because we connect the thermodynamic entities ahead, '1', and behind, '2', the shock analytically, we must connect also the specific gas constants R_1 and R_2 with each other, see, e.g., [7]. This is made with the so called compressibility factor $Z = Z(\rho, T)$:

$$R_2 = Z(\rho_2, T_2) R_1. \tag{10.8}$$

The compressibility factor in the frame of our considerations is, in the few cases where needed, a problematic issue. Details are given in Sub-Section 10.1.6.

In the classical way it follows then from eqs. (10.2) and (10.3):

$$v_1^2 = \left(\frac{\rho_2}{\rho_1}\right) \frac{p_2 - p_1}{\rho_2 - \rho_1}, \tag{10.9}$$

and

$$v_2^2 = \left(\frac{\rho_1}{\rho_2}\right) \frac{p_2 - p_1}{\rho_2 - \rho_1}, \tag{10.10}$$

and finally:

$$v_2^2 = \left(\frac{\rho_1}{\rho_2}\right)^2 v_1^2. \tag{10.11}$$

Putting eqs. (10.9) and (10.10) into eq. (10.7), we obtain the so called Hugoniot relation which combines the thermodynamic parameters ahead and behind the shock wave:

$$h_2 - h_1 = \frac{\gamma_2}{\gamma_2 - 1} \frac{p_2}{\rho_2} - \frac{\gamma_1}{\gamma_1 - 1} \frac{p_1}{\rho_1} = \frac{1}{2}\left(\frac{1}{\rho_1} + \frac{1}{\rho_2}\right)(p_2 - p_1). \tag{10.12}$$

From this the pressure ratio across the shock wave is obtained:

$$\frac{p_2}{p_1} = \left(\frac{\gamma_2 - 1}{\gamma_1 - 1}\right) \frac{\frac{\rho_2}{\rho_1}(\gamma_1 + 1) - (\gamma_1 - 1)}{(\gamma_2 + 1) - \frac{\rho_2}{\rho_1}(\gamma_2 - 1)}. \tag{10.13}$$

Mach number M_1 and speed of sound a_1 are defined by:

$$M_1 = \frac{v_1}{a_1}, \tag{10.14}$$

$$a_1 = \sqrt{\gamma_1 R_1 T_1} = \sqrt{\gamma_1 \frac{p_1}{\rho_1}}. \tag{10.15}$$

Combining eq. (10.14) with eq. (10.9) yields the connection of the pressure and the density ratio across the shock with the Mach number M_1 ahead of it:

$$\frac{1}{\gamma_1}\left(\frac{\rho_2}{\rho_1}\right) \frac{\frac{p_2}{p_1} - 1}{\frac{\rho_2}{\rho_1} - 1} = M_1^2. \tag{10.16}$$

We substitute here p_2/p_1 with eq. (10.13) and obtain the relation for the density ratio:

$$\left(\frac{\rho_2}{\rho_1}\right)^2\left[\gamma_1 M_1^2(\gamma_2-1)+2\gamma_1\frac{\gamma_2-1}{\gamma_1-1}\right]+ \\ +\left(\frac{\rho_2}{\rho_1}\right)\left[-2\gamma_1 M_1^2\gamma_2-2\gamma_2\right]+\left[\gamma_1 M_1^2(\gamma_2+1)\right]=0. \tag{10.17}$$

If we denote the square brackets with A, B, and C, respectively, we get a solution with:

$$\frac{\rho_2}{\rho_1}|_{1,2}=\frac{-B\pm\sqrt{B^2-4AC}}{2A} \quad \left(=\frac{v_1}{v_2}|_{1,2}\right). \tag{10.18}$$

A compact-form solution, as we have it for a constant γ in the whole flow domain, see, e.g., [6], [1], is not possible. The positive sign of the square root is valid. The result is then used to determine numerically the other ratios across the shock wave (see below).

Of course, choosing $\gamma_2 \equiv \gamma_1 = \gamma$ yields the familiar compact-form solutions for this and all other parameter ratios. If $\gamma_1 \neq \gamma_2$, the case $M_1 = 1$ is not included, because we assume iso-enthalpic flow.

For $M_1 \to \infty$ we get further:

$$\frac{\rho_2}{\rho_1}|_{M_1\to\infty}=\frac{\gamma_2+1}{\gamma_2-1}, \tag{10.19}$$

and also the other limiting cases, given for instance in [1].

The temperature ratio T_2/T_1 is determined from:

$$\frac{T_2}{T_1}=\frac{p_2}{p_1}\frac{\rho_1}{\rho_2}\frac{R_1}{R_2}. \tag{10.20}$$

The speed of sound behind the shock a_2 is defined by:

$$a_2=\sqrt{\gamma_2 T_2 R_2}=\sqrt{\gamma_2 p_2/\rho_2}. \tag{10.21}$$

The Mach number $M_2 = v_2/a_2$ is found with the help of eqs. (10.11) and (10.21) after some manipulation in the following way:

$$M_2^2=\frac{v_2^2}{a_2^2}=\left(\frac{\rho_1}{\rho_2}\right)^2 v_1^2\frac{\rho_2}{\gamma_2 p_2}=\frac{\gamma_1}{\gamma_2}M_1^2\frac{1}{\frac{\rho_2}{\rho_1}\frac{p_2}{p_1}}. \tag{10.22}$$

It remains the determination of total pressure p_{t_2}, density ρ_{t_2}, and temperature[4] T_{t_2} behind the shock wave.

[4] The total enthalpy is constant, but the total temperature not because of eq. (10.7).

The total pressure p_{t_2} is found from:

$$\frac{p_{t_2}}{p_{t_1}} = \frac{p_{t_2}}{p_2}\frac{p_2}{p_1}\frac{p_1}{p_{t_1}}, \tag{10.23}$$

with p_2/p_1 from eq. (10.13) after solution of eq. (10.17), and:

$$\frac{p_{t_2}}{p_2} = \left(1 + \frac{\gamma_2 - 1}{2} M_2^2\right)^{\frac{\gamma_2}{\gamma_2 - 1}}, \quad \frac{p_{t_1}}{p_1} = \left(1 + \frac{\gamma_1 - 1}{2} M_1^2\right)^{\frac{\gamma_1}{\gamma_1 - 1}}. \tag{10.24}$$

Similarly we find for the total density ρ_{t_2}:

$$\frac{\rho_{t_2}}{\rho_{t_1}} = \frac{\rho_{t_2}}{\rho_2}\frac{\rho_2}{\rho_1}\frac{\rho_1}{\rho_{t_1}}, \tag{10.25}$$

with

$$\frac{\rho_{t_2}}{\rho_2} = \left(1 + \frac{\gamma_2 - 1}{2} M_2^2\right)^{\frac{1}{\gamma_2 - 1}}, \quad \frac{\rho_{t_1}}{\rho_1} = \left(1 + \frac{\gamma_1 - 1}{2} M_1^1\right)^{\frac{1}{\gamma_1 - 1}}. \tag{10.26}$$

The total temperature T_{t_2} is to be determined with:

$$\frac{T_{t_2}}{T_{t_1}} = \frac{p_{t_2}}{p_{t_1}}\frac{\rho_{t_1}}{\rho_{t_2}}\frac{R_1}{R_2}, \tag{10.27}$$

or directly with the help of eq. (10.4), respectively eq. (10.7):

$$T_{t_2} = \frac{\gamma_2 - 1}{\gamma_2}\frac{1}{R_2}\left(\frac{\gamma_2}{\gamma_2 - 1}\frac{p_2}{\rho_2} + \frac{1}{2}v_2^2\right). \tag{10.28}$$

Looking back, we observe, that the specific gas constant behind the shock wave, R_2, is needed only if we want to determine the temperatures T_2 and/or T_{t_2}. This is a restriction, which we can not overcome with an adequately simple approximation of the compressibility factor $Z(\rho, T) = Z(\gamma_{eff})$, Sub-Section 10.1.6.

Because we don't get a compact-form solution for ρ_2/ρ_1, we also don't have such solutions for the other parameter ratios. We note in Table 10.1 the input into the respective equations for the numerical determination of these ratios, for the temperatures with the caveat in view of the compressibility factor.

10.1.4 Oblique Shock Wave

For the oblique shock wave, Fig. 10.1, the equations for the conservation of the three entities mass, momentum (now two components: normal and tangential to the oblique shock wave), and enthalpy read:

- mass:

$$\rho_1 \tilde{u}_1 = \rho_2 \tilde{u}_2, \tag{10.29}$$

Table 10.1. Flow parameters ratios across a normal shockwave with γ_{eff}, and the equations and the input for their numerical determination.

Parameter Ratio	Symbol	Eq.	Input
Density	ρ_2/ρ_1	(10.17)	$\gamma_1, \gamma_2, M_1^2$
Pressure	p_2/p_1	(10.13)	$\gamma_1, \gamma_2, \rho_2/\rho_1$
Temperature	T_2/T_1	(10.20)	$p_2/p_1, \rho_2/\rho_1, R_2$
Velocity	v_2/v_1	(10.2)	ρ_2/ρ_1
Mach number	M_2/M_1	(10.22)	$\gamma_1, \gamma_2, \rho_2/\rho_1, p_2/p_1$
Total pressure	p_{t_2}/p_{t_1}	(10.23)	$p_{t_2}/p_1, p_2/p_1, p_{t_1}/p_1$
Total density	ρ_{t_2}/ρ_{t_1}	(10.25)	$\rho_{t_2}/\rho_1, \rho_2/\rho_1, \rho_{t_1}/\rho_1$
Total temperature	T_{t_2}/T_{t_1}	(10.27)	$p_{t_2}/p_{t_1}, \rho_{t_2}/\rho_{t_1}, R_2$

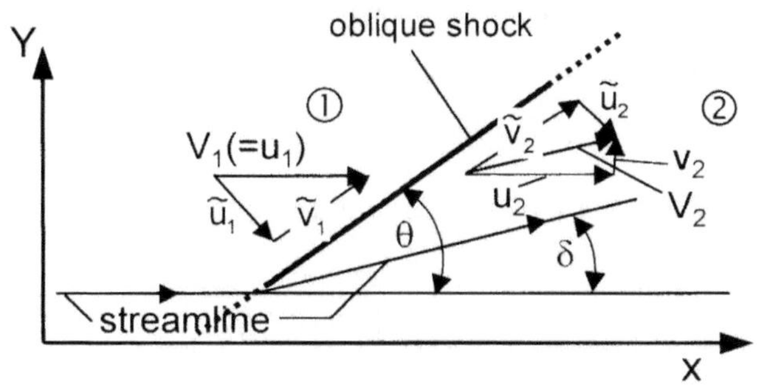

Fig. 10.1. Schematic of an oblique shock wave and notation [1].

- normal momentum component:

$$\rho_1 \tilde{u}_1^2 + p_1 = \rho_2 \tilde{u}_2^2 + p_2, \tag{10.30}$$

- tangential momentum component:

$$\rho_1 \tilde{u}_1 \tilde{v}_1 = \rho_2 \tilde{u}_2 \tilde{v}_2, \tag{10.31}$$

- enthalpy:

$$(h_{t_1} =)\ h_1 + \frac{1}{2}(\tilde{u}_1^2 + \tilde{v}_1^2) = h_2 + \frac{1}{2}(\tilde{u}_2^2 + \tilde{v}_2^2)\ \ (= h_{t_2}). \tag{10.32}$$

We note especially, that the resultant velocity behind the shock wave, V_2, is connected to that ahead of it, V_1, by:

$$V_2^2 = V_1^2 \left[\left(\frac{\rho_1}{\rho_2} \right)^2 \sin^2\theta + \cos^2\theta \right]. \tag{10.33}$$

We proceed as before and find the Hugoniot relation in the same form:

$$h_2 - h_1 = \frac{\gamma_2}{\gamma_2 - 1}\frac{p_2}{\rho_2} - \frac{\gamma_1}{\gamma_1 - 1}\frac{p_1}{\rho_1} = \frac{1}{2}\left(\frac{1}{\rho_1} + \frac{1}{\rho_2}\right)(p_2 - p_1). \tag{10.34}$$

Now, however, the relevant Mach number is that of the flow component normal to the shock wave:

$$\tilde{M}_1 = \sin\theta\, M_1 = \frac{\tilde{u}_1}{a_1}. \tag{10.35}$$

Proceeding further like for the normal shock wave, we obtain:

$$\frac{1}{\gamma_1}\left(\frac{\rho_2}{\rho_1}\right)\frac{\frac{p_2}{p_1} - 1}{\frac{\rho_2}{\rho_1} - 1} = \tilde{M}_1^2 = \sin^2\theta\, M_1^2. \tag{10.36}$$

The temperature T_2 again is:

$$\frac{T_2}{T_1} = \frac{p_2}{p_1}\frac{\rho_1}{\rho_2}\frac{R_1}{R_2}. \tag{10.37}$$

and the speed of sound behind the shock wave:

$$a_2 = \sqrt{\gamma_2 T_2 R_2} = \sqrt{\gamma_2 p_2/\rho_2}. \tag{10.38}$$

The Mach number $M_2 = v_2/a_2$ now is found with the help of eqs. (10.33) and (10.38) to be:

$$M_2^2 = \frac{V_2^2}{a_2^2} = \left(\frac{\rho_1}{\rho_2}\right)^2 V_1^2 \frac{\rho_2}{\gamma_2 p_2} = \frac{\gamma_1}{\gamma_2} M_1^2 \left[\left(\frac{\rho_1}{\rho_2}\right)^2 \sin^2\theta + \cos^2\theta\right] \frac{1}{\frac{\rho_1}{\rho_2}\frac{p_2}{p_1}}. \tag{10.39}$$

The shock angle θ is an unknown. From Fig. 10.1 we find with:

$$\tan\theta = \frac{\tilde{u}_1}{\tilde{v}_1}, \quad \tan(\theta - \delta) = \frac{\tilde{u}_2}{\tilde{v}_2}, \tag{10.40}$$

and eq. (10.29) the relation:

$$\frac{\tan\theta}{\tan(\theta - \delta)} = \frac{\tilde{u}_1}{\tilde{v}_1}\frac{\tilde{v}_2}{\tilde{u}_2} = \frac{\tilde{u}_1}{\tilde{u}_2} = \frac{\rho_2}{\rho_1}. \tag{10.41}$$

After some manipulation with eqs. (10.34), (10.36), and (10.41), like for the normal shock wave, we obtain two coupled equations for the two unknowns ρ_2/ρ_1 and $\tan\theta$ as functions of the free-stream Mach number M_1, the deflection angle δ, and the two ratios of specific heats γ_1 and γ_2:

$$\begin{aligned}&\left(\frac{\rho_2}{\rho_1}\right)^2 \left[\gamma_1 \sin^2\theta\, M_1^2(\gamma_2 - 1) + 2\gamma_1 \frac{\gamma_2 - 1}{\gamma_1 - 1}\right] + \\ &+ \left(\frac{\rho_2}{\rho_1}\right)\left[-2\gamma_1 \sin^2\theta\, M_1^2 \gamma_2 - 2\gamma_2\right] + \left[\gamma_1 \sin^2\theta\, M_1^2(\gamma_2 + 1)\right] = 0,\end{aligned} \tag{10.42}$$

and

$$\tan^2\theta\,[\tan\delta] + \tan\theta\,[1 - \frac{\rho_2}{\rho_1}] + [\frac{\rho_2}{\rho_1}\tan\delta] = 0. \tag{10.43}$$

The latter equation written in the form:

$$\tan\delta = \frac{\frac{\rho_2}{\rho_1} - 1}{\frac{\rho_2}{\rho_1} + \tan^2\theta}\tan\theta \tag{10.44}$$

leads for $\gamma_2 \equiv \gamma_1 = \gamma$ to the familiar compact-form relation for $\tan\delta$. The case $\gamma_1 \neq \gamma_2$ for $\delta = 0°$ is not included, again because we assume iso-enthalpic flow.

We proceed further and find the pressure p_2 like for the normal shock with the help of:

$$\frac{p_2}{p_1} = \left(\frac{\gamma_2 - 1}{\gamma_1 - 1}\right)\frac{\frac{\rho_2}{\rho_1}(\gamma_1 + 1) - (\gamma_1 - 1)}{(\gamma_2 + 1) - \frac{\rho_2}{\rho_1}(\gamma_2 - 1)}. \tag{10.45}$$

The velocity components normal to and behind the shock wave read:

$$\tilde{u}_2 = \frac{1}{\frac{\rho_2}{\rho_1}}V_1\sin\theta, \tag{10.46}$$

and, because of eq. (10.31),

$$\tilde{v}_1 = \tilde{v}_2 = V_1\cos\theta. \tag{10.47}$$

The resultant velocity behind the shock wave, Fig. 10.1, is:

$$V_2 = \sqrt{\tilde{u}_2^2 + \tilde{v}_2^2}. \tag{10.48}$$

Total pressure p_{t_2}, total density ρ_{t_2} and total temperature T_{t_2} can be found like for the normal shock wave. Again we need the specific gas constant R_2, if we wish to determine T_2 and/or T_{t_2}.

For the numerical determination of all parameters a scheme similar to that given in Table 10.1 can be devised, which however has to include eq. (10.43) for θ.

10.1.5 Bow Shock Total Pressure Loss: Restitution of Parameters of a One-Dimensional Surface Flow

With extended algebraic effort it can be shown that the governing equations for inviscid fluid flow around a body can be reduced exactly from their three-dimensional or two-dimensional form *off* the body surface, of course including the surface, to the two-dimensional or one-dimensional form only *on* the surface. We take this for granted and consider now, how for a two-dimensional body from a given pressure distribution along the surface other flow parameters can be determined, while taking into account the total pressure loss due to the flow-normal portion of the bow shock.

This is of interest, if for instance with Newton's method the pressure on the surface of a body was found, if a consideration like in Sub-Section 6.1.1 is made, or if experimental pressure data are given, and further investigations are intended.

We have to make one assumption, viz. that the streamline hitting the forward stagnation point crossed the locally normal bow shock surface, [1]. Usually at large angles of attack this is not the case, but in general it is an acceptable assumption.

With this assumption, we can obtain[5], with $\gamma_{eff} = \gamma_2 =$ const., the total pressure p_{t_2} and the total density ρ_{t_2} at the forward stagnation point on the body surface from eqs. (10.23) and (10.25). Both are constant along the body surface as long as we can assume that no embedded shock waves are present there.

Starting at the stagnation point, we denote the coordinate along the surface with x. If total pressure, total density, and the pressure $p\,(x)$ are given, we find with the help of Bernoulli's equation the velocity $v_2\,(x)$:

$$v_2(x) = \sqrt{\frac{2\gamma_2}{\gamma_2 - 1}\frac{p_{t2}}{\rho_{t_2}}\left[1 - \left(\frac{p_2(x)}{p_{t_2}}\right)^{\frac{\gamma_2-1}{\gamma_2}}\right]}. \tag{10.49}$$

Taking into account constant total enthalpy $h_{t_2} = h_{t_1}$, the temperature $T(x)$ is determined from:

$$h_2(x) = h_{t_1} - \frac{1}{2}\,v_2^2(x), \tag{10.50}$$

and

$$T_2(x) = \frac{h_2(x)}{c_{p2}} = \frac{\gamma_2 - 1}{\gamma_2}\frac{1}{R_2}\,h_2(x). \tag{10.51}$$

From this follows the Mach number $M_2(x)$:

$$M_2(x) = \frac{v_2(x)}{\sqrt{\gamma_2 R_2\,T_2(x)}}, \tag{10.52}$$

and finally the density $\rho_2(x)$ with:

$$\rho_2(x) = \frac{p_2(x)}{R_2\,T_2(x)}, \tag{10.53}$$

and, for checking purposes:

$$\rho_2(x) = \rho_{t_2}\left[1 + \frac{\gamma_2 - 1}{2}M_2^2(x)\right]^{\frac{-1}{\gamma_2-1}}. \tag{10.54}$$

---

5 We give, despite the problems with the compressibility factor, the derivation in terms of γ_{eff}.

The specific gas constant $R_2 = Z\, R_1$, appears quite often in these relations. This means that the compressibility factor plays a larger role here, than in the shock relations. In general it can not be recommended to use the γ_{eff} approach for the restitution of flow parameters, as described here, unless an exact numerical reference solution for a given problem permits to develop a simple approximation of Z valid for this problem, see the next sub-section.

10.1.6 The Compressibility Factor *Z*

Consider a mixture of n thermally perfect gases. The equation of state, with the universal gas constant R_0, is, see, e.g., [1]:

$$p = \sum_{i=1}^{n} p_i = \sum_{i=1}^{n} \rho_i \frac{R_0}{M_i} T = \rho T \sum_{i=1}^{n} c_i \frac{R_0}{M_i} = \rho T R, \tag{10.55}$$

with p_i being the partial pressure, ρ_i the partial density, and M_i the mass of the species i. R is the gas constant of the mixture. The mass fraction[6] c_i ($0 \leqq c_i \leqq 1$) of the species i is:

$$c_i = \frac{\rho_i}{\rho}, \tag{10.56}$$

and the mean molecular weight M:

$$\frac{1}{M} = \sum_{i=1}^{n} \frac{c_i}{M_i}. \tag{10.57}$$

If we consider a single diatomic gas, for instance N_2, it is obvious, that the specific gas constant R is twice as large for the fully dissociated gas ($c_{N_2} = 0$) than for the not dissociated gas ($c_{N_2} = 1$). If we express the specific gas constant in the dissociated state with R_2, it is related to that of the undissociated gas R_1 by:

$$R_2 = Z(\rho, T) R_1. \tag{10.58}$$

Z is the compressibility factor, see, e.g., [7]. It is found with eqs. (10.55) and (10.57):

$$Z = \frac{R_2}{R_1} = \frac{\sum_{i=1}^{n} c_i \frac{R_0}{M_i}|_2}{\sum_{i=1}^{n} c_i \frac{R_0}{M_i}|_1} = \frac{M_1}{M_2}. \tag{10.59}$$

For air in equilibrium in the temperature range $1{,}500.0K \leqq T \leqq 15{,}000.0K$ and the density range $1 \leqq log_{10}(\rho/\rho_0) \leqq -10$, with $\rho_0 = 1.293{\cdot}10^{-3}\ g/cm^3$ a graph is given in [7], which shows a highly non-linear behavior of $Z(\rho, T)$, with $1 \leqq Z \lessapprox 4$. For $log_{10}(\rho/\rho_0) = -5$, for instance, the compressibility factor is $Z \approx 1.3$ for $T = 4{,}000.0\ K$, $Z \approx 2$ for $T = 6{,}000.0\ K$, and $Z \approx 3.2$ for $T = 10{,}000.0\ K$.

[6] Other than in [1] we use c_i as symbol instead of ω_i.

We state expressively that all this is no problem in discrete numerical solutions of the governing equations of aerothermodynamics, where we don't need these data beforehand in the computation process, but can construct for instance Z or γ_{eff} a posteriori.

If we employ γ_{eff} in analytical methods of the kinds presented here, however, we need for the determination of some thermodynamic and flow entities the compressibility factor Z, which means – note that this is the critical point – its connection with γ_{eff}: $Z = Z(\gamma_{eff})$.

An expression of $Z = Z(\gamma_{eff})$ is not known. It is advisable therefore, to use the γ_{eff} approach only for the determination of data, where Z does not play a role. Across shock waves, these are almost all data with the exception of the temperatures T and T_t. For the restitution of parameters of one-dimensional or other flows, one can for a selected problem, of course, try approximations of the kind $Z = a + b\,\gamma_{eff}$. This is permissible, if it can be shown with an exact numerical reference solution, that this approximation is meaningful. Clearly then this approximation can only be employed in the class of problems, to which the selected problem belongs.

10.1.7 γ_{eff} Results Across Shocks in the Large M_1 Limit

For the normal shock we have found, eq. (10.19), that ρ_2/ρ_1 in the limit $M_1 \Rightarrow \infty$ depends on γ_2 only. This means that in that limit the state of the gas behind the shock wave does no more depend on the ratio of specific heats γ_1 ahead of the shock.

We study this in more detail in Fig. 10.2. The asymptotic behavior is seen well for ρ_2/ρ_1 and M_2. The limit is reached always the later the smaller γ_1 is. From the ordinary Rankine-Hugoniot relations, see, e.g., [1], we know that no large Mach number limits exist for p_2/p_1 and T_2/T_1. Hence we cannot assume, that these properties become independent of γ_1 in the limit. This is also well seen in the figure.

For the total pressure coefficient behind the shock wave, $c_{p_{t2}}$, we see finally, Fig. 10.3, that it becomes already for rather small M_1 independent of γ_1.

These results show on the one hand, that the γ_{eff} approach, which anyway must be handled with care, demands the use of the exact relations as we have derived them above. On the other hand, in certain cases, here for $c_{p_{t2}}$ in the range $M_1 \gtrapprox 10$, it is permitted to plug into the Rankine-Hugoniot conditions, also for the oblique shock, just the value of γ_{eff} ($\equiv \gamma_2$).

This means, that a gas with a given γ_{eff} can be handled like in the perfect-gas case, by employing a constant γ ahead and behind the shock. This would be the justification to work in a ground-simulation facility with gases other than air in order to study the influence of a chosen γ on the aerodynamic properties of a flight vehicle. An example is the investigation of the pitching moment anomaly of the Space Shuttle Orbiter, Section 3.5, in the LaRC 20" CF_4 (carbon tetrafluoride) tunnel, see, e.g., [8].

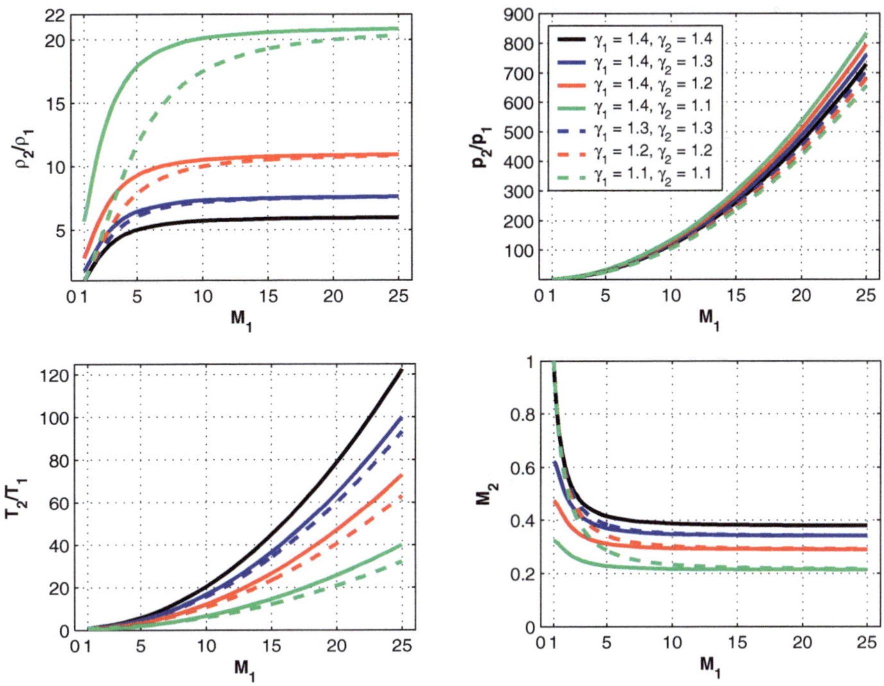

Fig. 10.2. Normal shock wave: dependence of density ratio (upper left), pressure ratio (upper right), temperature ratio (lower left) and Mach number behind the shock (lower right) on pairs of γ_1 and γ_2 as function of M_1.

10.2 Transport Properties

We list here only some power-law approximations for the viscosity and the thermal conductivity, valid for temperatures up to approximately 1,500.0 to 2,000.0 K. For details, the reader is referred to, e.g., [1].

Air

- Viscosity

  Sutherlands equation:

$$\mu = 1.458 \cdot 10^{-6} \frac{T^{1.5}}{T + 110.4}. \tag{10.60}$$

  Simple power-law approximation:

$$\mu = c T^{\omega}, \tag{10.61}$$

  where for $T \lessapprox 200.0\ K$: $c = 0.702 \cdot 10^{-7}$, $\omega = 1$,
  and for $T \gtrapprox 200.0\ K$: $c = 0.04644 \cdot 10^{-5}$, $\omega = 0.65$.

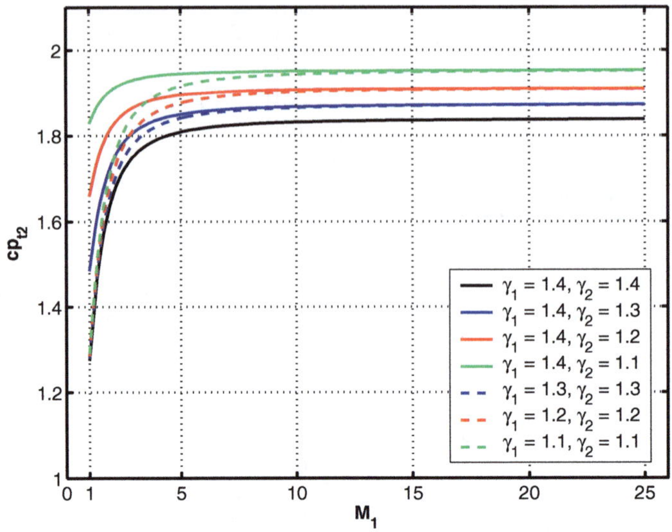

Fig. 10.3. Normal shock wave: dependence of the total pressure coefficient behind the shock wave, $c_{p_{t2}}$ on pairs of γ_1 and γ_2 as function of M_1.

- Thermal conductivity

The thermal conductivity of a molecular gas can be determined with the help of the Eucken formula, where c_p is the specific heat at constant pressure:

$$k = \left(c_p + \frac{5}{4}\frac{R_0}{M}\right)\mu. \tag{10.62}$$

The monatomic case is included, if for the specific heat $c_p = 2.5\,R_0/M$ is taken.

From the above the following relation for the Prandtl number can be derived:

$$Pr = \frac{\mu c_p}{k} = \frac{c_p}{c_p + 1.25 R_0/M} = \frac{4\gamma}{9\gamma - 5}. \tag{10.63}$$

Explicit relations for air are:

Hansen equation:

$$k = 1.993 \cdot 10^{-3} \frac{T^{1.5}}{T + 112.0}. \tag{10.64}$$

Simple power-law approximation:

$$k = c T^{\omega}, \tag{10.65}$$

where for $T \lessapprox 200.0\ K$: $c = 9.572 \cdot 10^{-5}$, $\omega = 1$,
and for $T \gtrapprox 200.0\ K$: $c = 34.957 \cdot 10^{-5}$, $\omega = 0.75$.

Molecular Nitrogen

- Viscosity

  Sutherlands equation for molecular Nitrogen:

$$\mu = 1.39 \cdot 10^{-6} \frac{T^{1.5}}{T + 102.0}. \tag{10.66}$$

- Heat conductivity

  The thermal conductivity of a molecular nitrogen can be determined, like for air, with the help of the Eucken formula, eq. (10.62).

10.3 Formulas for Stagnation Point Heating

At the early stages of aerothermodynamics there was nearly no possibility to measure or to theoretically determine the temperature gradient $\partial T/\partial y$ normal to the surface[7]. This temperature gradient determines by the Fourier law (the law of heat conduction in a material) the convective heat-flux to an isothermal wall in a non-reacting flow, i. e. the heat flux in the gas at the wall, [1]:

$$q_{gw} = -k \frac{\partial T}{\partial y}, \tag{10.67}$$

where k is the thermal conductivity[8].

Therefore, previously, the so called film formulation was generated, which reads:

$$q_{gw} = h_c(T_r - T_w), \tag{10.68}$$

with T_r the recovery and T_w the wall temperature, and h_c the film coefficient. The latter is not a constant, but depends on numerous variables, where the most important ones are the density and the velocity at the boundary layer edge as well as the transport properties. The film coefficient combines the effects of conduction and convection in the gas.

We list some of the most often used formulas for determining the stagnation point (sphere) heat-flux, which all are based on the film formulation concept, eq. (10.68). It is not the intention here to go in great detail, since applications of these formulas are numerously available in the literature. But we provide

---

[7] With modern numerical simulation methods approximating the full Navier-Stokes equations including thermochemical non-equilibrium, this situation has changed and most of the gradients in the boundary layer can be resolved properly.

[8] Note that usually for the wall heat flux the minus sign is omitted: $q_{gw} = k\, \partial T/\partial y$. If the y-coordinate is positive away from the surface, a heat flux towards the wall therefore has a positive value. This holds analogously also for the wall shear stress.

the reader with the possibility to have a first glance at the dependencies of forward stagnation point heat-fluxes for RV-W- and RV-NW-type vehicles during atmospheric re-entry, but also CAV/ARV-type vehicles flying through the atmosphere.

a) Formula of Van Driest, [9]

$$q_{gw} = k_{st} Pr^{-0.6} (\rho_e \mu_e)^{0.5} (h_r - h_w)_{st} \left(\frac{du_e}{dx}\right)^{0.5}_{st}, \tag{10.69}$$

with $k_{st} = 0.76$ for a sphere, Pr the Prandtl number, h_r the recovery enthalpy, h_w the fluid enthalpy at the wall, $du_e/dx|_{st}$ the tangential velocity gradient at the stagnation point, the subscript e denotes the boundary layer edge and st the stagnation point. Perfect gas is assumed. The velocity gradient, $du_e/dx|_{st}$, can be determined by Newton's pressure assumption at the stagnation point, see, e.g., [1], or from Euler equations at the edge of the boundary layer, using the relation:

$$\left(\frac{du_e}{dx}\right)_{st} = \frac{1}{R_N} \left(\frac{2(p_e - p_\infty)}{\rho_e}\right)^{0.5}, \tag{10.70}$$

with R_N being the nose radius.

b) Formula of Lees, [10]:

$$q_{gw} = 0.50 Pr^{-0.67} (\rho_{st} \mu_{st})^{0.5} h_{w,st} \sqrt{2} \left(\frac{du_e}{dx}\right)^{0.5}_{st}. \tag{10.71}$$

Perfect gas is assumed. This formula was used during the APOLLO project work for convective heat-flux prediction. For the heat-flux prediction away from the stagnation point on the APOLLO surface in the pitch plane the following relations were used exploiting the wall pressure distribution (e.g. known from wind tunnel experiment):

$$q_{gw} = 0.50 Pr^{-0.67} (\rho_{st} \mu_{st})^{0.5} h_{w,st} F(s), \tag{10.72}$$

$$F(s) = \frac{\dfrac{p_e}{p_{st}} \dfrac{\mu_e T_{st}}{\mu_{st} T_e} u_e r_o^k}{\left(2 \displaystyle\int_o^{s1} \dfrac{p_e}{p_{st}} \dfrac{\mu_e T_{st}}{\mu_{st} T_e} u_e r_o^{2k} ds\right)^{0.5}}, \tag{10.73}$$

with $k = 1$ for axisymmetric flow and $k = 0$ for plane flow, s denotes the contour length measured from the stagnation point, r_0 the radius of the nose section of the body of revolution, [11].

c) Formula of Fay and Riddell, [12]:

$$q_{gw} = 0.76 Pr^{-0.6} (\rho_w \mu_w)^{0.1} (\rho_e \mu_e)^{0.4} \cdot$$

$$\cdot \left[1 + (Le^{\phi} - 1) \frac{h_D}{h_e}\right] (h_e - h_w)_{st} \left(\frac{du_e}{dx}\right)^{0.5}_{st}, \tag{10.74}$$

with h_D being the enthalpy of dissociation of the gas mixture, $Le = Pr/Sc$ the Lewis number defining the ratio of heat transfer by mass diffusion to heat transfer by conduction [1], where, for example, for dissociating air the Prandtl number is $Pr = 0.71$ and the Schmidt number is $Sc = 0.5$, giving the Lewis number $Le = 1.42$. $Le = 0$ for frozen flow with non-catalytic wall, $\phi = 0.52$ for equilibrium flow and $\phi = 0.63$ for frozen flow with fully catalytic wall. This formula is likely the most often used and referenced relation for stagnation point heat-flux determination.

d) Formula of Cohen, [13]:

$$q_{gw} = 0.767 Pr^{-0.6} (\rho_w \mu_w)^{0.07} (\rho_e \mu_e)^{0.43} (h_e - h_w)_{st} \left(\frac{du_e}{dx} \right)_{st}^{0.5} . \qquad (10.75)$$

This equation is very similar to eq. (10.74) of Fay and Riddell for $Le = 0$ (frozen flow with non-catalytic wall) and is used in [14] as a basic relation in an engineering method for predicting the heat transfer at blunted cones.

e) Simple formula, [15], [16]:

$$q_{gw} = \frac{5.1564 \cdot 10^{-5}}{\sqrt{R_N}} \rho_\infty^{0.5} v_\infty^{3.15} . \qquad (10.76)$$

The input dimensions are kg/m^3 for the density ρ_∞, m/s for the flight speed v_∞. The resulting q_{gw} has the dimension W/m^2.

This relation is a correlation of data from several sources, based on eq. (2.5). This is a "cheap" and fair approximation of the heat flux in the gas at the wall at the stagnation point of a sphere. It serves typically in trajectory determination and optimization with the forward stagnation point as reference location, Chapter 2.

10.4 Flat Surface Boundary Layer Parameters Based on the Reference-Temperature/Enthalpy Concept

The reference-temperature/enthalpy concept permits to take into account temperature and compressibility effects approximately and in a simple way for the determination of boundary layer parameters, [17]. As reference-enthalpy concept it enables to treat flows with high-enthalpy, [18]. The reference-temperature/ enthalpy concept is discussed in some detail in, e.g., [1]. It is not an exact, but a well-proven approximate concept. Basically it works with boundary layer relations established for incompressible flow. These are applied with the inviscid flow data at the body surface, which are interpreted as being those at the boundary layer edge. Density and viscosity are interpreted as function of an appropriate reference temperature or enthalpy. In [19], for instance, the viability of the approach is demonstrated. Generalized formulations, which are valid for attached laminar and turbulent flow can be found in [20].

10.4.1 Reference-Temperature/Enthalpy Concept

The characteristic Reynolds number for a high speed boundary layer like flow is postulated to read:

$$Re_x^* = \frac{\rho^* v_e x}{\mu^*}. \tag{10.77}$$

Density ρ^* and viscosity μ^* are reference data, characteristic of the boundary layer. They are determined with the local pressure p and the reference enthalpy h^* or with the reference temperature T^*. v_e is the external inviscid flow.

The reference temperature T^* is empirically composed of the boundary layer edge temperature T_e (without hypersonic viscous interaction identical to T_∞), the wall temperature T_w, and the recovery temperature T_r, [17]:

$$T^* = 0.28 T_e + 0.5 T_w + 0.22 T_r. \tag{10.78}$$

The general reference enthalpy h^* is defined similarly, [18]:

$$h^* = 0.28 h_e + 0.5 h_w + 0.22 h_r. \tag{10.79}$$

The recovery data are found with:

$$T_r = T_e + r^* \frac{v_e^2}{2 c_p}, \tag{10.80}$$

or

$$h_r = h_e + r_h^* \frac{v_e^2}{2}. \tag{10.81}$$

Here r^*, or r_h^*, is the recovery factor, which is a function of the Prandtl number Pr. The Prandtl number depends rather weakly on the temperature, or the enthalpy, respectively, up to $T \approx 5{,}000.0\ K$. Usually it is sufficient to assume $r^* = r = \text{const.}$ For laminar flow the recovery factor is $r = \sqrt{Pr}$, and for turbulent flow $r = \sqrt[3]{Pr}$. With the Prandtl number at low temperatures, $Pr \approx 0.74$, see, e.g., [1], we get $r_{lam} \approx 0.86$, and $r_{turb} \approx 0.90$.

Introducing the boundary layer edge data as reference flow data into eq. (10.77) yields:

$$Re_x^* = \frac{\rho_e v_e x}{\mu_e} \frac{\rho^*}{\rho_e} \frac{\mu_e}{\mu^*} = Re_{e,x} \frac{\rho^*}{\rho_e} \frac{\mu_e}{\mu^*}, \tag{10.82}$$

with $Re_{e,x} = \rho_e v_e x / \mu_e$.

This relation can be simplified. If we apply it to boundary layer like flows, we can write, because $p = p_e = p_w$:

$$\frac{\rho^*}{\rho_e} = \frac{T_e}{T^*}. \tag{10.83}$$

If, for simplicity, we further assume a power-law expression for the viscosity, we obtain:

$$\frac{\mu^*}{\mu_e} = \frac{(T^*)^{\omega^*}}{(T_e)^{\omega_e}}. \tag{10.84}$$

Only if T^* and T_e are both in the same temperature ω^* and ω_e are equal, see, e.g., [1], and we see, e.g., [1], and we get:

$$\frac{\mu^*}{\mu_e} = \left(\frac{T^*}{T_e}\right)^{\omega}. \tag{10.85}$$

For air and for $T \gtrapprox 200.0\ K$, the exponent is $\omega = 0.65$, and if $T \lessapprox 200.0\ K$, even more simply $\omega = 1$, Section 10.2.

Introducing eqs. (10.83) and (10.85) into eq. (10.82) reduces the latter to:

$$Re_x^* = Re_{e,x}\left(\frac{T_e}{T^*}\right)^{1+\omega}. \tag{10.86}$$

10.4.2 Boundary Layer Thicknesses at Flat Surfaces

In this and the following sub-sections we give the flat-plate relations in generalized form, [20], see also [1]. The exponent in the following relations is $n = 0.5$ for laminar, and $n = 0.2$ for turbulent flow. If, see above, T^* and T_e are both in the same temperature interval, the exponents ω^* and ω_e are equal, and the given relations can be further reduced.

The boundary layer thickness δ reads:

$$\delta = C\frac{x}{(Re_{e,x}^*)^n}, \tag{10.87}$$

with $C = 5$ for laminar and $C = 0.37$ for turbulent flow.

Specializing for the reference temperature, we obtain with $Re^u = \rho_e\, v_e/\mu_e$:

$$\delta = C\frac{x^{1-n}}{(Re_e^u)^n}\left(\frac{\rho_e}{\rho^*}\frac{\mu^*}{\mu_e}\right)^n. \tag{10.88}$$

An alternative formulation for the thickness of laminar compressible boundary layers is, [21]:

$$\delta_{lam} = x^{0.5}\sqrt{\frac{C*}{Re_e^u}}\left(5 + 2.21\frac{\gamma - 1}{2}M_\infty^2 + 1.93\frac{T_w - T_r}{T_e}\right), \tag{10.89}$$

with:

$$C^* = \frac{\mu^* T_e}{\mu_e T^*}. \tag{10.90}$$

The relations for the displacement thickness δ_1 are similar. For the laminar case we have:

$$\begin{aligned}\delta_{1,lam} =& \delta_{1,lam,ic}\left(-0.122 + 1.122\frac{T_w}{T_\infty} +\right.\\ &\left. + 0.333\frac{\gamma_\infty - 1}{2}M_\infty^2\right)\left(\frac{T^*}{T_\infty}\right)^{0.5(\omega-1)},\end{aligned} \tag{10.91}$$

with that for laminar incompressible flow being:

$$\delta_{1,lam,ic} = 1.7208 \frac{x}{(Re_{\infty,x})^{0.5}}. \tag{10.92}$$

For turbulent flow the relation reads:

$$\begin{aligned}\delta_{1,turb,c} =& \delta_{1,turb,ic} \left(0.129 + 0.871 \frac{T_w}{T_\infty} + \right. \\ & \left. + 0.648 \frac{\gamma_\infty - 1}{2} M_\infty^2 \right) \left(\frac{T^*}{T_\infty}\right)^{0.2(\omega-4)},\end{aligned} \tag{10.93}$$

with that for turbulent incompressible flow:

$$\delta_{1,turb,ic} = 0.0504 \frac{x}{(Re_{\infty,x})^{0.2}}. \tag{10.94}$$

The relation for the momentum thickness δ_2 is:

$$\delta_2 = C_2 \frac{x^{1-n}}{1-n} \left(\frac{\rho^* \mu^*}{\rho_e \mu_e}\right)^n \left(\frac{\rho^*}{\rho_e}\right)^{1-2n} \left(\frac{1}{Re_e^u}\right)^n, \tag{10.95}$$

with $C_2 = 0.332$ for laminar flow, and $C_2 = 0.0296$ for turbulent flow.
For the thickness of the viscous sub-layer we have:

$$\delta_{vs} = 29.06 \frac{x^{0.1}}{(Re_e^u)^{0.9}} \left(\frac{\rho_e}{\rho^*} \frac{\mu^*}{\mu_e}\right)^{0.9}, \tag{10.96}$$

whereas the turbulent scaling thickness, reads:

$$\delta_{vs} = 33.78 \frac{x^{0.2}}{(Re_e^u)^{0.8}} \left(\frac{\rho_e}{\rho^*} \frac{\mu^*}{\mu_e}\right)^{0.8}. \tag{10.97}$$

10.4.3 Wall Shear Stress and Thermal State at Flat Surfaces

For the wall shear stress over a flat plate we get in generalized form, with $C = 0.332$ for laminar flow, and $C = 0.0296$ for turbulent flow:

$$\tau_w = C \mu_e v_e x^{-n} \left(\frac{T_e}{T^*}\right)^{1-n} \left(\frac{\mu^*}{\mu_e}\right)^n (Re_e^u)^{1-n}. \tag{10.98}$$

This can also be written as:

$$\frac{\tau_w}{0.5 \rho_e v_e^2} = c_f = 2 C x^{-n} \left(\frac{T_e}{T^*}\right)^{1-n} \left(\frac{\mu^*}{\mu_e}\right)^n (Re_e^u)^{-n}. \tag{10.99}$$

The heat flux in the gas at the wall reads, again with $C = 0.332$ for laminar flow, and $C = 0.0296$ for turbulent flow:

$$q_{gw} = C x^{-n} k_e Pr^{1/3} (T_r - T_w) \left(\frac{T_e}{T^*}\right)^{1-n} \left(\frac{\mu^*}{\mu_e}\right)^n (Re_e^u)^{1-n}. \tag{10.100}$$

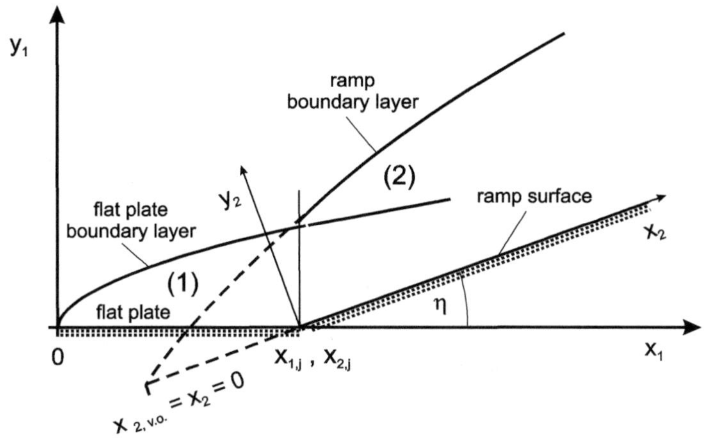

Fig. 10.4. Illustration of the virtual origin of a boundary layer at a junction demonstrated by means of a flat plate/ramp configuration.

The (implicit) relation for the radiation-adiabatic wall temperature is:

$$T_{ra} = \left[C x^{-n} \frac{k_e}{\sigma \varepsilon} Pr^{1/3} (T_r - T_{ra}) \left(\frac{T_e}{T^*} \right)^{1-n} \left(\frac{\mu^*}{\mu_e} \right)^n (Re_e^u)^{1-n} \right]^{0.25} . \tag{10.101}$$

10.4.4 Virtual Origin of Boundary Layers at Junctions

If the simple relations provided above are to be applied on consecutive surfaces with a change of characteristic properties, a virtual origin at the junction(s) must be constructed[9].

We show this with the help of the flat-plate/ramp configuration given in Fig. 10.4. With a flat-plate flow Mach number $M_1 > 1$, we get at the junction of the two planar surfaces a jump of the flow parameters, and especially of the unit Reynolds number, Fig. 6.18, depending on the ramp angle η and on the flat-plate flow Mach number M_1. Of course, in the frame of the simple approach we cannot describe local strong interaction phenomena, only the asymptotic properties on the ramp, Section 6.3.

If Re^u changes at the junction, the boundary layer on the ramp surface (2), Fig. 10.4, has other properties than that on the flat plate (1). It cannot be assumed, that simply a continuation takes place. In [22] therefore a matching procedure is proposed, which essentially leads to a ramp boundary layer with an origin – the virtual origin – different from that of the flat plate.

Proposed in [22] is the matching of the momentum deficit of the two boundary layers on both sides of the junction $x_{1,j} = x_{2,j}$:

[9] This holds, for instance, also for the correlation of experimental data.

$$(\rho u^2 \delta_2)|_2 = (\rho u^2 \delta_2)|_1, \tag{10.102}$$

with δ_2 being the momentum-loss thickness, see above.

The procedure is the following:

- determine $\delta_2|_1$, eq. (10.95), at the junction $x_1 = x_{1,j}$ with the flow parameters of the flat plate (1),
- determine $\delta_2|_2$ from eq. (10.102) with the ramp flow parameters (2),
- find the virtual junction coordinate $x_{2,j}$ of the ramp boundary layer with the inverted eq. (10.95). The effective ramp coordinate is then in terms of x_1 and the ramp angle η: $x_2 = x_{2,j} + (x_1 - x_{1,j})/\cos\eta$. The virtual origin $x_{2,v\,o}$ of the ramp boundary layer lies at $x_2 =$ - $x_{2,j}$.

This approach, in analogous ways, can be applied to cone/cylinder, blunt-nose/cylinder et cetera configurations, see, e.g., [22], [23], [20], including laminar-turbulent transition, for instance on a flat surface, where the prescribed transition point would be the junction point. When employing discrete numerical methods for the solution of the boundary layer or the Navier-Stokes equations for the flow past such configurations, of course such an approach is not necessary.

References

1. Hirschel, E.H.: Basics of Aerothermodynamics. Progress in Astronautics and Aeronautics, AIAA, Reston, Va, vol. 204. Springer, Heidelberg (2004)
2. Adams, J.C., Martindale, W.R., Mayne, A.W., Marchand, E.O.: Real Gas Scale Effects on Hypersonic Laminar Boundary-layer Parameters Including Effects of Entropy-Layer Swallowing. AIAA-Paper 76-358 (1976)
3. Schwane, R., Muylaert, J.: Design of the Validation Experiment Hyperboloid-Flare. ESA Doc. YPA/1256/RS, ESTEC (1992)
4. Wüthrich, S., Sawley, M.L., Perruchoud, G.: The Coupled Euler/ Boundary-Layer Method as a Design Tool for Hypersonic Re-Entry Vehicles. Zeitschrift für Flugwissenschaften und Weltraumforschung (ZFW) 20(3), 137–144 (1996)
5. Brauckmann, G.J., Paulson Jr., J.W., Weilmuenster, K.J.: Experimental and Computational Analysis of Shuttle Orbiter Hypersonic Trim Anomaly. Journal of Spacecraft and Rockets 32(5), 758–764 (1995)
6. Ames Research Staff. Equations, Tables, and Charts for Compressible Flow. NACA R-1135 (1953)
7. Vincenti, W.G., Kruger, C.H.: Introduction to Physical Gas Dynamics. John Wiley, New York (1965); Reprint edition, Krieger Publishing Comp., Melbourne, Fl. (1975)
8. Paulson Jr., J.W., Brauckmann, G.J.: Recent Ground-Facility Simulations of Space Shuttle Orbiter Aerodynamics. In: Throckmorton, D. A. (ed.), Orbiter Experiments (OEX) Aerothermodynamics Symposium. NASA CP-3248, Part 1, pp. 411–445 (1995)
9. Van Driest, E.R.: On Skin Friction and Heat Transfer Near the Stagnation Point. NACA Report, AL-2267 (1956)

10. Lees, L.: Laminar Heat Transfer over Blunt-Nosed Bodies at Hypersonic Flight Speeds. Jet Propulsion 26(4), 259–269 (1956)
11. Bertin, J.J.: The Effect of Protuberances, Cavities, and Angle of Attack on the Wind Tunnel Pressure and Heat Transfer Distribution for the Apollo Command Module. NASA TM X-1243 (1966)
12. Fay, J.A., Riddell, F.R.: Theory of Stagnation Point Heat Transfer in Dissociated Air. Journal of Aeronautical Sciences 25(2), 73–85 (1958)
13. Cohen, N.B.: Boundary Layer Similar Solutions and Correlation Equations for Laminar Heat Transfer Distribution in Equilibrium Air at Velocities Up to 41,100 Feet Per Second. NASA TR R-118 (1961)
14. Zoby, E.V., Moss, J.N., Sutton, K.: Approximate Convective Heating Equations for Hypersonic Flows. Journal of Spacecraft 18(1) (1981)
15. Detra, R.W., Kemp, N.H., Riddell, F.R.: Addendum to Heat Transfer to Satellite Vehicles Reentering the Atmosphere. Jet Propulsion 27(12), 1256–1257 (1957)
16. Riley, C.J., DeJearnette, F.R.: Engineering Aerodynamic Heating Method for Hypersonic Flow. Journal of Spacecraft and Rockets 29(3) (1992)
17. Rubesin, M.W., Johnson, H.A.: A Critical Review of Skin Friction and Heat Transfer Solutions of the Laminar Boundary Layer of a Flat Plate. Trans. ASME 71, 385–388 (1949)
18. Eckert, E.R.G.: Engineering Relations of Friction and Heat Transfer to Surfaces in High-Velocity Flow. J. Aeronautical Sciences 22(8), 585–587 (1955)
19. Simeonides, G., Walpot, L.M.G., Netterfield, M., Tumino, G.: Evaluation of Engineering Heat Transfer Prediction Methods in High Enthalpy Flow Conditions. AIAA-Paper 96-1860 (1996)
20. Simeonides, G.: Generalized Reference-Enthalpy Formulation and Simulation of Viscous Effects in Hypersonic Flow. Shock Waves 8(3), 161–172 (1998)
21. Simeonides, G.: Hypersonic Shock Wave Boundary Layer Interactions over Compression Corners. Doctoral Thesis, University of Bristol, U.K (1992)
22. Hayes, J.R., Neumann, R.D.: Introduction to the Aerodynamic Heating Analysis of Supersonic Missiles. In: Mendenhall, M. R. (ed.), Tactical Missile Aerodynamics: Prediction Methodology. Progress in Astronautics and Aeronautics, AIAA, Reston, Va, pp. 63–110 (1992)
23. Streit, T., Martin, S., Eggers, T.: Approximate Heat Transfer Methods for Hypersonic Flow in Comparison with Results Provided by Numerical Navier-Stokes Solutions. DLR FB 94-36 (1994)

11 The Governing Equations of Aerothermodynamics

In this chapter we give a complete set of the governing equations of aerothermodynamics. For a discussion of the general background see, e.g. [1], [2]. In Section 11.1 we describe the general governing equations for viscous, turbulent and chemical non-equilibrium flow in a way which is convenient and flexible for numerical approximations. In this context we define a separate conservative flux-vector formulation for each of the three sets, namely the Navier-Stokes equations, the equations for a basic $k - \omega$ turbulence model and the species continuity equations. The laminar, perfect gas, and the equilibrium gas case are each included, and can simply be extracted.

Experience shows, Sub-Section 5.5.1, that the prediction of heat transfer at higher altitudes ($H \gtrapprox 80\ km$) may require the consideration of the thermal non-equilibrium state of the gas. Further for entry speeds $v_e \gtrapprox 10.0\, km/s$ (entry from planetary mission) the gas radiation part reaches values of 50 per cent and more of the total wall heat transfer. Therefore we give in Section 11.2 the equations for the energy conservation of vibrationally excited molecules and of electronic excitations including gas radiation. Finally, also these equations are incorporated in the vector form of the Navier-Stokes equations.

Throughout we assume continuum flow with zero slip and no temperature jump at the body surface. Shock waves are considered to be captured, such that also the classical shock-slip effects are not described. We also don't provide the necessary and possible conditions at all boundaries, in particular at surfaces, and refer the reader instead to, e.g., [3].

11.1 Governing Equations for Chemical Non-Equilibrium Flow

a) The Navier-Stokes equations for turbulent flow in Favre averaged formulation read, [1], [4][1]:

---

[1] The comprehensive compilation of the governing equations for chemical non-equilibrium flow, presented in [4], is taken over here.

$$\frac{\partial \underline{Q}}{\partial t} + \frac{\partial \underline{E}}{\partial x} + \frac{\partial \underline{F}}{\partial y} + \frac{\partial \underline{G}}{\partial z} = \frac{\partial \underline{E}_{visc}}{\partial x} + \frac{\partial \underline{F}_{visc}}{\partial y} + \frac{\partial \underline{G}_{visc}}{\partial z} + \underline{S}_{NS}, \tag{11.1}$$

with

$$\underline{Q} = \begin{pmatrix} \rho \\ \rho u \\ \rho v \\ \rho w \\ \rho e_t \end{pmatrix}, \quad \underline{S}_{NS} = \begin{pmatrix} 0 \\ 0 \\ 0 \\ 0 \\ 0 \end{pmatrix}, \tag{11.2}$$

$$\underline{E} = \begin{pmatrix} \rho u \\ \rho u^2 + p \\ \rho v u \\ \rho w u \\ (\rho e_t + p)\, u \end{pmatrix}, \quad \underline{F} = \begin{pmatrix} \rho v \\ \rho u v \\ \rho v^2 + p \\ \rho w v \\ (\rho e_t + p)\, v \end{pmatrix}, \quad \underline{G} = \begin{pmatrix} \rho w \\ \rho u w \\ \rho v w \\ \rho w^2 + p \\ (\rho e_t + p)\, w \end{pmatrix}, \tag{11.3}$$

$$\underline{E}_{visc} = \begin{pmatrix} 0 \\ \tau_{xx} \\ \tau_{xy} \\ \tau_{xz} \\ u\tau_{xx} + v\tau_{xy} + w\tau_{xz} - q_x + \mu' \frac{\partial k}{\partial x} \end{pmatrix},$$

$$\underline{F}_{visc} = \begin{pmatrix} 0 \\ \tau_{yx} \\ \tau_{yy} \\ \tau_{yz} \\ u\tau_{yx} + v\tau_{yy} + w\tau_{yz} - q_y + \mu' \frac{\partial k}{\partial y} \end{pmatrix}, \tag{11.4}$$

$$\underline{G}_{visc} = \begin{pmatrix} 0 \\ \tau_{zx} \\ \tau_{zy} \\ \tau_{zz} \\ u\tau_{zx} + v\tau_{zy} + w\tau_{zz} - q_z + \mu' \frac{\partial k}{\partial z} \end{pmatrix}.$$

The meaning of the single terms is[2]:

$\tau_{ij} = \mu \left(\frac{\partial v_i}{\partial x_j} + \frac{\partial v_j}{\partial x_i} - \frac{2}{3}\delta_{ij} \frac{\partial v_k}{\partial x_k} \right) - \frac{2}{3}\rho k \delta_{ij}$: the viscous stress tensor[3] with

$v_{i,j,k} \hat{=} u, v, w, \; x_{i,j,k} \hat{=} x, y, z$: and

$k = \frac{1}{2\rho} \left(\overline{\rho u'^2} + \overline{\rho v'^2} + \overline{\rho w'^2} \right)$: the turbulent kinetic energy,

$e_t = e + 0.5 \left(u^2 + v^2 + w^2 \right) + k$: the total energy,

[2] The subscript "lam" denotes the molecular diffusive part of a transport entity ϕ, "turb" the apparent turbulent one, for instance $\phi_{eff} = \phi_{lam} + \phi_{turb}$.

[3] For convenience we do not keep the classical notation of the continuum mechanics, which puts a negative sign at the elements of the stress tensor.

$\mu = \mu_{lam} + \mu_{turb}$: the effective viscosity,

$\mu' = \mu_{lam} + \sigma * \mu_{turb}$: the specific viscosity,

$\kappa = \kappa_{lam} \left(1 + \frac{Pr_{lam}\, \mu_{turb}}{Pr_{turb}\, \mu_{lam}}\right)$: the thermal conductivity[4],

$\underline{q} = -\kappa \operatorname{grad} T - \sum_i \rho D_i h_i \operatorname{grad} c_i$: the vector of heat flow[5],

$D_i = D_{i,lam} + \frac{\mu_{turb}}{Sc_{turb}\, \rho}$: the total diffusion coefficient,

$\underline{j_i} = \rho_i \underline{v_i^d}$: the diffusion mass flux vector,

$\underline{v_i^d} = -D_i \frac{\operatorname{grad} c_i}{c_i}$: the diffusion velocity,

ρ : the density,

u, v, w : the Cartesian velocity components,

p : the pressure,

T : the translational-rotational temperature,

e : the internal energy,

$c_i = \frac{\rho_i}{\rho}$: the mass fraction,

$Pr_{lam} = \frac{\mu_{lam} c_p}{\kappa_{lam}}, \quad Sc_{lam} = \frac{\mu_{lam}}{\rho D_{i,lam}}$: the molecular Prandtl and Schmidt numbers,

Pr_{turb}, Sc_{turb} : the turbulent Prandtl and Schmidt numbers, e.g. [3].

b) Two equation turbulence model on Wilcox' standard $k-\omega$ model basis[6], [4], [5]:

$$\frac{\partial \underline{Q}_T}{\partial t} + \frac{\partial \underline{E}_T}{\partial x} + \frac{\partial \underline{F}_T}{\partial y} + \frac{\partial \underline{G}_T}{\partial z} = \frac{\partial \underline{E}_{visc,T}}{\partial x} + \frac{\partial \underline{F}_{visc,T}}{\partial y} + \frac{\partial \underline{G}_{visc,T}}{\partial z} + \underline{S}_T, \tag{11.5}$$

with

$$\underline{Q}_T = \begin{pmatrix} \rho k \\ \rho\omega \end{pmatrix}, \quad \underline{S}_T = \begin{pmatrix} P_k - \beta^* \rho k \omega \\ \alpha \frac{\rho}{\mu_{turb}} P_k - \beta \rho \omega^2 \end{pmatrix}, \tag{11.6}$$

---

[4] We use here κ for the thermal conductivity instead of k as otherwise in the book.

[5] The components of the diffusion mass flux vector depending on the pressure and the temperature gradients, see, e.g., [3], are neglected.

[6] For local Mach numbers $M > 3-4$ compressibility corrections of the type tested in e.g. [6] have to be added.

$$\underline{E}_T = \begin{pmatrix} \rho u k \\ \rho u \omega \end{pmatrix}, \quad \underline{F}_T = \begin{pmatrix} \rho v k \\ \rho v \omega \end{pmatrix}, \quad \underline{G}_T = \begin{pmatrix} \rho w k \\ \rho w \omega \end{pmatrix}, \tag{11.7}$$

$$\underline{E}_{visc,T} = (\mu_{lam} + \sigma^* \mu_{turb}) \begin{pmatrix} \dfrac{\partial k}{\partial x} \\ \dfrac{\partial \omega}{\partial x} \end{pmatrix},$$

$$\underline{F}_{visc,T} = (\mu_{lam} + \sigma^* \mu_{turb}) \begin{pmatrix} \dfrac{\partial k}{\partial y} \\ \dfrac{\partial \omega}{\partial y} \end{pmatrix}, \tag{11.8}$$

$$\underline{G}_{visc,T} = (\mu_{lam} + \sigma^* \mu_{turb}) \begin{pmatrix} \dfrac{\partial k}{\partial z} \\ \dfrac{\partial \omega}{\partial z} \end{pmatrix}.$$

The meaning of the single terms is :

$\mu_{turb} = \rho \dfrac{k}{\omega}$: the apparent turbulent viscosity,

$P_k = \tau_{ij} \dfrac{\partial v_i}{\partial x_j}$: the production term of turbulent kinetic energy,

ω : the turbulent dissipation,

$\alpha, \beta, \beta^*, \sigma^*$: model constants.

In compressible flow it can happen that the normal stresses becomes lower than zero which must be prevented. To overcome this problem either a compressible dissipation or a realizability correction is employed, see, e.g., [6], [7].

c) Equations of species continuity, [4], [8]:

$$\frac{\partial \underline{Q}_S}{\partial t} + \frac{\partial \underline{E}_S}{\partial x} + \frac{\partial \underline{F}_S}{\partial y} + \frac{\partial \underline{G}_S}{\partial z} = \frac{\partial \underline{E}_{visc,S}}{\partial x} + \frac{\partial \underline{F}_{visc,S}}{\partial y} + \frac{\partial \underline{G}_{visc,S}}{\partial z} + \underline{S}_S, \tag{11.9}$$

with

$$\underline{Q}_S = \begin{pmatrix} \rho_1 \\ \rho_2 \\ \vdots \\ \rho_N \end{pmatrix}, \quad \underline{S}_S = \begin{pmatrix} \dot{\omega}_1 \\ \dot{\omega}_2 \\ \vdots \\ \omega_N \end{pmatrix}, \tag{11.10}$$

$$\underline{E}_S = \begin{pmatrix} \rho_1 u \\ \rho_2 u \\ \vdots \\ \rho_N u \end{pmatrix}, \quad \underline{F}_S = \begin{pmatrix} \rho_1 v \\ \rho_2 v \\ \vdots \\ \rho_N v \end{pmatrix}, \quad \underline{G}_S = \begin{pmatrix} \rho_1 w \\ \rho_2 w \\ \vdots \\ \rho_N w \end{pmatrix}, \tag{11.11}$$

$$\underline{E}_{visc,S} = \begin{pmatrix} \rho D_1 \dfrac{\partial c_1}{\partial x} \\ \rho D_2 \dfrac{\partial c_2}{\partial x} \\ \vdots \\ \rho D_N \dfrac{\partial c_N}{\partial x} \end{pmatrix}, \quad \underline{F}_{visc,S} = \begin{pmatrix} \rho D_1 \dfrac{\partial c_1}{\partial y} \\ \rho D_2 \dfrac{\partial c_2}{\partial y} \\ \vdots \\ \rho D_N \dfrac{\partial c_N}{\partial y} \end{pmatrix},$$

$$\underline{G}_{visc,S} = \begin{pmatrix} \rho D_1 \dfrac{\partial c_1}{\partial z} \\ \rho D_2 \dfrac{\partial c_2}{\partial z} \\ \vdots \\ \rho D_N \dfrac{\partial c_N}{\partial z} \end{pmatrix}. \tag{11.12}$$

The meaning of the single terms is :

$\dot{\omega}_i = M_i \dfrac{d\,[X_i]}{dt}$	: the source term of chemical reactions,
$[X_i] = \dfrac{\rho_i}{M_i}$	: the species concentration,
$X_i = \dfrac{M}{M_i} c_i$	: the mol fraction with $\sum\limits_i X_i = 1$,
$\dfrac{d\,[X_i]}{dt}$	: the reaction rate of species i,
M_i	: the mol mass of species i,
$D_{i,lam} = \dfrac{1 - X_i}{\sum\limits_j \frac{X_j}{D_{ij}}}$	: the molecular diffusion coefficient.

μ_{lam}, κ_{lam} and the binary diffusion coefficient D_{ij}[7] are often approximated by polynomial forms. For example, one defines:

$$\mu_{lam} = 0.5 \left(\sum_i X_i Z_i^{\mu} + 1 / \sum_i \frac{X_i}{Z_i^{\mu}} \right)$$

with the polynomial form $Z_i^{\mu} = exp\left(d_{1i} + d_{2i} T_L + d_{3i} T_L^2 + d_{4i} T_L^3\right)$ and $T_L = lnT$, where the coefficients $d_{ki},\ k = 1,..,4$ are experimentally determined values depending on the composition of the gas considered, see [4].

Note:
$i = 1, \ldots, N_{mol}, N_{mol} + 1, \ldots, N$ the total number of species.

[7] For a binary gas D_{AB} is written instead of D_{ij}, see, e.g., [3].

For thermodynamic equilibrium the set of equations is considerably simpler. In **a)** the term $\sum_i \rho D_i h_i \text{grad} c_i$ can be dropped and the species continuity equations **c)** can be cancelled.

Since in equilibrium every thermodynamic variable is completely described by two other thermodynamic variables, it is convenient to use the density ρ and the internal energy e as independent variables with $p = p\,(e, \rho)$, $T = T\,(e, \rho)$, $\mu_{lam} = \mu_{lam}\,(e, \rho)$ and $\kappa_{lam} = \kappa_{lam}\,(e, \rho)$. These variables can be determined either by the computer program of Gordon & McBride [9] or by a table look-up approach, [10], [11].

The speed of sound, classically defined by $a^2 = \left(\frac{\partial p}{\partial \rho}\right)_s$, often used in the solution process for the convective part of the equations, has now to be defined in terms of the independent thermodynamic variables ρ and e.

From

$$Tds = de + pd\frac{1}{\rho}\ ,$$

$$dp = \left(\frac{\partial p}{\partial \rho}\right)_s d\rho + \left(\frac{\partial p}{\partial s}\right)_\rho ds \quad \text{for} \quad p = p\,(\rho, s)\ ,$$

$$dp = \left(\frac{\partial p}{\partial \rho}\right)_e d\rho + \left(\frac{\partial p}{\partial e}\right)_\rho de \quad \text{for} \quad p = p\,(\rho, e),$$

we obtain with $ds = 0$:

$$\frac{de}{d\rho} = \frac{p}{\rho^2},\quad \frac{dp}{d\rho} = \left(\frac{\partial p}{\partial \rho}\right)_s,$$

and finally the speed of sound for thermo-chemical equilibrium flow:

$$a^2(\rho, e) = \frac{p}{\rho^2}\left(\frac{\partial p}{\partial e}\right)_\rho + \left(\frac{\partial p}{\partial \rho}\right)_e, \tag{11.13}$$

where again the derivatives of p with respect to e and ρ can be extracted from [9] or [10], [11].

11.2 Equations for Excitation of Molecular Vibrations and Electron Modes

11.2.1 Excitation of Molecular Vibrations

The vibrational energy equation describing the thermal non-equilibrium of each molecular species i' has the form, see, e.g., [1], [12]:

$$\frac{\partial}{\partial t}(\rho_{i'} e_{v,i'}) + \frac{\partial}{\partial x_j}(\rho_{i'} e_{v,i'} v_j) = -\frac{\partial}{\partial x_j}(\rho_{i'} e_{v,i'} v^d_{i',j}) - \frac{\partial q_{v,i',j}}{\partial x_j} + \tag{11.14}$$
$$+ Q_{T-v_{i'}} + Q_{e-v_{i'}} + Q_{D_{i'}} + Q_{v-v_{i'}}\,.$$

The meaning of the single terms is:

$\underline{j}_{i'} = \rho_{i'} \underline{v}_{i'}^{d} = -\rho D_{i'} \, \text{grad} \, c_{i'}$: the diffusion mass flux vector,
$c_{i'} = \rho_{i'}/\rho$: the mass fraction,
$\underline{q}_{v,i'} = -\kappa_{v,i'} \text{grad} \, T_{v,i'}$: the conduction of vibrational energy due to vibrational temperature gradients,
$e_{v,i'}$: the vibrational energy,
$T_{v,i'}$: the vibrational temperature of species i',
$\kappa_{v,i'}$: the frozen thermal conductivity of vibrational energy of species i',

$Q_{T-v_{i'}} = \rho_{i'} \dfrac{\bar{e}_{v,i'}(T) - e_{v,i'}}{< \tau_{i'} >}$: the energy exchange between translational and vibrational modes, T the translational-rotational temperature, $\bar{e}_{v,i'}(T)$ the vibrational energy at temperature T, $< \tau_{i'} >$ the characteristic relaxation time for translational-vibrational energy exchange,

$Q_{e-v_{i'}} = \rho_{i'} \dfrac{\bar{\bar{e}}_{v,i'}(T_e) - e_{v,i'}}{< \tau_{e,i'} >}$: the energy exchange between electronic and vibrational modes, T_e the electronic temperature, $\bar{\bar{e}}_{v,i'}(T_e)$ the vibrational energy at temperature T_e, $< \tau_{e,i'} >$ the characteristic relaxation time for electron-vibrational energy exchange,

$Q_{D_{i'}} = \dot{\omega}_{i'} \hat{e}_{v,i'}$: the energy removed at dissociation or gained at recombination, $\dot{\omega}_{i'}$ the mass rate of formation of species i', $\hat{e}_{v,i'}$ the vibrational energy of species i' generated or lost with mass rate $\dot{\omega}_{i'}$ ($\hat{e}_{v,i'}$ can be higher than the vibrational energy $e_{v,i'}$),

$Q_{v-v_{i'}}$: the energy exchange between different vibrational modes.

Note:
$i' = 1, \ldots, N_{mol}$ the number of molecular species.

For a vibrational model with one averaged temperature T_v the energy equation reads:

$$\frac{\partial}{\partial t}(\rho e_v) + \frac{\partial}{\partial x_j}(\rho e_v v_j) = \frac{\partial}{\partial x_j}\left(\rho \sum_{i'=1}^{N_{mol}} e_{v,i'} D_{i'} \frac{\partial c_{i'}}{\partial x_j}\right) - \frac{\partial q_{v,j}}{\partial x_j} + \qquad (11.15)$$
$$+ Q_{T-v} + Q_{e-v} + Q_D + Q_{v-v},$$

with
$$\rho e_v = \sum_{i'=1}^{N_{mol}} \rho_{i'} e_{v,i'}, \quad \underline{q}_v = -\kappa_v \text{grad} \, T_v \,,$$

$$Q_{T-v} = \sum_{i'=1}^{N_{mol}} \rho_{i'} \frac{\bar{e}_{v,i'}(T) - e_{v,i'}}{< \tau_{i'} >}, \quad Q_{e-v} = \sum_{i'=1}^{N_{mol}} \rho_{i'} \frac{\bar{\bar{e}}_{v,i'}(T_e) - e_{v,i'}}{< \tau_{e,i'} >} \,,$$

$$Q_D = \sum_{i'=1}^{N_{mol}} Q_{D_{i'}}, \quad Q_{v-v} = \sum_{i'=1}^{N_{mol}} Q_{v-v_{i'}}.$$

11.2.2 Electron Modes

The energy equation for electron modes is given by, e.g., [1], [13], [14]:

$$\frac{\partial}{\partial t}(\rho e_e) + \frac{\partial}{\partial x_j}((\rho e_e + p_e)v_j) = \frac{\partial}{\partial x_j}\left(\rho\sum_{i=1}^{N} e_{e,i} D_i \frac{\partial c_i}{\partial x_j}\right) - \frac{\partial q_{e,j}}{\partial x_j} + \quad (11.16)$$
$$+ u_j \frac{\partial p_e}{\partial x_j} + Q_{T-e} + Q_{e-v} + Q_I + q_{rad,g}.$$

The meaning of the single terms is:

$\rho e_e = \sum\limits_{i=1}^{N} \rho_i e_{e,i}$: the electronic energy,

$q_{e,j} = -\kappa_e \frac{\partial T_e}{\partial x_j}$: the conduction of electronic energy due to electron temperature gradient,

$q_{rad,g}$: the energy amount due to radiation caused by electron transition,

κ_e : the frozen thermal conductivity for electronic energy,

p_e : the pressure of electrons,

ρ_e : the electron density,

R_0 : the universal gas constant,

ν_e : the collision frequency for electrons and heavy particles,

M_i : the mole mass of species i,

$\dot{n}_{e,i''}$: the number density of species i'' produced by electron impact ionization,

$I_{i''}$: the ionization potential of species i'',

$Q_{T-e} = 2\rho\frac{3}{2}R_0\,(T - T_e)\sum\limits_{i=1}^{N-1}\frac{\nu_{e,i}}{M_i}$: the energy exchange due to elastic collisions between electrons and heavy particles,

$Q_{e-v} = \sum\limits_{i=1}^{N} Q_{e-v_i}$: the energy exchange between electronic and vibrational modes,

$Q_I = \sum\limits_{i''=1}^{N_{ions}} \dot{n}_{e,i''} I_{i''}$: the energy amount due to ionization by electron impact.

Note: the index i'' counts up the ions.

11.3 Vector Form of Navier-Stokes Equations Including Thermal Non-Equilibrium

For thermal non-equilibrium including electron modes the vectors of eqs. 11.2, 11.3, 11.4, 11.14 and 11.16 have the form:

$$\underline{Q} = \begin{pmatrix} \rho \\ \rho u \\ \rho v \\ \rho w \\ \rho e_t \\ \rho_1 e_{v,1} \\ \rho_2 e_{v,2} \\ \vdots \\ \rho_{N_{mol}} e_{v,N_{mol}} \\ \rho e_e \end{pmatrix}, \quad \underline{S}_{NS} = \begin{pmatrix} 0 \\ 0 \\ 0 \\ 0 \\ 0 \\ Q_{v,1} \\ Q_{v,2} \\ \vdots \\ Q_{v,N_{mol}} \\ Q_e \end{pmatrix}, \tag{11.17}$$

$$\underline{E} = \begin{pmatrix} \rho u \\ \rho u^2 + p \\ \rho v u \\ \rho w u \\ (\rho e_t + p)\, u \\ \rho_1 e_{v,1} u \\ \rho_2 e_{v,2} u \\ \vdots \\ \rho_{N_{mol}} e_{v,N_{mol}} u \\ (\rho e_e + p_e) u \end{pmatrix}, \quad \underline{F} = \begin{pmatrix} \rho v \\ \rho u v \\ \rho v^2 + p \\ \rho w v \\ (\rho e_t + p)\, v \\ \rho_1 e_{v,1} v \\ \rho_2 e_{v,2} v \\ \vdots \\ \rho_{N_{mol}} e_{v,N_{mol}} v \\ (\rho e_e + p_e) v \end{pmatrix},$$

$$\underline{G} = \begin{pmatrix} \rho w \\ \rho u w \\ \rho v w \\ \rho w^2 + p \\ (\rho e_t + p)\, w \\ \rho_1 e_{v,1} w \\ \rho_2 e_{v,2} w \\ \vdots \\ \rho_{N_{mol}} e_{v,N_{mol}} w \\ (\rho e_e + p_e) w \end{pmatrix}, \tag{11.18}$$

$$
E_v = \begin{pmatrix} 0 \\ \tau_{xx} \\ \tau_{xy} \\ \tau_{xz} \\ u\tau_{xx} + v\tau_{xy} + w\tau_{xz} - q_x + \mu' \dfrac{\partial k}{\partial x} \\ \kappa_{v,1}\dfrac{\partial T_{v,1}}{\partial x} + \rho D_1 \dfrac{\partial c_1}{\partial x}\, e_{v,1} \\ \kappa_{v,2}\dfrac{\partial T_{v,2}}{\partial x} + \rho D_2 \dfrac{\partial c_2}{\partial x}\, e_{v,2} \\ \vdots \\ \kappa_{v,N_{mol}}\dfrac{\partial T_{v,N_{mol}}}{\partial x} + \rho D_{N_{mol}} \dfrac{\partial c_{N_{mol}}}{\partial x}\, e_{v,N_{mol}} \\ \kappa_e \dfrac{\partial T_e}{\partial x} + \rho \sum_{i=1}^{N} D_i \dfrac{\partial c_i}{\partial x}\, e_{e,i} \end{pmatrix},
$$

$$
F_v = \begin{pmatrix} 0 \\ \tau_{yx} \\ \tau_{yy} \\ \tau_{yz} \\ u\tau_{yx} + v\tau_{yy} + w\tau_{yz} - q_y + \mu' \dfrac{\partial k}{\partial y} \\ \kappa_{v,1}\dfrac{\partial T_{v,1}}{\partial y} + \rho D_1 \dfrac{\partial c_1}{\partial y}\, e_{v,1} \\ \kappa_{v,2}\dfrac{\partial T_{v,2}}{\partial y} + \rho D_2 \dfrac{\partial c_2}{\partial y}\, e_{v,2} \\ \vdots \\ \kappa_{v,N_{mol}}\dfrac{\partial T_{v,N_{mol}}}{\partial y} + \rho D_{N_{mol}} \dfrac{\partial c_{N_{mol}}}{\partial y}\, e_{v,N_{mol}} \\ \kappa_e \dfrac{\partial T_e}{\partial y} + \rho \sum_{i=1}^{N} D_i \dfrac{\partial c_i}{\partial y}\, e_{e,i} \end{pmatrix}, \qquad (11.19)
$$

$$
G_v = \begin{pmatrix} 0 \\ \tau_{zx} \\ \tau_{zy} \\ \tau_{zz} \\ u\tau_{zx} + v\tau_{zy} + w\tau_{zz} - q_z + \mu' \dfrac{\partial k}{\partial z} \\ \kappa_{v,1}\dfrac{\partial T_{v,1}}{\partial z} + \rho D_1 \dfrac{\partial c_1}{\partial z}\, e_{v,1} \\ \kappa_{v,2}\dfrac{\partial T_{v,2}}{\partial z} + \rho D_2 \dfrac{\partial c_2}{\partial z}\, e_{v,2} \\ \vdots \\ \kappa_{v,N_{mol}}\dfrac{\partial T_{v,N_{mol}}}{\partial z} + \rho D_{N_{mol}} \dfrac{\partial c_{N_{mol}}}{\partial z}\, e_{v,N_{mol}} \\ \kappa_e \dfrac{\partial T_e}{\partial z} + \rho \sum_{i=1}^{N} D_i \dfrac{\partial c_i}{\partial z}\, e_{e,i} \end{pmatrix}.
$$

The meaning of the terms is:

$\underline{q} = -\kappa \operatorname{grad} T - \sum_{i'=1}^{N_{mol}} \kappa_{v,i'} \operatorname{grad} T_{v,i'} - \kappa_e \operatorname{grad} T_e - \sum_{i=1}^{N} \rho h_i D_i \operatorname{grad} c_i$: the vector of heat flow,

$e_t = \sum_{i=1}^{N} \frac{\rho_i e_i}{\rho} + 0.5(u^2 + v^2 + w^2) + k$: the total energy,

$Q_{v,i'} = Q_{T-v_{i'}} + Q_{e-v_{i'}} + Q_{D,i'} + Q_{v-v_{i'}},$: the vibrational coupling terms,

$Q_e = u_j \frac{\partial p_e}{\partial x_j} + Q_{T-e} + Q_{e-v} + Q_I - q_{rad,g},$: the electron coupling terms,

e_i : the internal energy of species i in thermal equilibrium.

References

1. Gnoffo, P.A., Gupta, R.N., Shinn, J.L.: Conservation Equations and Physical Models for Hypersonic Air Flows in Thermal and Chemical Nonequilibrium. NASA-TN-2867 (1989)
2. Sarma, G.S.R.: Physico-Chemical Modelling in Hypersonic Flow Simulation. Progress in Aerospace Sciences 36(3-4), 281–349 (2000)
3. Hirschel, E.H.: Basics of Aerothermodynamics. Progress in Astronautics and Aeronautics, AIAA, Reston, Va., vol. 204. Springer, Heidelberg (2004)
4. Gross, A.: Numerische Untersuchung abgelöster Düsenströmungen (Numerical Investigations of Separated Nozzle Flows). Doctoral Thesis, Department of Combustion and Propulsion, RWTH Aachen, Germany (2000)
5. Wilcox, D.C.: Turbulence Modelling for CFD. DCW Industries, La Cañada, CAL., USA (1998)
6. Weber, C., Behr, R., Weiland, C.: Investigation of Hypersonic Turbulent Flow over the X-38 Crew Return Vehicle. AIAA-Paper 2000-2601 (2000)
7. Moore, J.G., Moore, J.: Realizability in Two-Equation Turbulence Models. AIAA-Paper 99-3779 (1999)
8. Park, C.: Nonequilibrium Hypersonic Flow. John Wiley & Sons, New York (1990)
9. Gordon, S., McBride, B.J.: Computer Program for Calculation of Complex Chemical Equilibrium Compositions, Rocket Performances, Incident and Reflected Shocks, and Chapman-Jouguet Detonations. NASA SP-273, interim (1976)
10. Srinivasan, S., Tannehill, J.C., Weilmuenster, K.J.: Simplified Curve Fits for the Thermodynamic Properties of Equilibrium Air. ISU-ERI-Ames-86401, Iowa State University (1986)
11. Mundt, C., Keraus, R., Fischer, J.: New, Accurate. Vectorized Approximations of the State Surfaces for the Thermodynamic and Transport Properties of Equilibrium Air. Zeitschrift für Flugwissenschaften and Weltraumforschung (ZfW) 15(3), 179–184 (1991)
12. Candler, G.V., MacCormack, R.W.: The Computation of Hypersonic Ionized Flows in Chemical and Thermal Nonequilibrium. AIAA-Paper 88-0511 (1988)
13. Otsu, H., Suzuki, K., Fujita, K., Abe, T.: Radiative Heating Analysis around the MUSES-C Reentry Capsule at a Superorbital Speed. AIAA-Paper 98-2447 (1998)
14. Surzhikov, S.T.: Computing System for Solving Radiative Gasdynamic Problems of Entry and Re-Entry Space Vehicles. ESA-SP-533 (2003)

12 The Earth Atmosphere

The aerothermodynamic design problems considered in this book concern hypersonic flight predominantly in the Earth atmosphere of RV-W-type, RV-NW-type and CAV-type vehicles. We provide in this chapter some basic features and properties of the Earth atmosphere, and some velocity-altitude maps of flight parameters and re-entry vehicle trajectories.

The Earth atmosphere consists of several layers, Fig. 12.1. The weather phenomena occur mainly in the troposphere, there the fluctuations mix and disperse introduced contaminants. These fluctuations are only weakly present at higher altitudes.

The stratosphere is characterized by a temperature plateau of around 220.0 K to 230.0 K, in the mesosphere it becomes colder, in the thermosphere the temperature rises fast with altitude, Table 12.1. Ecologically important is the altitude between 18.0 and 25.0 km with the vulnerable ozone layer.

The pressure p decreases fast with increasing altitude, and also the density ρ. Because the temperature T does not change much, the curves of p and ρ look similar. The dynamic viscosity μ is a function of temperature and is determined by the Sutherland formula, see, e.g., [1]. Sutherland's formula becomes inaccurate for the conditions present at high altitudes, [2]. Therefore the data are restricted to altitudes $H \lessapprox 86.0\, km$. An empirical relation comparable to Sutherland's formula is used for the calculation of the thermal conductivity k, [2]. Again the data are restricted to altitudes $H \lessapprox 86.0\, km$.

The composition of the atmosphere can be considered as constant in the homosphere, up to approximately 100.0 km altitude. In the heterosphere, above 100.0 km altitude, it changes with altitude. This is important especially for computational simulations of aerothermodynamics. Note that also around 100.0 km altitude the continuum domain ends, see, e.g., [1], where also atmospheric gas models are discussed.

It should be mentioned, that these numbers are average numbers, which partly depend strongly on geographical latitude, and that they are changing with time (seasons, atmospheric tides, sun-spot activities). A large number of reference and standard atmosphere models is discussed in [3], where also model uncertainties and limitations are noted.

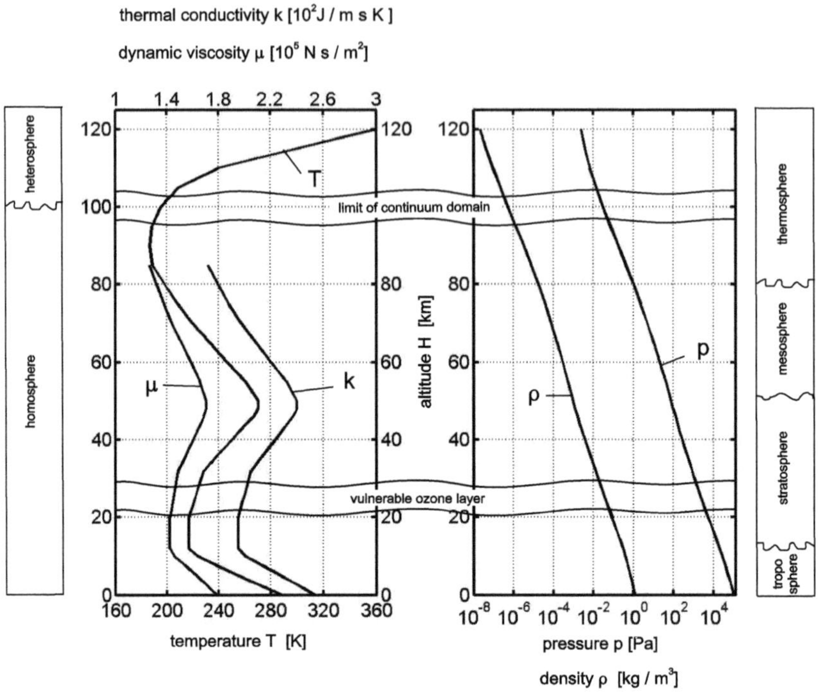

Fig. 12.1. Atmospheric properties as function of the altitude.

In aerothermodynamics we work usually with the U. S. standard atmosphere [2], in order to determine static pressure (p_∞), density (ρ_∞), static temperature (T_∞) et cetera as function of the altitude, Table 12.1. The 15°C standard atmosphere assumes a temperature of 15°C at sea level.

A very simple approximation for the atmospheric density is given by [4], [5]:

$$\frac{\rho(H)}{\rho(0)} = \exp^{-\beta H}, \tag{12.1}$$

with $\beta = 1.40845 \cdot 10^{-4}/m$, which assumes an isothermal atmosphere with $T(H) = T(0)$. With the perfect gas equation we have:

$$\frac{\rho(H)}{\rho(0)} = \frac{p(H)}{p(0)}.$$

To illustrate the influence of the atmospheric properties on the aerodynamic forces during re-entry of a RV-W-type vehicle, we show in Fig. 12.2 as an example for the X-38 vehicle the drag force D as function of the altitude H. At approximately 7.0 km exists a strong peak in D which is accompanied by the total maxima of the Reynolds number Re and the dynamic pressure q_∞, Fig. 12.2. The drag coefficient C_D has a spike-like local maximum. The reason for the peak lies in the momentum flux $\rho_\infty v_\infty^2$, which has a global maximum.

Table 12.1. Properties of the 15°C U. S. standard atmosphere as function of the altitude, [2].

Altitude H [km]	Temperature T [K]	Pressure p [Pa]	Density ρ [kg/m^3]	Dynamic viscosity μ [$N\ s/m^2$]	Thermal conductivity k [$W/(m\ K)$]
0.0	288.150	$1.013 \cdot 10^5$	$1.225 \cdot 10^0$	$1.789 \cdot 10^{-5}$	$2.536 \cdot 10^{-2}$
1.0	281.651	$8.988 \cdot 10^4$	$1.112 \cdot 10^0$	$1.758 \cdot 10^{-5}$	$2.485 \cdot 10^{-2}$
2.0	275.154	$7.950 \cdot 10^4$	$1.007 \cdot 10^0$	$1.726 \cdot 10^{-5}$	$2.433 \cdot 10^{-2}$
3.0	268.659	$7.012 \cdot 10^4$	$9.092 \cdot 10^{-1}$	$1.694 \cdot 10^{-5}$	$2.381 \cdot 10^{-2}$
4.0	262.166	$6.166 \cdot 10^4$	$8.193 \cdot 10^{-1}$	$1.661 \cdot 10^{-5}$	$2.329 \cdot 10^{-2}$
5.0	255.676	$5.405 \cdot 10^4$	$7.364 \cdot 10^{-1}$	$1.628 \cdot 10^{-5}$	$2.276 \cdot 10^{-2}$
6.0	249.187	$4.722 \cdot 10^4$	$6.601 \cdot 10^{-1}$	$1.595 \cdot 10^{-5}$	$2.224 \cdot 10^{-2}$
7.0	242.700	$4.110 \cdot 10^4$	$5.900 \cdot 10^{-1}$	$1.561 \cdot 10^{-5}$	$2.170 \cdot 10^{-2}$
8.0	236.215	$3.565 \cdot 10^4$	$5.258 \cdot 10^{-1}$	$1.527 \cdot 10^{-5}$	$2.117 \cdot 10^{-2}$
9.0	229.733	$3.080 \cdot 10^4$	$4.671 \cdot 10^{-1}$	$1.493 \cdot 10^{-5}$	$2.063 \cdot 10^{-2}$
10.0	223.252	$2.650 \cdot 10^4$	$4.135 \cdot 10^{-1}$	$1.458 \cdot 10^{-5}$	$2.009 \cdot 10^{-2}$
12.0	216.650	$1.940 \cdot 10^4$	$3.119 \cdot 10^{-1}$	$1.421 \cdot 10^{-5}$	$1.953 \cdot 10^{-2}$
14.0	216.650	$1.417 \cdot 10^4$	$2.279 \cdot 10^{-1}$	$1.421 \cdot 10^{-5}$	$1.953 \cdot 10^{-2}$
16.0	216.650	$1.035 \cdot 10^4$	$1.665 \cdot 10^{-1}$	$1.421 \cdot 10^{-5}$	$1.953 \cdot 10^{-2}$
18.0	216.650	$7.565 \cdot 10^3$	$1.216 \cdot 10^{-1}$	$1.421 \cdot 10^{-5}$	$1.953 \cdot 10^{-2}$
20.0	216.650	$5.529 \cdot 10^3$	$8.891 \cdot 10^{-2}$	$1.421 \cdot 10^{-5}$	$1.953 \cdot 10^{-2}$
22.0	218.574	$4.047 \cdot 10^3$	$6.451 \cdot 10^{-2}$	$1.432 \cdot 10^{-5}$	$1.969 \cdot 10^{-2}$
24.0	220.560	$2.972 \cdot 10^3$	$4.694 \cdot 10^{-2}$	$1.443 \cdot 10^{-5}$	$1.986 \cdot 10^{-2}$
26.0	222.544	$2.188 \cdot 10^3$	$3.426 \cdot 10^{-2}$	$1.454 \cdot 10^{-5}$	$2.003 \cdot 10^{-2}$
28.0	224.527	$1.616 \cdot 10^3$	$2.508 \cdot 10^{-2}$	$1.465 \cdot 10^{-5}$	$2.019 \cdot 10^{-2}$
30.0	226.509	$1.197 \cdot 10^3$	$1.841 \cdot 10^{-2}$	$1.475 \cdot 10^{-5}$	$2.036 \cdot 10^{-2}$
32.0	228.490	$8.891 \cdot 10^2$	$1.355 \cdot 10^{-2}$	$1.486 \cdot 10^{-5}$	$2.053 \cdot 10^{-2}$
34.0	233.743	$6.634 \cdot 10^2$	$9.887 \cdot 10^{-3}$	$1.514 \cdot 10^{-5}$	$2.096 \cdot 10^{-2}$
36.0	239.282	$4.985 \cdot 10^2$	$7.258 \cdot 10^{-3}$	$1.543 \cdot 10^{-5}$	$2.142 \cdot 10^{-2}$
38.0	244.818	$3.771 \cdot 10^2$	$5.366 \cdot 10^{-3}$	$1.572 \cdot 10^{-5}$	$2.188 \cdot 10^{-2}$
40.0	250.350	$2.871 \cdot 10^2$	$3.996 \cdot 10^{-3}$	$1.601 \cdot 10^{-5}$	$2.233 \cdot 10^{-2}$
42.0	255.878	$2.200 \cdot 10^2$	$2.995 \cdot 10^{-3}$	$1.629 \cdot 10^{-5}$	$2.278 \cdot 10^{-2}$
44.0	261.403	$1.695 \cdot 10^2$	$2.259 \cdot 10^{-3}$	$1.657 \cdot 10^{-5}$	$2.323 \cdot 10^{-2}$
46.0	266.925	$1.313 \cdot 10^2$	$1.714 \cdot 10^{-3}$	$1.685 \cdot 10^{-5}$	$2.367 \cdot 10^{-2}$
48.0	270.650	$1.023 \cdot 10^2$	$1.317 \cdot 10^{-3}$	$1.704 \cdot 10^{-5}$	$2.397 \cdot 10^{-2}$
50.0	270.650	$7.978 \cdot 10^1$	$1.027 \cdot 10^{-3}$	$1.704 \cdot 10^{-5}$	$2.397 \cdot 10^{-2}$
55.0	260.771	$4.252 \cdot 10^1$	$5.681 \cdot 10^{-4}$	$1.654 \cdot 10^{-5}$	$2.318 \cdot 10^{-2}$
60.0	247.021	$2.196 \cdot 10^1$	$3.097 \cdot 10^{-4}$	$1.584 \cdot 10^{-5}$	$2.206 \cdot 10^{-2}$
65.0	233.292	$1.093 \cdot 10^1$	$1.632 \cdot 10^{-4}$	$1.512 \cdot 10^{-5}$	$2.093 \cdot 10^{-2}$
70.0	219.585	$5.221 \cdot 10^0$	$8.283 \cdot 10^{-5}$	$1.438 \cdot 10^{-5}$	$1.978 \cdot 10^{-2}$

Table 12.1. *(continued)*

Altitude H $[km]$	Temperature T $[K]$	Pressure p $[Pa]$	Density ρ $[kg/m^3]$	Dynamic viscosity μ $[N \cdot s/m^2]$	Thermal conductivity k $[J/(m \cdot s \cdot K)]$
75.0	208.399	$2.388 \cdot 10^{0}$	$3.992 \cdot 10^{-5}$	$1.376 \cdot 10^{-5}$	$1.883 \cdot 10^{-2}$
80.0	198.639	$1.052 \cdot 10^{0}$	$1.846 \cdot 10^{-5}$	$1.321 \cdot 10^{-5}$	$1.800 \cdot 10^{-2}$
85.0	188.893	$4.457 \cdot 10^{-1}$	$8.220 \cdot 10^{-6}$	$1.265 \cdot 10^{-5}$	$1.716 \cdot 10^{-2}$
90.0	186.870	$1.836 \cdot 10^{-1}$	$3.416 \cdot 10^{-6}$		
95.0	188.420	$7.597 \cdot 10^{-2}$	$1.393 \cdot 10^{-6}$		
100.0	195.080	$3.201 \cdot 10^{-2}$	$5.604 \cdot 10^{-7}$		
105.0	208.840	$1.448 \cdot 10^{-2}$	$2.325 \cdot 10^{-7}$		
110.0	240.000	$7.104 \cdot 10^{-3}$	$9.708 \cdot 10^{-8}$		
115.0	300.000	$4.010 \cdot 10^{-3}$	$4.289 \cdot 10^{-8}$		
120.0	360.000	$2.538 \cdot 10^{-3}$	$2.222 \cdot 10^{-8}$		

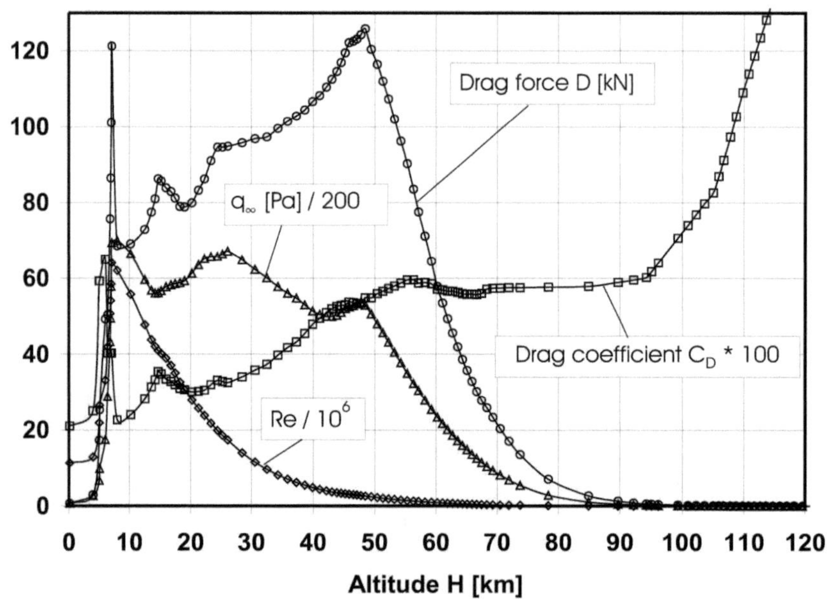

Fig. 12.2. Behavior of aerodynamic drag force D, drag coefficient C_D, dynamic pressure q_∞ and Reynolds number $Re/10^6$ along the X-38 re-entry trajectory as function of the altitude H.

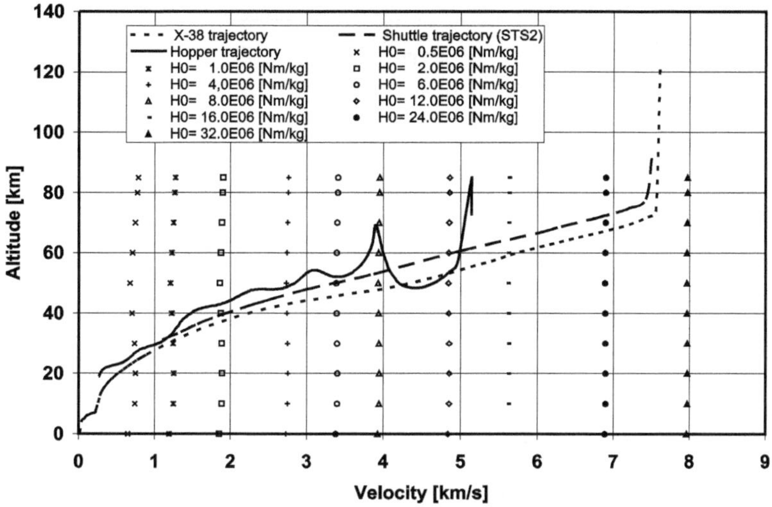

Fig. 12.3. Velocity-altitude map for three RV-W-type vehicles with constant total enthalpy lines (symbols).

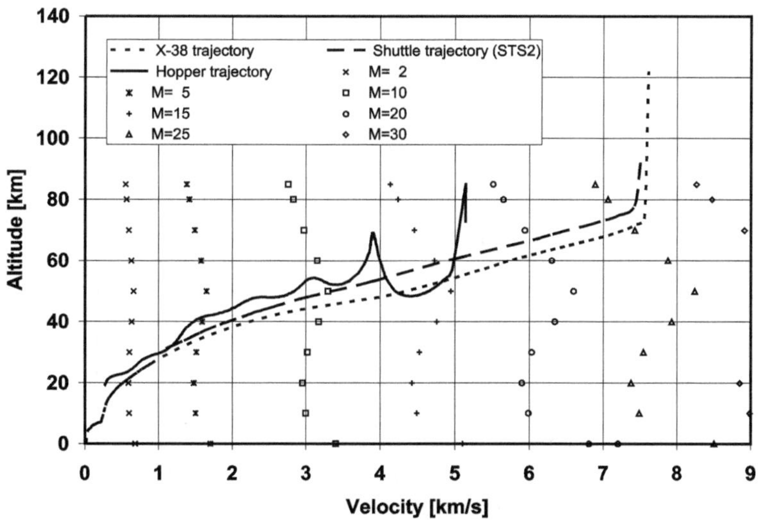

Fig. 12.4. Velocity-altitude map for three RV-W-type vehicles with constant Mach number lines (symbols).

The global maximum of the drag force D is obtained at $H = 48.0\ km$. This is due to the weak local maximum of the dynamic pressure combined with the beginning plateau of the drag coefficient. For the flight mechanical considerations of the vehicle the aerodynamic forces are very important at

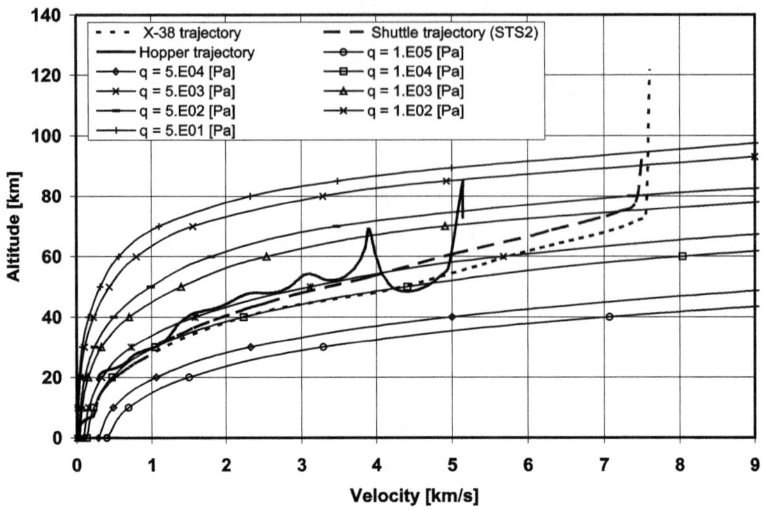

Fig. 12.5. Velocity-altitude map for three RV-W-type vehicles with constant dynamic pressure lines.

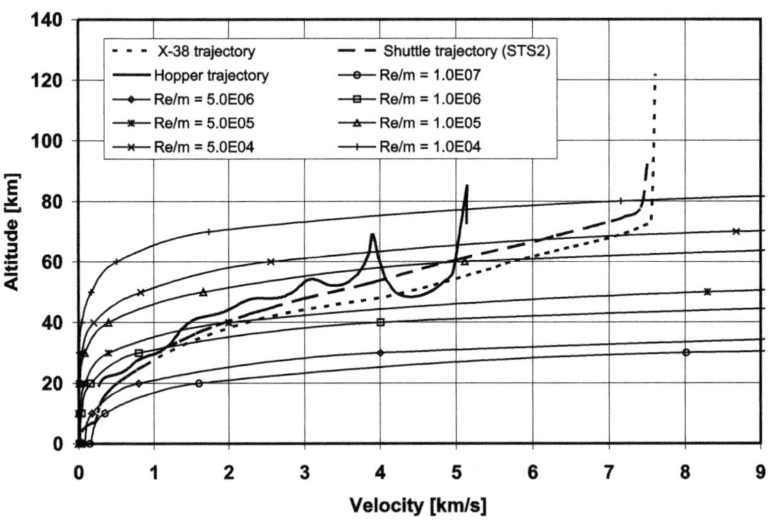

Fig. 12.6. Velocity-altitude map for three RV-W-type vehicles with constant Reynolds number lines.

altitudes up to approximately $60.0\,km$, but play only a minor role at altitudes $H \gtrapprox 80.0\ km$. The aerodynamic drag value for instance at $H = 48.0\ km$ is 460 times larger than at $H = 100.0\ km$.

Sometimes it is helpful to know for a quick estimate how some variables develop during the re-entry flight of a RV-W- or RV-NW-type vehicle. Therefore in the next four figures, Figs. 12.3 to 12.6, velocity-altitude maps of the trajectories of the Space Shuttle Orbiter (STS-2), the X-38 and the HOPPER vehicles[1], Section 3.3, are shown, combined with lines of constant total enthalpy, Mach number, dynamic pressure and Reynolds number.

References

1. Hirschel, E.H.: Basics of Aerothermodynamics. Progress in Astronautics and Aeronautics, AIAA, Reston, Va, vol. 204. Springer, Heidelberg (2004)
2. N.N.: U.S. Standard Atmosphere. Government Printing Office, Washington, D.C. (1976)
3. N.N.: Guide to Reference and Standard Atmosphere Models. ANSI/AIAA G-003A-1996 (1997)
4. Hankey, W.L.: Re-Entry Aerodynamics. AIAA Education Series, Washington, D.C. (1988)
5. Regan, F.J.: Re-Entry Vehicle Dynamics. AIAA Education Series, Washington, D.C. (1984)

[1] As one can see, Hopper has not reached the velocity necessary for orbital flight. On the re-entry flight path some skipping can be observed which has the effect of aerobraking.

13 Constants, Units and Conversions

In this book units are in general the SI units (Système International d'unités), see [1], [2], where also the constants can be found. The basic units, the derived units, and conversions[1] to US units are given in Section 13.2.

We give first some general constants and air properties and then the most important units and their conversions.

13.1 Constants and Air Properties

Molar universal gas constant: $R_0 = 8.314472 \cdot 10^3\ kg\, m^2/s^2\, kmol\, K = 4.97201 \cdot 10^4\ lb_m\, ft^2/s^2\, lb_m\text{-}mol\, ^\circ R$

Stephan-Boltzmann constant: $\sigma = 5.670400 \cdot 10^{-8}\ W/m^2\, K^4 = 1.7123 \cdot 10^{-9}\ Btu/hr\, ft^2\, ^\circ R^4$

Standard gravitational acceleration of earth at sea level: $g_0 = 9.80665\ m/s^2 = 32.174\ ft/s^2$

Change of gravitational acceleration of earth with altitude H, [3]: $g = g_0 (R_E/(R_E + H))^2$

Mean earth diameter: $R_E = 6.378 \cdot 10^6\ m = 20.925 \cdot 10^6\ ft$

Circular speed: $v_c = \sqrt{g_0 R_E} = 7.9086 \cdot 10^3\ m/s = 25.9470 \cdot 10^3\ ft/s$

---

[1] Details can be found, for instance, at `http://www.chemie.fu-berlin.de/chemistry/general/si_en.html` and at `http://physics.nist.gov/cuu/Units/units/html`.

Table 13.1. Molecular weights, gas constants, and intermolecular force parameters of air constituents for the low temperature domain, [4], [5]. * is the U. S. standard atmosphere value, + the value from [5].

Gas	Molecular weight M $[kg/kmol]$	Specific gas constant R $[m^2/s^2K]$	Collision diameter $\sigma \cdot 10^{10}$ [m]	2nd Lennard-Jones parameter ϵ/k $[K]$
air	28.9644* (28.97+)	287.06	3.617	97.0
N_2	28.02	296.73	3.667	99.8
O_2	32.00	259.83	3.430	113.0
NO	30.01	277.06	3.470	119.0
N	14.01	593.47	2.940	66.5
O	16.00	519.65	2.330	210.0
Ar	39.948	208.13	3.432	122.4
He	4.003	2077.06	2.576	10.2

Table 13.2. Characteristic rotational, vibrational, and dissociation temperatures of air molecules, [6].

Gas:	N_2	O_2	NO
Θ_{rot} $[K]$	2.9	2.1	2.5
Θ_{vibr} $[K]$	3,390.0	2,270.0	2,740.0
Θ_{diss} $[K]$	$113,000.0$	$59,500.0$	$75,500.0$

13.2 Units and Conversions

SI basic units and SI derived units are listed of the major flow, transport, and thermal entities. In the left column name and symbol are given and in the right column the unit (dimension), with → the symbol used in Chapter 14, and in the line below its conversion.

SI Basic Units

length, L — $[m]$, $\rightarrow$ $[L]$
$1.0\ m = 100.0\ cm = 3.28084\ ft$
$1{,}000.0\ m = 1.0\ km$

mass, m — $[kg]$, $\rightarrow$ $[M]$

	$1.0\ kg = 2.20462\ lb_m$
time, t	$[s]\ (= [sec]),\ \rightarrow [t]$
temperature, T	$[K],\ \rightarrow [T]$ $1.0\ K = 1.8\ ^{\circ}R$ $\Rightarrow T_{Kelvin} = (5/9)\ (T_{Fahrenheit} + 459.67)$ $\Rightarrow T_{Kelvin} = T_{Celsius} + 273.15$
amount of substance, *mole*	$[kmol],\ \rightarrow [mole]$ $1.0\ kmol = 2.20462\ lb_m\text{-}mol$

SI Derived Units

area, A	$[m^2],\ \rightarrow L^2]$ $1.0\ m^2 = 10.76391\ ft^2$
volume, V	$[m^3],\ \rightarrow L^3]$ $1.0\ m^3 = 35.31467\ ft^3$
speed, velocity, v, u	$[m/s],\ \rightarrow [M/t]$ $1.0\ m/s = 3.28084\ ft/s$
force, F	$[N] = [kg\, m/s^2],\ \rightarrow [M\, L/t^2]$ $1.0\ N = 0.224809\ lb_f$
pressure, p	$[Pa] = [N/m^2],\ \rightarrow [M/L\, t^2]$ $1.0\ Pa = 10^{-5}\ bar = 9.86923{\cdot}10^{-6}\ atm =$ $= 0.020885\ lb_f/ft^2$
density, ρ	$[kg/m^3],\ \rightarrow [M/L^3]$ $1.0\ kg/m^3 = 0.062428\ lb_m/ft^3 =$ $= 1.94032{\cdot}10^{-3}\ lb_f s^2/ft^4$
(dynamic) viscosity, μ	$[Pa\, s] = [N\, s/m^2],\ \rightarrow [M/L\, t]$ $1.0\ Pa\, s = 0.020885\ lb_f\, s/ft^2$
kinematic viscosity, ν	$[m^2/s],\ \rightarrow [M^2/t]$ $1.0\ m^2/s = 10.76391\ ft^2/s$
shear stress, τ	$[Pa] = [N/m^2],\ \rightarrow [M/L\, t^2]$ $1.0\ Pa = 0.020885\ lb_f/ft^2$
energy, enthalpy, work, quantity of heat	$[J] = [N\, m],\ \rightarrow [M\, L^2/t^2]$ $1.0\ J = 9.47813{\cdot}10^{-4}\ BTU =$ $= 23.73036\ lb_m ft^2/s^2 = 0.737562\ lb_f/s^2$

(mass specific) internal energy, enthalpy, e, h	$[J/kg] = [m^2/s^2]$, $\rightarrow [L^2/t^2]$ $1.0\ m^2/s^2 = 10.76391\ ft^2/s^2$
(mass) specific heat, c_v,c_p specific gas constant, R	$[J/kg\,K] = [m^2/s^2\,K]$, $\rightarrow [L^2/t^2\,T]$ $1.0\ m^2/s^2\,K = 5.97995\ ft^2/s^2\,{}^\circ R$
power, work per unit time	$[W] = [J/s] = [N\,m/s]$, $\rightarrow [M\,L^2/t^3]$ $1.0\ W = 9.47813{\cdot}10^{-4}\ BTU/s =$ $= 23.73036\ lb_m\,ft^2/s^3$
thermal conductivity, k	$[W/m\,K] = [N/s^2\,K]$, $\rightarrow [M\,L/t^3\,T]$ $1.0\ W/m\,K = 1.60496{\cdot}10^{-4}\ BTU/s\,ft\,R =$ $= 4.018342\ lb_m\,ft/s^3\,R$
heat flux, q	$[W/m^2] = [J/m^2\,s]$, $\rightarrow [M/t^3]$ $1.0\ W/m^2 = 0.88055{\cdot}10^{-4}\ BTU/s\,ft^2 =$ $= 2.204623\ lb_m/s^3$
(binary) mass diffusivity, D_{AB}	$[m^2/s]$, $\rightarrow [L^2/t]$ $1.0\ m^2/s = 10.76391\ ft^2/s$
thermo diffusivity, D_A^T	$[kg/m\,s]$, $\rightarrow [M/L\,t]$ $1.0\ kg/m\,s = 0.67197\ lb_m/ft\,s$
diffusion mass flux, j	$[kg/m^2\,s]$, $\rightarrow [M/L^2\,t]$ $1.0\ kg/m^2\,s = 0.20482\ lb_m/ft^2\,s$

References

1. Taylor, B.N. (ed.): The International System of Units (SI). US Dept. of Commerce, National Institute of Standards and Technology, NIST Special Publication 330, 2001, US Government Printing Office, Washington, D.C. (2001)
2. Taylor, B.N.: Guide for the Use of the International System of Units (SI). US Dept. of Commerce, National Institute of Standards and Technology, NIST Special Publication 811, 1995, US Government Printing Office, Washington, D.C. (1995)
3. Regan, F.J.: Re-Entry Vehicle Dynamics. AIAA Education Series, Washington, D.C. (1984)
4. Hirschfelder, J.O., Curtiss, C.F., Bird, R.B.: Molecular Theory of Gases and Liquids. John Wiley, New York (1964) (corrected printing)
5. Bird, R.B., Stewart, W.E., Lightfoot, E.N.: Transport Phenomena, 2nd edn. John Wiley & Sons, New York (2002)
6. Vincenti, W.G., Kruger, C.H.: Introduction to Physical Gas Dynamics. John Wiley, New York (1965); Reprint edition, Krieger Publishing Comp., Melbourne (1975))
7. Neufeld, P.D., Jansen, A.R., Aziz, R.A.: J. Chem. Phys. 57, 1100–1102 (1972)

14 Symbols

Only the important symbols are listed. If a symbol appears only locally or infrequent, it is not included. In general the page number is indicated, where a symbol appears first or is defined. Dimensions are given in terms of the SI basic units: length $[L]$, time $[t]$, mass $[M]$, temperature $[T]$, and amount of substance $[mole]$, Chapter 13. For actual dimensions and their conversions see Section 13.2.

14.1 Latin Letters

A, Y, N	body axis forces, p. 22 , $[ML/t^2]$
A	stream-tube area, p. 153, $[L^2]$
A_1, A_2	arbitrary constants, p. 251, $[-]$
A_{ref}	reference area, p. 14, $[L^2]$
a	acceleration, p. 127, $[L/t^2]$
a	speed of sound, p. 424, $[L/t]$
a	function, p. 251, $[-]$
a_1, a_2, a_3	action lines of forces, p. 94, $[-]$
a_i	constants ($i = 1...7$), p. 183, $[-]$
B	force entity, p. 127, $[ML/t^2]$
$\underline{B}$	transformation vector, p. 376, $[-]$
$\underline{b}$	vector of body forces, p. 376, $[-]$
b	$= \omega_d$, p. 251, $[1/t]$
C	damping matrix, p. 377, $[-]$
C	heat capacity matrix, p. 378, $[-]$
C	damping parameter, p. 251, $[-]$
C_D	drag coefficient, p. 14, $[-]$
C_{FGX}	axial thrust force coefficient, p. 43 , $[-]$
CG, cog	center of gravity
C_L	lift coefficient, p. 14, $[-]$
C_{gw}	constant , p. 20, $[-]$
C_m	pitching moment coefficient, p. 75, $[-]$
$C_{mq} + C_{m\dot{\alpha}}$	dynamic derivative of pitch motion, p. 242, $[-]$

C_X, C_Z	force coefficients, p. 221, $[-]$
C_{yr}	number of cycles damped, p. 251, $[-]$
c_f	skin-friction coefficient, p. 134, $[-]$
c_p	(mass) specific heat at constant pressure, $[L^2/t^2T]$
c_p	pressure coefficient, $[-]$
c_{p_B}	base pressure coefficient, p. 139 , $[-]$
c_{p_e}	boundary-layer edge pressure coefficient, p. 281, $[-]$
$c_{p_{stag}}$	stagnation pressure coefficient, p. 63 , $[-]$
c_{p_t}	total pressure coefficient, p. 63 , $[-]$
$c_{p_{vac}}$	vacuum pressure coefficient, p. 70 , $[-]$
$c_{p_{wall}}$	wall pressure coefficient, p. 100 , $[-]$
$c^*_{p_{wall}}$	normalized wall pressure coefficient, p. 116, $[-]$
D, C, L	air-path axis forces, p. 22, $[ML/t^2]$
D	diameter, $[L]$
D	drag, p. 14, $[ML/t^2]$
D	elasticity matrix, p. 376, $[-]$
D_{trim}	drag of trimmed flight, p. 23, $[ML/t^2]$
$\underline{D}$	drag vector, p. 50 $[-]$
dS	differential surface element, p. 376, $[-]$
E	air-path energy, p. 13, $[ML^2/t^2]$
$\underline{\hat{E}}, \underline{\hat{F}}, \underline{\hat{G}}$	flux vectors for non-orthogonal coordinates, p. 373, $[-]$
F	force, $[ML/t^2]$
F	surface, p. 152, $[L^2]$
$\underline{F}$	force vector, $[-]$
$\underline{F}_A$	aerodynamic force vector, p. 47, $[-]$
$\underline{F}_{aero}$	aerodynamic force vector, p. 94, $[-]$
$\underline{F}_{bf}$	aerodynamic bodyflap force vector, p. 94, $[-]$
$F_{aero,trim}$	resultant aerodynamic force of trimmed flight, p. 23, $[ML/t^2]$
F_{centr}	centrifugal force, p. 23, $[ML/t^2]$
F_{mass}	force, p. 23, $[ML/t^2]$
F_{net}	installed net thrust, p. 169, $[ML/t^2]$
$\underline{F}$	load vector, p. 377, $[-]$
f	frequency, p. 251, $[1/t]$
f_1, f_2	functions defined by eq. 4.10, p. 183, $[-]$
f_i	unit surface forces ($i = x, y, z$), p. 357, $[M/Lt^2]$
f_a	aerodynamic loads, p. 380, $[ML/t^2]$
$\underline{G}$	gravity vector, p. 47 $[-]$
$\mathbf{g}$	g frame, p. 49
g	gravitational acceleration, p. 13, $[L/t^2]$
g_{eff}	effective gravitational acceleration, p. 129 , $[L/t^2]$
g_0	gravitational acceleration at sea level, p. 26 , $[L/t^2]$
H	altitude, p. 2, $[L]$
H_i	initial altitude, p. 13, $[L]$
H_0	total (mass-specific) enthalpy, p. 144 , $[L^2/t^2]$
h	stream-tube height, p. 165, $[L]$

h	enthalpy, $[L^2/t^2]$
h	film coefficient, p. 435, $[M/Tt^3]$
h_t	total enthalpy, p. 423, $[L^2/t^2]$
h^*	reference enthalpy, p. 438, $[L^2/t^2]$
h_∞	free-stream (mass-specific) enthalpy, p. 20 , $[L^2/t^2]$
I_{ij}	components of moment of inertia, p. 246, $[ML^2]$
I_{sp}	specific impulse, p. 127, $[t]$
J	determinant of Jacobi matrix of geometry, p. 374, $[-]$
K	stiffness matrix, p. 377, $[-]$
K	restoring parameter, p. 251, $[-]$
K_t	transfer matrix, p. 380, $[-]$
K_c	conduction matrix, p. 378, $[-]$
K_h	convection matrix, p. 378, $[-]$
Kn	Knudsen number, p. 145, $[-]$
$\mathbf{k}$	k frame, p. 50
k	thermal conductivity, p. 433, $[ML/t^3T]$
k	reduced frequency, p. 251, $[-]$
L	body length, p. 28, $[L]$
$\underline{L}$	lift vector, p. 50 $[-]$
L	lift, p. 14, $[ML/t^2]$
L_{eff}	effective lift, p. 16, $[ML/t^2]$
L_{ref}	reference length, p. 73, $[L]$
L_{trim}	lift of trimmed flight, p. 23, $[ML/t^2]$
L, M, N	body axis moments, p. 23, $[ML^2/t^2]$
L^A, M^A, N^A	body axis moments, p. 22 , $[ML^2/t^2]$
L_a, M_a, N_a	air-path axis moments, p. 22 , $[ML^2/t^2]$
L/D	lift-to-drag ratio, p. 14, $[-]$
l_F	surface length, p. 153, $[L]$
l	wing length, p. 150, $[L]$
l	vehicle length, $[L]$
l_{aero}	lever arm of aerodynamic force, p. 95, $[L]$
l_{bf}	lever arm of flap force, p. 95, $[L]$
$\underline{M}$	torque vector, p. 355, $[-]$
M	magnitude of torque vector, p. 355, $[ML^2/t^2]$
M	Mach number, p. 75, $[-]$
M	mass matrix, p. 377, $[-]$
M	pitching moment, $[ML^2/t^2]$
M	transformation matrix, p. 49, $[-]$
M_e	boundary-layer edge Mach number, p. 60, $[-]$
M_i	molecular weight of species i, p. 431 , $[M/mole]$
M_w	(inviscid) wall Mach number, p. 115, $[-]$
M_∞	flight Mach number, p. 17, $[-]$
M'_∞	saturation Mach number, p. 108, $[-]$
$\tilde{M}, \tilde{C}, \tilde{K}$	fictitious mass, damping and stiffness matrices, p. 380, $[-]$
m	mass, p. 13, $[M/L^3]$

m	exponent, p. 20, $[-]$
m_1	constant, p. 183, $[-]$
m_v	vehicle mass, p. 127, $[M]$
$\dot{m}$	mass-flow rate, p. 165, $[M/t]$
$\underline{N}$	vector of shape functions at nodal points , p. 376, $[-]$
$\underline{n}$	unit vector, p. 380, $[-]$
n	exponent, p. 20, $[-]$
n	exponent in boundary-layer relations, p. 439, $[-]$
n_z	normal load factor, p. 21, $[-]$
$\mathbf{O}$	O frame, p. 48
P_h	heat input from heat transfer, p. 378, $[ML^2/t^3]$
P_q	heat input from heat convection, p. 378, $[ML^2/t^3]$
P_r	heat input from heat radiation, p. 378, $[ML^2/t^3]$
P_Ω	heat input from internal heat generation, p. 378, $[ML^2/t^3]$
Pr	Prandtl number, p. 434, $[-]$
p	pressure, $[M/Lt^2]$
$\mathbf{p}$	p frame, p. 47
p, q, r	components of angular velocity vector, p. 246, $[1/t]$
p_e	boundary-layer edge pressure, p. 281, $[M/Lt^2]$
p_i	partial pressure of species i, p. 431 , $[M/Lt^2]$
p_{stag}	stagnation pressure, p. 62, $[M/Lt^2]$
p_t	total pressure, p. 63, $[M/Lt^2]$
p_{wall}	wall pressure, p. 98, $[M/Lt^2]$
p_∞	free-stream pressure, p. 62 , $[M/Lt^2]$
$\underline{\hat{Q}}$	solution vector for non-orthogonal coordinates, p. 373, $[-]$
$\underline{Q}$	solution vector for Cartesian coordinates, p. 373, $[-]$
$\underline{Q^a}$	vector of aerodynamic moments, p. 246, $[-]$
Q_∞	total heat flux, p. 20, $[M/t^3]$
q	heat flux, $[M/t^3]$
$\underline{q}$	vector of heat conduction, p. 378, $[-]$
q	dynamic pressure, $[M/Lt^2]$
q_e	boundary-layer edge dynamic pressure, p. 281, $[M/Lt^2]$
q_{gw}	heat flux in the gas at the wall, p. 411 , $[M/t^3]$
q_i	modal coordinate i, p. 389, $[-]$
q_w	heat flux into the wall, p. 411 , $[M/t^3]$
q_{rad}	thermal radiation heat flux, p. 412 , $[M/t^3]$
q_∞	free-stream dynamic pressure, p. 18 , $[M/Lt^2]$
R	gas constant, p. 466, $[L^2/t^2T]$
R	radius, p. 20, $[L]$
R_E	mean radius of earth, p. 21, $[L]$
R_N	nose radius, p. 14, $[L]$
R_0	universal gas constant, p. 465 , $[ML^2/t^2moleT]$
Re	Reynolds number, $[-]$
Re^u	unit Reynolds number, p. 130, $[1/L]$
Re^u_e	boundary-layer edge unit Reynolds number, p. 281 , $[1/L]$

R_∞	flight path radius, 23, $[L]$
r	radius, p. 24, $[L]$
r	radius of gyration, p. 242, $[L]$
$\underline{r}$	position vector, p. 47, $[-]$
r	magnitude of position vector, p. 49, $[L]$
r	recovery factor, p. 438, $[-]$
r, θ	conical coordinates, p. 177, $[L, °]$
r_c	local cone radius, p. 181, $[L]$
S	transformation matrix, p. 374, $[-]$
Sr	Strouhal number, p. 243, $[-]$
St	Stanton number, p. 172, $[-]$
s	half span of wing, p. 150, $[L]$
T	tensor of moment of inertia, p. 246, $[-]$
T	thrust, p. 127, $[M/Lt^2]$
T	temperature, $[T]$
T_e	boundary-layer edge temperature, p. 281 , $[T]$
T_{gw}	temperature of the gas at the wall, p. 411 , $[T]$
T_r	recovery temperature, p. 135, $[T]$
T_{ra}	radiation-adiabatic temperature, p. 412, $[T]$
T_t	total temperature, p. 71, $[T]$
T_w	wall temperature, p. 411, $[T]$
T_0	total temperature, p. 144, $[T]$
T_∞	free-stream temperature, p. 133, $[T]$
T^*	reference temperature, p. 438, $[T]$
t	time, $[t]$
$\underline{t}$	vector of tractions, p. 376, $[-]$
U, V, W	contravariant velocity components, p. 374 $[L/t]$
u, v, w	Cartesian velocity components, $[L/t]$
u, v, w	components of displacement vector, p. 374, $[L]$
$\underline{u}$	vector of displacements in elements, p. 376, $[-]$
$\underline{\tilde{u}}$	vector of displacements at nodes, p. 376, $[-]$
$\widetilde{u}, \widetilde{v}$	velocity components normal and tangential to an oblique shock wave, p. 427 , $[L/t]$
u_∞, v_∞	free-stream velocity, flight speed, $[L/t]$
$\underline{V}$	flight velocity vector of space vehicle, 48, $[-]$
V	magnitude of flight velocity vector, p. 24, $[L/t]$
V_e	speed at (re-)entry, p. 32, $[L/t]$
V_m	maximum velocity, p. 71, $[L/t]$
V_1, V_2	resultant velocities ahead and behind an oblique shock wave, p. 427 , $[L/t]$
$\overline{V}$	viscous interaction parameter, p. 145 , $[-]$
V'	viscous interaction parameter, p. 145 , $[-]$
v_c	circular (orbital) speed, p. 26, $[L/t]$
v_e	boundary-layer edge velocity, p. 130, $[L/t]$
v_i	initial speed, p. 13, $[L/t]$

v_0	circular (orbital) speed, p. 129, $[L/t]$
W	weight, p. 14, $[ML/t^2]$
W_e	engine weight, p. 127, $[ML/t^2]$
W_v	vehicle weight, p. 127, $[ML/t^2]$
w	stream-tube width, p. 165, $[L]$
X, Y, Z	body axis forces, p. 22, $[ML/t^2]$
X_a, Y_a, Z_a	air-path axis forces, p. 22 , $[ML/t^2]$
X_N	neutral point location, p. 195, $[L]$
x_N	neutral point definition, p. 360, $[L]$
x, y, z	Cartesian coordinates, $[L]$
x, y, z	body axis coordinates, p. 22, $[L]$
x_a, y_a, z_a	air-path axis coordinates, p. 22 , $[L]$
x_{bc}, z_{bc}	blunt cone coordinates, p. 109 , $[L]$
x_{cog}	x-location of center of gravity, p. 73 , $[L]$
$x_{ref}, y_{ref}, z_{ref}$	x-, y-, z-location of moment reference point, p. 223, $[L]$
$\underline{x}$	vector of numerical mesh, p. 380, $[-]$
Z	body axis coordinate, p. 100, $[L]$
Z	compressibility factor, p. 431, $[-]$
z	state variable in Lagrange equation, p. 385, $[-]$
z	surface function, p. 183, $[-]$
z_{cog}	z-location of center of gravity, p. 221 , $[L]$

14.2 Greek Letters

$\underline{\alpha}$	vector of thermal expansion coefficients , p. 376, $[-]$
α	angle of attack, p. 13 , $[°]$
α_{eff}	effective angle of attack, p. 153 , $[°]$
α_m	lift parameter, p. 25 , $[M/Lt^2]$
α_w	lift parameter, p. 14 , $[M/Lt^2]$
$\overline{\alpha}$	inclination angle, p. 71 , $[°]$
$\overline{\alpha}_0$	inclination angle at zero angle of attack, p. 155 , $[°]$
β	sideslip (yaw) angle, p. 18, $[°]$
β	exponent, p. 31, $[1/L]$
β_w	ballistic parameter, p. 14 , $[M/Lt^2]$
β_m	ballistic parameter, p. 25 , $[M/Lt^2]$
Γ	interface boundary, p. 380, $[-]$
γ	flight path angle, p. 23 , $[°]$
γ	ratio of specific heats, p. 63 , $[-]$
γ_e	flight path angle at entry, p. 32 , $[°]$
γ_{eff}	effective ratio of specific heats, p. 64 , $[-]$
Δ	thrust-vector angle, p. 43 , $[°]$
ΔF	magnitude of force increment, p. 357, $[ML/t^2]$
Δ_x	down range, p. 29, $[L]$
Δ_y	cross range, p. 29, $[L]$
$\delta\underline{\varepsilon}$	vector of virtual strains , p. 376, $[-]$

$\delta\underline{u}$	vector of virtual displacements , p. 376, [−]
δ	flow (ordinary) boundary layer thickness, p. 133 , $[L]$
δ	ramp angle, p. 277 , [°]
δ	wedge and cone angle, p. 177, 180 , [°]
δ_i	Lagrange multiplier i, p. 385, [−]
δ_{lam}	laminar boundary-layer thickness, p. 413 , $[L]$
δ_{turb}	turbulent boundary-layer thickness, p. 413 , $[L]$
δ_{sc}	turbulent scaling thickness, p. 413 , $[L]$
δ_t	power setting, p. 44 , [−]
δ_{vs}	viscous sub-layer thickness, p. 413 , $[L]$
δ_1	boundary-layer displacement thickness ($\delta_1 \equiv \delta^*$), p. 133 , $[L]$
$\underline{\varepsilon}$	strain vector, p. 374, [−]
$\underline{\varepsilon}_I$	initial strain vector, p. 374, [−]
$\underline{\varepsilon}_T$	thermal strain vector, p. 374, [−]
ε	emissivity coefficient, p. 412 , [−]
ε_{eff}	effective (fictitious) emissivity coefficient, p. 411 , [−]
η	flap/control surface deflection angle, p. 73 , [°]
η_{bf}	body flap deflection angle, p. 78 , [°]
η_{el}	elevon deflection angle, p. 79 , [°]
θ	longitude angle, p. 48, [°]
θ	shock angle, p. 135 , [°]
λ	thermal conductivity, p. 378, $[ML/t^3T]$
μ	viscosity, p. 432 , $[M/Lt]$
μ_a	bank angle, p. 16 , [°]
μ_e	boundary-layer edge viscosity, p. 130 , $[M/Lt]$
ξ, η, ζ	arbitrary non-orthogonal time-dependent coordinates, p. 373, [−]
ξ	dynamic stability coefficient, p. 242 , [−]
ρ	density, $[M/L^3]$
ρ_e	boundary-layer edge density, p. 130 , $[M/L^3]$
ρ_i	partial density of species i, p. 431 , $[M/L^3]$
ρ_0	atmospheric density at sea level, p. 31 , $[M/L^3]$
ρ_∞	free-stream density, p. 12 , $[M/L^3]$
$\underline{\sigma}$	stress vector, p. 374, [−]
σ	shock angle, p. 177, [°]
$\underline{\underline{\tau}}$	viscous stress tensor, p. , $[M/t^2L]$
τ_{ij}	components of viscous shear stress tensor, p. 374, [−]
τ_w	wall shear stress, skin friction, p. 20 , $[M/Lt^2]$
ϕ	latitude angle, p. 48, [°]
ϕ	semi-apertural cone angle, p. 211, [°]
ϕ	angular displacement, p. 253, [°]
ϕ, θ, ψ	Euler angles (heading, pitching, banking) , p. 246, [°]
Φ_f	equivalence ratio, p. 44, [−]
φ	sweep angle of leading edge or cylinder, p. 196 , [°]
φ_0	constant, p. 183, [°]
χ	flight path azimuth angle, p. 24 , [°]

χ, $\overline{\chi}$	viscous interaction parameter, p. 296 , [−]
ψ	state of structural field, p. 385, [−]
ψ_S	state of structural interface boundary, p. 384, [−]
$\underline{\bar{\Omega}}$	angular velocity vector, p. 47, [−]
$\underline{\Omega}$	angular velocity vector, p. 51, [−]
Ω	internal heat generation, p. 378, $[M/Lt^3]$
ω	relaxation coefficient, p. 384, [−]
ω	angular velocity, p. 24 , $[°/t]$
ω	exponent in the power-law equations of viscosity and heat conductivity, p. 433 , [−]
ω	oscillation frequency, p. 253, $[1/t]$
ω_d	damped natural frequency, p. 251, $[1/t]$
ω_i	mass fraction of species i, p. 431 , [−]
ω_n	undamped natural frequency, p. 251, $[1/t]$

14.3 Indices

14.3.1 Upper Indices

a	aerodynamic
a	nodal point
e	individual element
T	transposed
t	tare
u	unit
ϕ,θ	spherical angles
γ,χ	flight trajectory angles
p,O	frames
$*$	reference-temperature/enthalpy value
$+$	dimensionless sub-layer entity

14.3.2 Lower Indices

a	air path axis system
a	nodal point
$aero$	aerodynamic
B	base
b	nodal point
bc	blunt cone
bf	body flap
bl	boundary layer
c	centrifugal
$circ$	circular orbit
cog	center of gravity
cp	center of pressure

D	drag
E	Earth
e	boundary-layer edge, external (inviscid flow)
e	entry
e	experimental system
e	individual element
eff	effective
el	elevon
env	environment, ambient
exp	experiment
F	fluid
f	body axis system
f	non-inertial coordinate system
fix	fixed
g	gas radiation
g	inertial coordinate system
g,k,p,O	frames
gw	gas at the wall
Han	Hansen
is	incipient separation
j	arbitrary reference point
k	thermal conductivity
L	lift
L	length
L_α	lift change due to angle of attack
l	lee side
lam	laminar
$m\dot{\alpha}$	pitching moment change due to angle of attack movement
$m\alpha$	pitching moment change due to angle of attack
max	maximum
min	minimum
mq	pitching moment change due to pitch movement
net	net
$offset$	geometrical offset from centerline or given reference line
pw	wall pressure
q	pitch velocity
r,θ	conical coordinates
R	resultant
ra	radiation adiabatic
rad	radiation
ram	ramjet
ref	reference
rel	relative
res	resultant
S	structural

SL	sea level
$Suth$	Sutherland
s	stagnation point
$stag$	stagnation
s	surface
sp	specific
T	thermal
TA	terminal area
t	total
$trans$	translational
$trim$	trimmed
$turb$	turbulent
$turbo$	turbojet
v	vacuum conditions
v	vehicle
vac	vacuum
$visc$	viscous
$vibr$	vibrational
vs	viscous sub-layer
w	wall
w	wind-on conditions
x,y,z	Cartesian coordinates
ya	side force in air-path system
$\dot{\alpha}$	angle of attack velocity
$\dot{\theta}$	pitch velocity
μ	viscosity
τ_w	wall shear stress
τ	time
0	inlet
1	ahead of the shock wave
2	behind the shock wave
9	jet
∞	infinity

14.4 Other Symbols

O( )	order of magnitude
∞	infinity
$\overline{q}$	time-integrated value of q
$\underline{v}$	vector
$\underline{\underline{t}}$	tensor
$\times$	vector product
$\cdot$	scalar product
$\sim$	proportional
$\widehat{=}$	corresponds to

15 Glossary, Abbreviations, Acronyms

15.1 Glossary

- **Catalytic surface recombination**: in flow with dissociated gas species the recombination of atoms to molecules promoted by the "catalytic" property of the surface material of the vehicle.
- **Cayley's design paradigm**: product definition and development approach with loose couplings of the various systems and sub-systems (first aspect), differentiation of disciplines (second aspect).
- **Compressibility (pressure) effects**: increase of pressure, density and temperature across – especially – bow shocks of hypersonic vehicles, accompanied by high-temperature real-gas effects.
- **Entropy layer**: flow field with an entropy gradient from streamline to streamline (non-homentropic flow) behind a curved shock, especially a blunt-nosed vehicle's bow shock. The entropy layer flow field is a rotational inviscid flow field (Crocco's theorem).
- **Entropy layer swallowing**: at a blunt-nosed vehicle grows the boundary layer (thickness) in down-stream direction and "swallows" the entropy layer. A better wording would be "growing of the boundary layer into the entropy layer".
- **High-temperature real-gas effects**: real-gas effects – in contrast to van der Waals effects – due to high gas temperatures: vibrational excitation and dissociation of molecules, ionization of atoms and molecules (thermo-chemical flow properties).
- **Hypersonic viscous interaction**: strong displacement effect (due to a large boundary-layer displacement thickness) of attached viscous flow (with oblique shock formation) at high Mach and small Reynolds numbers with strong increase of wall pressure and heat flux.
- **Low-density (rarefaction) effects**: low-density, also rarefaction or Knudsen effects, appear, if the domain of continuous flow is left. The effects are slip flow and temperature jump at the surface, thick boundary layers

(effects similar to hypersonic viscous interaction effects), strengthening of high-temperature real-gas effects, change of transport properties.

- **Mechanical loads**: surface pressure and wall shear stress, the latter directly influenced by the wall temperature (decrease of wall shear stress with increase of wall temperature, especially in turbulent flow).
- **Necessary and permissible surface properties**: thermal surface effects are in any case governed also by the properties of the vehicle's surface, for instance by surface emissivity (should be high), surface catalycity (should be low), and surface roughness, waviness, steps et cetera (must be sub-critical, especial in view of laminar-turbulent transition and turbulent flow). Demands on surface properties can be different for the different vehicle classes.
- **Non-convex effects**: Radiation cooling is reduced, if vehicles' surfaces "look" at each other. Non-convex effects are highly non-linear because of the mutual influence of boundary-layer thickness and wall temperature.
- **Radiation adiabatic temperature**: resulting surface temperature in presence of radiation cooling, if the heat flux into the wall is zero. Also called radiation equilibrium temperature.
- **Radiation cooling**: thermal surface radiation cooling is the major (passive) cooling mode of hypersonic vehicles. It demands a high surface-material emissivity (surface coating) and is (locally) strongly influenced by the boundary-layer thickness.
- **Strong interaction**: mutual interaction of inviscid and viscous (attached) flow portions with flow separation, imbedded shock waves et cetera. Strong interaction is critical if accompanied with strong increase of surface pressure and wall heat transfer.
- **Thermal loads**: heat flux in the gas at the wall towards the wall (several mechanisms) and the wall temperature. Thermal loads are governed by surface radiation cooling.
- **Thermal state of the surface**: the wall temperature and the heat flux in the gas at the wall. It governs both thermal surface effects and thermal loads, but also mechanical (wall shear stress) loads.
- **Thermal surface effects**: the influence of the thermal state of the surface on surface and surface-near flow properties.
- **Thermal surface effects – thermo-chemical**: the influence of the thermal state of the surface on especially thermo-chemical flow properties.
- **Thermal surface effects – viscous**: the influence of the thermal state of the surface on especially viscous flow properties.
- **Viscosity effects**: in attached viscous flow the wall shear stress and heat transfer, further the displacement of the inviscid flow (boundary-layer displacement thickness), the loss of longitudinal flow momentum (boundary-layer momentum (loss) thickness) et cetera. Viscosity effects are in general stronger in turbulent than in laminar flow.
- **Weak interaction**: displacement of the inviscid flow by the attached viscous flow (small boundary-layer displacement thickness) with small or negligible changes of surface pressure, wall shear stress and heat transfer.

15.2 Abbreviations, Acronyms

Indicated is the page where the acronym is used for the first time.

AEDC	Arnold Engineering Development Center, p. 7
AFE	Aeroassisted Flight Experiment, p. 217
AOTV	aeroassisted orbital transfer vehicle, p. 2
ARA	Aircraft Research Association, p. 255
ARD	Atmospheric Re-entry Demonstrator, p. 215
ARV	ascent and re-entry vehicle, p. 2
ASTRA	Selected Systems and Technologies for Future Space Transportation Systems Applications, p. 394
BDW	blunt delta wing, p. 69
CAV	cruise and acceleration vehicle, p. 2
CIRA	National Aerospace Research Center Italy, p. 88
CTV	crew transport vehicle, p. 216
DLR	German Aerospace Center, p. 85
ESA	European Space Agency, p. 76
FESTIP	Future European Space Transportation Investigations Programme, p. 282
FFA	National Aerospace Research Center Sweden, p. 255
GETHRA	general thermal radiation, p. 312
GEO	geostationary Earth orbit, p. 210
HALIS	high alpha inviscid solution, p. 71
IMENS	Integrated Multidisciplinary Design of Hot Structures for Space Vehicles, p. 403
ISS	International Space Station, p. 74
LEO	low Earth orbit, p. 210
LH_2	liquid hydrogen, p. 142
LOX	liquid oxygen, p. 142
MSRO	Mars Sample Return Orbiter, p. 218
MSTP	Manned Space Transportation Programme, p. 216
NASA	National Aeronautics and Space Administration, p. 85
NASP	National Aerospace Plane, p. 137
NS	Navier-Stokes, p. 64
ONERA	National Aerospace Research Center France, p. 386
PRORA-USV	Programma di Ricerca Aerospaziale, Unmanned Space Vehicle, p. 88
RCS	reaction control system, p. 12
RHPM	Rankine-Hugoniot-Prandtl-Meyer, p. 16
RV-W	winged re-entry vehicle, p. 2
RV-NW	non-winged re-entry vehicle, p. 2
SERN	single expansion ramp nozzle, p. 132
SSTO	single-stage-to-orbit, p. 2
STS-X	Space Shuttle Flight No. X, p. 7
TA	terminal area, p. 17

TETRA	Technologies for Future Space Transportation Systems, p. 85
TSTO	two-stage-to-orbit, p. 2
TPS	thermal protection system, p. 13

16 Solution Guide and Solutions of the Problems

We give here the solution guides and the solutions of the problems defined at the end of the various chapters.

16.1 Problems of Chapter 2

Problem 2.1

a) We find from Table 12.1 with linear interpolation $\rho_\infty = 1.64 \cdot 10^{-5}\ kg/m^3$ and $T_\infty = 196.7\ K$. From $q_\infty = 0.5\ \rho_\infty v_\infty^2$ we obtain $v_\infty = 7{,}648.1\ m/s$. The speed of sound is $a_\infty = (\gamma\, R T_\infty)^{0.5}$. With $R = 287.06\ [m^2/s^2 K]$, Table 13.1, and $\gamma = 1.4$, we get $a_\infty = 281.1\ m/s$, and $M_\infty = 27.2$.
b) We find from Table 12.1 $\rho_\infty = 9.89 \cdot 10^{-5}\ kg/m^3$ and $T_\infty = 222.3\ K$ and obtain $v_\infty = 6{,}198.6\ m/s$, $a = 298.9\ m/s$, and $M_\infty = 20.7$.

Problem 2.2

From eq. (9.2) we find $T_{ra} = 1{,}085.8\ K$.

Problem 2.3

At 70.0 km altitude we have $\rho_\infty = 8.283 \cdot 10^{-5}\ kg/m^3$, Table 12.1. With eq. (10.76) we find $q_{gw} = 411.7\ kW/m^2$ and from eq. (9.2) $T_{ra} = 1{,}709.6\ K$. This temperature is within the limits. The nose cone of the Orbiter is made from reinforced carbon-carbon (RCC). This material performs up to approximately 2,000.0 K, see , e. g., [1]. We observe moreover, that the radiation-adiabatic temperature is larger than the actual temperature due to heat conduction into and within the nose cap material, see for example Sub-Section 8.4.3.

Problem 2.4

Proceed as with Problem 2.3. The temperature is above the limit.

Problem 2.5

a) Using the equations

$$V_{circ} = [(R_E + H)g(H)]^{1/2},$$
$$V_E = \omega(R_E + H),$$
$$g = g_0 \left(\frac{R_E}{R_E + H}\right)^2.$$

and setting $V_{circ} = V_E$, we find $H = 35.8096 \cdot 10^6$ m.

b) By applying the eqs. (2.55) we have to use:

$$V^0_{|g} = 0.0\ m/s,$$

and by eqs. (2.56)

$$V^g_{|g} = V_{circ} = 3,076.5\ m/s.$$

16.2 Problems of Chapter 3

Problem 3.1

We find from Table 12.1 for each altitude the temperature T_∞ and the density ρ_∞. With the speed of sound $a_\infty = (\gamma R T_\infty)^{0.5}$, $R = 287.06\ [m^2/s^2K]$, Table 13.1, and $\gamma = 1.4$, we obtain the velocities v_∞. Finally with $q_\infty = 0.5\ v^2_\infty$ we obtain 1) $q_\infty = 0.51\ kPa$, 2) $q_\infty = 1.79\ kPa$, 3) $q_\infty = 3.79\ kPa$, 4) $q_\infty = 5.92\ kPa$, 5) $q_\infty = 7.48\ kPa$, 6) $q_\infty = 9.69\ kPa$. At these altitudes the dynamic pressures are below the maximum dynamic pressure $q_\infty \approx 14.0\ kPa$.

Problem 3.2

The total enthalpy is at the ambient temperatures of the atmosphere T_∞:

$$h_t = h_\infty + \frac{1}{2}\, v^2_\infty = \frac{\gamma}{\gamma - 1}\, R T_\infty + \frac{1}{2}\, v^2_\infty.$$

We find with the data from Problem 3.1: 1) $h_t = 0.199{\cdot}10^6 + 27.819{\cdot}10^6 = 28.019{\cdot}10^6\ m^2/s^2$, $\triangle_k = 99.3$ per cent, 2) $h_t = 21.771{\cdot}10^6\ m^2/s^2$, $\triangle_k = 99.0$ per cent, 3) $h_t = 12.483{\cdot}10^6\ m^2/s^2$, $\triangle_k = 98.0$ per cent, 4) $h_t = 6.042{\cdot}10^6\ m^2/s^2$, $\triangle_k = 95.5$ per cent, 5) $h_t = 2.123{\cdot}10^6\ m^2/s^2$, $\triangle_k = 88.1$ per cent, 6) $h_t = 0.753{\cdot}10^6\ m^2/s^2$, $\triangle_k = 69.8$ per cent.

Problem 3.3

From eq. (2.8) we find for small and constant γ:

$$m = \frac{L}{g - \frac{v^2_\infty}{R_\infty}}.$$

The earth acceleration at $H = 80.0\ km$ altitude is, Section 13.1: $g = 9.565\ m/s^2$. The radius $R_\infty = R_E + H$ is $R_\infty = 6.458{\cdot}10^6\ m$. With $q_\infty = 0.513\ kPa$ and $v_\infty = 7{,}459.1\ m/s$, see Problem 3.1, we get a) $m = 118{,}709.73\ kg$, b) $v_\infty^2/R_\infty = 8.61\ m/s^2 \mathrel{\widehat{=}} 90$ per cent, c) $W = 1{,}164{,}144.8\ N$.

Problem 3.4

From eq. (2.8) we find for small and constant γ:

$$v_\infty = \sqrt{\frac{g}{\frac{C_L A_{ref} \rho_\infty}{2m} + \frac{1}{R_\infty}}}.$$

At 80.0 km altitude we have $\rho_\infty = 1.846{\cdot}10^{-5}\ kg/m^3$, Table 12.1. With $R_\infty = 6.458{\cdot}10^6\ m$ we find $v_\infty = 7{,}569.97\ m/s$ and $M_\infty = 26.8$.

Problem 3.5

The dynamic pressure at $M_\infty = 24$, $H_\infty = 70.0\ km$ is $q_\infty = 2.105\ kPa$. The force acting on the boattailing surface is $F_N = \triangle c_{p_w}\ A_{bt}\ q_\infty$. The delta force is $\triangle F_N = 157{,}106.67\ N - 179{,}958.55\ N = -\ 22{,}851.88\ N$ and the delta normal force coefficient $\triangle C_N = \triangle F_N/(q_\infty\ A_{ref}) = -\ 0.0434$. The delta moment is $\triangle M = -\ \triangle F_N\ (0.91 - 0.665)\ L_{ref} = 194{,}678.82\ Nm$ and the delta pitching moment $\triangle C_m = \triangle M/(\bar{c}\ A_{ref}\ q_\infty) = 0.0307$.

Problem 3.6

a) With the help of eq. (7.23) we obtain $C_N = 1.205$ and $C_X = 0.0608$.
b) The plan area of the Orbiter is $A_{plan} = 361.3\ m^2$. Subtracting the boattailing area $A_{bt} = 135.7\ m^2$, we find what we call the main surface to $A_{ms} = 225.6\ m^2$. With the approximate normal force coefficient:

$$C_N \approx \frac{c_{p_{w,ms}} A_{ms} + c_{p_{w,bt}} A_{bt}}{A_{ref}},$$

we obtain, case 2.a: $C_N = 1.145$ and case 2.c: $C_N = 1.093$.
c) C_X is much smaller than C_N, because the surface pressure at the windward side acts predominantly in normal force direction, Fig. 3.37. C_N of case 2.c is smaller than that of case 2.a because of the Mach number and high-temperature real-gas effects. The level of these coefficients is smaller than that found by the transformation of the data base coefficients, because in this simple approximation the higher c_{p_w} at the nose part of the vehicle was not taken into account.

Problem 3.7

The dynamic pressure at $M_\infty = 22.1$ and $H = 70.0\ km$ is $q_\infty = 1.79\ kPa$. a) We assume, that the stagnation point streamline penetrates the bow shock surface orthogonally. Then we have $c_{p_{w,perfect,stag}} = 1.8394$. b) The force on the stagnation point area is $\triangle F_N = (c_{p_{w,real,stag}} - c_{p_{w,perfect,stag}})\, q_\infty \,\triangle A_{nose} =$ 180.07 N. $\triangle C_N = \triangle F_N/(q_\infty\, A_{ref}) = 4.02{\cdot}10^{-4}$. c) $\triangle M = \triangle F_N\, x_l =$ 3,835.6 Nm, $\triangle C_M = (\triangle F_N\, x_l)/(q_\infty\, A_{ref}\, \overline{c}) = 7.1{\cdot}10^{-4}$.

Problem 3.8

a) The boattailing force is $F_{bt} =$ 153,028.9 N. The normal force is $F_{bt} = N =$ 539,043.4 N and hence $F_{ms} =$ 386,014.5 N. From eq. (3.10) we get:

$$x_{ms} = \frac{F_{res}\, x_{cp} - F_{bt}\, x_{bt}}{F_{ms}}.$$

The location of x_{cp} is found with the help of the moment and the normal force coefficient: $\triangle x =$ - $M/N =$ - $C_M\, \overline{c}/C_N = 0.41034$ and $x_{cp} = x_{cog} + \triangle x$ $= 21.71\ m$. With $x_{bt} = 29.82\ m$ we get finally $x_{ms} = 18.4936\ m$.

b) The boattailing force is now $F'_{bt} =$ 133,596.6 N. With eq. (3.11) we have:

$$x'_{cp} = \frac{x_{ms} + x_{bt}\frac{F'_{bt}}{F_{ms}}}{1 + \frac{F'_{bt}}{F_{ms}}},$$

and $x'_{cp} = 21.41\ m$.

c) The forward shift of the center of pressure is $|\triangle\, x_{cp}| \approx 0.30\ m$.

16.3 Problems of Chapter 4

Problem 4.1

We employ the flat surface relations given in Section 10.4. The recovery temperature is:

$$T_r = T_e(1 + r\frac{\gamma - 1}{2} M_e^2).$$

The viscosity is determined with eq. (10.61) for $T \gtrapprox 200.0\ K$. We find $T_{r,lam}$ = 2,069.25 K, $T_{r,turb} =$ 2,154.74 K, $T^*_{r,lam} = 803.03\ K$, $T^*_{r,turb} =$ 1,016.84 K. Because all temperatures are above $T = 200.0\ K$, eq. (10.99) can be written:

$$c_f = 2\,C\,x^{-n} \left(\frac{T_e}{T^*}\right)^{1-n(1+\omega)} (Re_e^u)^{-n}.$$

The result must be re-normalized with the dynamic pressure q_∞. We obtain a) $c_{f,lam} = 1.10{\cdot}10^{-4}$, $c_{f,turb} = 1.61{\cdot}10^{-3}$.

The relation for the heat flux in the gas at the wall, eq. (10.100), is used in the form:

$$q_{gw} = C\,x^{-n}\, k_e\, Pr^{1/3}(T_r - T_w) \left(\frac{T_e}{T^*}\right)^{1-n(1+\omega)} (Re_e^u)^{1-n}.$$

We obtain b) $q_{gw,lam} = 2.69\ kW/m^2$, $q_{gw,turb} = 31.73\ kW/m^2$.

The agreement with the data in Fig. 4.8 is quite good.

Problem 4.2

We write eq. (10.88) in the form:

$$\delta = C \frac{x^{1-n}}{(Re_e^u)^n} \left(\frac{T^*}{T_e}\right)^{n\,\omega},$$

and obtain $\delta_{lam} = 0.043\ m$ (with the alternative relation for δ_{lam}, eq. (10.89), the thickness is smaller). With the other relations from Sub-Section 10.4.2 we find $\delta_{1,lam} = 0.022\ m$, $\delta_{turb} = 0.66\ m$, $\delta_{1,turb} = 0.187\ m$, $\delta_{vs} = 0.002\ m$.

Problem 4.3

We obtain $c_{f,lam} = 0.57{\cdot}10^{-4}$, $c_{f,turb} = 0.58{\cdot}10^{-3}$, $q_{gw,lam} = 1.35\ kW/m^2$, $q_{gw,turb} = 11.02\ kW/m^2$, $\delta_{lam} = 0.078\ m$, $\delta_{1,lam} = 0.043\ m$, $\delta_{turb} = 0.84\ m$, $\delta_{1,turb} = 0.255\ m$, $\delta_{vs} = 0.005\ m$.

The skin-friction coefficients and the heat flux in the gas at the wall are smaller, the thicknesses are larger. The latter reflect directly the smaller Reynolds number, which is the flight Reynolds number, whereas the former are influenced by the changed flow parameters at the lower side of the forebody.

Problem 4.4

We find $c_{f,turb} = 0.612{\cdot}10^{-3}$, $q_{gw,turb} = 0.152\ kW/m^2$, $\delta_{turb} = 0.54\ m$, $\delta_{1,turb} = 0.128\ m$. a) The heat flux in the gas at the wall is small because of the chosen large wall temperature. b) The boundary layer thicknesses are smaller than before, mainly because of the much larger unit Reynolds number. c) The skin friction increases.

Problem 4.5

From eq. (4.8) we obtain for small angles $\overline{\alpha}$ and $\gamma = 1.4$:

$$\theta|_{M_\infty \to \infty, \overline{\alpha} \ll 90^\circ} = \frac{\gamma - 1}{\gamma + 1} = 0.1667.$$

Problem 4.6

Newton's theory gives $C_L = 2\sin^2\overline{\alpha}\cos\overline{\alpha}$ and $C_D = 2\sin^3\overline{\alpha}$.

We find for $\overline{\alpha} = 9.4°$: $C_L = 0.0526$, $C_D = 0.00871$, for $\overline{\alpha} = 8.4°$: $C_L = 0.0422$, $C_D = 0.00623$, and for $\overline{\alpha} = 10.4°$: $C_L = 0.0641$, $C_D = 0.0118$.

The changes are for $\overline{\alpha} = 8.4°$: $\triangle C_L = -\ 19.8$ per cent, $\triangle C_D = -\ 28.5$ per cent, and for $\overline{\alpha} = 10.4°$: $\triangle C_L = +\ 21.9$ per cent, $\triangle C_D = +\ 35.5$ per cent.

The changes are, in view of Fig. 4.44, very large. This means that, in order to arrive at a reliable conclusion, the performance changes of the whole vehicle must be considered, see the remarks at the end of Sub-Section 4.5.4.

Problem 4.7

In spherical coordinates rot $\underline{v}$ reads:

$$\text{rot}\underline{v} = \begin{pmatrix} \dfrac{1}{r}\dfrac{\partial v_\varphi}{\partial \theta} - \dfrac{1}{r\sin\theta}\dfrac{\partial v_\theta}{\partial \varphi} + \dfrac{\cot\theta}{r} v_\varphi \\ \dfrac{1}{r\sin\theta}\dfrac{\partial v_r}{\partial \varphi} - \dfrac{\partial v_\varphi}{\partial r} - \dfrac{1}{r} v_\varphi \\ \dfrac{\partial v_\theta}{\partial r} - \dfrac{1}{r}\dfrac{\partial v_r}{\partial \theta} + \dfrac{1}{r} v_\theta \end{pmatrix} = 0.$$

For conical and axisymmetric flows we have $\dfrac{\partial}{\partial r} = 0$, $\dfrac{\partial}{\partial \varphi} = 0$, $v_\varphi = 0$ and from the third component of rot $\underline{v}$ we find $v_\theta = \dfrac{\partial v_r}{\partial \theta}$.

Problem 4.8

With $L_{ref} = 80.0\ m$ we have $\Delta x = x_N - x_{cp} = 40.0\ cm$. From eqs. 7.18 and 7.23 we find with the 1 per cent condition:

$$\begin{aligned} \Delta z|^{\alpha=0^\circ} &= 1.60\,cm, \\ \Delta z|^{\alpha=3^\circ} &= 2.40\,cm, \\ \Delta z|^{\alpha=10^\circ} &= 2.88\,cm. \end{aligned}$$

16.4 Problems of Chapter 5

Problem 5.1

1. The coefficient of surface pressure is $c_{p_w} = 2\sin^2(\alpha + \theta)$.
2. The unit-surface force F_{aero} acting on a surface element $d\,S$ is only a pressure force and hence $F_{aero} = p_w = c_{p_w} q_\infty$ acts normal to the (windward) surface element. The resultant force R, also acting normal to the windward surface $A = w\,L/\cos\theta$, is $R = c_{p_w} q_\infty A = c_{p_w} q_\infty\, 0.1\, L^2/\cos\theta$ and its coefficient $C_R = R/(q_\infty A_{ref}) = c_{p_w}/\cos\theta$.
3. The normal force is $Z = \cos\theta R$, the axial force $X = \sin\theta\, R$, and their coefficients are $C_Z = c_{p_w}$ and $C_X = c_{p_w}\tan\theta$.
4. The pitching moment is $M =$ - $RL/(2\cos\theta)$ and its coefficient $C_m = M/(q_\infty A_{ref} L_{ref}) =$ - $c_{p_w}/\cos^2\theta$.
5. The surface element dS is connected to dx by $dS = w\,dx/\cos\theta$ and to dz by $dS =$ - $w\,dz/\sin\theta$. The components of the unit-surface force F_{aero} are $F_{x,aero} = F_{aero}\sin\theta$ and $F_{z,aero} = F_{aero}\cos\theta$.
6. The moment M is split according to eq. (7.4). We find with eqs. (7.9) $C_m|_0^x =$ - $(w/(q_\infty A_{ref} L_{ref}))\int_0^L c_{p_w} q_\infty x dx =$ - c_{p_w} and $C_m|_0^z =$ - $(w/(q_\infty A_{ref} L_{ref}))\int_{-L\tan\theta}^0 c_{p_w} q_\infty z dz =$ - $c_{p_w}\tan^2\theta$.
7. The coordinates of the center-of-pressure are finally with eqs. (7.5) and (7.7) $x_{cp} = L_{ref} C_m|_0^x / C_Z =$ - $0.5\ L$ and $z_{cp} =$ - $L_{ref} C_m|_0^z / C_X = + \ 0.5\ L\tan\theta$.

8. The x-coordinate of the center-of-gravity for trim is like eq. (7.11) $x_{cog} = x_{cp} - \frac{C_X}{C_Z} z_{cp}$, Fig. 7.5.

Problem 5.2

The pressure coefficient on the windward side comes to:

$$c_p^w = 2\sin^2(\phi_w + \alpha),$$

and on the leeward side to:

$$c_p^l = 2\sin^2(\phi_w - \alpha).$$

The lift coefficient is given by

$$\begin{aligned} C_L &= C_L^w - C_L^l = \\ &= 2\sin^2(\phi_w + \alpha)\cos(\phi_w + \alpha) - 2\sin^2(\phi_w - \alpha)\cos(\phi_w - \alpha). \end{aligned}$$

a) $\phi_w = 45°,\ \alpha = 45° \Longrightarrow C_L = 0.$

b) The function $C_L(\phi_w, \alpha)$ possesses a turning point in $\alpha = 0°$ with:

$$\begin{aligned} \frac{dC_L}{d\alpha} &= -2\left(\sin^3(\phi_w + \alpha) + \sin^3(\phi_w - \alpha)\right) + \\ &\quad + 4\left(\sin(\phi_w + \alpha)\cos^2(\phi_w + \alpha) + \sin(\phi_w - \alpha)\cos^2(\phi_w - \alpha)\right) = \\ &= 0, \end{aligned}$$

and from $dC_L/d\alpha|_{\alpha=0} = 0$ one finds:

$$\phi_w = \arcsin\sqrt{\frac{2}{3}} = 54.732°.$$

With that the inequality holds:

$$\begin{aligned} &\sin^2(\arcsin\sqrt{\frac{2}{3}} + \alpha)\cos(\arcsin\sqrt{\frac{2}{3}} + \alpha) < \\ &< \sin^2(\arcsin\sqrt{\frac{2}{3}} - \alpha)\cos(\arcsin\sqrt{\frac{2}{3}} - \alpha) \end{aligned}$$

for any $\alpha \leq 45°$, which means that C_L is negative.

Problem 5.3

From eq. 7.2 we have:

$$\begin{aligned} L_{ref}\ C_{m(1)} - (x_{(1)} - x_{cp})C_z + (z_{(1)} - z_{cp})C_x = 0, \\ L_{ref}\ C_{m(0)} - (x_{(0)} - x_{cp})C_z + (z_{(0)} - z_{cp})C_x = 0, \end{aligned}$$

and

$$L_{ref}\, C_{m(1)} = L_{ref}\, C_{m(0)} + (x_{(1)} - x_{(0)})\, C_z - (z_{(1)} - z_{(0)})\, C_x.$$

With $C_{m(0)} = 0.13$, $C_x = 1.45$, $C_z = 0.42$ we find:

$$C_{m(1)} = -\,0.0539.$$

Problem 5.4

For the quarter circle we find, Fig. 16.1:

$$\begin{aligned} F_x|^{circ} &= q_\infty 2r \int_0^{\pi/2} \cos^2(\varphi - \alpha) \sin(\frac{\pi}{2} + \varphi)\, d\varphi \\ &= q_\infty 2r \left(\frac{1}{3} + \frac{1}{3} \cos\alpha\, (\cos\alpha + 2\sin\alpha)\right) \\ &= q_\infty A_{ref} C_x|^{circ}, \end{aligned}$$

$$\begin{aligned} F_z|^{circ} &= q_\infty 2r \int_0^{\pi/2} \cos^2(\varphi - \alpha) \sin\varphi\, d\varphi \\ &= q_\infty 2r \left(\frac{1}{3} + \frac{1}{3} \sin\alpha\, (\sin\alpha + 2\cos\alpha)\right) \\ &= q_\infty A_{ref} C_z|^{circ}, \end{aligned}$$

$$\begin{aligned} C_m|_o^{x,circ} &= -\frac{1}{q_\infty A_{ref} L_{ref}}\, q_\infty 2r^2 \int_0^{\pi/2} \cos^2(\varphi - \alpha)\ \sin\varphi\, (1 - \cos\varphi)\, d\varphi \\ &= -\frac{1}{q_\infty A_{ref} L_{ref}}\, q_\infty 2r^2 \left(\frac{1}{12} + \sin\alpha \left(\cos\alpha\, \left(\frac{2}{3} - \frac{\pi}{8}\right) + \frac{1}{3}\sin\alpha\right)\right), \end{aligned}$$

$$\begin{aligned} C_m|_o^{z,circ} &= -\frac{1}{q_\infty A_{ref} L_{ref}}\, q_\infty 2r^2 \int_0^{\pi/2} \cos^2(\varphi - \alpha)\ \cos\varphi\, (-\sin\varphi)\, d\varphi \\ &= -\frac{1}{q_\infty A_{ref} L_{ref}}\, q_\infty 2r^2 \left(\frac{1}{4} + \frac{\pi}{8} \cos\alpha \sin\alpha\right), \end{aligned}$$

$$z_{cp}|^{circ} = L_{ref} \frac{C_m|_o^{z,circ}}{C_x|^{circ}} = -3r \frac{\frac{1}{4} + \frac{\pi}{8} \cos\alpha \sin\alpha}{1 + \cos\alpha\, (\cos\alpha + 2\sin\alpha)},$$

$$x_{cp}|^{circ} = -L_{ref} \frac{C_m|_o^{x,circ}}{C_z|^{circ}} = 3r \frac{\frac{1}{12} + \sin\alpha \left(\cos\alpha \left(\frac{2}{3} - \frac{\pi}{8}\right) + \frac{1}{3}\sin\alpha\right)}{1 + \sin\alpha\, (\sin\alpha + 2\cos\alpha)}.$$

For the complete contour we have:

$$
\begin{aligned}
F_x\big|^{complete} &= F_x\big|^{circ}, \\
F_z\big|^{complete} &= F_z\big|^{circ} + 4r\sin^2\alpha, \\
C_m\big|_o^{x,complete} &= C_m\big|_o^{x,circ} - 8r^2\sin^2\alpha, \\
C_m\big|_o^{z,complete} &= C_m\big|_o^{z,circ}, \\
z_{cp}\big|^{complete} &= z_{cp}\big|^{circ}, \\
x_{cp}\big|^{complete} &= 3r\frac{\frac{1}{12} + \sin\alpha\cos\alpha\left(\frac{2}{3} - \frac{\pi}{8}\right) + \frac{13}{3}\sin^2\alpha}{1 + 7\sin^2\alpha + 2\sin\alpha\cos\alpha}.
\end{aligned}
$$

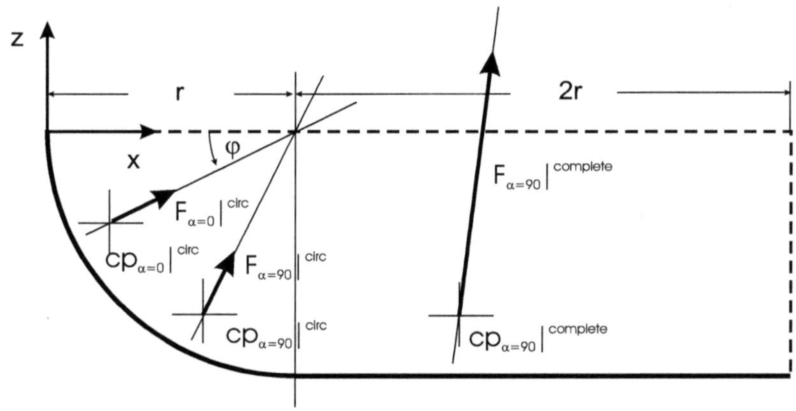

Fig. 16.1. Positions of center-of-pressure, as well as directions and magnitude of aerodynamic force vectors on the generic contour for $\alpha = 0°$ and $90°$

16.5 Problems of Chapter 6

Problem 6.1

The viscosities and the Chapman-Rubesin constant are $\mu_w = 1.88 \cdot 10^{-5}$ $kg/(m\,s)$, $\mu_1 = 0.62 \cdot 10^{-5}$ $kg/(m\,s)$, $C_1 = 0.9$, Reynolds number and hypersonic viscous interaction parameter are $Re_{1,l_{onset}} = 1.037 \cdot 10^5$, $\chi_1 = 8.3$, and finally the incipient separation angle is $\theta_{is} = 16.3°$.

Problem 6.2

Because $M_1 \gtrapprox 4.5$ we obtain $p_{2,is}/p_1 = 43.9$. From the relations for the pressure jump across an oblique shock and for the connection of shock angle θ and ramp angle $\delta \equiv \varphi$, see, e. g., [2]:

$$
\frac{p_2}{p_1} = \frac{2\gamma M_1^2 \sin^2\theta - (\gamma - 1)}{\gamma + 1},
$$

and

$$\tan\delta = \frac{2\cot\theta(M_1^2\sin^2\theta - 1)}{2 + M_1^2(\gamma + 1 - 2\sin^2\theta)},$$

we obtain $\theta = 41.8°$ and $\varphi_{is} = 32.1°$.

Problem 6.3

The pressure coefficient c_{p_2} is defined by:

$$c_{p_2} = \frac{p_2 - p_1}{q_1}.$$

From it we get with $p_1/q_1 = 0.5\gamma M_1^2$:

$$\frac{p_2 - p_1}{p_1} = c_{p2}\frac{\gamma}{2}M_1^2,$$

and with eq. (6.9) finally:

$$\frac{p_2 - p_1}{p_1} = \frac{\gamma(\gamma + 1)}{2}M_1^2\eta^2.$$

Problem 6.4

Because $p_2/p_1 = 1$ at $\eta = 0°$ in both figures, we write:

$$\frac{p_2}{p_1} - 1 = \frac{p_2 - p_1}{p_1} = \triangle\left(\frac{p_2}{p_1}\right) = C\,\eta^2.$$

a) In Fig. 6.13 we find $p_2/p_1 \approx 1.3$ at $\eta = 5°$, and with η in radian measure $C = 39.4$. Applying this we see that despite the small Mach number $M_{e,onset}$ $= 2$ a reasonable agreement can be found in the interval $0° \leqq \eta \lessapprox 7°$.

b) In Fig. 6.17 we find $p_2/p_1 \approx 3$ at $\eta = 5°$, and with η in radian measure $C = 262.6$. Applying this we see that with the larger Mach number $M_{e,onset} =$ 10 a reasonable agreement can be found in the somewhat larger interval $0° \leqq$ $\eta \lessapprox 10°$.

Problem 6.5

From $log(p/p_0) \approx$ - 1.8 we get $p/p_0 = p/p_\infty = 0.0158$ and with

$$c_p = \frac{2}{\gamma M_\infty^2}\left(\frac{p}{p_\infty} - 1\right)$$

and $M_\infty = 15$, $\gamma = 1.4$, finally $c_p =$ - 0.00625.

The vacuum pressure coefficient is $c_{p_{vac}} =$ - 0.00635. The base pressure is close to vacuum.

16.6 Problems of Chapter 7

Problem 7.1

1. $C_D = 1.140$, $C_L = 0.404$.
2. With eq. (7.2) we have, with the subscript 0 denoting the original data, and 1 the ones after the shift:

$$L_{ref}\, C_{m(1)} = L_{ref}\, C_{m(0)} + \left(x_{ref(1)} - x_{ref(0)}\right) C_z - \left(z_{ref(1)} - z_{ref(0)}\right) C_x,$$

and $C_{m(1)} =$ - 0.07756.
3. The z_{ref} shift in positive direction leads always to a pitch down increment. The effect on the pitching moment of a positive x_{ref} shift depends on the sign of C_Z, which can be either positive or negative. In this case C_Z is slightly negative, so that the pitching moment experiences a small pitch down increment. Note, that $L_{ref} \equiv D_1$.

16.7 Problems of Chapter 8

Problem 8.1

Euler's equation along streamlines reads:

$$V dV + \frac{dp}{\rho} = 0, \qquad (1)$$

with $V^2 = u^2 + v^2 + w^2$ being the resultant velocity along a streamline. Since the inviscid flow along streamlines behaves isentropic (in flow fields without shocks), we use the thermodynamic relation for an isentropic process of a perfect gas:

$$\frac{p}{\rho^\gamma} = constant. \qquad (2)$$

Substitution of eq. (2) in eq. (1) and integration leads to:

$$\frac{1}{2}V^2 + \frac{\gamma}{\gamma - 1}\frac{p}{\rho} = constant,$$

or with $p = \rho R T$ to:

$$\frac{1}{2}V^2 + c_p T = c_p T_o = constant,$$

where T_o is the total temperature.

The maximum velocity is then reached for $T \longrightarrow 0$. With $c_p = \dfrac{\gamma}{\gamma - 1} R$ and $R = 287.06\, m^2/s^2K$ we find, [2]:

$$V_{max} = \sqrt{2 c_p T_o} = 1736.12\, \frac{m}{s}.$$

Problem 8.2

The first component of the unsteady, two-dimensional inviscid part of eq. (8.2) has the form:

$$\frac{\partial}{\partial \tau}\left(J^{-1}\rho\right) + \frac{\partial}{\partial \xi}\left(J^{-1}\rho(\xi_t + \xi_x u + \xi_y v)\right) + \frac{\partial}{\partial \eta}\left(J^{-1}\rho(\eta_t + \eta_x u + \eta_y v)\right) = 0, \quad (1)$$

and of eq. (8.8)

$$\frac{\partial}{\partial t}\left(J^{-1}\rho\right) + J^{-1}\frac{\partial}{\partial x}(\rho(u - x_\tau)) + J^{-1}\frac{\partial}{\partial y}(\rho(v - y_\tau)) = 0. \quad (2)$$

The first terms of both equations are identical, so that we have to consider in each case only the second and the third terms.

From the geometrical Jacobian matrix $\mathbf{J} = \dfrac{\partial(\xi, \eta, \tau)}{\partial(x, y, t)}$ and its inverse $\mathbf{J^{-1}} = \dfrac{\partial(x, y, t)}{\partial(\xi, \eta, \tau)}$ we find for the components:

$$\xi_x = \frac{1}{J^{-1}}\, y_\eta, \qquad \xi_y = -\frac{1}{J^{-1}}\, x_\eta, \qquad \xi_\tau = \frac{1}{J^{-1}}(y_\tau x_\eta - x_\tau y_\eta),$$

$$\eta_x = -\frac{1}{J^{-1}}\, y_\xi, \qquad \eta_y = \frac{1}{J^{-1}}\, x_\xi, \qquad \eta_\tau = \frac{1}{J^{-1}}(x_\tau y_\xi - y_\tau x_\xi),$$

with the determinant $J^{-1} = x_\xi y_\eta - y_\xi x_\eta$.

Inserting these relations in eq. (1) and using the chain rules $\dfrac{\partial}{\partial x} = \xi_x \dfrac{\partial}{\partial \xi} + \eta_x \dfrac{\partial}{\partial \eta}$, $\dfrac{\partial}{\partial y} = \xi_y \dfrac{\partial}{\partial \xi} + \eta_y \dfrac{\partial}{\partial \eta}$ in eq. (2), confirms that the identities of eq. (1) and (eq. 2) are valid, if $\dfrac{\partial^2 x}{\partial \xi \partial \eta} = \dfrac{\partial^2 x}{\partial \eta \partial \xi}$ et cetera.

Problem 8.3

We write:

$$T_{ra} = C\, R^{-0.125},$$

and find with the value $T_w = T_{ra} = 1{,}490.0\ K$ from Fig. 8.18 for the radius $R = 64.0\ mm$ the constant to be $C = 2{,}505.87\ K/mm^{-0.125}$. We determine with that the radiation-adiabatic temperatures for some nose radii: a) $R = 32.0\ mm \Rightarrow T_{ra} = 1{,}624.8\ K$, b) $R = 16.0\ mm \Rightarrow T_{ra} = 1{,}771.9\ K$, c) $R = 4.0\ mm \Rightarrow T_{ra} = 2{,}107.2\ K$. In Fig. 8.18 we read for these radii approximately a) $T_w = 1{,}585.0\ K$, b) $T_w = 1{,}650.0\ K$, c) $T_w = 1{,}680.0\ K$. Our conclusion is: the smaller the nose radius, the stronger the tangential heat flux away from the nose point and the stronger the reduction of the wall temperature, which then is no more the radiation-adiabatic temperature.

Problem 8.4

The cowl lip of the inlet: due to the small lip radius!

The nozzle expansion surface: the combustion gases radiate heat towards the nozzle expansion surface (SERN) and prevent partly the cooling heat radiation away from that surface!

References

1. Schwartz, M.: New Materials, Processes, and Methods Technology. Taylor and Francis, Boca Raton (2006)
2. Hirschel, E.H.: Basics of Aerothermodynamics. Progress in Astronautics and Aeronautics, AIAA, Reston, Va, vol. 204. Springer, Heidelberg (2004)

Permissions

Figures reproduced with permission by

the American Astronomical Society (AAS): Fig. 2.1,
the American Control Conference (ACC): Fig. 2.13,
the American Institute of Aeronautics and Astronautics (AIAA), Inc.: Figs. 3.39, 4.24, 4.25, 4.45, 4.55, 4.67, 5.29, 5.30, 8.5 to 8.7, 8.25 to 8.27,
Elsevier: Figs. 6.12, 6.33, 6.34,
the European Space Agency (ESA): Figs. 4.60, 4.64, 6.49,
ESA and Ö. Karatekin, J.-M. Charbonnier, F. Wang and V. Dehant: Fig. 5.31,
ESA and M. Haupt: Figs. 8.17, 8.18,
METS, Inc.: Fig. 4.10,
RTO/AGARD: Fig. 4.5, 4.14 to 4.18,
F. W. Spaid and RTO/AGARD: Fig. 6.37,
Springer Science and Business Media: Figs. 2.14, 4.4, 4.37, 4.38, 6.31, 6.40 to 6.43, 8.8 to 8.13,
Wiley-VCH: Fig. 4.46,
Zeitschrift für Flugwissenschaften und Weltraumforschung (ZfW): Figs. 2.16 to 2.19, 3.3, 6.19.

The permission to reproduce figures, which did originate in the German Hypersonics Technology Programme, was given by the programmes former director, Dr. H. Kuczera.

The permissions to reprint all other figures have been provided directly from the authors, see Acknowledgements at the beginning of the book.

Name Index

Subject Index

MIX
Papier aus verantwortungsvollen Quellen
Paper from responsible sources
FSC® C105338

If you have any concerns about our products,
you can contact us on
ProductSafety@springernature.com

In case Publisher is established outside the EU,
the EU authorized representative is:
Springer Nature Customer Service Center GmbH
Europaplatz 3, 69115 Heidelberg, Germany

Printed by Libri Plureos GmbH
in Hamburg, Germany